The Control Principle of Wind Power Generation System

Hongwei Ma · Yongdong Li · Lie Xu · Jianyun Chai

The Control Principle of Wind Power Generation System

机械工业出版社
CHINA MACHINE PRESS

Springer

Hongwei Ma
School of Automation
Beijing Institute of Technology
Beijing, China

Yongdong Li
School of Electrical Engineering
Tsinghua University
Beijing, China

Lie Xu
School of Electrical Engineering
Tsinghua University
Beijing, China

Jianyun Chai
School of Electrical Engineering
Tsinghua University
Beijing, China

ISBN 978-981-99-6082-8 ISBN 978-981-99-6083-5 (eBook)
https://doi.org/10.1007/978-981-99-6083-5

Jointly published with China Machine Press Co., Ltd.
The print edition is not for sale in China (Mainland). Customers from China (Mainland) please order the print book from: China Machine Press Co., Ltd.

This Springer imprint is published by the registered company Springer Nature Singapore Pte Ltd.
The registered company address is: 152 Beach Road, #21-01/04 Gateway East, Singapore 189721, Singapore

Preface

Nowadays, environmental pollution and energy transition have become two important issues for social development. The massive consumption of traditional fossil energy represented by coal and oil has not only seriously polluted the living environment of human beings, but also depleted their recoverable reserves. Therefore, the large-scale development and use of clean and non-polluting renewable energy sources and the shift to a sustainable energy strategy have gradually become the consensus of all countries in the world. Renewable energy is an inexhaustible and automatically renewing form of energy obtained from a continuous natural cycle, which is generally environmentally friendly, widely distributed and suitable for local exploitation. At present, renewable energy mainly includes wind, solar, hydro, geothermal, biomass, ocean energy and other forms of energy.

China's renewable energy resources are spread over a wide area, with huge reserves, and the conditions for developing the renewable energy industry are unique. Since the issuance of the "China Renewable Energy Exploitation Law" in 2006, China's wind energy industry has developed rapidly, successively catching up with pioneering countries such as Denmark, Spain, Germany and the United States, and has now become the world's largest country in wind power generation. By the end of 2018, the cumulative installed capacity of grid-connected wind turbines in China attained 210 GW, with an annual power generation capacity of 366 TWh, accounting for 11% and 5% of China's power system respectively, saving approximately 150 million tonnes of standard coal, reducing carbon dioxide emissions by over 360 million tonnes, sulphur dioxide emissions by over 10.98 million tonnes and nitrogen oxide emissions by over 5.49 million tonnes per year, providing more than 500,000 jobs, and bringing good economic, environmental and social benefits.

At present, wind power is considered to be the most technically mature form of renewable energy technology with great potential for large-scale development and commercialization, and its lowest cost of generating electricity is already significantly lower than that of many other major forms of energy such as fossil, nuclear, gas and solar. The rapid expansion of wind power generation capacity is based on improvements in the technology and performance of wind turbine systems. Wind turbine systems have gone through a process of development from single-unit operation to multi-unit grid connection, from fixed speed to variable speed, from fixed pitch to variable pitch, and from onshore generation to offshore generation, with the individual capacity of wind turbines constantly expanding. In order to make full use of wind resources and further improve the economics of wind power generation, it is foreseeable that in the coming future, the individual capacity of wind turbines tends to be larger, while offshore wind farms will gain more space for development.

Since the 1980s, wind power technology has flourished and attracted the attention and interest of many researchers, accumulating a large amount of literature and extremely fruitful results. Especially in recent years, countries around the world have invested heavily in wind power technology, and China has become the world's largest manufacturing base and market for wind power. The Power Electronics and Motor Control Laboratory of the Department of Electrical Engineering at Tsinghua University has been engaged in research on wind power generation systems together with relevant national entities since the 1990s and has achieved some valuable research results and promoted a number of more mature technologies, which have attracted the attention of relevant national departments and application entities.

It is hoped that this book will, on the one hand, provide a systematic introduction to the working principles of wind power systems, laying the foundation for further in-depth study of the technology, and on the other hand, provide a more comprehensive introduction to modelling and control methods for such systems, so that the reader can make use of them in solving practical problems. In addition, full attention is given to newer international research topics, such as medium voltage permanent magnet synchronous wind power systems, and it is hoped that readers will pay attention to these topics, thus promoting the further development of wind power system researches.

The book is divided into nine chapters. Chapter 1 briefly describes the history of renewable energy and wind power, the basic principles and some relevant issues. Chapter 2 focuses on the mathematical modelling of the doubly-fed wind power system and modelling methods. Chapter 3 focuses on the control strategies for the doubly-fed wind power system under ideal grid conditions, which have been widely used and studied, including vector control of the doubly-fed wind turbine and vector control of the grid-side converter. Direct power control strategies and predictive control strategies for the doubly-fed generator are also presented in this chapter. Chapter 4 gives the operational characteristics of the doubly-fed wind power system under non-ideal grid conditions, including the types of grid faults and their impact on the wind power system. Chapter 5 presents the low-voltage ride-through technology for the doubly-fed wind power system and its control and protection strategies. Chapter 6 presents the control strategies for the doubly-fed wind power system under asymmetric grid fault conditions, including model predictive control and multi-objective control. Chapter 7 introduces the mathematical model and control principles of the permanent magnet synchronous wind power system and its operating characteristics under weak grid conditions. Chapter 8 introduces the medium/high-voltage permanent magnet synchronous wind power system based on three-level converters and the transformerless medium/high-voltage permanent magnet synchronous wind power system. Chapter 9 introduces offshore wind power systems based on HVDC transmission, including MMC-based and MFT-based HVDC transmission systems for offshore wind farms.

This book summarizes the research results of our laboratory over the years in the theory and practice of wind power generation. Many of my colleagues and postgraduate students have made important contributions to the content of this book. They are Prof. Jianyun Chai, who started the research work on new generation of wind power generation systems in the laboratory at very early time, and each Ph.D.'s thesis work could not have been finished without Chai's patient inspiration and enthusiastic guidance; Associate Professor Xinjian Jiang, who did a lot of work in the field of general control and engineering; Associate Professor Lie Xu, who started the researches on the brand new field of wind power generation from the very beginning of his return to Tsinghua from the UK; Dr. Guofeng Yuan, the first Ph.D. in the laboratory to work on wind power generation, has done a lot of pioneering work; Dr. Yanwen Zheng, has provided unique insights into the operational characteristics of wind power systems under grid faults, especially low-voltage ride-through technology; Dr. Xibo Yuan, has pioneered the research field of permanent magnet synchronous wind power systems, and has an enthusiasm for weak grid operation and medium voltage systems; Dr. Ling Peng, who has completed researches on graphical modelling and LVRT control of the doubly-fed wind power system; Dr. Firas, who has completed researches on offshore wind power systems based on high-voltage DC transmission; and Dr. Hongwei Ma, who has completed researches on operation control, predictive control and multi-objective control of wind power systems under asymmetric grid conditions. The organizing and writing of this book were done by Dr. Ma Hongwei, who is currently working at Beijing Institute of Technology. The outline layout and the unification of the whole book were done by Prof. Yongdong Li.

During the topic-selection and publication process of this book, I have received great support from China Machine Press, for which I am deeply grateful. I would also like to express my deep gratitude to the M.Sc. and Ph.D. students who have worked on related researches in my laboratory and have now graduated. Finally, I would also like to thank my wife and children who have given me their selfless support.

Due to the limitations of our knowledge and the time constraint, there must be a lot of content in the field of wind power technology that has not been reflected. Inevitably, there are also inappropriate and erroneous points in the book, and I would appreciate comments and criticism from our respectable readers.

Beijing, China
November 2019

Dr. Yongdong Li

Contents

Chapter 1
Introduction

1.1 Renewable Energy and Wind Power

Nowadays, energy shortage and environmental pollution have become two important problems of human survival and development. Traditional fossil energy sources (represented by coal, oil, natural gas, etc.) not only cause serious environmental pollution, but also their reserves have become increasingly depleted [1]. The search for clean and pollution-free renewable energy sources and the adoption of sustainable energy strategies have gradually become a consensus among countries around the world. In December 2018, the 24th Conference of the United Nations Framework Convention on Climate Change was held in Poland, which once again obliged countries to vigorously develop renewable energy technologies and form sustainable energy structures with high efficiency and low emissions in clean and pollution-free ways as soon as possible [2].

1.1.1 Development of Renewable Energy

Renewable energy refers to the inexhaustible and self-renewing energy obtained from natural processes that are continuously replenished,[1] generally with the characteristics of no harm or minimal damage to the environment, huge and extensive distribution of resources, and suitability for local development and utilisation. At present, renewable energy mainly includes wind energy, solar energy, hydro energy, geothermal energy, biomass energy, ocean energy and other forms of energy.

In recent years, renewable energy technologies have been seeing huge investment and rapid development. In 2017, global renewable energy received investment funds of more than \$300 billion, providing 18.1% of global energy consumption. In the field

[1] This definition originates from the International Energy Agency's (IEA) Renewable Energy Group.

H. Ma et al., *The Control Principle of Wind Power Generation System*,
https://doi.org/10.1007/978-981-99-6083-5_1

of power generation, the new installed capacity of renewable energy power generation in 2018 was 181 GW, accounting for nearly 2/3 of the new installed capacity of global power generation, and the total installed capacity of renewable energy power generation had reached 2378 GW, accounting for 1/3 of the total installed capacity of global power generation. wherein, hydro power generation occupied the largest share, reaching 1172 GW, wind power and solar power generation occupied most of the remaining part, reaching 564 GW and 480 GW respectively (Figs. 1.1 and 1.2).

China has huge renewable energy reserves, rapid development and vast industrial space. In 2018, China invested a total of $91.2 billion in renewable energy, accounting for 32% of global capital investment in renewable energy (more than half of which was invested in wind power), and China's installed capacity of renewable energy power generation reached 32.4% of the total installed capacity of global renewable energy power generation, while 1870 TWh grid-connected power was generated, accounting for 26.7% of China's total power generation capacity, saving about 750 million tons of standard coal, reducing carbon dioxide emissions by more than 1.86 billion tons, sulfur dioxide emissions by more than 56.1 million tons, and nitrogen oxide emissions by more than 28 million tons, providing more than 11 million jobs and bringing excellent environmental, economic and social benefits (Fig. 1.3).

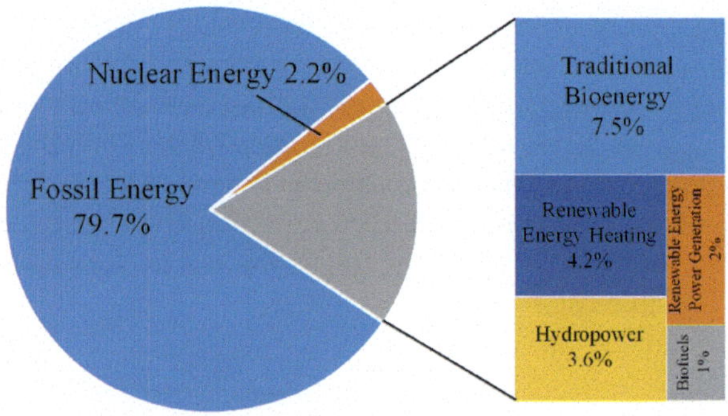

Fig. 1.1 Global energy consumption (2018)

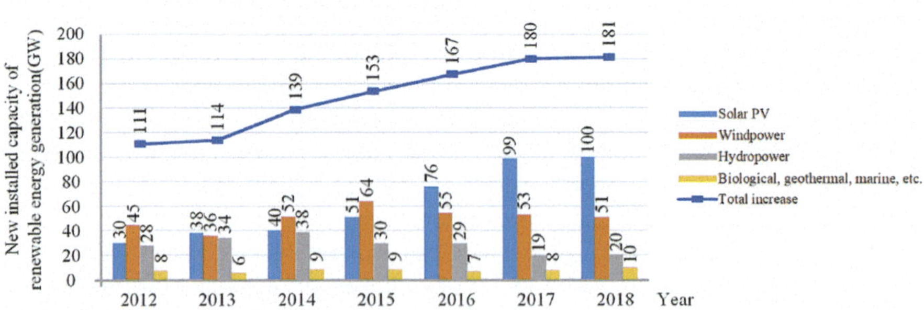

Fig. 1.2 New installed capacity of global renewable power generation (2012–2018)

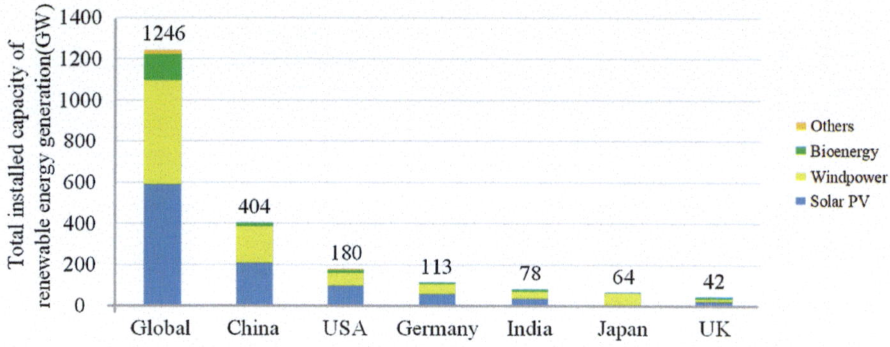

Fig. 1.3 Total installed capacity of global renewable energy generation (2018)

1.1.2 Development of Wind Power

Approximately 2% of solar radiation received on the earth is converted into wind energy, of which the available wind energy amounts to 130 TW [3]. At the same time, wind power generation has significant environmental effects, and for each kilowatt hour of electricity generated by wind power, it can generally reduce the consumption of 0.31–0.34 kg of standard coal, and reduce the emission of 0.853–0.953 kg of CO_2, as well as a certain amount of NO_2, SO_2, and dust and ash, etc. CO_2 will cause the greenhouse effect, NO_2 and SO_2 in the air will produce acid rain, dust will pollute the air, and ash will require land for disposal [4]. Nowadays, when countries around the world are emphasizing environmental protection and energy conservation, wind power generation has a very positive effect on improving the ecological environment of the earth and reducing air pollution, so countries are paying more and more attention on the research and development of wind power generation technologies.

Currently, wind power is considered as the most technologically mature form of renewable energy with the greatest potential for large-scale development and commercialization, and its minimum cost of power generation has been significantly lower than that of many major energy forms such as fossil energy, nuclear energy, gas, and solar energy [5]. In the last decade, the installed capacity of wind power grew at an average annual rate of 14% worldwide roughly, with an average of 43 GW new capacity installed each year, and commercial applications in at least 103 countries worldwide, 33 of which have a total installed capacity of more than 1 GW. According to the Global Wind Energy Council (GWEC) [6], global wind power generation is expected to exceed 2157 TWh by 2020, reaching 9% of global electricity demand; by 2020, global wind power generation will likely reach 5546 TWh, accounting for 18–20% of global electricity demand; by 2050, global wind power generation will likely reach 15258 TWh, accounting for 36–41% of global electricity demand, bringing excellent environmental and economic benefits (Figs. 1.4 and 1.5).

China's wind power resources are very rich, including about 253 GW exploitable wind energy reserves on land and 750 GW exploitable wind energy reserves at sea.

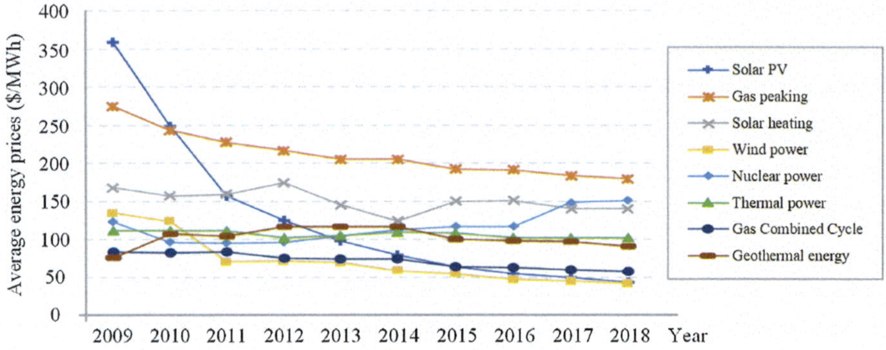

Fig. 1.4 Average global energy prices (2009–2018)

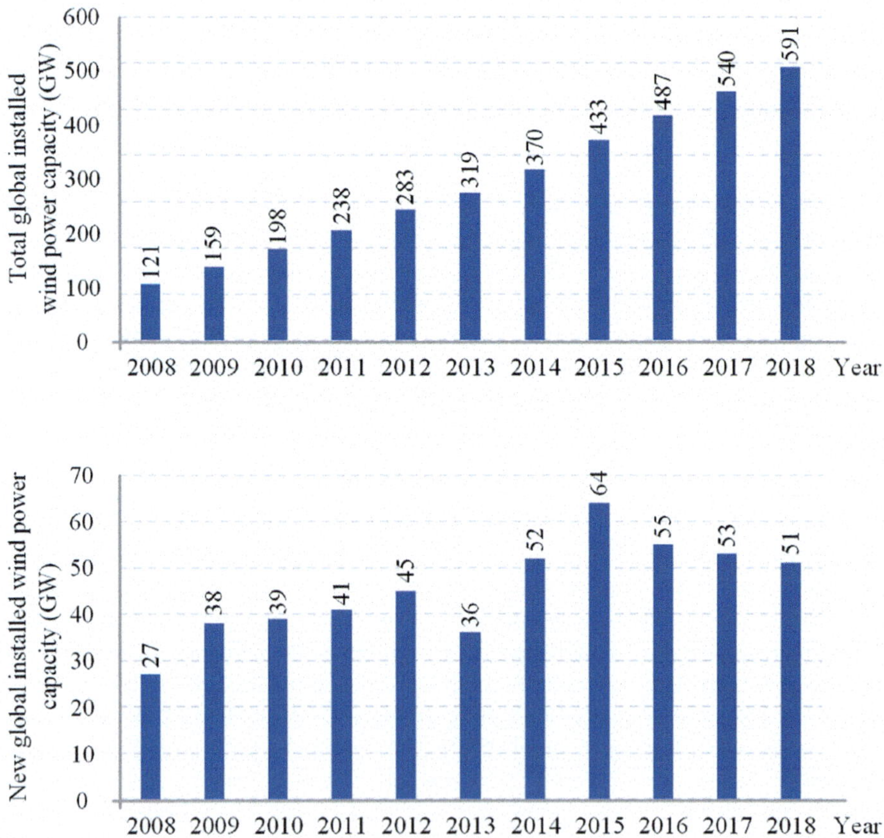

Fig. 1.5 Global installed capacity of wind power (2009–2018)

The rich wind power resources provide the necessary foundation for the development of China's wind power technologies and related industries, while the national planning guidelines and various favorable policies have further promoted the rapid development of China's wind power industry [7, 8]. In 2006, China promulgated the Renewable Energy Law, and in the following year, the National Development and Reform Commission announced the "Medium and Long-term Development Plan for Renewable Energy", which explicitly required China's installed wind power capacity to reach 30 GW by 2020, and this requirement was further increased to 210 GW in the "Thirteenth Five-Year Plan for Wind Power Development" released in 2016. In 2009, China became the second largest country in the world in terms of total installed wind power capacity, with a total installed capacity of 25.8 GW [9]; in 2010, China became the first country in the world in terms of total installed wind power capacity, with a total installed capacity of 41.8 GW [10]. In the past decade, China's total installed wind power capacity grew at an average annual rate of 68%, with an average annual addition of about 18.5 GW, and reached 210 GW in 2018, and China became the first country in the world with more than 200 GW total installed wind power capacity. The China Wind Power Development Roadmap 2050, jointly published by China's National Development and Reform Commission and the International Energy Agency (IEA), predicts that China's cumulative installed wind power capacity will reach 400 GW by 2030, and 1 TW by 2050 to meet 17% of China's electricity demand and become one of the top five energy sources in China (Fig. 1.6).

The rapid expansion of wind power capacity is based on improvements in technology and performance of wind turbine systems. Wind turbine systems have undergone a process of change from the stand-alone mode to the generator-set mode, from the fixed-speed mode to the variable-speed mode, from the fixed-pitch mode to the variable-pitch mode, and from the onshore systems to the offshore systems, and the stand-alone capacity has been expanding. From the 1980s to the beginning of this century, the stand-alone capacity increased from 50 to 5 MW, and considering the maturity and stability of the technology, until the beginning of this century, wind turbines with 1.5–2 MW stand-alone capacity began to become the mainstream models in the market. GWEC data shows that the average power of new turbines worldwide in 2018 is 2.45 MW, while the average power of new turbines in the UK and Germany is close to 4 MW, which indicates that 3–5 WM wind turbines are becoming the mainstream models. In June 2018, Vestas announced that its 9.5 MW stand-alone turbine passed the type approval, making it the world's largest wind turbine in terms of stand-alone capacity in full commercial operation. With the further increase in wind power capacity and the need to reduce the cost per kilowatt of wind power, it is foreseeable that the stand-alone turbine capacity will still be larger in the coming period, while offshore wind farms will gain more space for development (Fig. 1.7).

Through a series of approaches such as state support, independent research and technology import, China's wind power industry is developing rapidly, and a number of high-level wind power equipment manufacturing enterprises, such as Goldwind Science and Technology, Envision Energy, Guangdong Mingyang, United Power,

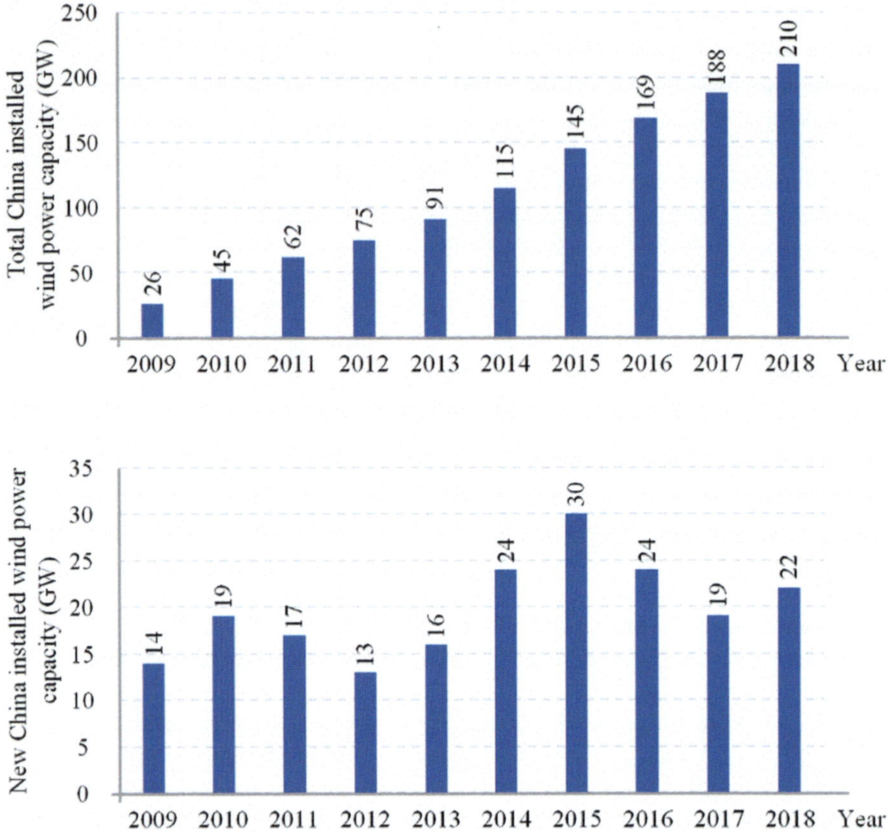

Fig. 1.6 China installed capacity of wind power (2009–2018)

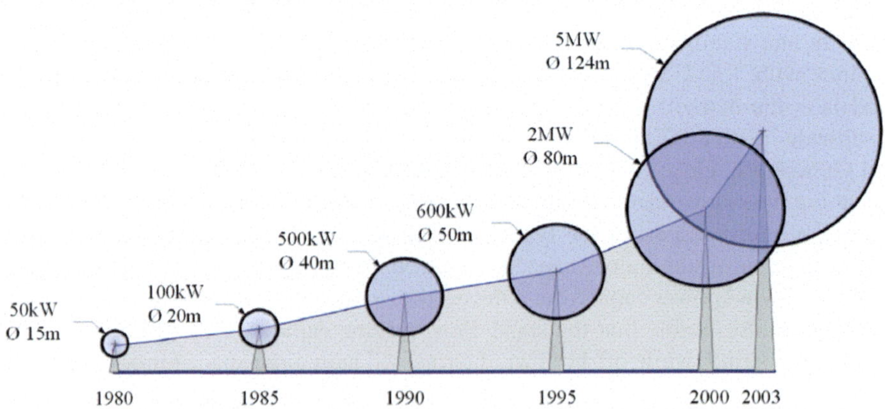

Fig. 1.7 Development of stand-alone wind turbine capacity (1980–2003)

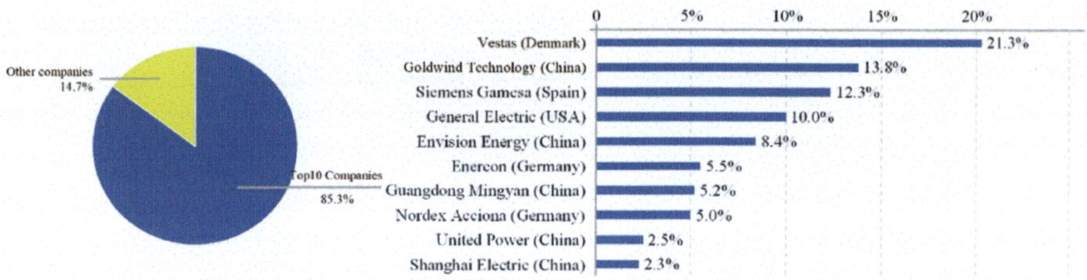

Fig. 1.8 Market share of top global wind power manufacturers (in 2018)

Shanghai Electric, have emerged one after another, occupying the majority of the market share in the domestic market, and already contributing to the international market. At present, China's enterprises already have the ability to independently research, develop and manufacture key components of wind power systems such as tower, nacelle cover, blades, gearboxes and generators, and have formed the related industrial chain. Though more and more control systems, converters, and other core electronic control equipment have also been localized, but China's research and development in this area started late, and gaps still exist in key technologies compared to some other countries. Especially in the design, manufacture and control of large wind turbines and offshore wind farms, there is still a lot of research and engineering work waiting to be completed (Fig. 1.8).

1.2 Overview of Wind Power Systems

Wind energy has a long history, and the large-scale use of wind energy for grid-connected power generation began after the energy crisis in the 1980s. In the past 20 years, with the development of aerodynamics, materials science, electrical engineering, power electronics, computers and control technology, wind power technology made great progress: the stand-alone capacity grown from 200 kW in the early days to 9.5 MW in commercial operation; the power generation form developed from the initial stand-alone operation of a small individual generator to the grid-connected operation of a large set of generators; the control technology evolved from fixed pitch stall control to variable-pitch variable-speed control using modern power electronics and control technology; the operational reliability also increased from 50% at the beginning to over 98% at present. Nowadays, wind power is also being used in a wider range of applications, moving from onshore platforms to offshore platforms.

a. A vertical-shaft wind turbine b. A horizontal-shaft wind turbine

Fig. 1.9 Two types of wind turbines

1.2.1 Fundamentals of Wind Power Systems

A wind power generation system mainly consists of two major components: a wind turbine and a generator, the former converting wind energy into mechanical energy for the generator and the latter converting mechanical energy into electrical energy for the grid. The wind turbine can be classified into the vertical-shaft wind turbine and the horizontal-shaft wind turbine due to the different axial direction of the rotating impeller [11, 12]. The vertical-shaft wind turbine appeared earlier and can capture wind energy in any direction, but its efficiency is low, so most of wind turbine applications currently in use employ the horizontal-shaft structure, generally in the form of three blades (Fig. 1.9).

When the wind passes through the turbine, the wind speed will drop and part of the kinetic energy of the wind is transferred to the wind turbine, at which point the power generated by the turbine is [11]

$$P = \frac{1}{2}\rho C_p S v^3 \tag{1.1}$$

where P is the power from the wind turbine, C_p is the wind energy utilisation coefficient, ρ is the air density, S is the swept area of wind turbine impeller and v is the wind speed.

The wind energy utilisation coefficient is used to express the efficiency of wind energy conversion to mechanical energy in a wind turbine and is mainly related to impeller design, pitch angle and tip-speed ratio.[2] At a given wind speed, the

[2] Tip-speed ratio, is defined as the ratio of blade tip speed to wind speed, which can be expressed as $\lambda = 2\pi Rn/v = \omega R/v$, where R is the impeller radius, n is the impeller revolution speed, and ω is the impeller angular speed.

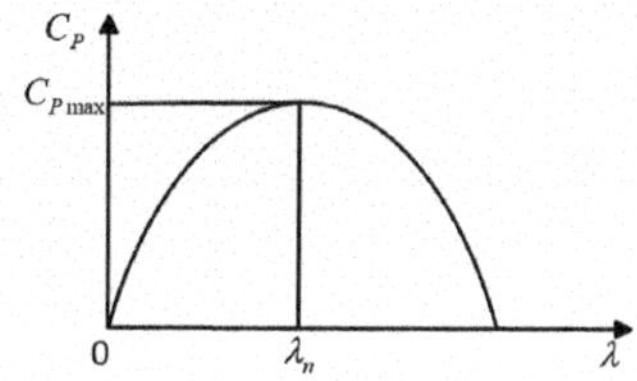

Fig. 1.10 Relation between wind energy utilisation coefficient and tip-speed ratio (pitch angle remained constant)

wind energy utilisation coefficient will decrease as the pitch angle increases and will reach its maximum at a particular tip-speed ratio. Theoretically, wind energy utilisation cannot be increased indefinitely, and the German aerodynamicist Albert Betz pointed out that the limit is 0.593, i.e. the Betz limit. With the development of wind power technology, the design of the impeller has been continuously optimized, and the wind energy utilisation coefficient of the wind turbine has already been high and close to the Betz limit, leaving limited potential for further improvement. Therefore, at present, the main way to increase the output power of the wind turbine is to increase the swept area of the impeller and select a high-wind-speed installation location (Fig. 1.10).

The regulation of the amount of wind energy captured by the wind turbine will help the generator achieve stable operation in various complex wind conditions. According to the different regulation methods for the wind turbine to capture the wind energy, the current grid-connected wind turbine can be classified into three main types: the fixed-pitch fixed-speed (FPFS) wind turbine, the variable-pitch fixed-speed (VPFS) wind turbine, and the variable-pitch variable-speed (VPVS) wind turbine. Among them, the variable-pitch wind turbine is the current dominant model [3].

The pitch angle of the FPFS wind turbine is constant, and the wind power captured by this type of wind turbine will not be regulated within the normal operating wind speed range. At the same time, the FPFS wind turbine solves the problem of safety shutdown when the turbine suddenly unloads. With the advantages of reliable performance and simple structure, the FPFS wind turbine were firstly commercialized in wind farms, but its low energy conversion rate and poor start-up performance have restricted its further development.

Based on the aerodynamic point of view, it is difficult for the FPFS wind turbine to maintain a constant power output from the turbine by the stall characteristics of blades. For this reason, the VPFS wind turbine has been developed, which adjusts the pitch angle through its hydraulic or electric system to regulate the aerodynamic torque of the turbine and achieve a stable power output from the turbine, and by changing the pitch angle, the VPFS wind turbine can obtain a larger starting torque during its starting process, which is more suitable for a large-capacity wind power system.

The proposal of the VPFS wind turbine is a great progress in wind power control technology, but the performance of this turbine under the rated wind speed is still unsatisfactory. For this reason, by closely combining the wind power generation system with power electronics technology, the VPVS wind turbine is further developed: under the rated wind speed, it can track the best tip-speed ratio by controlling

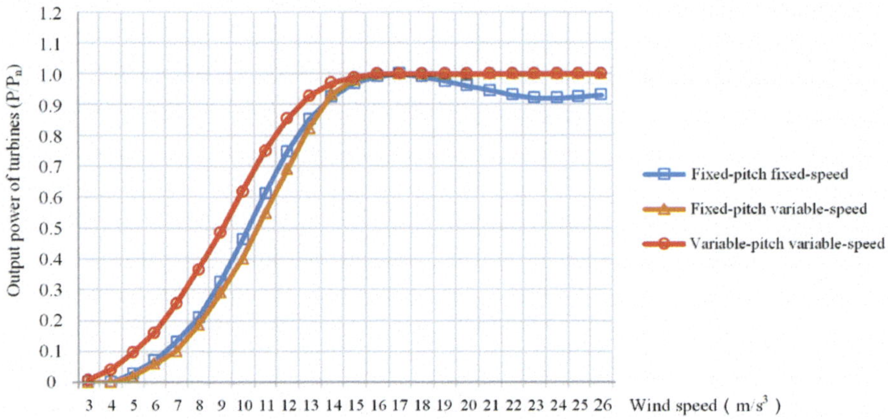

Fig. 1.11 Power characteristics of different types of turbines

the generator speed, so as to achieve the best power curve; above the rated wind speed, it can increase the flexibility of the drive train by adjusting the pitch angle, so as to make the output power smooth and improve the quality of power generation. The VFVS wind turbine can better respond to various complex wind environments, and it is generally employed in a MW-class turbine and above (Fig. 1.11).

1.2.2 Typical Solutions for Wind Power Systems

A wind power generation system is centered on a generator and a power electronic converter and consists of supporting equipment such as a wind turbine and a gearbox (optional). Wherein, the generator can be a common one such as a squirrel-cage asynchronous generator, a wire-wound asynchronous generator, an electrically excited synchronous generator, and a permanent magnet synchronous generator. And the power electronic converter can take the form of no converter, partial-scale converter and full-scale converter [13, 14].

A constant-speed and constant-frequency wind power generation system does not require a converter, generally uses a squirrel-cage asynchronous generator and maintain the generator speed through fixed-pitch stall regulation or active stall regulation, the structure of which is shown in Fig. 1.12. Since the generator is directly connected to the grid, it will absorb reactive power from the grid to provide excitation for the generator during grid-connected operation, so capacitors need to be connected in parallel between the generator and the grid for reactive power compensation to improve the power factor on the grid side. In addition, a soft starter is required to reduce the starting current of the generator. The advantages of this system are the simple structure, low cost, and high reliability. However, because the generator speed can only run within a narrow range above the rated speed, the wind energy utilisation

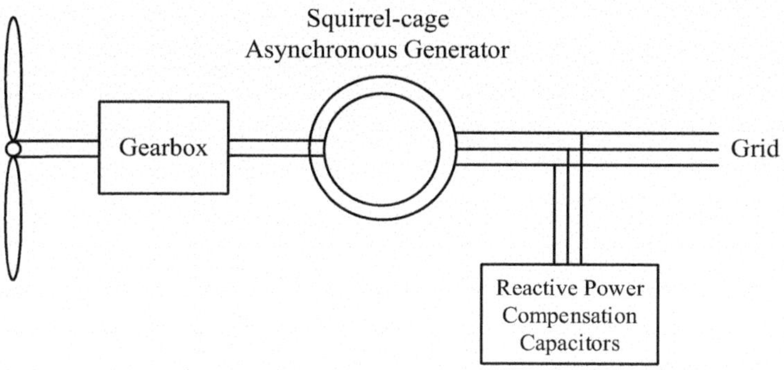

Fig. 1.12 A constant-speed wind power generation system based on a squirrel-cage asynchronous generator

coefficient here is very low. To improve the wind energy conversion efficiency, two squirrel-cage asynchronous generators with different number of pole pairs can be combined, where the generator with more pole pairs works at low speed when the wind is weak and the generator with fewer pole pairs works at high speed when the wind is strong. However, since the two generators are still operating at constant speed at their respective rated speed points, their bearings and gearboxes are exposed to higher mechanical stress and are prone to damage (Fig. 1.12).

A variable-speed and constant-frequency wind power generation system generally uses a wire-wound asynchronous generator. An OptiSlip structure is shown in Fig. 1.13, where a variable resistor is connected to the rotor of the generator to realize a finite variable-speed and constant-frequency wind power generation system. By dynamically changing the resistance of the rotor circuit through a power electronic converter, the generator speed can be adjusted to achieve the variable-speed and constant-frequency operation. The biggest advantage of this system is the simplicity of its converter structure and control strategy; however, the generator can only work above the synchronous speed, and the speed regulation range is not wide. Also, the system efficiency is relatively low because all the differential power is consumed by the variable resistor. In addition, such a system with a limited variable-speed range also needs to be equipped with a soft starter and a reactive power compensator.

To widen the operating speed range of the generator, a four-quadrant power electronic converter can be used instead of the variable resistor to construct a variable-speed and constant-frequency doubly-fed wind power system. In this way, energy on the rotor side will not be consumed, but can be fed into or out of the grid through the converter. The converter of the doubly-fed system is available in various forms such as an AC/DC converter, an AC/DC converter, a matrix converter, a parallel converter, and a resonant converter [15, 16]. Here, the most widely used structure is the back-to-back voltage converter, as shown in Fig. 1.14. As the converter is in the rotor circuit of the generator in the doubly-fed system, and only the differential power needs to be controlled, the capacity of the converter is only a part of the total system capacity. Generally, the speed of the doubly-fed wind generator operates between ±30% of

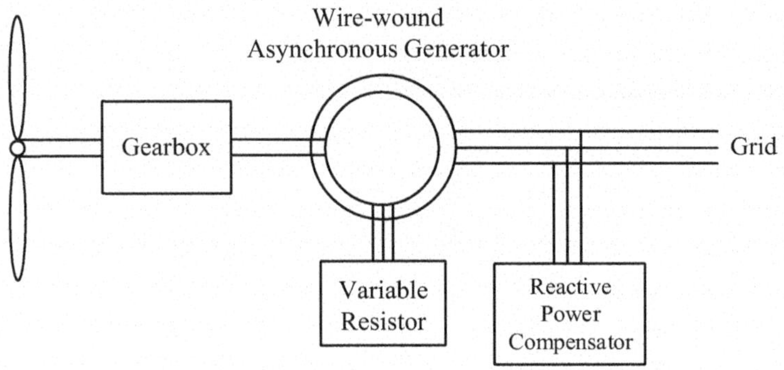

Fig. 1.13 A limited-range variable-speed wind power generation system based on a wire-wound asynchronous generator

its synchronous speed. Therefore, the capacity of the power electronic converter is generally about 1/3 of the whole machine capacity, which greatly reduces the volume, the weight and the cost, and leads to low losses and high efficiency of the whole system. This partial-scale converter structure of the double-fed wind power generation system gives it an advantage in cost and efficiency, which, together with the good power output quality, makes the double-fed wind power generation system as the mainstream model at present. The main shortcoming of such a system is that it works with gear box, slip ring and brush. The use of these structures reduces the reliability to a certain extent and increases the maintenance cost, meanwhile, the direct coupling between the stator of the double-fed generator and the grid makes it very sensitive to grid failure, and the capacity of the partial-scale converter will further limit its ability to operate uninterruptedly in case of grid failures.

A brushless doubly-fed wind power generation system [17] as shown in Fig. 1.15 can be used in order to maintain the advantage of small converter capacity of the

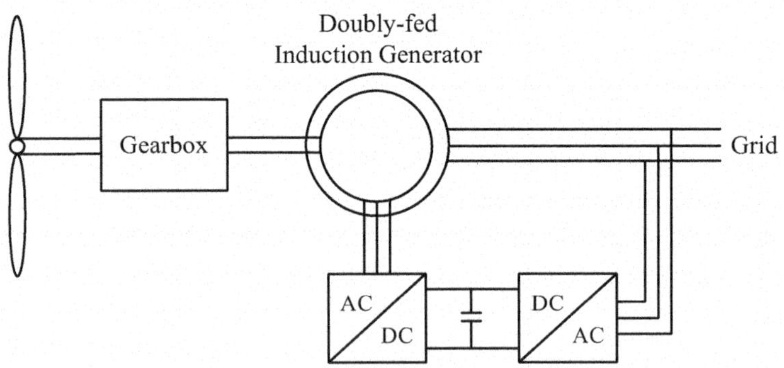

Fig. 1.14 A variable-speed constant-frequency wind power generation system based on a doubly-fed induction generator

doubly-fed wind power generation system and to have the structure of a squirrel-cage asynchronous generator without brushes and slip rings. The brushless doubly-fed generator has two sets of windings with different pole pairs in the stator, where the power winding is connected to the grid and the control winding is connected to the converter, and the rotor pole pairs are the sum of the pole pairs of the two stator windings. The advantages of this system are small converter capacity, low cost and high efficiency, as well as the simple generator structure and high reliability. However, the design of a high-power brushless doubly-fed generator is still difficult, so the system is still in the stage of small-power test prototype and has not met the commercial standard.

To realize variable-speed and constant-frequency operation of a squirrel cage asynchronous generator, a full-scale converter can be used to construct a variable-speed squirrel cage asynchronous wind power generation system, as shown in Fig. 1.16. Since the generator is separated from the grid by the converter, the rotor speed is no longer limited by the grid frequency, variable-speed operation can be achieved, and no additional reactive power compensator is required. However, the use of the full-scale converter increases the cost and reduces the efficiency of the whole system.

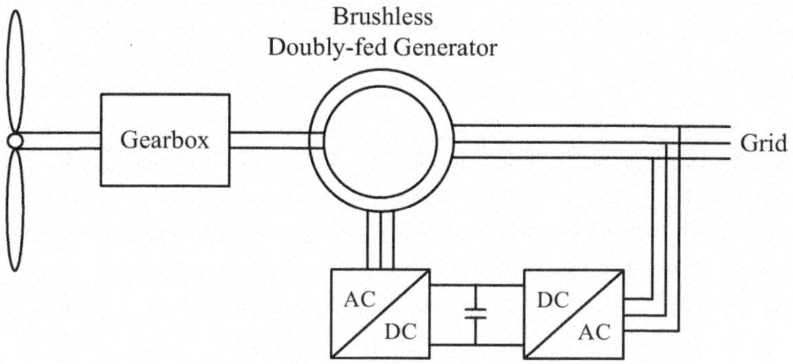

Fig. 1.15 A brushless doubly-fed wind power generation system

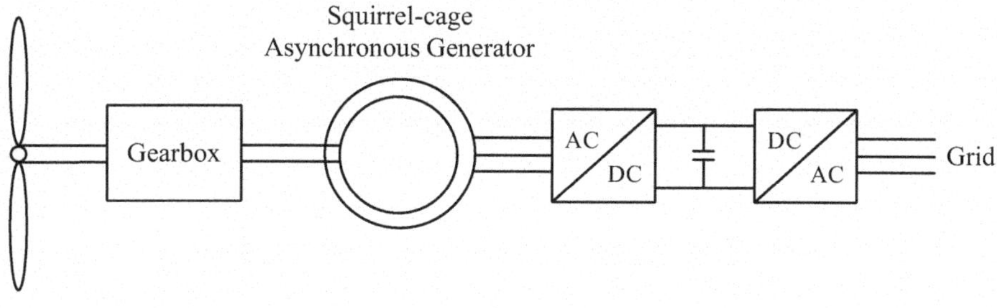

Fig. 1.16 A variable-speed wind power generation system based on a squirrel-cage asynchronous generator

In all the above systems based on asynchronous generators, because of the high rated speed of the generator, it is required to use a multi-stage booster gearbox to increase the speed of the wind turbine before connecting it to the generator. The multi-stage gearbox is complex, costly and prone to damage. According to statistics, downtime due to gearbox failure accounts for 20% of total downtime, which is the highest percentage of all failures. At the same time, the gearbox increases the amount of routine maintenance [18]. To eliminate this weak link in the wind power system, improve reliability and reduce maintenance cost and operational noise, the direct-drive wind power system can be constructed using a full-scale converter and a multi-pole synchronous generator. A direct-drive system based on an electrically excited synchronous generator is shown in Fig. 1.17. The rectifier on the rotor side provides excitation for the generator, while the stator-side alternating current (AC), which varies in frequency, is first rectified and then inverted to AC output at the same frequency as the grid. The system allows for a full speed range operation and a flexible connection to the grid. The main disadvantages are the high cost and the high losses of the converter when a full-scale converter is used. At the same time, the size and weight of the generator increases as the number of pole pairs increases, which results in higher strength requirements for supporting components such as the tower.

By using a permanent magnet synchronous generator (PMSG) instead of an electrically excited synchronous generator, a permanent magnet direct-drive wind power system can be obtained, as shown in Fig. 1.18. This system has the same advantages as the direct-drive wind power system based on an electrically excited synchronous generator and has lower losses and higher efficiency thanks to the elimination of the excitation winding as no external excitation is required. In addition, the full-scale converter isolates the generator from the grid, improving its ability to cope with grid faults [19, 20]. The disadvantages are that the permanent magnet material is more expensive and can be demagnetized, and the use of the full-scale converter increases the size, losses and cost of the system. With the increase in the capacity of wind power systems, the size and weight of multi-pole PMSGs also increase significantly,

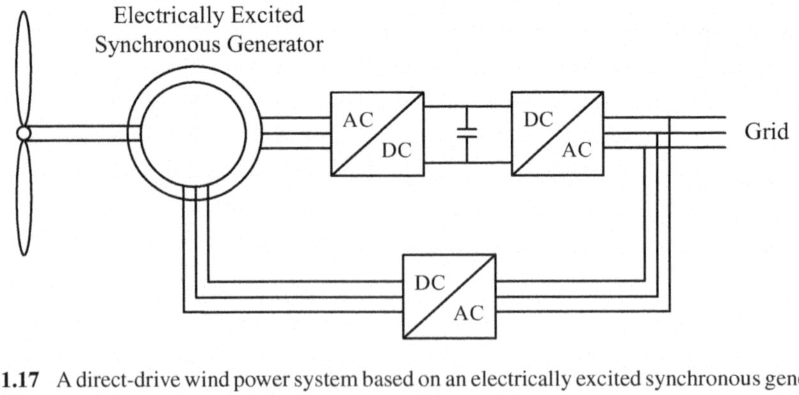

Fig. 1.17 A direct-drive wind power system based on an electrically excited synchronous generator

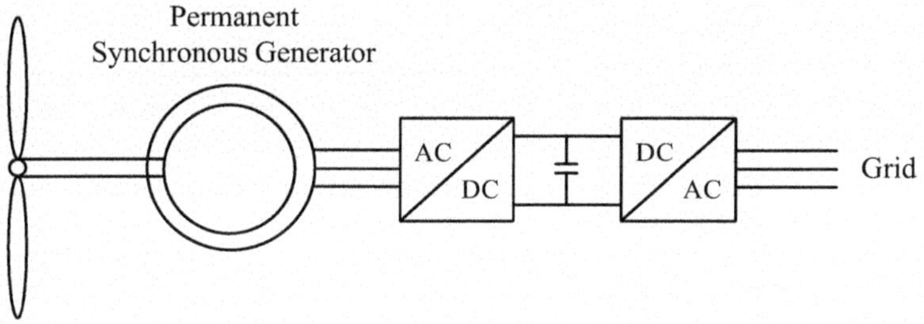

Fig. 1.18 A direct-drive wind power system based on a permanent magnet synchronous generator

and their manufacturing process becomes more complex. A semi-direct-drive wind power system based on a compromise idea has therefore emerged [21]. It consists of a single-stage booster gearbox and a medium-speed permanent magnet synchronous generator or a medium-speed doubly-fed induction generator, and combines the benefits of both direct-drive and doubly-fed wind power systems, with a reduced generator size and lower manufacturing costs compared to the direct-drive wind power system, and a reduced number of gearbox stages and lower operation and maintenance costs compared to the double-fed wind power system.

The features of various major wind power systems and their comparison are shown in Table 1.1.

In addition, some new generators, such as switched reluctance generators [22], transverse field generators [23] and high voltage generators [24] etc., have been used in wind power systems. New models such as the wind power system based on mechanical and electrical variable transmission have also emerged [25].

Currently, a doubly fed induction generator (DFIG) with a partial-scale converter[3] and a permanent magnet synchronous generator (PMSG) with a full-scale converter[4] are the two most widely used solutions for the large-capacity variable-pitch variable-speed wind power system, both of which have already been developed into the megawatt class and are progressing to a higher power class. Whereas the DFIG-based wind power system has the biggest market share and the largest installed capacity, and the PMSG-based wind power system has also grown very rapidly in recent years, with its market share increasing year on year.

---

[3] Unless otherwise specified, it will be referred to as the DFIG-based wind power system in subsequent parts of the book.

[4] Unless otherwise specified, it will be referred to as the PMSG-based wind power system in subsequent parts of the book.

Table 1.1 Features of major wind power systems

Converter	Generator	Advantages	Disadvantages
No converter	Squirrel-cage asynchronous generator	• Simple structure • Low cost • High reliability	• Constant-speed operation • Low wind energy utilisation • Requirement of reactive power compensator and soft stater
Partial-scale converter	Wire-wound asynchronous generator	• Simple structure • Simple control	• Narrow speed range • Low system efficiency • Requirement of reactive power compensator and softer starter
	Doubly-fed induction generator	• Variable-speed and constant-frequency operation • Small converter capacity • Low cost • High efficiency	• Complex control • More maintenance due to brushes, slip rings and gearboxes
Full-scale converter	Squirrel-cage asynchronous generator	• Variable-speed and constant-frequency operation • Simple motor structure • High reliability	• High converter cost • Low system efficiency • Requirement of gearbox
	Electrically excited or permanent magnet synchronous generator	• No gearbox • Full speed range operation • No external excitation by using permanent magnet materials	• High system cost • Low converter efficiency • Large size and weight of generators • High price of permanent magnet materials

1.3 Some Key Issues in Current Wind Power Technologies

1.3.1 MPPT Control

According to the basic principle of the wind power system, when the wind speed is constant and the output power of the wind turbine does not reach its rated power, there exists a maximum wind energy utilisation coefficient C_{pmax}, which makes the wind turbine capture the maximum wind energy under the current working condition. The wind energy utilisation coefficient C_p is mainly related to the impeller design, pitch angle and tip-speed ratio, therefore, the maximum wind energy utilisation coefficient can be obtained through impeller design optimization, pitch control and speed regulation, to make the wind power system output power as high as possible (not higher than the rated power) under the same wind speed condition. In the operation process of the wind power system, the impeller design can not change since the impeller has been manufactured, and the pitch system is not suitable for frequent and rapid adjustment of the pitch angle, so the wind power system will regulate the tip-speed ratio to maintain its maximum wind energy utilisation coefficient over a wide range of wind speed, namely Maximum Power Point Tracking (MPPT).

The control of the turbine system is mainly to make it output the appropriate mechanical power according to the wind conditions, by making the generator speed change with the wind speed properly. For this purpose, a power-speed characteristic curve (as shown in Fig. 1.19) can be pre-designed, according to the capacity and operating speed range of the turbine, as an operating criterion for this system [26, 27].

As can be seen from the above figure, the wind power system operates mainly in four different zones:

Zone 1 where the wind power system is starting—when the generator speed (driven by the wind) is higher than the cut-in speed, the generator starts to run.

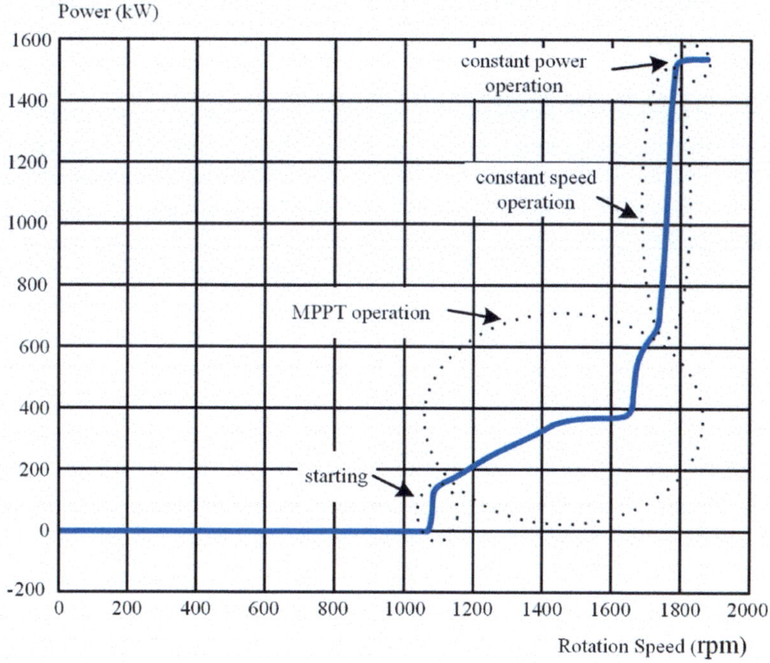

Fig. 1.19 Typical power-speed curve of a turbine system

Zone 2 (with the widest range of operating speed) where the wind power system is operating in MPPT mode—the generator speed is controlled to track the changing wind speed so that the tip-speed ratio is always at an optimum value, thus making that the system absorbs the maximum amount of wind energy in this zone.

Zone 3 where the wind power system is in constant-speed operation—the generator speed is controlled as constant and the turbine output power is increased with the wind speed to near its rated value.

Zone 4 where the wind power system is in constant-power operation—once the system output power has reached its rated value, the wind energy absorbed by the wind turbine is limited by pitch control until the generator speed rises to the cut-out speed and then the system is allowed to exit operation.

It can be noticed that the wind power system operates over the widest range of speed in Zone 2 and the corresponding MPPT control plays the most important role in the overall operational efficiency of the system.

Currently, MPPT control mainly includes methods such as the look-up table method, the estimation method [28] and the peak power seeking method [29], some of which will be described in this book.

1.3.2 Uninterrupted Operation Under Fault/Non-ideal Grid Conditions[5]

The main grid fault for wind power systems is the grid voltage dip, which refers to a sudden drop in grid voltage amplitude to 10–90% of the rated condition for a period of 0.5 industrial frequency cycles to a few seconds [30]. In particular, about 60% of the voltage dip comes from insulator flashover [31] and line discharge to ground caused by lightning strikes, and a portion of the voltage dip mainly comes from transmission line faults, including three-phase short circuits, single-phase grounding, two-phase grounding and phase-to-phase short circuits, where the voltage dip caused by three-phase short circuits is symmetrical and the voltage dip caused by the other three line faults is asymmetrical, and the latter three occupy a major part of the actual system faults [32].

[5] Faulty grid conditions and non-ideal grid conditions are defined mainly based on relevant national grid standards in China, e.g., the "China Emergency Grid Operation Regulations" requires power equipment to be capable of stable long-term operation in the grid condition with a steady-state asymmetry of no more than 2% and a transient asymmetry of no more than 5%. Therefore, in terms of asymmetric grid conditions, this book considers a grid with a non-zero steady-state asymmetry of no more than 2% and a non-zero transient asymmetry of more than 5% to be a non-ideal one, and a grid with a steady-state asymmetry of greater than 2% or a transient asymmetry of greater than 5% to be a faulty one. In practice it can be considered that a faulty grid is a special case of a non-ideal grid, known as a heavily non-ideal grid, and the two concepts will not be clearly distinguished in the subsequent part of this book.

In a DFIG-based wind power system, the stator of the DFIG is directly connected to the grid via a grid-connected switch, making the system very sensitive to grid voltage faults. When the grid voltage dip occurs, it introduces an intense electromagnetic process in the DFIG system [33], which in turn causes overcurrent and overvoltage on the rotor side with the possibility of exceeding the current and voltage levels of the switching devices in the power electronics converter, thus causing permanent damage to the system hardware. Traditional vector control strategies for the DFIG-based wind power system are modelled and derived based on the three symmetrical grid voltage, which facilitates the simplification of control algorithms, but cannot effectively cope with the effects of the asymmetrical grid voltage on system operation, leading to a series of serious problems:

- On the generator side, the asymmetrical grid voltage will introduce a pulsation at twice the grid frequency in the stator flux linkage, which will result in pulsations in the generator output power and electromagnetic torque, causing shocks to the grid and mechanical systems. Besides, the asymmetrical grid voltage will also cause asymmetrical stator and rotor currents, uneven heating, rotor-side over-current and over-voltage, etc.
- On the grid side, the slip power with the pulsation at twice the grid frequency surges into the DC bus, causing the DC-bus voltage to pulsate at twice the grid frequency, which may lead to DC-bus overvoltage, where frequent charging and discharging will also shorten the lifetime of DC-bus capacitors. Besides, the asymmetrical grid voltage will also lead to the asymmetrical grid-side current, heating of the grid-side filter and the pulsation of the grid-side power, etc.

Initially, the wind power system was immediately disconnected from the grid to protect its own hardware system in the event of a grid fault. However, with the increasing individual capacity of the wind power system and the scale of the wind farm, the proportion of wind power in the grid has been increasing, so that large-scale wind power going off-grid will further deprive the grid of its ability to support voltage and frequency, causing further deterioration of grid faults and even triggering more serious chain reactions, which seriously threaten the stability of the grid. Therefore, new grid operation standards have imposed more strict requirements on grid-connected wind power, which require the wind power system to have uninterrupted operation within a certain range of the grid voltage dip, including the fault ride-through (FRT) capability of the wind farm at the point of common coupling (PCC) and the low voltage ride-through (LVRT) capability of the wind power generator.

The LVRT requires that: when the grid voltage dips, the wind power system must remain connected to the grid within the time frame required by the grid, and can even provide some reactive power support for the grid to help the grid voltage recover; when the grid voltage is recovered, the wind power system can quickly resume normal operation and provide normal power output to maintain the stable operation of the power system. Figure 1.20 shows the LVRT standards of some countries with the strong wind power technology, where the horizontal axis is time and the vertical axis is the residual voltage amplitude at the end of the generator. The wind power system is only allowed to disconnect from the grid when the operating point of the system

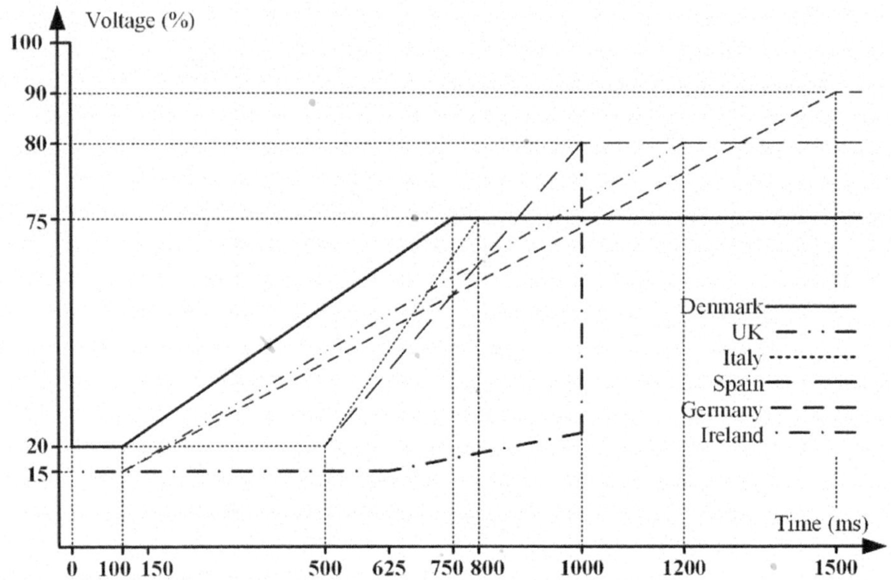

Fig. 1.20 LVRT standard curves of selected countries

is below each curve in the figure, otherwise it must remain connected to the grid to achieve uninterrupted operation. The LTVRT capability can avoid the wind power system from going off and reconnecting to the grid frequently due to grid faults, reducing the impact on the grid, and enable the wind power generator to provide the necessary frequency and voltage support to the grid during grid faults, just like the conventional synchronous generator, improving the stability of the power system.

Currently, most of LVRT standard curves are given for the three-phase symmetrical voltage fault. In practical grid operation, however, the asymmetrical voltage fault rate is much higher than the symmetrical voltage fault rate. This situation has been noticed by researchers, and some countries and companies have already started or are in the process of improving their national standards and corporate standards, respectively, to incorporate the asymmetrical grid voltage fault into the original LVRT requirements in order to cope with more complex grid conditions and to achieve more effective and widespread grid fault ride-through capability. In recent years, the technology for uninterrupted operation of the wind power system under the asymmetrical grid fault has become a new research hotspot in the field of wind power generation [20, 30].

Compared to the DFIG-based wind power system, the PMSG-based wind power system uses a full-scale converter to isolate its permanent magnet synchronous generator from the grid to a certain extent, reducing the impact of grid faults on generator

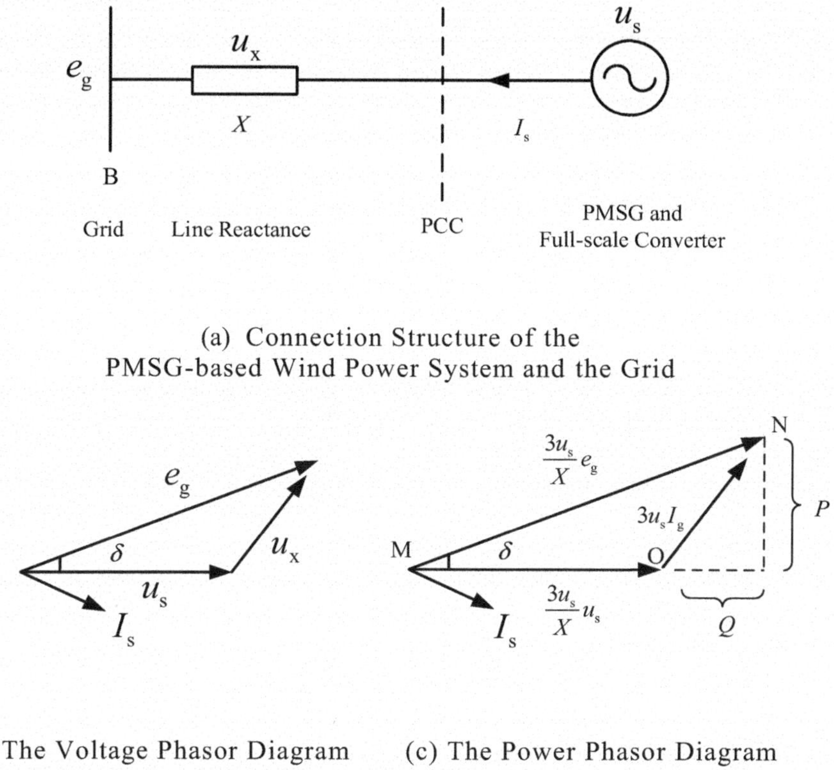

Fig. 1.21 Connection structure of the PMSG-based wind power system and the grid

operation, and its uninterrupted operation under grid faults is easier to achieve [20], but it still has to cope with another non-ideal grid condition—the weak grid.[6]

Figure 1.21a illustrates the interface between the PMSG-based wind power system and the grid. As the wind farm is usually located far from the electricity consuming areas (economically developed areas), the electricity generated by the wind farm needs to be transmitted over a long distance to the load side, where the reactance on the transmission line is high. The wind power system in the weak grid will have the following characteristics:

Firstly, the output power of the wind power system will be limited. The relation between the output power and the line reactance is shown in Eq. (1.2).

$$\begin{aligned} P &= \frac{3U_g}{X}U_c \cdot \sin\delta \\ Q &= \frac{3 \cdot U_g}{X}(U_c \cdot \cos\delta - U_g) \end{aligned} \tag{1.2}$$

[6] A weak grid is a grid where the voltage at the PCC is significantly affected by the output power of the wind power system.

where P and Q are the active and reactive power output from the wind power system respectively, U_g is the grid voltage, U_c is the converter output voltage and δ is the angle of the converter output voltage leading to the grid voltage. It can be seen that: if the wind power converter is to inject active power into the grid, the converter output voltage must be leading to the grid voltage; if the wind power converter is to inject reactive power into the grid, the amplitude of the converter output voltage must be greater than the grid voltage. Also, as can be seen in Fig. 1.21, because of the line reactance, when the output power of the wind power system varies, it can lead to different line reactance voltage drops, resulting in fluctuations in the grid voltage at the PCC, which can affect not only the surrounding electrical equipment, but even the stability of the grid at the access point.

Secondly, with the increasing share of wind power in the grid, it is expected that the wind power generator can stabilize the grid voltage amplitude and frequency just like a conventional synchronous generator, which is especially important in a weak grid. In this case the grid-side converter can no longer operate as a current source, but as a voltage source, to regulate the grid voltage amplitude and frequency. This operating mode also applies to the case where wind power is the only power supply (stand-alone mode). Stand-alone mode can occur in remote areas where wind power is the only power supply when the conventional grid is inaccessible, or when the transmission line in Fig. 1.21a fails and wind power is still needed to maintain the local power supply.

The regulation of the grid voltage frequency and amplitude by the wind power system is achieved by feeding active and reactive power into the grid, respectively. For active power, the wind power captured by the turbine and the load power on the grid side need to be balanced. At this point, the turbine is not operating at its maximum power point but tracking the load power. When the wind power captured by the turbine is greater than the power required by the load, the captured wind power can be reduced by adjusting the speed of the generator and the pitch angle of the turbine. As wind speed varies randomly, when the maximum wind power captured still cannot meet the load active power demand, the energy storage system need to be used to provide energy. For reactive power, the ability of the system to generate reactive power depends mainly on the current capacity of the grid-side converter, as the reactive power mainly circulates within the grid-side converter.

The following aspects of wind power system operation in weak grid conditions need to be taken into account: (1) The detection of grid status, including grid faults, and the determination of strong and weak grids. (2) The control method for grid-side converter operating as a voltage source. The converter operating as a voltage source has been widely used in uninterruptible power supplies (UPS). Output voltage amplitude and frequency regulation of converter has been studied in the literature for linear, non-linear, symmetrical and asymmetrical loads. (3) Parallel operation control of multiple wind power converters. When multiple wind turbines are operated in parallel, the load power needs to be distributed according to the different capacities of the converters. Droop control is an effective method to achieve equal power distribution, and its basic principle is to simulate the adjustment characteristics of the output voltage amplitude and frequency of a conventional synchronous generator. (4) The problem of active control of active power. As mentioned earlier, the power captured by the turbine needs to be balanced with the load power to achieve active power regulation. In terms of the topology of the full-power converter, this is translated into the problem of controlling the converter DC bus voltage, and if the DC bus voltage control is relatively stable, it implies that the power captured by the turbine is well-balanced with the load power.

1.3.3 Topologies and Control Methods of Large-Capacity and High-Voltage[7] Wind Power Systems

With the gradual increase in stand-alone wind power generator capacity, the capacity of the power electronic converter supporting the generator also needs to be increased accordingly. A common low-voltage (690 V) two-level power electronic converter is difficult to meet the requirements of a stand-alone large-capacity wind power system. At present, the methods for high-capacity wind power converters mainly include:

(1) The Method Based on Parallel Converters

It is achieved by increasing the current capacity of the system through the parallel connection of converters, without increasing the voltage level of the system, and by breaking up a large-capacity converter into several small-capacity converters, as shown in Fig. 1.22.

The parallel-converter scheme in Fig. 1.22 uses a rated voltage of 690 V. For an onshore wind power system, the step-up transformer connected to the grid is usually placed at the bottom of the tower and the electricity is transmitted to the transformer at a low voltage and high current. If the 690 V rated voltage is used, the rated current will reach 1673 A for a 2 MW system and 4183 A for a 5 MW system, which will bring a series of problems: The excessive transmission current increases the cost and manufacturing difficulty of the power cables and supporting switchgear in the tower and causes extra line losses (I^2R) and voltage drops. At the same time, the electromagnetic interference in the nacelle and tower is serious, and the high-power low-voltage power electronic conversion devices have higher switching losses due to the higher current flowing through the devices. In addition, the use of the grid-side step-up transformer increases costs and reduces the operational efficiency of the system. It can be seen that, as far as the stand-alone large-capacity wind power system is concerned, the 690 V low-voltage electrical system has been unable to meet its economic and rational requirements, and the shift from low-voltage to high-voltage electrical system configuration has become an inevitable development trend.

[7] In this book, it is referred to as 690 V for low voltage systems and 3–35 kV for high voltage systems.

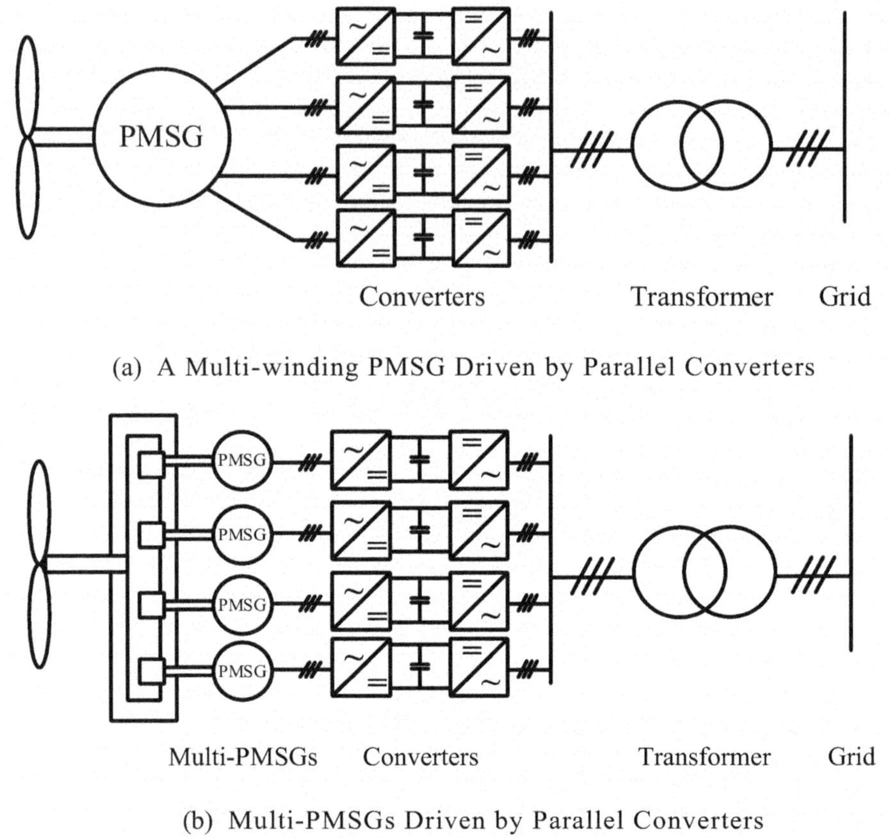

Fig. 1.22 Large-capacity wind power systems based on parallel converters

(2) The Method Based on Multi-level Converters

A more suitable converter topology for large capacity and high voltage is the multi-level converter. There are two main types of high-voltage multi-level converters in widespread use today, the neutral point clamped (NPC) converter and the H-bridge cascaded converter. The advantages of the multi-level converter compared to the conventional two-level converter are: (in a multi-level converter) the lower voltage withstand level required for the same DC-bus voltage, and the improved harmonics and dv/dt of the output voltage due to the increased levels of the output voltage waveform, reducing the impact of the resulting EMI on the surrounding circuits.

A topology using a PMSG and a back-to-back diode-clamped three-level converter (a typical NPC converter) is illustrated in Fig. 1.23. In the converter, each bridge arm has four switching devices, each of which is subjected to the half voltage of a conventional two-level converter, so that the topology can increase the dc bus voltage as well as the output voltage level with the same withstand voltage of the devices. At the same time, the output phase voltage is 3-level and the line voltage is 5-level, thus

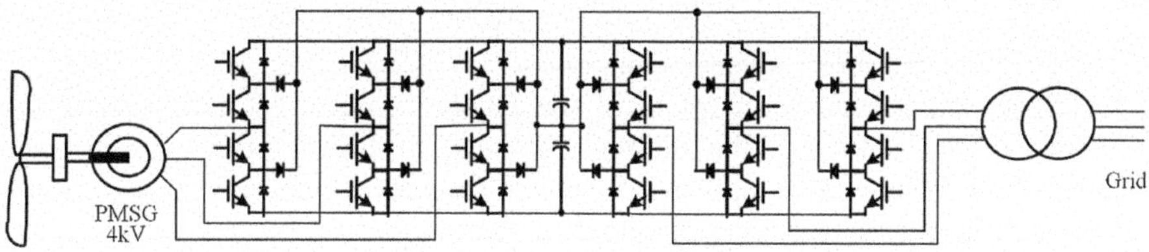

Fig. 1.23 A wind power topology based on NPC 3-level converters

reducing the harmonics of the output voltage and the voltage stress on the devices. If a 4.5 kV power electronic device, such as an IGCT, is used, the DC bus voltage can be set at 5.4 kV, the output voltage can reach 3300 V and the output power can reach 10 MW. For the same capacity wind power converter, a high-voltage system at 3300 V can reduce the rated current to 1/5 of that of a low-voltage system at 690 V. Related research results show that in wind power converters above 2 MW, the three-level high-voltage converter has a greater cost advantage compared to the two-level converter in parallel structure. To increase the output voltage level further, a five-level converter can be used, where the midpoint voltage does not have stability problems due to the back-to-back structure.

Currently, research on the diode-clamped converter for wind power generation is focusing on:

I. Pulse width modulation (PWM) algorithm. For a three-level converter in Fig. 1.23 (consisting of 3 bridge arms connected to the grid) there are 27 switching states and for a five-level converter there are up to 125 switching states. It is a complicated problem to choose the proper voltage vector according to the operating conditions of the load and the requirements of the converter itself. If the traditional space vector pulse width modulation (SVPWM) method is used, the determination of the vector triangle, the selection of the basic vector and the calculation of the vector action time will greatly increase the computational complexity of the algorithm. Although the selection of vectors can be simplified by using methods such as 60° and 120° coordinate transformations, such transformations again increase the computation and the selection of redundant vectors is still obtained by looking up tables. A simple modulation method that corresponds to SVPWM is the carrier PWM (CPMW), which obtains the proper switching states simply by comparing the designed modulation wave and carrier wave. Since CPWM is based on a three-phase stationary coordinate frame and the actual switching devices of a converter are also in the stationary coordinate frame, the use of CPWM simplifies the modulation algorithm. The key issue here is how to obtain the required modulation wave and carrier wave. Different modulation waves are generated by various control algorithms, while for a single modulation wave, different carriers will influence its PWM harmonic characteristics. Essentially, SVPWM and CPWM have equivalence when the zero-sequence component is considered. Moreover, the selection of the redundant vector in SVPWM corresponds in essence to the selection of the zero-sequence component.

II. Voltage balance at the midpoint of the DC-bus capacitors. The problem of controlling (balancing) the midpoint voltage of the DC-bus capacitors in Fig. 1.23 is inherent to a diode-clamped converter topology. In a back-to-back structure, both rectifier and inverter can coordinate to control the DC-bus midpoint voltage. If SVPWM is used, the midpoint voltage of DC-bus capacitors can be balanced by selecting proper redundant vectors while outputting the desired voltage. As mentioned earlier, if CPWM is used, it is necessary to find the quantitative relation between the DC-bus midpoint voltage and the zero-sequence component to achieve balance control of the midpoint voltage. In addition, the switching loss and common mode voltage suppression of a three-level converter are also hot topics of interest to current researchers.

III. Fault detection and protection. Compared to the two-level converter, the three-level converter has twice as many switching devices and a more complex topology. For a wind power system, especially an offshore case, the high cost of system maintenance and troubleshooting places high demands on the reliability of the converter and therefore requires a thorough study of the fault and protection of the three-level converter for wind power system. So far, only a few articles have explored the fault and protection of a multi-level converter, mainly focusing on overcurrent fault protection and uninterrupted operation during faults.

A converter with a similar structure to the diode-clamped converter is the capacitor-clamped converter, where the clamping diodes in Fig. 1.23 are replaced by clamping capacitors. This topology is not widely used, considering the large number of its capacitors, as well as lifetime issues and control complexity [1].

A new converter topology using a multi-winding permanent magnet synchronous wind generator and a modular cascaded converter is given in Fig. 1.24. The basic idea is that individual windings inside the generator, which are insulated to each other, provide isolated power sources to the individual modules of the cascaded converter. Each module has a rectifier, a booster and an inverter with a single direction of power flow. The single-phase high-voltage converter consists of a group of low-voltage modules connected in series at the output ports. Three single-phase high-voltage converters are connected in a star connection to make a three-phase converter. The output of the 3-phase converter can be directly connected to the 10.5 kV high-voltage AC network via a grid-connected reactor, eliminating the requirement for the grid-side transformer. This is the only multi-level converter topology that can achieve an output of over 10 kV at current power electronics device withstand voltage level.

In a motor drive system, the modular cascaded converter is also known as the H-bridge cascade structure or Robincon structure, where it usually requires a zigzag transformer with multiple secondary windings to provide multiple isolated power sources, connects each isolated power source to a rectifier and an inverter sequentially to form a low-voltage module, and then cascades the low-voltage modules to achieve a high-voltage output. In contrast, in a wind power system, the PMSG

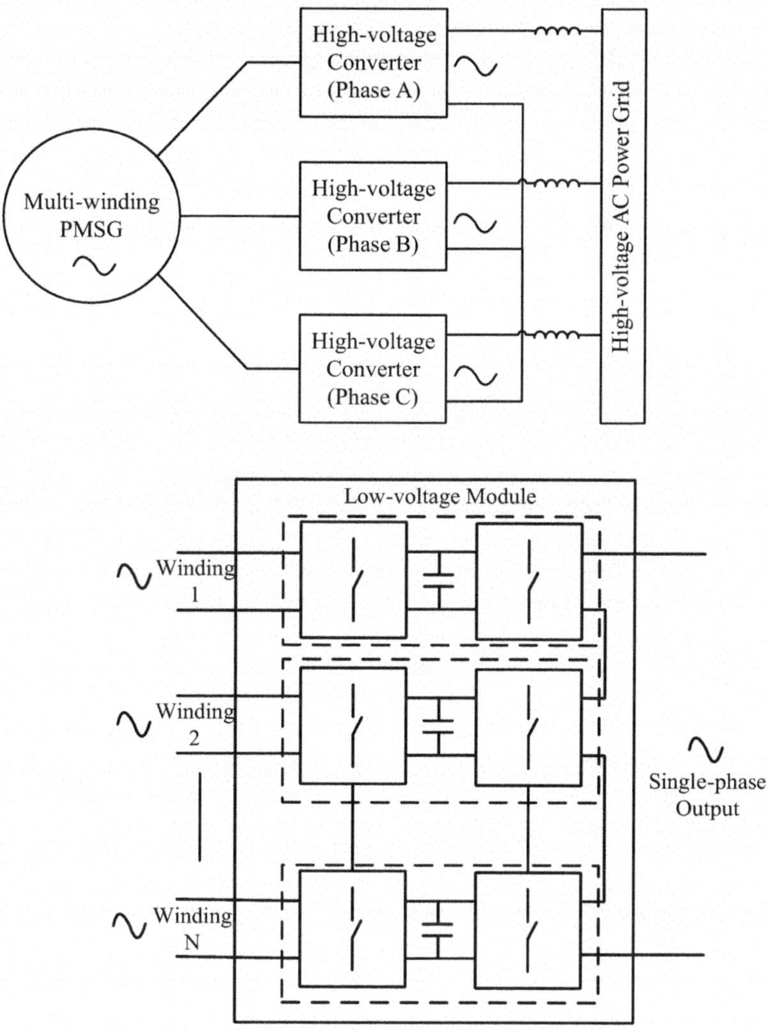

Fig. 1.24 A novel modular cascaded converter topology for megawatt-class high-voltage PMSG-based wind power system

usually has multiple poles, multiple phases and multiple windings. These multiple generator windings can be used separately as multiple isolated power sources, thus eliminating the requirement for the complex input transformer. At the same time, an important advantage of modulization is that, if one module fails, the others can function normally. This uninterrupted operation capability and redundancy is particularly suitable for wind power applications where the reliability of the converter is critical. Another advantage of this topology is that it makes it possible to eliminate the need for a large step-up transformer on the grid side. For an onshore wind power system, the high-voltage output means reduced output current as well as the corresponding transmission loss, voltage drop and cable cost. For an offshore wind power system, the conventional topology usually places the bulky step-up transformer at the top of the tower, which significantly increases the mechanical stress on the tower. The novel topology here, which is connected directly to the grid via a high-voltage converter, can eliminate the step-up transformer and thus reduce the mechanical stress on the tower. Research on similar topologies is currently being carried out by Tsinghua University in China and Durham University in the UK.

At present, research on the new cascaded converter for wind power generation is mainly focused on: (1) The study of the low-voltage module. The low-voltage module is the basic component of the high-voltage multi-level converter, which has the structure of AC/DC/AC conversion. And the topology and control strategy of each part of the low-voltage module need to be designed in consideration of the output voltage variation of the multi-winding PMSG. (2) The analysis and trade-off of many factors in the whole system such as the number of series stages of the low-voltage modules, the volt-ampere capacity of the high-power power electronics devices, inductors and capacitors, the number and the withstand voltage of the PMSG stator armature windings and the fault-tolerant operating capability of the generator system. (3) The design of the overall control strategy for the new topology.

(3) The Method Based on High-voltage Generator

An alternative method of achieving the large-capacity conversion of wind power is to use a high-voltage generator. At present, some well-known international wind power companies have already started research and attempts to develop high-voltage wind power generators.

Semikron has proposed a large-capacity wind power system based on a high-voltage PMSG, a high-voltage DC bus (5–10 kV) and an inverter made up of low-voltage modules connected in series, as shown in Fig. 1.25. In this system, the three-phase high-voltage PMSG is connected to the 5–10 kV DC bus via the diode rectifier. The inverter consists of series-connected low-voltage modules on the DC side (the low-voltage module structure is shown in the dashed rectangular box in Fig. 1.25). L is the parasitic inductance of each wire, the switching tube S is used to bypass the corresponding low-voltage module, the inverter circuit of each module is a low-voltage three-phase PWM inverter, the AC output side of each module is the low AC voltage, and the system is connected to the high-voltage (35 kV) grid through a high-capacity transformer with several low-voltage primary three-phase windings and a high-voltage (35 kV) secondary three-phase winding. As can be seen, the system still requires a high-capacity step-up transformer and the control is relatively complex.

ABB has also proposed a high-voltage PMSG-based offshore wind power system (also known as Windformer), which is shown in Fig. 1.26. It is characterized by: firstly, the use of circular cross-section cable windings with a high insulation class, which allows the generator windings to withstand voltages of 20–40 kV; secondly, the use of the high-voltage DC bus to connect multiple rectified generators in parallel, which is then connected to a high-voltage inverter to convert the DC energy into

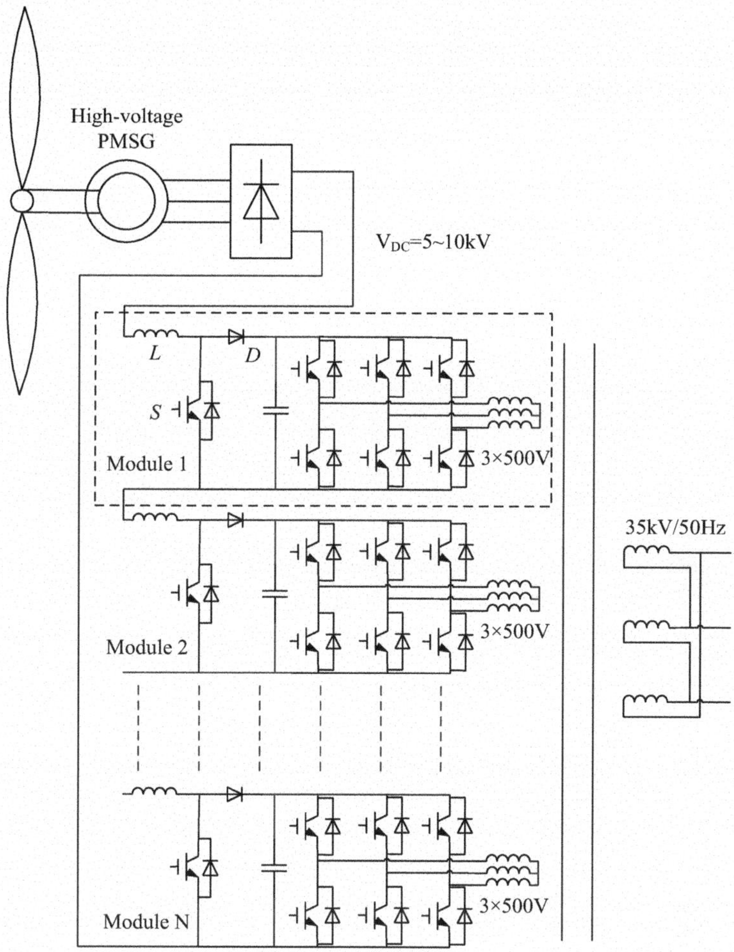

Fig. 1.25 A high-voltage PMSG-based wind power system proposed by Semikron Company

appropriate AC power. The technical challenge of this wind power system, based on a light DC transmission concept, also remains the high-voltage power electronic conversion.

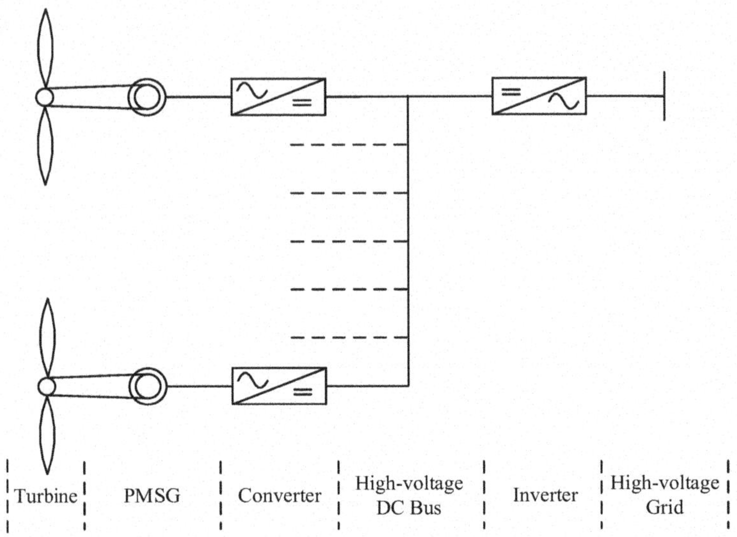

Fig. 1.26 A high-voltage PMSG-based wind power system proposed by ABB Company

1.3.4 Topologies and Control Methods of Large-Scale Offshore Wind Power Systems

Although the total installed capacity of wind power has been increasing continuously over the past decade, the growth rate of onshore wind power installed capacity has started to slow down since 2015, while the growth rate of offshore wind power installed capacity has been steadily increasing, up to 23.5% in 2018. Offshore wind power features huge reserves, stable wind energy propagation, no occupation of land resources, no consumption of water resources, high power utilisation hours, and low ecological impact, etc. Compared with onshore wind power, offshore wind power is more continuous, especially offshore wind farms have higher power output efficiency and are close to traditional load centers, facilitating grid consumption and avoiding a series of problems such as long-distance transmission, which makes offshore wind power attract widespread attention from various countries, and large-scale offshore wind power systems have become an important trend in the future development of wind power.

Currently, there are three main forms of grid-connected offshore wind power transmission: High Voltage Alternate Current (HVAC), Line Commutated Converter—High Voltage Direct Current (LCC-HVDC) and Voltage Source Converter—High Voltage Direct Current (VSC-HVDC). Among them, the HVAC is the most mature, but is easily affected by the transmission cable capacitive resistance and inductive resistance (especially the submarine cable to ground capacitive resistance), which can cause system oscillations and require large amounts of reactive power compensation, increasing the system difficulty and cost, therefore the HVAC is only suitable for short-distance offshore wind power transmission. By contrast, the HVDC is suitable for long-distance transmission due to the absence issue of reactive power compensation, and the HVDC-based system generally requires fewer cables than the HVAC-based system, resulting lower cable cost. In HVDC technology, the VSC-HVDC can complete all the functions of LCC-HVDC, and is characterized by flexible control, bi-directional power flow, active and reactive power decoupling control, etc., thereby being highly advantageous in terms of voltage stabilization, fault ride-through and weak grid support, etc. The VSC-HVDC is becoming the most important form of transmission for offshore wind power, with its various topologies, control methods, fault ride-through methods and other related technologies being the current research hotspots of large-scale offshore wind power systems.

Depending on the connection between stand-alone wind power units, large-capacity offshore wind power topologies can be classified into series, parallel and hybrid topologies.

Figures 1.27 and 1.28 illustrate two typical parallel VSC-HVDC topologies. Figure 1.27 shows an AC parallel topology, where the AC side of each stand-alone wind power unit is connected in parallel to the AC bus and converges on the offshore HVDC converter, and then, via a subsea DC cable, the DC power outputted by the offshore HVDC converter is transmitted to the onshore HVDC converter, where the DC power is converted to the AC power and integrated into the grid. It can be seen that the offshore part of this topology requires a larger number of high-capacity transformers, thus increasing the size and weight of the system and bringing inconvenience to offshore wind power construction. Some researchers have proposed to use high-frequency AC current in the AC bus of the offshore part, so as to significantly reduce the size and weight of the transformers. Figure 1.28 shows a DC parallel topology, where the DC side of each stand-alone wind power unit is connected in parallel to the DC bus and converges on the offshore HVDC converter, and then, via a subsea DC cable, the DC power outputted by the offshore HVDC converter is transmitted to the onshore HVDC converter, where the DC power is converted to the AC power and integrated into the grid. This topology eliminates the transformers on the offshore platform and obtains high DC voltage directly through DC boost process but may require a multi-step boost conversion. The above-mentioned parallel VSC-HVDC topologies both require a large-capacity centralized converter platform to be built offshore and require a multi-step conversion (by transformers or by multi-step DC/DC converters), which increases the difficulty and cost of construction, while the low-voltage and high-current working mode also increases the loss of the system.

Figure 1.29 illustrates a typical series VSC-HVDC topology. In this case, the DC side of each stand-alone wind power unit is connected in series to directly obtain the high DC voltage required by the offshore transmission, and, via a subsea DC cable, the high-voltage DC power is transmitted to the onshore VSC-HVDC converter, where the high-voltage DC power is converted to the AC power and integrated into the grid. This topology eliminates the transformers and the large-capacity centralized converter platform in the offshore part, thus decreasing cost, reducing complexity

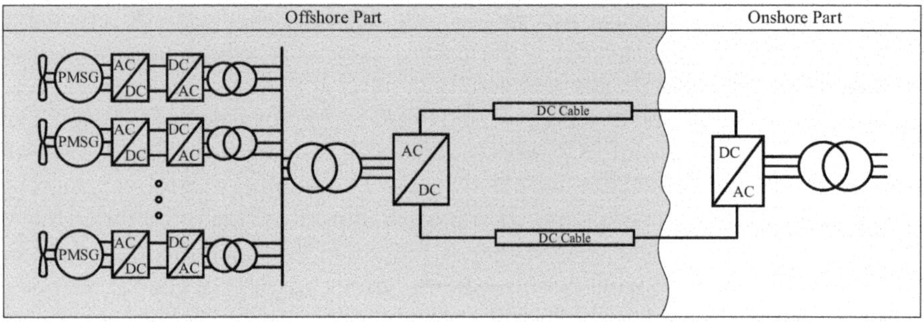

Fig. 1.27 A large-scale offshore wind power topology based on the paralleled VSC-HVDC structure with the paralleled AC bus

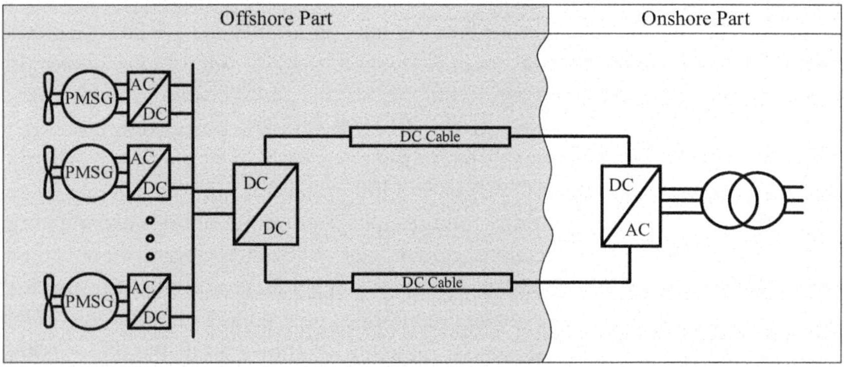

Fig. 1.28 A large-scale offshore wind power topology based on the paralleled VSC-HVDC structure with the paralleled DC bus

and improving efficiency, reliability and maintainability of the system, but the series-connected DC bus poses certain difficulties in dealing with the faults of stand-alone wind power units and transmission cables.

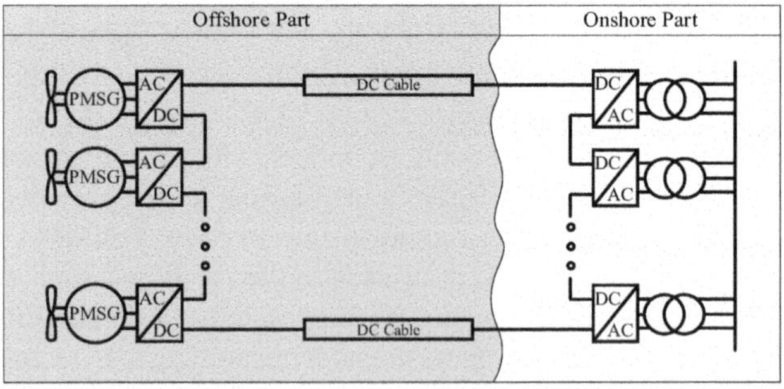

Fig. 1.29 A large-scale offshore wind power topology based on the series VSC-HVDC structure

Figures 1.30 and 1.31 show two hybrid VSC-HVDC topologies. In Fig. 1.30, each stand-alone wind power unit is connected in series and then in parallel. While maintaining the advantages of a series topology, it limits the impact caused by a stand-alone wind power unit failure to the branch where the faulty stand-alone wind power unit is located, avoiding the impact on the whole system and providing a certain economic value, but this topology may also cause current loops between the branches, thus reducing system efficiency. In Fig. 1.31, each stand-alone wind power unit is connected in a matrix. In which, by controlling controllable switches, the faulty unit or DC fault point can be bypassed and the faulty branch can be merged into its adjacent branch, thus reducing the impact caused by the stand-alone wind power unit fault on the whole system and avoiding the branch loop, but this topology is complex, uses many switch devices and increases the size and cost more.

In recent years, with the research and maturity of converter topologies such as Matrix Converter (MC) and Modular Multi-level Converter (MMC) as well as their control methods, more novel converter topologies have been applied to large-scale offshore wind power systems, and the research on related topologies and control methods is flourishing.

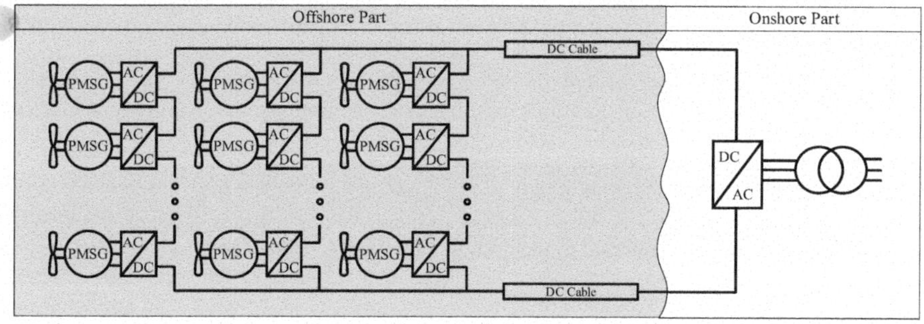

Fig. 1.30 A large-scale offshore wind power topology based on the hybrid VSC-HVDC structure

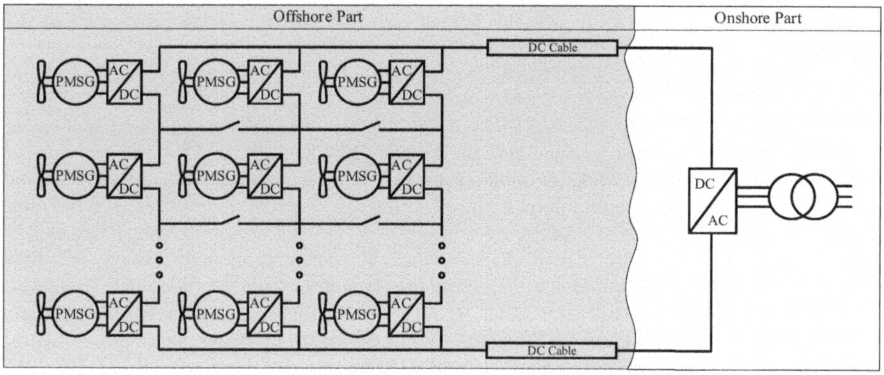

Fig. 1.31 A large-scale offshore wind power topology based on the matrixed VSC-HVDC structure

References

1. Y.D. Li, X. Xiao, Y. Gao, *Large-Capacity Multi-level Converters: Principles, Control and Applications* (Science Press, Beijing, 2005)
2. H.Y. Ye, *Control Technologies of Wind Power Generators* (China Machine Press, Beijing, 2002)
3. Renewable Energy Policy Network for 21st Century. Renewables 2017: Global Status Report, 2017
4. Global Wind Energy Council. Global Wind Energy Outlook 2016, 2016
5. Lazard Freres. Lazard Levelized Cost of Energy 2017, 2017
6. Renewable Energy Policy Network for 21st Century. Renewables 2019: Global Status Report, 2019
7. J. Baroudi, V. Dinavahi, A. Knight, A review of power converter topologies for wind generators. Renew. Energy **32**, 2369–2385 (2007)
8. X.J. Wu, J.Y. Chai, X.H. Wang, Overview of AC excitation for variable speed constant frequency double fed wind power generator systems. Autom. Electr. Power Syst. **28**(23), 92–96 (2004)
9. Y.K. He, M.M. He, R.D. Zhao et al., Analysis of the frequency converter served for the AC excitation of wind-power DFIG. Autom. Electric Power Syst. **30**(4), 105–112 (2006)
10. J. Ribrant, L. Bertling, Survey of failures in wind power systems with focus on Swedish wind power plants during 1997–2005. IEEE Trans. Energy Convers. **22**(1), 167–173 (2007)
11. H. Polinder, F. Pijl, G. Vilder et al., Comparison of direct-drive and geared generator concepts for wind turbines. IEEE Trans. Energy Convers. **21**(3), 725–733 (2006)
12. S. Aimani, B. Francois, B. Robyns et al., Modeling and simulation of doubly fed induction generators for variable speed wind turbines integrated in a distribution network, in *EPE 2003* (2003)
13. S.Q. Sun, *Voltage Fluctuations and Flickers* (China Electric Power Systems, Beijing, 1998)
14. G. Wu, Y. Teng, Y.G. Pan et al., Research on voltage sag of power systems. North China Electric Power **4**, 1–4 (2004)
15. M. Tazil, V. Kumar, R.C. Bansal et al., Three-phase doubly fed induction generators: an overview. IET Electr. Power Appl.Electr. Power Appl. **4**(2), 75–89 (2010)
16. M. Chinchilla, S. Arnalte, J.C. Burgos, Active and reactive power limits of three-phase PWM voltage source inverter connected to the grid, in *EPE-PEMC* (2002), pp. 1–10
17. J.L. Li, S.J. Hu, X.B. Fu et al., High-power direct-driven wind power system topologies. Electric Power Autom. Equip. **28**(7), 73–76 (2008)
18. X. Zeng, Z. Chen, F. Blaabjerg, Design and comparison of full-size converters for large variable-speed wind turbines, in *EPE* (2007), pp. 1–10
19. A. Faulstich, J.K. Stinke, F. Wittwer, Medium voltage converter for permanent magnet wind power generators up to 5 MW, in *EPE* (2005), pp. 9–17
20. Y. Gao. Research on the stability domain of diode-clamped multilevel inverter voltage balancing control. Ph.D. Thesis, Tsinghua University, Beijing, 2007
21. D.G. Holmes, T.A. Lipo, *Pulse Width Modulation for Power Converters* (Wiley, New York, 2003)
22. C. Wang, Y. Li, X. Xiao, A unified SVM algorithm for multilevel converter and analysis of zero sequence voltage components, in *IEEE IECON* (2006), pp. 2020–2024
23. N. Celanovic, D. Boroyevich, A comprehensive study of neutral-point voltage balancing problem in three-level neutral-point-clamped voltage source PWM inverters. IEEE Trans. Power Electron. **15**(2), 242–249 (2000)
24. M.A. Perez, J.R. Espinoza, J.R. Rodriguez et al., Regenerative medium-voltage AC drive based on a multicell arrangement with reduced energy storage requirements. IEEE Trans. Ind. Electron. **52**(1), 171–180 (2005)
25. J.M. Carrasco, L.G. Franquelo, J.T. Bialasiewicz et al., Power electronic systems for the grid integration of renewable energy sources: a survey. IEEE Trans. Ind. Electron. **53**(4), 1002–1016 (2006)

26. E. Cengelci, P. Enjeti, Modular PM generator/converter topologies, suitable for utility interface of wind/micro turbine and flywheel type electromechanical energy conversion systems, in *IEEE IAS* (2000), pp. 2269–2276
27. Y. Guofeng, Research on megawatt variable speed constant frequency doubly fed induction generator and its experimental systems. Postdoctoral Exit Research Report, Tsinghua University, Beijing, China, 2008
28. C.H. Ng, M.A. Parker, L. Ran et al., A multilevel modular converter for a large, light weight wind turbine generator. IEEE Trans. Power Electron. **23**(3), 1062–1074 (2008)
29. D. Schreiber, Power converter circuit for generators with dynamically varying output power. US Patent US6680856, 2004
30. M. Dahlgren, H. Frank, M. Leijon, Windformer wind power goes large-scale. ABB Rev. **3**, 31–37 (2000)
31. X. Li, Research on technology of grid integration and power transmission for large capacity of offshore wind turbines. Ph.D. Thesis, North China Electric Power University, Beijing, 2014
32. D.Q. He. Series-connected DC based offshore wind farm and its control. Ph.D. Thesis, Shanghai Jiaotong University, Shanghai, 2013
33. C. Shadi, T. Ahmadreza, Matrix interconnected topology for DC collector systems of offshore wind farms, in *IET Conference on Renewable Power Generation (RPG 2011)*, Edinburgh, UK, 6–8 September 2011, pp. 1–4

Chapter 2
Mathematical Models and Modelling Methods for the DFIG-Based Wind Power System

In the DFIG-based wind power system, the stator of the generator is connected directly to the grid through a grid-connected switch, while the rotor of the generator is connected to the grid through a generator-side converter and a grid-side converter, which is illustrated in Fig. 2.1. Here, the generator-side converter is connected to the rotor and regulates the rotor current for generator control,[1] and the grid-side converter regulates the DC-bus voltage, feeds the slip power into the grid and realizes the decoupled control of active and reactive power. A filter structure is usually found between the grid-side converter and the grid to suppress the harmonic currents caused by the grid-side converter.

In this chapter, the dynamic mathematical models of the DFIG and the grid-side converter in various coordinate frames are first introduced, and then common system modelling methods are discussed, focusing on two graphical modelling methods—the Causal Ordering Graph (COG) and the Energetic Macroscopic Representation (EMR), followed by an example of the mechanical part of the DFIG-based wind power system to illustrate how it can be modelled using the COG and the EMR and how its basic control methods can be derived.

2.1 Mathematical Model of the DFIG

The DFIG is a multi-variable, strong-coupled, non-linear, time-varying and high-order system for which the following assumptions are usually made to facilitate analysis [1]:

[1] In this paragraph, the generator control includes speed regulation, MPPT, active and reactive power decoupled control, etc.

H. Ma et al., *The Control Principle of Wind Power Generation System*,
https://doi.org/10.1007/978-981-99-6083-5_2

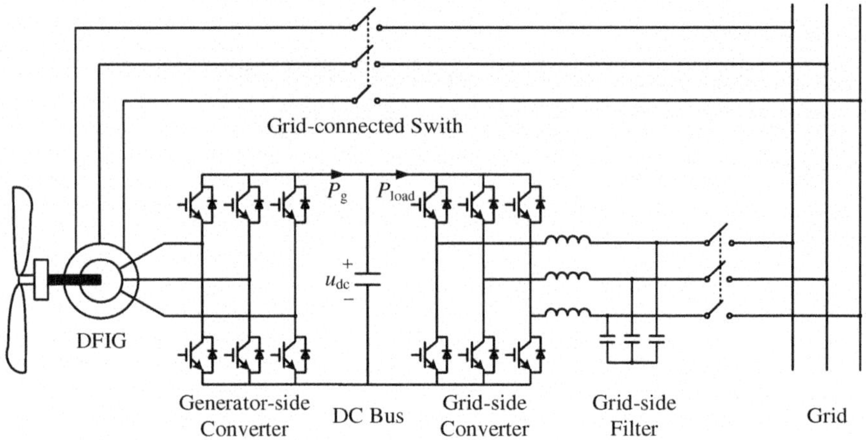

Fig. 2.1 Typical topology of the DFIG-based wind power system

(1) Neglecting the space harmonics, the three-phase windings are considered symmetrical with a mutual difference of 120° electrical angle in space, which produces the magnetic motive force distributed sinusoidally along the air gap.
(2) Neglecting magnetic circuit saturation, the self- and mutual inductances of each winding are considered to be constant.
(3) The iron core losses are not taken into account.
(4) The effect of frequency variation and temperature variation on winding resistance is not considered.
(5) The parameters of the rotor winding are converted to the stator side with the same number of turns in the stator and the rotor windings after conversion.

Under the above assumptions, the windings of the DFIG can be equated to the winding model shown in Fig. 2.2. In this case, the stator three-phase winding axes **A**, **B** and **C** are fixed in space, with the **A**-axis as the reference axis; the rotor three-phase winding axes **a**, **b** and **c** rotate with the rotor, and the angle between the rotor **a**-axis and the stator **A**-axis is θ_r in electrical angle. The positive direction of voltage, current and flux in each winding is specified according to a general motor model (but not a general generator model) and the right-hand spiral rule [2].

The mathematical model of the DFIG shown in Fig. 2.2 can be represented by the following voltage equation, flux linkage equation, torque equation and motion equation.

(1) **Voltage Equation**

As can be seen in Fig. 2.2, the voltage balance equation for the three-phase stator winding is

Fig. 2.2 Winding model of the DFIG in 3-phase stationary coordinate frame

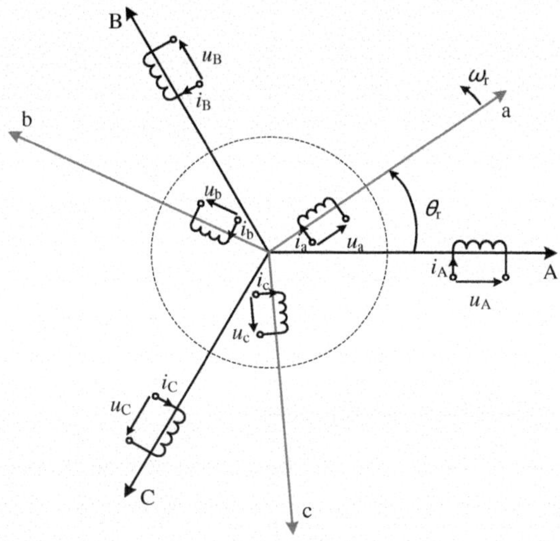

$$\begin{bmatrix} u_A \\ u_B \\ u_C \end{bmatrix} = \begin{bmatrix} R_s & 0 & 0 \\ 0 & R_s & 0 \\ 0 & 0 & R_s \end{bmatrix} \begin{bmatrix} i_A \\ i_B \\ i_C \end{bmatrix} + p \begin{bmatrix} \psi_A \\ \psi_B \\ \psi_C \end{bmatrix} \tag{2.1}$$

where p is the differential operator, equivalent to d/dt; u_A, u_B, u_C are the instantaneous stator phase voltages; i_A, i_B, i_C are the instantaneous stator phase currents; ψ_A, ψ_B, ψ_C are the instantaneous stator phase flux linkages; R_s is the stator phase resistance.

Similarly, after winding conversion of the rotor winding (to the stator side), the voltage equation based on the rotation axes **a**, **b** and **c** is

$$\begin{bmatrix} u_a \\ u_b \\ u_c \end{bmatrix} = \begin{bmatrix} R_r & 0 & 0 \\ 0 & R_r & 0 \\ 0 & 0 & R_r \end{bmatrix} \begin{bmatrix} i_a \\ i_b \\ i_c \end{bmatrix} + p \begin{bmatrix} \psi_a \\ \psi_b \\ \psi_c \end{bmatrix} \tag{2.2}$$

where u_a, u_b, u_c are the instantaneous rotor phase voltages; i_a, i_b, i_c are the instantaneous rotor phase currents; ψ_a, ψ_b, ψ_c are the instantaneous rotor phase flux linkages; R_r is the rotor phase resistance. All the above physical quantities have been converted (only by winding conversion) to the stator side, and for convenience, the upper corner mark "′" indicating the converted quantities has been ignored in subsequent parts of this book.

From Eqs. (2.1) and (2.2), the voltage equation of the DFIG is

$$\begin{bmatrix} u_A \\ u_B \\ u_C \\ u_a \\ u_b \\ u_c \end{bmatrix} = \begin{bmatrix} R_s & 0 & 0 & 0 & 0 & 0 \\ 0 & R_s & 0 & 0 & 0 & 0 \\ 0 & 0 & R_s & 0 & 0 & 0 \\ 0 & 0 & 0 & R_r & 0 & 0 \\ 0 & 0 & 0 & 0 & R_r & 0 \\ 0 & 0 & 0 & 0 & 0 & R_r \end{bmatrix} \begin{bmatrix} i_A \\ i_B \\ i_C \\ i_a \\ i_b \\ i_c \end{bmatrix} + p \begin{bmatrix} \psi_A \\ \psi_B \\ \psi_C \\ \psi_a \\ \psi_b \\ \psi_c \end{bmatrix} \tag{2.3}$$

It can also be vectorized as

$$\boldsymbol{u} = \boldsymbol{R}\boldsymbol{i} + p\boldsymbol{\psi} \tag{2.3a}$$

(2) **Flux Linkage Equation**

As shown in Fig. 2.3, the flux linkage in the stator A-phase winding consists of the self-induced flux linkage ψ_{AA} (including the A-phase mutual induced flux linkage ψ_{Am} and the leakage flux linkage $\psi_{\sigma 1}$) produced by the A-phase current and the mutual induced flux linkage (including ψ_{AB}, ψ_{AC}, ψ_{Aa}, ψ_{Ab}, ψ_{Ac}) produced by the B-phase, C-phase, a-phase, b-phase and c-phase currents, i.e. there are

$$\psi_{AA} = \psi_{Am} + \psi_{\sigma 1} \tag{2.4}$$

$$\psi_A = \psi_{AA} + \psi_{AB} + \psi_{AC} + \psi_{Aa} + \psi_{Ab} + \psi_{Ac} \tag{2.5}$$

Similarly, the stator and the rotor flux linkages of each phase in the DFIG can be obtained as following:

$$\begin{bmatrix} \psi_A \\ \psi_B \\ \psi_C \\ \psi_a \\ \psi_b \\ \psi_c \end{bmatrix} = \begin{bmatrix} \psi_{AA} + \psi_{AB} + \psi_{AC} + \psi_{Aa} + \psi_{Ab} + \psi_{Ac} \\ \psi_{BA} + \psi_{BB} + \psi_{BC} + \psi_{Ba} + \psi_{Bb} + \psi_{Bc} \\ \psi_{CA} + \psi_{CB} + \psi_{CC} + \psi_{Ca} + \psi_{Cb} + \psi_{Cc} \\ \psi_{aA} + \psi_{aB} + \psi_{aC} + \psi_{aa} + \psi_{ab} + \psi_{ac} \\ \psi_{bA} + \psi_{bB} + \psi_{bC} + \psi_{ba} + \psi_{bb} + \psi_{bc} \\ \psi_{cA} + \psi_{cB} + \psi_{cC} + \psi_{ca} + \psi_{cb} + \psi_{cc} \end{bmatrix}$$

$$= \begin{bmatrix} L_{AA} & L_{AB} & L_{AC} & L_{Aa} & L_{Ab} & L_{Ac} \\ L_{BA} & L_{BB} & L_{BC} & L_{Ba} & L_{Bb} & L_{Bc} \\ L_{CA} & L_{CB} & L_{CC} & L_{Ca} & L_{Cb} & L_{Cc} \\ L_{aA} & L_{aB} & L_{aC} & L_{aa} & L_{ab} & L_{ac} \\ L_{bA} & L_{bB} & L_{bC} & L_{ba} & L_{bb} & L_{bc} \\ L_{cA} & L_{cB} & L_{cC} & L_{ca} & L_{cb} & L_{cc} \end{bmatrix} \begin{bmatrix} i_A \\ i_B \\ i_C \\ i_a \\ i_b \\ i_c \end{bmatrix} \tag{2.6}$$

where L_{AA}, L_{BB}, L_{CC}, L_{aa}, L_{bb}, L_{cc} are the self-inductance of each respective winding, the remaining inductance parameters are the mutual inductance between each winding, and i_A, i_B, i_C, i_a, i_b, i_c are the phase currents of each winding respectively.

Fig. 2.3 Space distribution of flux linkages in stator a-phase winding of the DFIG[2]

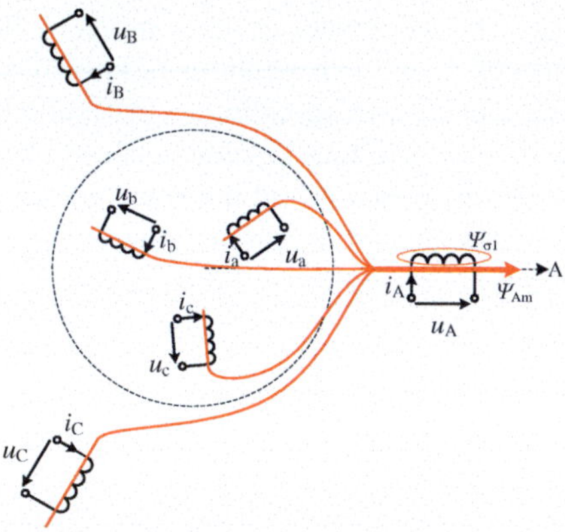

(a) Self-induced Flux Linkages

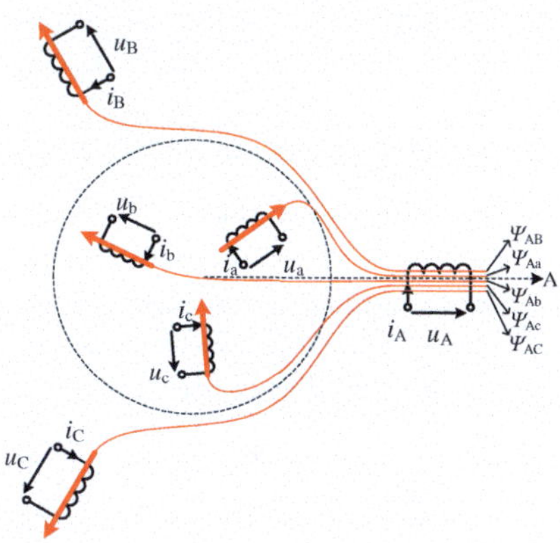

(b) Mutual-induced Flux Linkages

It can be seen from Eq. (2.4) that the self-induced flux linkage in each winding is composed of two parts: one is the part of the mutual induced flux linkage, coupling with other phases, (e.g. ψ_{Am} in Fig. 2.3a), and the other is the part of the leakage flux linkage coupling with the winding itself (e.g. $\psi_{\sigma 1}$ in Fig. 2.3a), the former being the major one. The inductance corresponding to each phase of the stator leakage flux is known as stator leakage inductance $L_{\sigma 1}$, and due to the symmetry of the windings, the leakage inductance of each stator winding is equal; similarly, the inductance corresponding to each phase of the rotor leakage flux is rotor leakage inductance

[2] This figure is only a schematic illustration that shows the coupling of flux linkages rather than the actual space distribution of flux linkages.

$L_{\sigma 2}$. The inductance corresponding to the mutual flux linkage produced by the stator current in its own winding is known as stator mutual inductance L_{m1}, while the inductance corresponding to the mutual flux linkage produced by the rotor current in its own winding is known as rotor mutual inductance L_{m2}. Since the number of turns of the stator and the rotor windings are equal after winding conversion, and the resistance of the mutual flux linkage through the air gap is the same between the windings, the following is obtained.

$$L_{m1} = L_{m2} = L_m \tag{2.7}$$

Further, from Eq. (2.4) it follows that

$$\psi_{AA} = L_{AA} i_A = \psi_{Am} + \psi_{\sigma 1} = L_m i_A + L_{\sigma 1} i_A \Rightarrow L_{AA} = L_m + L_{\sigma 1} \tag{2.8}$$

Since the stator and the rotor windings are symmetric with respect to each other, there is

$$\begin{cases} L_{AA} = L_{BB} = L_{CC} = L_m + L_{\sigma 1} \\ L_{aa} = L_{bb} = L_{cc} = L_m + L_{\sigma 2} \end{cases} \tag{2.9}$$

The mutual inductance between windings can be classified into two types:

(1) Type-I mutual inductance refers to the mutual inductance between the stator windings and the mutual inductance between the rotor windings. Sine the relative positions of the stator three-phase windings to each other and the rotor three-phase windings to each other are fixed, the Type-I mutual inductance can be considered as constant.
(2) Type-II mutual inductance refers to the mutual inductance between the stator winding and the rotor winding. Since the relative position of the stator winding and the rotor winding to each other is variable, the Type-II mutual inductance can be considered as a function of the angular position θ_r of the rotor winding with respect to the stator winding.

For the Type-I mutual inductance, the position angle between the stator three-phase winding axes or the rotor three-phase winding axes is an electrical angle of $\pm 120°$, and under the assumption that the air gap flux is sinusoidally distributed, the value of mutual inductance is $L_m \cos 120° = L_m \cos(-120°) = -L_m/2$, i.e. there is

$$\begin{cases} L_{AB} = L_{BC} = L_{CA} = L_{BA} = L_{CB} = L_{AC} = -L_m/2 \\ L_{ab} = L_{bc} = L_{ca} = L_{ba} = L_{cb} = L_{ac} = -L_m/2 \end{cases} \tag{2.10}$$

For the Type-II mutual inductance, the position angle between the stator three-phase winding axes and the rotor three-phase winding axes is a function of the electrical angle θ_r (as shown in Figs. 2.2 and 2.3), therefore there is

$$\begin{cases} L_{\mathrm{Aa}} = L_{\mathrm{aA}} = L_{\mathrm{Bb}} = L_{\mathrm{bB}} = L_{\mathrm{Cc}} = L_{\mathrm{cC}} = L_{\mathrm{m}} \cos\theta_{\mathrm{r}} \\ L_{\mathrm{Ab}} = L_{\mathrm{bA}} = L_{\mathrm{Bc}} = L_{\mathrm{cB}} = L_{\mathrm{Ca}} = L_{\mathrm{aC}} = L_{\mathrm{m}} \cos(\theta_{\mathrm{r}} + 120°) \\ L_{\mathrm{Ac}} = L_{\mathrm{cA}} = L_{\mathrm{Ba}} = L_{\mathrm{aB}} = L_{\mathrm{Cb}} = L_{\mathrm{bC}} = L_{\mathrm{m}} \cos(\theta_{\mathrm{r}} - 120°) \end{cases} \tag{2.11}$$

Substituting Eqs. (2.7)–(2.11) into Eq. (2.6) leads to a further expression of the flux linkage equation as

$$\begin{bmatrix} \boldsymbol{\psi}_{\mathrm{s}} \\ \boldsymbol{\psi}_{\mathrm{r}} \end{bmatrix} = \boldsymbol{L}_{\theta\mathrm{r}} \begin{bmatrix} \boldsymbol{i}_{\mathrm{s}} \\ \boldsymbol{i}_{\mathrm{r}} \end{bmatrix} = \begin{bmatrix} \boldsymbol{L}_{\mathrm{ss}} & \boldsymbol{L}_{\mathrm{sr}}(\theta_{\mathrm{r}}) \\ \boldsymbol{L}_{\mathrm{rs}}(\theta_{\mathrm{r}}) & \boldsymbol{L}_{\mathrm{rr}} \end{bmatrix} \begin{bmatrix} \boldsymbol{i}_{\mathrm{s}} \\ \boldsymbol{i}_{\mathrm{r}} \end{bmatrix} \tag{2.12}$$

where $\boldsymbol{\psi}_{\mathrm{s}} = \left[\psi_{\mathbf{A}}\ \psi_{\mathbf{B}}\ \psi_{\mathbf{C}}\right]^{\mathbf{T}}$, $\boldsymbol{\psi}_{\mathrm{r}} = \left[\psi_{\mathbf{a}}\ \psi_{\mathbf{b}}\ \psi_{\mathbf{c}}\right]^{\mathbf{T}}$, $\boldsymbol{i}_{\mathrm{s}} = \left[i_{\mathbf{A}}\ i_{\mathbf{B}}\ i_{\mathbf{C}}\right]^{\mathbf{T}}$, $\boldsymbol{i}_{\mathrm{r}} = \left[i_{\mathbf{a}}\ i_{\mathbf{b}}\ i_{\mathbf{c}}\right]^{\mathbf{T}}$, and

$$\boldsymbol{L}_{\mathrm{ss}} = \begin{bmatrix} L_{\mathbf{m}} + L_{\sigma\mathbf{1}} & -L_{\mathbf{m}}/2 & -L_{\mathbf{m}}/2 \\ -L_{\mathbf{m}}/2 & L_{\mathbf{m}} + L_{\sigma 1} & -L_{\mathbf{m}}/2 \\ -L_{\mathbf{m}}/2 & -L_{\mathbf{m}}/2 & L_{\mathbf{m}} + L_{\sigma 1} \end{bmatrix} \tag{2.12a}$$

$$\boldsymbol{L}_{\mathbf{rr}} = \begin{bmatrix} L_{\mathbf{m}} + L_{\sigma\mathbf{2}} & -L_{\mathbf{m}}/2 & -L_{\mathbf{m}}/2 \\ -L_{\mathbf{m}}/2 & L_{\mathbf{m}} + L_{\sigma 2} & -L_{\mathbf{m}}/2 \\ -L_{\mathbf{m}}/2 & -L_{\mathbf{m}}/2 & L_{\mathbf{m}} + L_{\sigma 2} \end{bmatrix} \tag{2.12b}$$

$$\boldsymbol{L}_{\mathrm{rs}}(\theta_{\mathbf{r}}) = L_{\mathbf{m}} \begin{bmatrix} \cos\theta_{\mathbf{r}} & \cos(\theta_{\mathbf{r}} - 120°) & \cos(\theta_{\mathbf{r}} + 120°) \\ \cos(\theta_{\mathbf{r}} + 120°) & \cos\theta_{\mathbf{r}} & \cos(\theta_{\mathbf{r}} - 120°) \\ \cos(\theta_{\mathbf{r}} - 120°) & \cos(\theta_{\mathbf{r}} + 120°) & \cos\theta_{\mathbf{r}} \end{bmatrix} \tag{2.12c}$$

$$\boldsymbol{L}_{\mathrm{sr}}(\theta_{\mathbf{r}}) = \boldsymbol{L}_{\mathrm{rs}}(\theta_{\mathbf{r}})^{\mathbf{T}} \tag{2.12d}$$

(3) **Torque Equation and Motion Equation**

The electromagnetic torque of the DFIG can be derived from the rate of change of the magnetic co-energy with respect to the mechanical angular position as [2]:

$$\begin{aligned} T_{\mathrm{e}} = n_{\mathrm{p}} L_{\mathrm{m}} [& (i_{\mathrm{A}} i_{\mathrm{a}} + i_{\mathrm{B}} i_{\mathrm{b}} + i_{\mathrm{C}} i_{\mathrm{c}}) \sin\theta_{\mathrm{r}} \\ & + (i_{\mathrm{A}} i_{\mathrm{b}} + i_{\mathrm{B}} i_{\mathrm{c}} + i_{\mathrm{C}} i_{\mathrm{a}}) \sin(\theta_{\mathrm{r}} + 120°) \\ & + (i_{\mathrm{A}} i_{\mathrm{c}} + i_{\mathrm{B}} i_{\mathrm{a}} + i_{\mathrm{C}} i_{\mathrm{b}}) \sin(\theta_{\mathrm{r}} - 120°)] \end{aligned} \tag{2.13}$$

The motion equation of the DFIG can thus be found as

$$T_{\mathrm{e}} = T_{\mathrm{L}} + \frac{J}{n_{\mathrm{p}}} \frac{d\omega_{\mathrm{r}}}{dt} \tag{2.14}$$

where T_{L} is the load torque, ω_{r} is the rotor electrical angular velocity and J is the load inertia.

Fig. 2.4 Relations between ABC-frame, αβ-frame and dq-frame

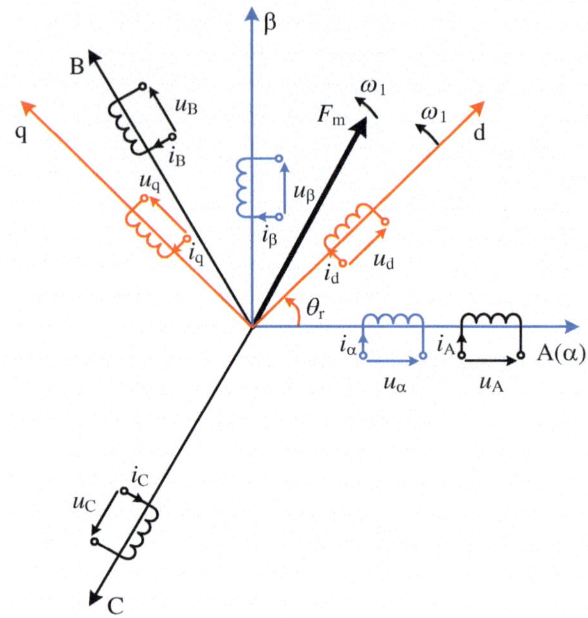

In summary, Eqs. (2.3), (2.6), (2.13) and (2.14) are the mathematical model of the DFIG in the ABC coordinate frame (ABC-frame).

To simplify the mathematical model, it is generally necessary to transform the above mathematical model of the DFIG into a two-phase stationary coordinate frame (αβ-frame) and a two-phase synchronous rotating coordinate frame (dq-frame), the relation between the coordinate frames being shown in Fig. 2.4.

To keep the magnetic motive force $\boldsymbol{F}_{\mathbf{m}}$ equivalent (with the same magnitude and position) in each coordinate frame and with the same power before and after the coordinate transformations,[3] the following coordinate transformation formulas are obtained.

$$C_{3s/2s} = \sqrt{\frac{2}{3}}\begin{bmatrix} 1 & -\frac{1}{2} & -\frac{1}{2} \\ 0 & \frac{\sqrt{3}}{2} & -\frac{\sqrt{3}}{2} \end{bmatrix} \tag{2.15a}$$

$$C_{2s/3s} = \sqrt{\frac{2}{3}}\begin{bmatrix} 1 & 0 \\ -\frac{1}{2} & \frac{\sqrt{3}}{2} \\ -\frac{1}{2} & -\frac{\sqrt{3}}{2} \end{bmatrix} \tag{2.15b}$$

[3] It is also reasonable to keep the amplitudes of the current and the voltage unchanged before and after the transformations, but the resulting coordinate transformation formulas will have different coefficients from those given in this book, and unless otherwise stated, the coordinate transformation formulas in this book are always for equal power transformations.

$$C_{2s/2r} = \begin{bmatrix} \cos\theta_r & \sin\theta_r \\ -\sin\theta_r & \cos\theta_r \end{bmatrix} \tag{2.15c}$$

$$C_{2r/2s} = \begin{bmatrix} \cos\theta_r & -\sin\theta_r \\ \sin\theta_r & \cos\theta_r \end{bmatrix} \tag{2.15d}$$

Applying the coordinate transformation represented by Eqs. (2.15a) to (2.2), (2.6) and (2.13) gives the voltage, the flux linkage and the torque equations of the DFIG in the $\alpha\beta$-frame as

$$\begin{cases} u_s^\alpha = R_s i_s^\alpha + p\psi_s^\alpha \\ u_s^\beta = R_s i_s^\beta + p\psi_s^\beta \\ u_r^\alpha = R_r i_r^\alpha + p\psi_r^\alpha + \omega_r \psi_r^\beta \\ u_r^\beta = R_r i_r^\beta + p\psi_r^\beta - \omega_r \psi_r^\alpha \end{cases} \tag{2.16}$$

$$\begin{cases} \psi_s^\alpha = L_s i_s^\alpha + L_M i_r^\alpha \\ \psi_s^\beta = L_s i_s^\beta + L_M i_r^\beta \\ \psi_r^\alpha = L_r i_r^\alpha + L_M i_s^\alpha \\ \psi_r^\beta = L_r i_r^\beta + L_M i_s^\beta \end{cases} \tag{2.17}$$

$$\begin{aligned} T_e &= n_p L_M \left(i_s^\beta i_r^\alpha - i_s^\alpha i_r^\beta\right) = n_p\left(\psi_s^\beta i_s^\alpha - \psi_s^\alpha i_s^\beta\right) \\ &= n_p (L_M/L_r)\left(i_s^\beta \psi_r^\alpha - i_s^\alpha \psi_r^\beta\right) = n_p\left(\psi_s^\beta i_r^\alpha - \psi_s^\alpha i_r^\beta\right) \\ &= n_p \frac{L_M}{\sigma L_s L_r}\left(\psi_s^\beta \psi_r^\alpha - \psi_s^\alpha \psi_r^\beta\right) \end{aligned} \tag{2.18}$$

where $L_M = 1.5L_m$, $L_s = 1.5L_m + L_{\sigma 1}$, $L_r = 1.5L_m + L_{\sigma 2}$, $\sigma = 1 - \left(L_M^2/L_s L_r\right)$. Here, Eqs. (2.16), (2.17), (2.18) and (2.14) are the mathematical model of the DFIG in the $\alpha\beta$-frame.

Similarly, by applying the coordinate transformation represented by Eqs. (2.15c) to (2.16), (2.17) and (2.18), the voltage, the flux linkage and torque equations of the DFIG in the dq-frame can be obtained as

$$\begin{cases} u_s^d = R_s i_s^d + p\psi_s^d - \omega_1 \psi_s^q \\ u_s^q = R_s i_s^q + p\psi_s^q + \omega_1 \psi_s^d \\ u_r^d = R_r i_r^d + p\psi_r^d - \omega_{sl} \psi_r^q \\ u_r^q = R_r i_r^q + p\psi_r^q + \omega_{sl} \psi_r^d \end{cases} \tag{2.19}$$

$$\begin{cases} \psi_s^d = L_s i_s^d + L_M i_r^d \\ \psi_s^q = L_s i_s^q + L_M i_r^q \\ \psi_r^d = L_r i_r^d + L_M i_s^d \\ \psi_r^q = L_r i_r^q + L_M i_s^q \end{cases} \tag{2.20}$$

$$T_e = n_p L_M \left(i_s^q i_r^d - i_s^d i_r^q\right) = n_p\left(\psi_s^q i_s^d - \psi_s^d i_s^q\right)$$

$$= n_\mathrm{p}(L_\mathrm{M}/L_\mathrm{r})\left(i_\mathrm{s}^\mathrm{q}\psi_\mathrm{r}^\mathrm{d} - i_\mathrm{s}^\mathrm{d}\psi_\mathrm{r}^\mathrm{q}\right) = n_\mathrm{p}\left(\psi_\mathrm{s}^\mathrm{q} i_\mathrm{r}^\mathrm{d} - \psi_\mathrm{s}^\mathrm{d} i_\mathrm{r}^\mathrm{q}\right)$$

$$= n_\mathrm{p}\frac{L_\mathrm{M}}{\sigma L_\mathrm{s} L_\mathrm{r}}\left(\psi_\mathrm{s}^\mathrm{q}\psi_\mathrm{r}^\mathrm{d} - \psi_\mathrm{s}^\mathrm{d}\psi_\mathrm{r}^\mathrm{q}\right) \tag{2.21}$$

where $\omega_\mathrm{sl} = \omega_1 - \omega_\mathrm{r}$ is the slip speed, Eqs. (2.19), (2.20), (2.21) and (2.14) are the mathematical model of the DFIG in the dq-frame.

Then the stator power of the DFIG injected into the grid can be written as

$$\begin{cases} P_\mathrm{s} = u_\mathrm{s}^\mathrm{d} i_s^\mathrm{d} + u_s^\mathrm{q} i_s^\mathrm{q} \\ Q_\mathrm{s} = u_s^\mathrm{q} i_\mathrm{s}^\mathrm{d} - u_\mathrm{s}^\mathrm{d} i_\mathrm{s}^\mathrm{q} \end{cases} \tag{2.22}$$

where P_s and Q_s represent the active and reactive power injected into the grid by the DFIG stator, respectively.

2.2 Mathematical Model of the Grid-Side Converter

The typical structure of the grid-side converter is shown in Fig. 2.1. In general, to simplify the modelling process, the effect of the capacitors in the grid-side filter can be neglected and the filter can be equated to an inductive one. The simplified power circuit of the grid-side converter is shown in Fig. 2.5.

Where e_ga, e_gb and e_gc indicate the a-phase, b-phase and c-phase grid voltages respectively, i_ga, i_gb and i_gc indicate the a-phase, b-phase and c-phase grid-side currents respectively, L_ga, L_gb and L_gc indicate the grid-side filter inductance in each phase respectively, R_ga, R_gb and R_gc indicate the equivalent resistance of the grid-side filter and wire in each phase respectively, u_ga, u_gb and u_gc indicate the a-phase, b-phase and c-phase equivalent phase voltages at the output of the grid-side converter respectively, i.e. the fundamental component of the output phase voltage of the grid-side converter after PWM, C indicates the DC-bus capacitance, u_dc and i_dc indicate the DC-bus voltage and the DC-bus charging current respectively.

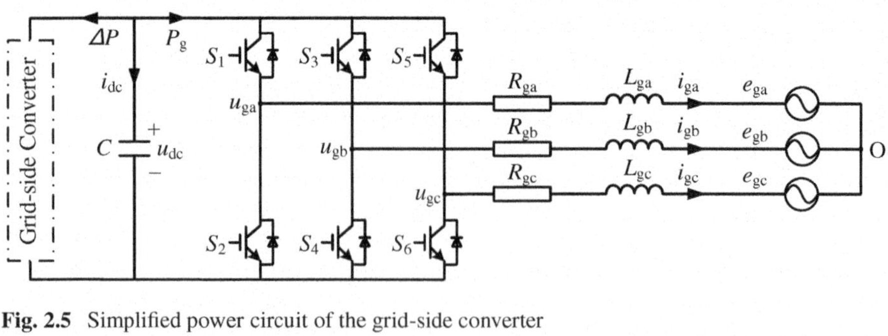

Fig. 2.5 Simplified power circuit of the grid-side converter

In the ABC-frame, the mathematical model of the AC side of the grid-side converter can be expressed as

$$\begin{cases} u_{\mathrm{ga}} = R_{\mathrm{ga}} i_{\mathrm{ga}} + L_{\mathrm{ga}} p i_{\mathrm{ga}} + e_{\mathrm{ga}} \\ u_{\mathrm{gb}} = R_{\mathrm{gb}} i_{\mathrm{gb}} + L_{\mathrm{gb}} p i_{\mathrm{gb}} + e_{\mathrm{gb}} \\ u_{\mathrm{gc}} = R_{\mathrm{gc}} i_{\mathrm{gc}} + L_{\mathrm{gc}} p i_{\mathrm{gc}} + e_{\mathrm{gc}} \end{cases} \tag{2.23}$$

Ideally, considering the system as tri-symmetric, i.e. $L_{\mathrm{ga}} = L_{\mathrm{gb}} = L_{\mathrm{gc}}$ and $R_{\mathrm{ga}} = R_{\mathrm{gb}} = R_{\mathrm{gc}}$, the expression of Eq. (2.23) in the dq-frame is obtained after the coordinate transformation as Eq. (2.24).

$$\begin{cases} u_{\mathrm{g}}^{\mathrm{d}} = R_{\mathrm{g}} i_{\mathrm{g}}^{\mathrm{d}} + L_{\mathrm{g}} p i_{\mathrm{g}}^{\mathrm{d}} - \omega_{\mathrm{e}} L_{\mathrm{g}} i_{\mathrm{g}}^{\mathrm{q}} + e_{\mathrm{g}}^{\mathrm{d}} \\ u_{\mathrm{g}}^{\mathrm{q}} = R_{\mathrm{g}} i_{\mathrm{g}}^{\mathrm{q}} + L_{\mathrm{g}} p i_{\mathrm{g}}^{\mathrm{q}} + \omega_{\mathrm{e}} L_{\mathrm{g}} i_{\mathrm{g}}^{\mathrm{d}} + e_{\mathrm{g}}^{\mathrm{q}} \end{cases} \tag{2.24}$$

where $u_{\mathrm{g}}^{\mathrm{d}}$ and $u_{\mathrm{g}}^{\mathrm{q}}$ indicate the d-axis and q-axis components of the equivalent output voltage of the grid-side converter respectively, $i_{\mathrm{g}}^{\mathrm{d}}$ and $i_{\mathrm{g}}^{\mathrm{q}}$ indicate the d-axis and q-axis components of the grid-side current respectively, $e_{\mathrm{g}}^{\mathrm{d}}$ and $e_{\mathrm{g}}^{\mathrm{q}}$ indicate the d-axis and q-axis components of the grid voltage respectively, R_{g} indicates the d-axis and q-axis equivalent resistance on the grid side, L_{g} indicates the d-axis and q-axis equivalent inductance of the grid-side filter, and ω_{e} indicates the synchronous angular frequency of the grid voltage.

Then the power of the grid-side converter injected into the grid can be written as

$$\begin{cases} P_{\mathrm{g}} = e_{\mathrm{g}}^{\mathrm{d}} i_{\mathrm{g}}^{\mathrm{d}} + e_{\mathrm{g}}^{\mathrm{q}} i_{\mathrm{g}}^{\mathrm{q}} \\ Q_{\mathrm{g}} = e_{\mathrm{g}}^{\mathrm{q}} i_{\mathrm{g}}^{\mathrm{d}} - e_{\mathrm{g}}^{\mathrm{d}} i_{\mathrm{g}}^{\mathrm{q}} \end{cases} \tag{2.25}$$

where P_{g} and Q_{g} represent the active and reactive power injected into the grid by the grid-side converter, respectively.

According to the power balance principle, the active power injected into the DC-bus capacitor under ideal conditions should be equal to the difference between the active power absorbed by the grid-side converter from the grid and the power consumed by the DC-side load, then the mathematical model of the DC bus of the grid-side converter is

$$u_{\mathrm{dc}} i_{\mathrm{dc}} = C u_{\mathrm{dc}} p u_{\mathrm{dc}} = -P_{\mathrm{g}} - \Delta P \tag{2.26}$$

where ΔP indicates the active power consumed by the load on the DC side.

2.3 Graphical Modelling Methods

A system model is the basis for its characteristic analysis, algorithm design, simulation research and so on. By the traditional modelling method, the mathematical model of a system is obtained by listing the mathematical equations of its components according to their physical laws and then arranging these equations in a particular way to represent the interactions between or among the components of the system. This mathematical modelling method has shown itself to be excellent in the study of relatively simple systems, to be powerful for the analysis and study in a variety of situations and to be strong in both qualitative analysis and quantitative calculations. However, if the system is very large and its structure is very complex, it becomes very difficult to study the system using its mathematical model and thus to derive its control methods, and it is also not convenient for the construction of computer simulation modules. This is why graphical modelling methods have been proposed as a complement to mathematical modelling methods.

2.3.1 *Overview of Graphical Modelling Methods*

One of the most widely used graphical modelling methods is the Bond Graph (BG), which was first proposed by the American scholar Paynter in the 1960s, and was further developed and improved in the 1970s. A bond graph can represent the energy flow among the parts of a physical system. The energy exchange between two subsystems is represented by an energy bond, as shown in Fig. 2.6, where the energy bond is represented by a half arrow indicating the direction of energy flow.

Where e represents the effort, f represents the flow and the instantaneous energy exchanged between subsystems can be expressed as the product of the effort and the flow.

There are three types of elements in a bode graph: (1) three passive elements representing the consumption of energy (R) or the storage of energy (I, C); (2) two active elements (Se, Sf) representing the source of energy; and (3) four connecting elements (0, 1, TF, GY) linking the active and passive elements according to the structure of the system.

In a bond graph, the causal relation is generally represented by a vertical line at the end of the half-arrow, as shown in Fig. 2.6. It indicates that Element A applies a force to Element B while Element B returns a flow to Element A. The vertical line representing the causal relation is therefore close to the Element B side. This causality is determined by the interaction between the elements, independent to the

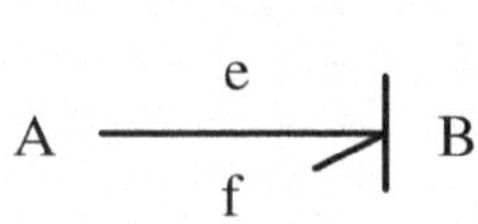

Fig. 2.6 Basic representation of the bond graph

direction of energy flow. This independence allows either an integral or a differential causality to be used, which is entirely determined by the structure of the system itself. It is worth noting that the integral causality is always given preference in the modelling process, although it is not the only option that is mandatory.

The biggest feature of the Bond Graph is that it most closely resembles the topology of an actual system, with the components in the model in one-to-one correspondence with the respective components of the actual system. It also provides a clear representation of the direction of energy flow within the actual system. However, the presence of differential causality makes it impossible to derive control methods for the actual system directly from its bond graph, and it is also difficult to use it for direct simulation studies.

Based on the Bond Graph, various graphical modelling methods have been developed, such as the Power-oriented Graph (POG), the Power Flow Diagram (PFD) and the Energetic Puzzle (EP).

The Power-oriented Graph [3] was proposed by the Italian scholar Zanasi in 1996 and is characterized by using two basic components instead of traditional transfer functions. These two basic components are shown in Fig. 2.7. The one on the left is the connection component, which represents the energy conversion, while the one on the right is the elaboration component, which represents the storage or consumption of energy. The limited number of representation components makes this method relatively easy to apply. The most important feature of this method is that its basic components correspond to the transfer functions of the respective components in a system, which allows it to correspond better to an actual physical system and therefore better represent the energy flow among the components of the system. As in the Bond Graph, the causality in the Power-oriented Graph can be either integral or differential, as determined by the mathematical model of the actual system. A major drawback of this method is that it does not show the coupling relations inside the system in structural terms and therefore does not clearly represent the structure of the original actual system.

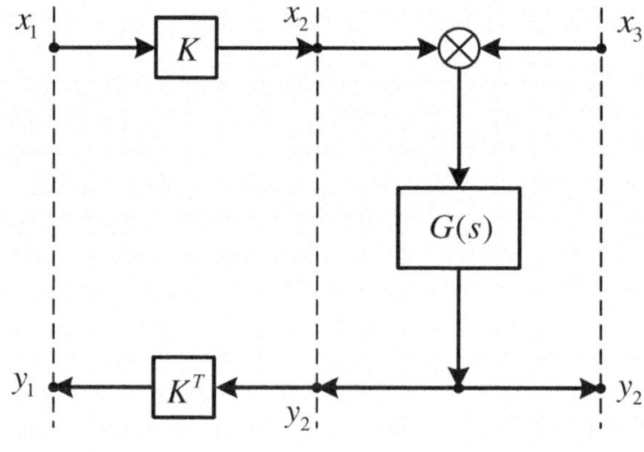

Fig. 2.7 Basic representation of the power-oriented graph

The Power Flow Diagram [4] is a method proposed by the German scholar Schonfeld in 2004 to study the energy flow of a system and its energy conversion efficiency. Unlike the Power-oriented Graph, the Power Flow Diagram defines its own unique graphical symbols for representing the source, conversion, storage and consumption of energy and the coupling of different components, so that it maintains the physical structure of a system to the greatest extent possible, as does the Bond Graph. A typical representation of this is shown in Fig. 2.8.

The diagram above shows, from left to right, the energy source, energy conversion and energy storage components. Unlike the Bond Graph, the causality of the Power Flow Diagram can only be integral and can therefore be used directly for simulation studies. Its greatest advantage is that it can be used to easily calculate the efficiency of the entire system, as the components representing energy consumption are specifically considered in the modelling process.

The Energetic Puzzle [5] was proposed at the beginning of this century by the Laplace Laboratory in Toulouse, France. This method was originally used to represent the energy conversion of a system, with the components represented by special puzzles and the different kinds of variables represented by the corresponding additional parts of the puzzles, as shown in Fig. 2.9.

Figure 2.9 shows the energy source, energy conversion and energy storage components from the left to the right. The triangles sticking out of the puzzle represent the output force, while the embedded squares represent the input flow. As can be seen, similar to the Bond Graph, the Energy Puzzle can also clearly represent the direction of energy flow in a system. As in the case of the Power Flow Diagram, the causality of the Energy Puzzle is also integral in nature. However, it can also be seen that the representation does not show the names of the system variables and therefore cannot be used to derive the control method of the system directly.

Although the graphical modelling methods introduced above all have their own unique graphical components, they all essentially use graphical components to represent the mathematical model of the corresponding subsystem to obtain the unified representation of a whole system. Compared to traditional mathematical modelling methods, the graphical modelling methods are advantageous in the study of complex systems consisting of many subsystems, including both mechanical and electrical

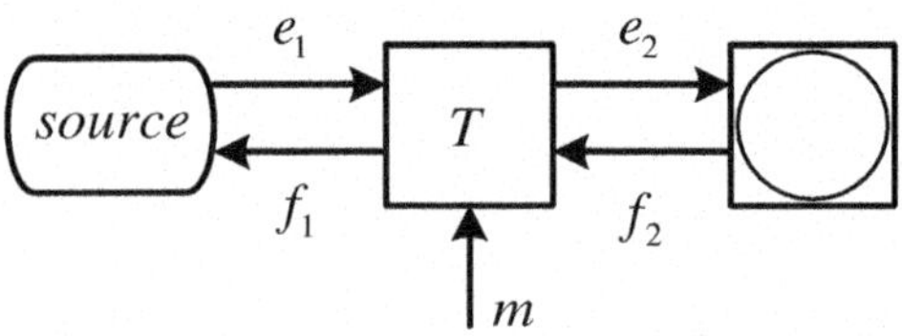

Fig. 2.8 Basic representation of the power flow diagram

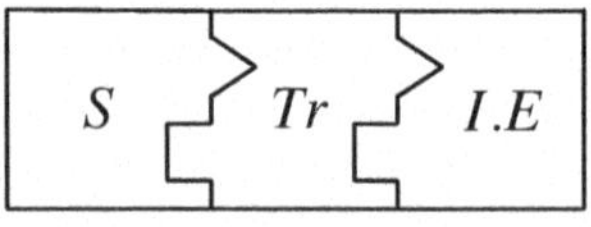

Fig. 2.9 Basic representation of the energetic puzzle

parts, such as a wind power system or an electric vehicle. This is because it allows a more intuitive and comprehensive model to be obtained that corresponds to the physical structure of a system. Furthermore, these methods can represent the causal relations among system variables as well as energy flow inside a system, making them more suitable for the qualitative analysis of a large and complex system. However, these methods do not allow for the direct derivation of control strategies, and new and more effective graphical representations of modelling and control methods must be further developed.

The next part of this section will focus on two graphical modelling methods, the Causal Ordering Graph (COG) and the Energetic Macroscopic Representation (EMR).

2.3.2 *Causal Ordering Graph*

The Causal Ordering Graph is a method proposed at the end of the last century by Professor Hautier of the L2EP laboratory in Lille, France, specifically for the study of electromechanical energy conversion systems [6]. It deals with the information inside a system in a graphical representation and is a transformation method for logical thinking in system analysis, thanks to an accurate description of the interrelations among the different components of a system and their own characteristics. The basic concept of the Causal Ordering Graph is based on a natural causal relation, and this causality can only be integral which must be followed when modelling any energy transformation process using the Causal Ordering Graph. This is because the integral causality is consistent with the fundamental physical law that the energy gained by any object accumulates over time. The general modelling process is through connecting graphical models of each part of a system, to ultimately obtain a more integrated representation of the system as a whole. The greatest advantage of this method is that the corresponding control methods can be derived directly from the graphical model according to specific inverse rules. The process of deriving the control methods must also follow integral causality.

2.3.2.1 Basic Components of Causal Ordering Graph

Similar to a bonding graph, the variables of a causal order graph are either potential energy variables (corresponding to the force variables) or kinetic energy variables (corresponding to the flow variables). In contrast, the relations between the input and output variables are physically characterized by an operator. This operator is the basic graphical representation of a physical object or set of objects and is used to establish the transformation relations between the input variables that impose their effects and the output variables that are affected by the input variables. Such relations can be classified into two categories: the static relation and the dynamic relation.

The static relation is one in which the relation between the output and input variables is independent of time, i.e. there is no natural causal relation, and can usually be expressed by an algebraic expression as follows.

$$R_{\mathrm{a}}\colon y(t) = R_{\mathrm{a}}(x(t)) \tag{2.27}$$

The operator representing the static relation is shown in Fig. 2.10, where the double arrow indicates that the externally determined causality is reversible.

It is often used to represent components with no definite causal relation, such as dissipative components like a resistor. All energy transferred to the resistor is consumed as heat, so that the voltage and current of the resistor show a static relation, which can be expressed in the following two forms.

$$R_{\mathrm{R1}}\colon u_{\mathrm{R}}(t) = r i_{\mathrm{R}}(t) \tag{2.28}$$

$$R_{\mathrm{R2}}\colon i_{\mathrm{R}}(t) = \frac{1}{r} u_{\mathrm{R}}(t) \tag{2.29}$$

Figure 2.11 shows a resistor component represented by the Causal Ordering Graph, corresponding to each of the two equations above.

As can be seen from the above analysis, the input and output variables in an operator representing a static relation are not definite and are determined by the other components directly connected to the operator.

In contrast, in the dynamic relation, the input and output variables are determined by their own causal relations, representing the causes and their resulting outcomes respectively. The dynamic relation is time-dependent and can usually be described by the differential equation:

$$\frac{dy(t)}{dt} = ax(t) + b \tag{2.30}$$

Therefore, the output variable is the integral of the input variable here, which can be expressed as

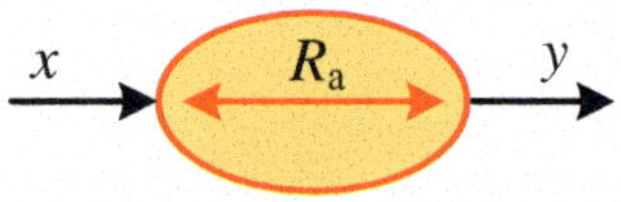

Fig. 2.10 Static relation represented by the causal ordering graph

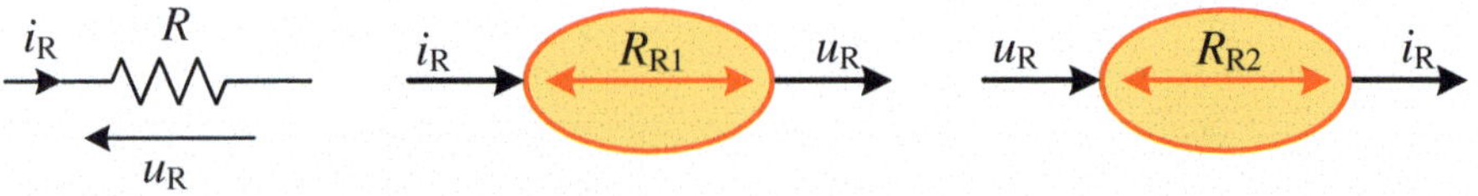

Fig. 2.11 A resistor component represented by the causal ordering graph

$$R_b\colon y(t) = R_b(x(t)) \tag{2.31}$$

The operator representing the dynamic relation is shown in Fig. 2.12, where the single arrow indicates that the integral causality is determined by itself, is not reversible and is not influenced by the other components connected to the operator.

It usually represents a component that can store energy, such as an inductor and a capacitor. The potential or kinetic energy stored in such a component is obtained by integrating the kinetic or potential energy that is input to the same component in the opposite pair.

According to the law of electromagnetic induction, the voltage drop across an inductor is

$$R_L\colon u_L(t) = \frac{d\psi(t)}{dt} = L\frac{di_L(t)}{dt} \tag{2.32}$$

Figure 2.13 shows an inductive component represented by the Causal Order Graph where the output current is the integral of the input voltage, which is determined by the physical characteristics of the inductor itself.

Similarly, the current flowing through a capacitor can be expressed as

$$R_C\colon i_C(t) = \frac{dQ(t)}{dt} = C\frac{du_C(t)}{dt} \tag{2.33}$$

Figure 2.14 shows a capacitor component represented by the Causal Order Graph, where the output voltage is the integral of the input current, which is also determined by the physical characteristics of the capacitor itself.

In summary, the causality between the input and output variables in an operator representing the dynamic relation is definite, which on the one hand ensures the physical characteristics of the system itself and on the other hand better explains the natural phenomena determined by the characteristics of the system. Moreover, it is these dynamic components that determine the causal relations of the static components directly connected to them.

Thus, the COG-based modelling method can be developed. Firstly, the corresponding graphical operators are obtained according to the physical law (mathematical model) of each part of a system, then the input and output variables of each

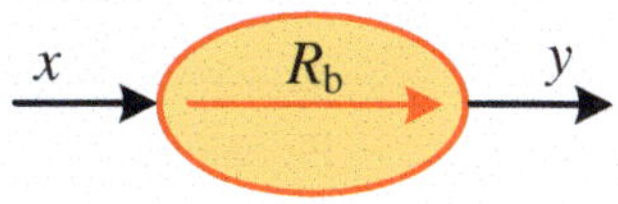

Fig. 2.12 Dynamic relation represented by the causal ordering graph

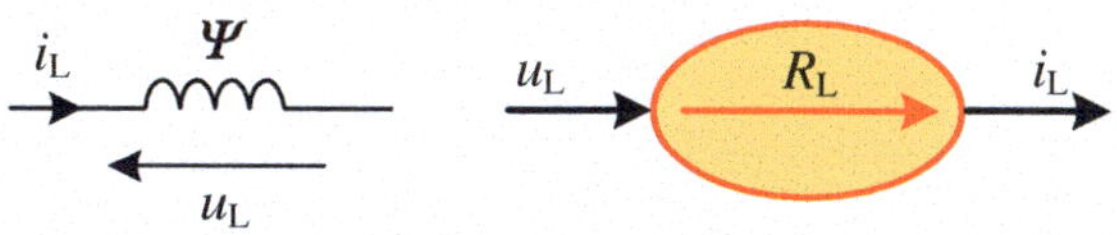

Fig. 2.13 An inductor component represented by the causal ordering graph

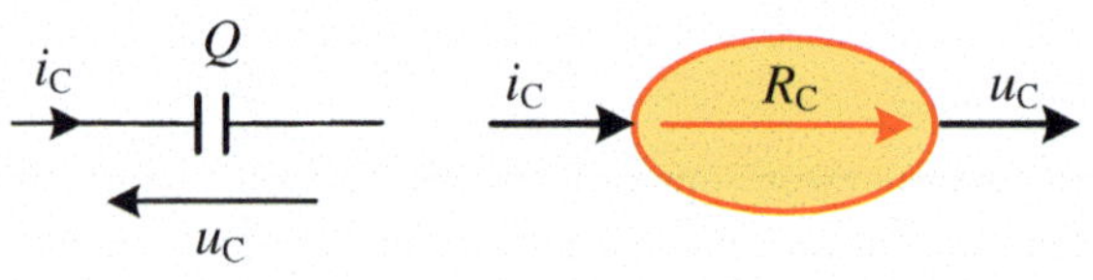

Fig. 2.14 A capacitor component represented by the causal ordering graph

operator are determined according to the causal relations of the dynamic components, and finally the corresponding graphical operators are connected according to the physical structure of the system to obtain the causal order graph of the whole system. As this method contains only two types of basic components, the resulting system model is more concise and uniform.

2.3.2.2 Control Methods Derived from Inverse Operation of COG

The most remarkable feature of the Causal Ordering Graph is that the basic control method of a system can be derived by inverse operation according to the specific inverse rules [6, 7]. The principle of such inverse operation can be summarized as follows: in order to get the desired output effect, a suitable input cause must be imposed on the system, provided that the causal relation of the system is known.

For a given basic operator, the control method of the model can be obtained by inverting its input–output relation. The controller thus transposes the input and output variables of the system and then obtains the model representing its inverse process. It should also be noticed that the inverse process still needs to follow the integral causality.

For an operator representing the static relation, since the relation between its input and output variables is not time-dependent, its controller can be obtained by a direct inverse operation if the model itself satisfies a one-to-one correspondence. The inverse operation for the static relation is shown in Fig. 2.15a, and its controller structure is similar to that of a model operator both of which use bidirectional arrow to represent the relation between the input and output variables.

Where R_{ac} indicates the control equation obtained from the inverse of the model equation R_a.

$$R_{ac}\colon x_{reg}(t) = R_{ac}(y_{ref}(t)) \tag{2.34}$$

Thus, in order to make the output equal to the reference value y_{ref}, the regulation value x_{ref} can be calculated directly using the above equation.

For an operator representing the dynamic relation, it is not so simple to obtain its controller. This is because the input–output relation is integral causality and its direct inverse results in a differential relation, which is not realizable in an actual physical system.

To solve this problem, a closed loop control method can be used to minimize the error between the actual measured output value $\widehat{y}$ and the reference value y_{ref}

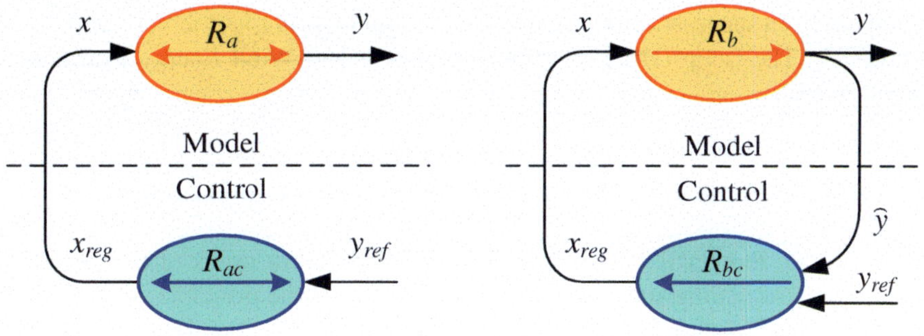

(a) The Static Relation Case (b) The Dynamic Relation Case

Fig. 2.15 Controller structure represented by the causal ordering graph

by designing a regulator C_{PI}. This control equation, obtained by indirect inverse operation, can be expressed as

$$R_{\mathrm{bc}}\colon x_{\mathrm{reg}}(t) = C_{\mathrm{PI}}(y_{\mathrm{ref}}(t) - \widehat{y}(t)) \tag{2.35}$$

The inverse operation of the dynamic relationship is shown in Fig. 2.15b. Unlike the static relationship, the controller is represented by a one-way arrow and the output needs to be measured in order to achieve closed-loop control.

From the above analysis, it is clear that the graphical representation of a system can be easily obtained from the mathematical model of the system using the Causal Ordering Graph, and the control structure can then be obtained by inversion, which is a proven method of modelling and controlling the system. However, when the system has a large number of components and a complex structure, the overall model obtained by this method becomes very complex and the readability of the system structure itself is also lost. Therefore, a more integrated graphical representation must be developed on this basis.

2.3.3 *Energetic Macroscopic Representation*

The Energetic Macroscopic Representation (EMR) is a graphical modelling method proposed at the beginning of this century by Professor Bouscayrol of the L2EP laboratory in Lille, France, in his study of complex electromechanical drive systems consisting of multiple motors and their inverters [8]. It is developed on the basis of the COG and enables a more integrated representation of complex electromechanical systems, highlighting their functions while maintaining the physical structures of the systems, so that the system models obtained using this method are more readable. The method has been successfully applied to the study of Hybrid Electric Vehicle (HEV), Hardware-In-The-Loop (HIL) real-time simulation and Wind Energy Conversion System (WECS).

2.3.3.1 Basic Components of Energetic Macroscopic Representation

Like the COG, the EMR is based on integral causality in terms of the interrelations between or among subsystems. It is also based on the principle of action and reaction, i.e. each action on a subsystem will produce a corresponding reaction. For different types of subsystems, the EMR uses three basic components to represent the energy source, transformation and storage parts respectively. It is also worth noting in particular that it specifically uses the linkage component to connect the different energy conversion channels in order to more clearly represent the energy distribution within a system. The specific graphical symbols in this method and the corresponding descriptions can be found in Table 2.1.

It can be seen that, different from the COG in which the components are used to replace the mathematical equations of subsystems, the EMR uses its basic components to represent subsystems. Therefore, the model obtained by the EMR provides the more integrated representation of a system and reflects its actual physical structure, as well as representing the interrelations and energy exchange between or among the subsystems.

2.3.3.2 Maximum Control Structure

A further advantage of the EMR is the ability to derive the Maximum Control Structure (MCS) for the corresponding system. The derivation process also requires the same specific inverse rules based on the integral relations as for the COG. The control diagram obtained using the inverse rules is called the maximum control structure because it contains the most information required for control operations and measurements. On this basis, a more practical control system can be obtained by making the necessary simplifications and estimating the unmeasurable variables [9].

The maximum control structure of a system is obtained by inverting its subsystems one by one sequentially. Since the components for energy transformation represent only the static relations, they can be inverted directly. For the transformation components containing regulation, two different types of the controller are possible: if the controller output is the regulation, the input effect variable is considered as the disturbance and its effect needs to be minimized; conversely, if the output is the effect variable, the regulation is considered as the disturbance to be measured. For the energy storage components, which inherently contain the integral causality, closed-loop controllers must be designed to indirectly invert them. Since an individual energy storage component generally contains two input effect variables, when one of them is selected as the output variable, the other will be measured as an interference or compensation term. For the linkage components, it is also possible to invert them

Table 2.1 Components of energetic macroscopic representation

Component	Symbol	Description	Instance
Energy source	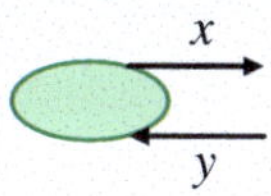	Source or recipient of energy (either kinetic or potential energy)	Power source
Energy storage	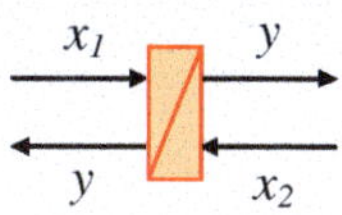	Energy storage component with integral causality	Inductor
			Capacitor
			Inertia shaft
Transformation	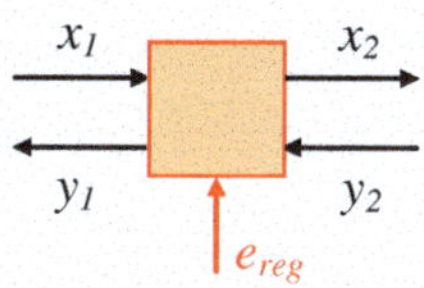	Component for electrical energy transformation without energy loss	Transformer
			Power converter
	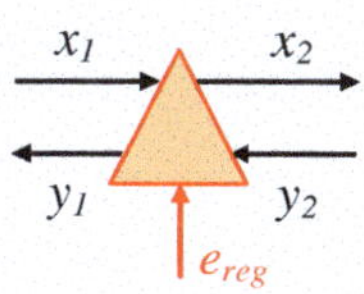	Component for mechanical energy transformation without energy loss	Gearbox
			Gear
	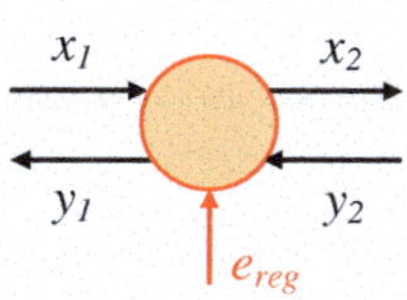	Component for energy transformation between electrical side and mechanical side	DC motor
Linkage	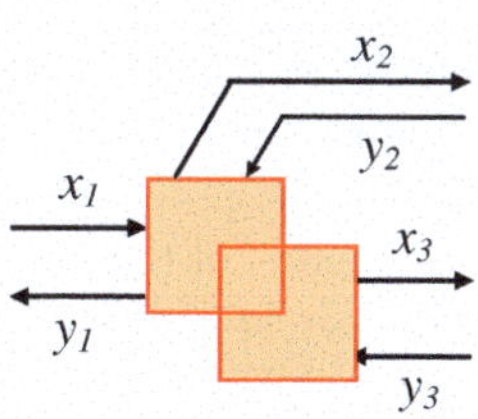	Electrical linkage component	Series connection
			Parallel connection

(continued)

Table 2.1 (continued)

Component	Symbol	Description	Instance
	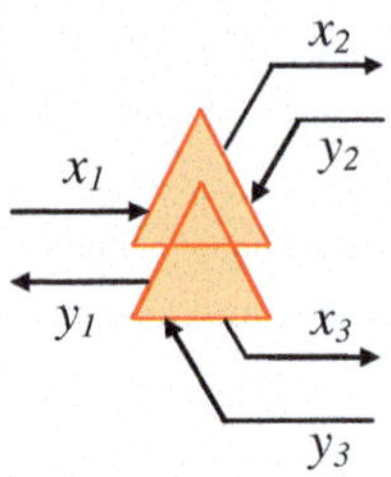	Mechanical linkage component	Pulley
	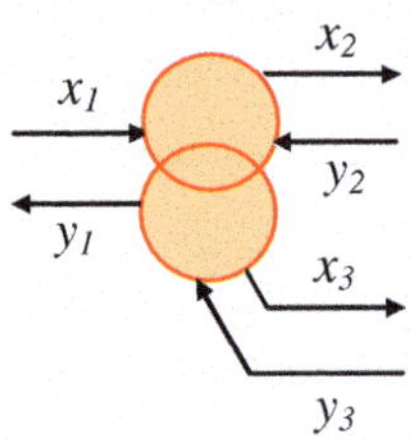	Electromechanical linkage component	Electrical machine

directly since they represent the distribution or aggregation of energy inside systems. The graphical symbols for the specific control components and the corresponding descriptions can be found in Table 2.2.

It follows that the maximum control structure corresponds to the optimal energy management of each subsystem. It is also necessary to determine the control strategy for the whole system in order to obtain the reference value of the output response quantity that the system needs to control, of which the graphical representation is shown in Table 2.2.

In summary, the EMR gives the comprehensive model of a whole system and enables the direct derivation of its maximum control structure. However, it is also necessary to use the Causal Ordering Graph when designing the structure of each controller in the control system. The specific parameters of the designed controllers must then depend on the transfer functions of their corresponding models. Therefore, the modelling and control of an electromechanical energy conversion system requires the combination of the tools described above, of which the process can be illustrated as shown in Fig. 2.16.

Table 2.2 Components of maximum control structure

Component	Symbol	Description
Inverse of transformation component	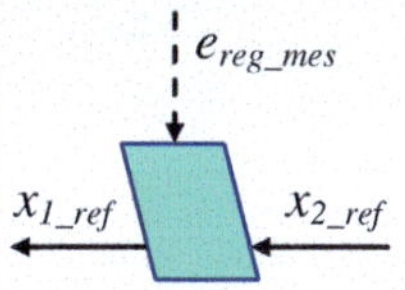	Inversed directly, considering e_{reg} as a disturbance
	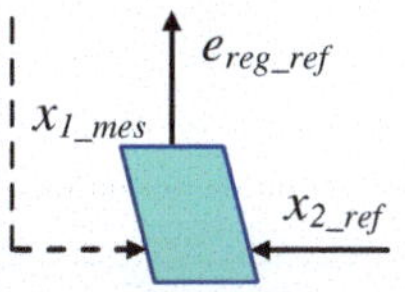	Inversed directly, considering x_1 as a disturbance
Inverse of storage component	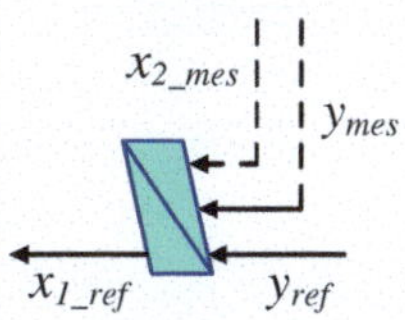	Inversed indirectly, requiring a controller
Inverse of linkage component	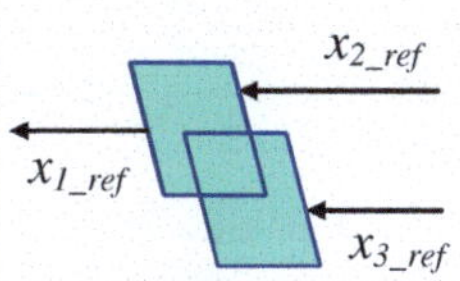	Inversed directly
Controller	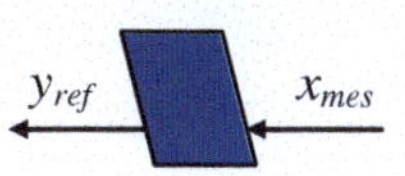	Determining the reference value of the controlled quantity

2.3.4 Graphical Modelling and Control Derivation of the Wind Turbine [10–12]

A wind power system is a device that converts wind energy into electrical energy, and it is the most important way to use wind energy. As a typical electromechanical energy conversion system, it mainly consists of the mechanical part that converts wind energy into mechanical energy and the electrical part that converts mechanical energy into electrical energy. The mechanical part, shown in Fig. 2.17, is made up of three main components: the wind turbine, the drive shaft and the gearbox. The wind turbine is the device that captures wind energy and converts it into mechanical energy, while the drive shaft and gearbox are the two mechanical transmission devices that transmit the mechanical energy output from the wind turbine to the generator.

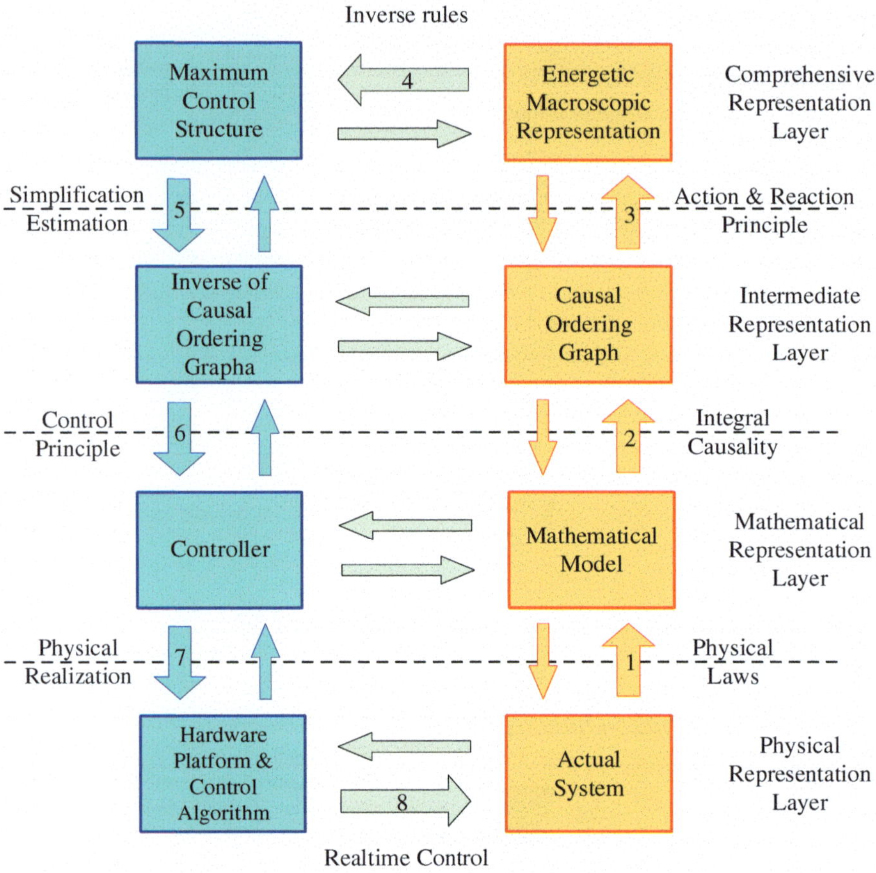

Fig. 2.16 Process of modelling and control of a system based on the EMR and the COG

2.3.4.1 Model of the Wind Turbine

The wind turbine is a device that converts wind energy into mechanical energy and its output can be expressed by Eq. (1.1).

The torque applying to the drive shaft of the wind turbine can be determined by its output mechanical power P and the angular speed of rotation Ω_t.

$$R_1 : T = \frac{P}{\Omega_t} = \frac{1}{2\Omega_t} C_p \rho S v^3 \tag{2.36}$$

The wind energy utilisation coefficient Cp in the above equation describes the proportion of mechanical energy absorbed by the wind turbine to the total wind energy and is the function of the tip-speed ratio λ and the pitch angle β. Where the tip-speed ratio λ is defined as the ratio of the blade tip linear speed to the wind speed

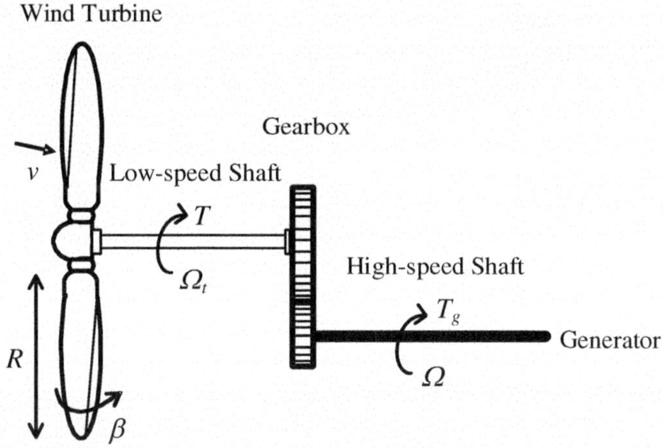

Fig. 2.17 Mechanical part of a wind power system

and is used to represent the state of the wind turbine at different wind speeds.

$$R_2\colon \lambda = \frac{R\Omega_t}{v} \tag{2.37}$$

An approximate expression of the wind energy utilisation coefficient Cp can be fitted based on the recorded data of the operation status of an actual 1.5 MW wind turbine [13].

$$\begin{aligned} R_3\colon C_p(\lambda, \beta) = (0.5 - 0.167(\beta - 2)) \sin\left[\frac{\pi(\lambda+0.1)}{18.5-0.3(\beta-2)}\right] \\ -0.00184(\lambda - 3)(\beta - 2) \end{aligned} \tag{2.38}$$

The aerodynamic characteristic of the wind turbine shows that at the same pitch angle, there is only one tip-speed ratio that makes it obtain the maximum wind energy utilisation coefficient, i.e. there is only one rotational speed point for the same wind speed that makes the wind turbine capture the maximum wind energy, and that point is the maximum power point in that state. The maximum power curve can be found by connecting the maximum power points at different wind speeds. In addition, as the pitch angle increases, the wind energy utilisation coefficient of the wind turbine decreases, which makes it possible to reduce the power absorbed by the wind turbine at a high wind speed by increasing the pitch angle, thus protecting the wind power system.

From Eqs. (2.36), (2.37) and (2.38), the causal ordering graph and the energetic macroscopic representation of the wind turbine can be obtained, as shown in Fig. 2.18.

As can be seen from the above figure, the causal ordering graph provides a clearer representation of the relations among the variables inside the model, whereas the energetic macroscopic representation focuses only on the input and output variables

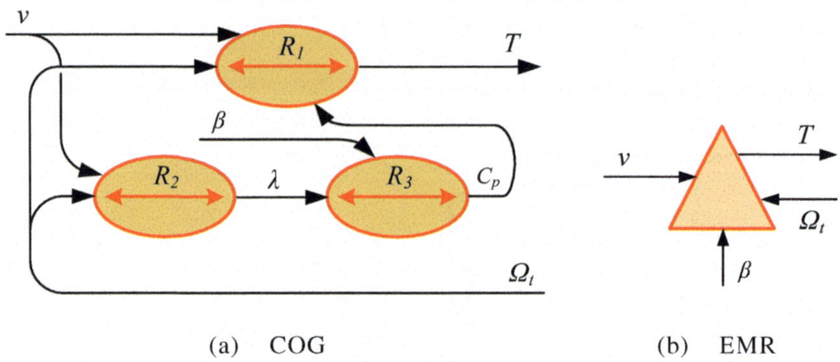

Fig. 2.18 Graphical model of the wind turbine

resulting from the interaction between the model and the outside environment, thus the system model obtained by the latter is more compact and integrated.

2.3.4.2 Model of the Shaft

The drive shaft of the wind power system includes the low-speed shaft of the wind turbine and the high-speed shaft of the generator. To simplify the modelling process, the high-speed shaft can be converted to the low-speed shaft side to obtain the concentrated mass block equivalent model of the drive shaft system.

$$R_4\colon J\frac{d\Omega_{\mathrm{t}}}{dt} = T_{\mathrm{t}} \tag{2.39}$$

where T_{t} is the total mechanical torque applied to the equivalent rotating shaft. In order to ensure that the input and output energy of the system does not change before and after the conversion, the corresponding parameters of the high-speed shaft need to be converted to the low-speed shaft side as well, from which the expression of the equivalent concentrated inertia J can be written as follows.

$$J = J_{\mathrm{t}} + \frac{J_{\mathrm{g}}}{G^2} \tag{2.40}$$

where J_{t} and J_{g} are the rotational inertia of the wind turbine and generator respectively, while G is the gearbox's speed ratio.

The total mechanical torque T_{t} applied to the equivalent shaft can be obtained from the wind turbine output torque T, the gearbox output torque T_{g} and the equivalent friction torque T_{vis}.

$$R_5\colon T_{\mathrm{t}} = T - T_{\mathrm{g}} - T_{\mathrm{vis}} \tag{2.41}$$

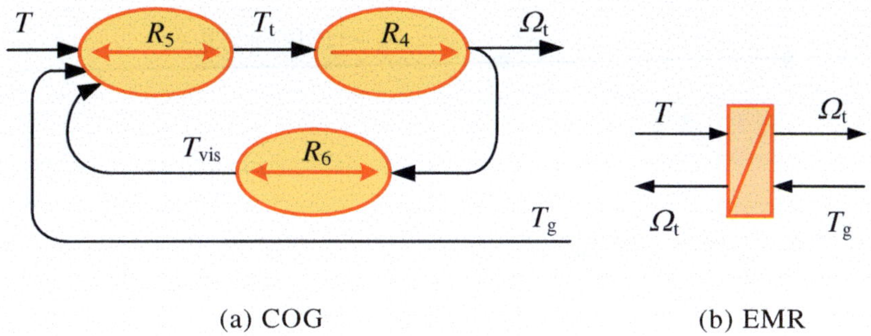

Fig. 2.19 Graphical model of the equivalent concentrated shaft

And the equivalent friction torque T_{vis} can be expressed as:

$$R_6\colon T_{vis} = f\Omega_t \tag{2.42}$$

where f is the equivalent damping factor and can be written as:

$$f = f_t + \frac{f_g}{G^2} \tag{2.43}$$

where f_t and f_g are the damping factors for the low-speed and high-speed shafts respectively.

As a result, the causal ordering graph and the energetic macroscopic representation of the equivalent concentrated shaft can be modelled, as shown in Fig. 2.19.

2.3.4.3 Model of the Gearbox

As the speed of the wind turbine is generally low, while the speed of the generator in normal operation is high, a mechanical speed-boosting device such as the gearbox is usually required to increase the low wind turbine speed to the high speed required for generator operation.

If the gearbox is assumed to be ideal and its own mechanical losses are neglected, it can be considered as a typical mechanical energy transformation component whose input and output energy remains constant. Therefore, as the gearbox increases its speed, its output torque changes accordingly.

Because the gearbox indicates the static relation, the relation between its input and output is determined by the equivalent concentrated shaft directly connected to it. Its mathematical model can therefore be written as:

$$R_7\colon \Omega = G\Omega_t \tag{2.44}$$

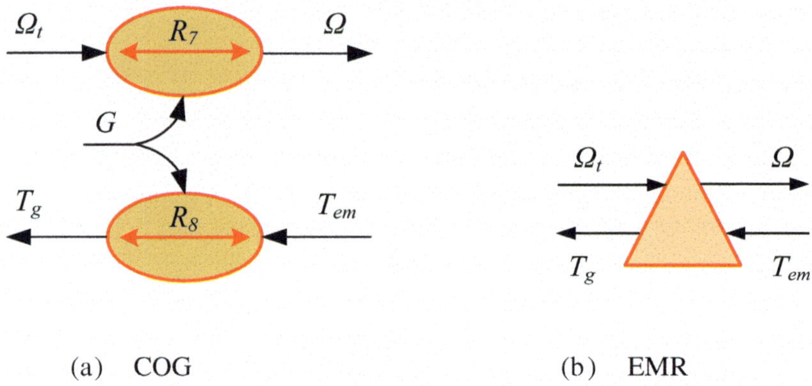

Fig. 2.20 Graphical model of the gearbox

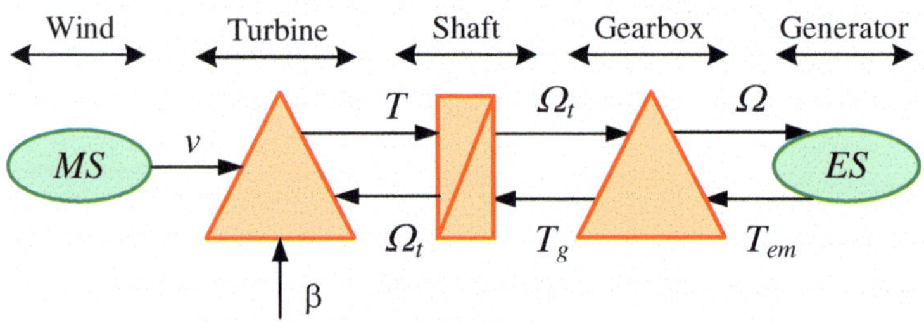

Fig. 2.21 EMR-based model of the wind turbine and its drive shaft

$$R_8 \colon T_g = G T_{em} \tag{2.45}$$

where Ω is the generator speed and T_{em} is the electromagnetic torque of the generator.

As a result, the model of the gearbox expressed by the COG and the EMR can be obtained, as shown in Fig. 2.20.

In summary, if the EMR-based components representing each part are connected, the graphical model of the entire mechanical part of the wind power system can be achieved, as shown in Fig. 2.21.

Similarly, the causal ordering graph, which captures the internal details of the system, can be represented as shown in Fig. 2.22.

According to the causal ordering graph, the system structure diagram of the mechanical part of the wind power system can be easily built, as shown in Fig. 2.23.

2.3.4.4 Control Strategy of the Wind Turbine

Taking the MPPT algorithm as an example, the wind turbine tracks its maximum power point, essentially controlling the wind turbine speed to track the wind speed

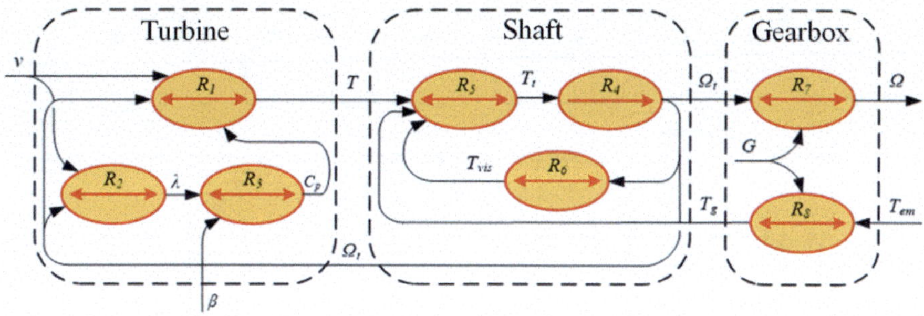

Fig. 2.22 COG-based model of the wind turbine and its drive shaft

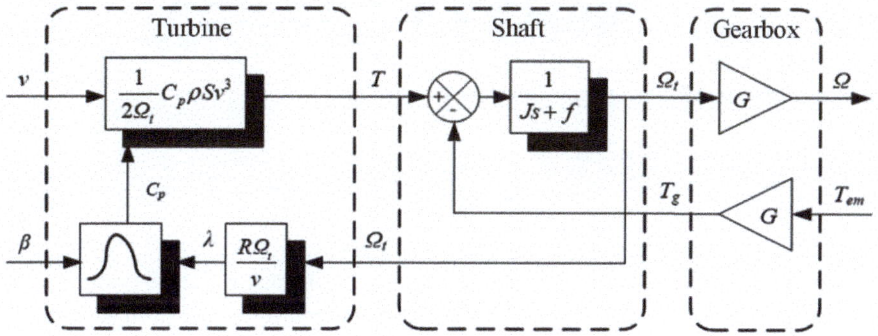

Fig. 2.23 Structure diagram of the wind turbine and its drive shaft

and achieve the optimum speed accordingly. From the EMR model of the wind turbine and its shaft, the MCS of the system can be easily derived, as shown in Fig. 2.24.

As the only energy storage in the model is the shaft, only one closed-loop controller is required at most in its control system. Assuming that all variables in the system can be measured, it is possible to achieve the MPPT of the wind turbine by the closed-loop control of its speed. Based on the mathematical model of the shaft, described by Eqs. (2.39), (2.40) and (2.41), and the inverse rules of the causal ordering graph, the expression of the speed controller can be derived as

$$T_{\text{g\_ref}} = C_{\text{PI}}(\Omega_{\text{t\_ref}} - \widehat{\Omega}_{\text{t}}) \tag{2.46}$$

where $T_{\text{g\_ref}}$ is the required torque reference, C_{PI} is the speed regulator, $\Omega_{\text{t\_ref}}$ is the wind turbine speed reference, and $\widehat{\Omega}_{\text{t}}$ is the measured wind turbine speed. However, the wind turbine speed is not generally measured in the practical system, so the wind turbine speed can be estimated from the generator speed by using Eq. (2.44) as:

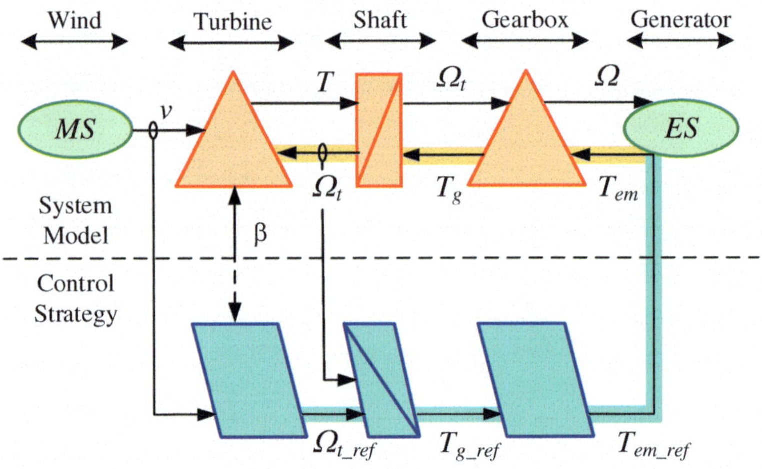

Fig. 2.24 EMR and MCS of the turbine and its drive shaft

$$R_{7c}: \tilde{\Omega}_{\mathrm{t}} = \frac{\widehat{\Omega}}{G} \tag{2.47}$$

where $\tilde{\Omega}_{\mathrm{t}}$ is the estimated wind turbine speed and $\widehat{\Omega}$ is the measured generator speed. This leads to the actual available wind turbine speed controller equation as:

$$R_{4c}: T_{\mathrm{g\_ref}} = C_{\mathrm{PI}}(\Omega_{\mathrm{t\_ref}} - \tilde{\Omega}_t) \tag{2.48}$$

Then the wind turbine speed reference can be obtained from the mathematical model of the wind turbine as Eq. (2.37) by directly inverting.

$$R_{2c}: \Omega_{\mathrm{t\_ref}} = \frac{\lambda_{\mathrm{Cp\_max}} \widehat{\upsilon}}{R} \tag{2.49}$$

where $\widehat{\upsilon}$ is the measured wind speed, and $\lambda_{\mathrm{Cp\_max}}$ is the value of the tip-speed ratio at which the MPPT is achieved. Since the pitch angle β is fixed to the minimum value in Zone 2 (shown as Fig. 1.19), the value of λ that maximizes the value of C_{p} can be found from the wind energy utilisation coefficient curve described by Eq. (2.38), which is the required $\lambda_{\mathrm{Cp\_max}}$ in Eq. (2.49).

Finally, the electromagnetic torque reference $T_{\mathrm{em\_ref}}$ required by the generator can be obtained from the reference value $T_{\mathrm{g\_ref}}$ outputted by the speed controller, according to Eq. (2.45).

$$R_{8c}: T_{\mathrm{em\_ref}} = \frac{T_{\mathrm{g\_ref}}}{G} \tag{2.50}$$

From the above, the control diagram, represented by the COG, can be obtained as shown in Fig. 2.25, and the corresponding structure diagram can be found in Fig. 2.26.

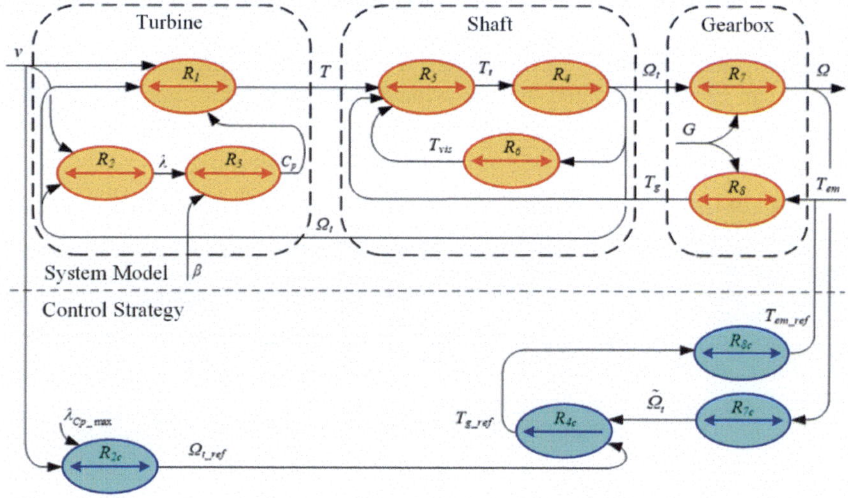

Fig. 2.25 COG of the wind turbine with the closed-loop speed control

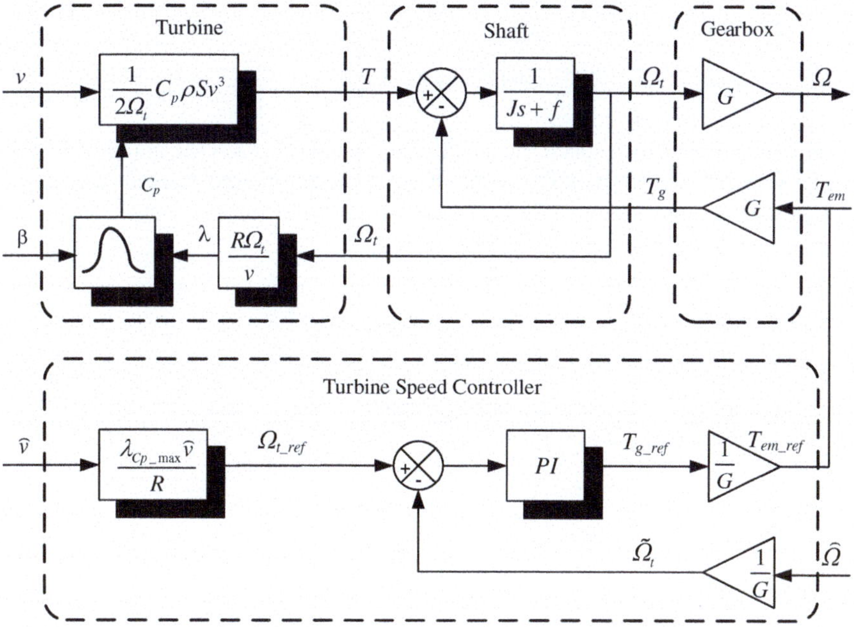

Fig. 2.26 Structure diagram of the turbine with the closed-loop speed control

In practice, however, it is difficult to measure the wind speed accurately. As a result, the wind turbine speed reference value, which is dependent on the wind speed measurement, may not be accurate. It is therefore necessary to investigate a MPPT control strategy that does not depend on the measured wind speed. Again, the MSC obtained in Fig. 2.24 can be simplified accordingly to produce a new control algorithm.

Assuming that the change in wind speed is not very drastic during the control period, it can be approximated that the wind turbine speed does not change during this period due to the very high rotational inertia of the megawatt-class turbine. According to Eqs. (2.39) and (2.41), it is possible to obtain:

$$J\frac{d\Omega_{\mathrm{t}}}{dt} = T - T_{\mathrm{g}} - T_{\mathrm{vis}} = 0 \tag{2.51}$$

Thus Eq. (2.39) degenerates from the differential equation to the algebraic one and the corresponding causality of the shaft also changes from the dynamic relation to the static one. If the frictional torque T_{vis} is neglected, it can be considered as follows.

$$R_{5c}\colon \tilde{T}_{\mathrm{g}} = \tilde{T} \tag{2.52}$$

That is, the estimated output torque of the wind turbine ($\tilde{T}$) is equal to the estimated output torque of the gearbox ($\tilde{T}_{\mathrm{g}}$). The required reference electromagnetic torque of the generator can thus be obtained from Eq. (2.45) as

$$R_{8c}\colon T_{\mathrm{em\_ref}} = \frac{\tilde{T}_{\mathrm{g}}}{G} \tag{2.53}$$

When the wind turbine speed tracks the wind speed always at the maximum power tracking point, the corresponding wind energy utilisation coefficient should also be equal to its maximum value. The estimate of the wind turbine output torque can therefore be derived from Eq. (2.36) as

$$R_{1c}\colon \tilde{T} = \frac{1}{2\tilde{\Omega}_{\mathrm{t}}} C_{\mathrm{p\_max}} \rho S \tilde{v}^3 \tag{2.54}$$

where the estimated wind turbine speed $\tilde{\Omega}_{\mathrm{t}}$ can also be obtained from Eq. (2.47), and the estimated wind speed can be obtained by directly inverting Eq. (2.41) as

$$R_{2c}\colon \tilde{v} = \frac{R\tilde{\Omega}_{\mathrm{t}}}{\lambda_{\mathrm{Cp\_max}}} \tag{2.55}$$

By combining the above equations, the wind turbine speed open-loop MPPT control strategy can be obtained, and the required generator electromagnetic torque reference value can finally be expressed as

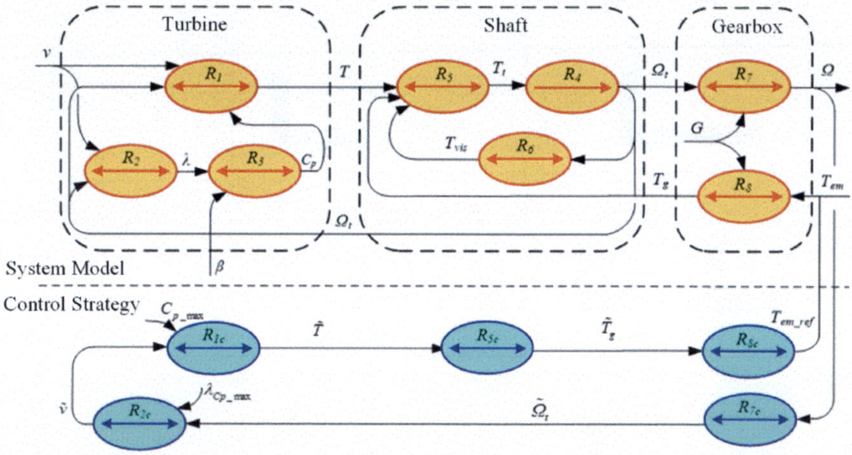

Fig. 2.27 COG of the turbine with the open-loop control

$$T_{\text{em\_ref}} = \frac{1}{2\lambda_{\text{Cp\_max}}^3 G^3} C_{\text{p\_max}} \rho \pi R^5 \widehat{\Omega}^2 \tag{2.56}$$

The model and its control system, represented by the COG, can be obtained from the above equations, as shown in Fig. 2.27, and the corresponding structure diagram can be found in Fig. 2.28.

And the simplified control diagram and the EMR of the wind turbine and its shaft system can be obtained accordingly as shown in Fig. 2.29.

With the help of simulation software such as Matlab, the above system model and the control methods can be implemented very quickly (Fig. 2.30).

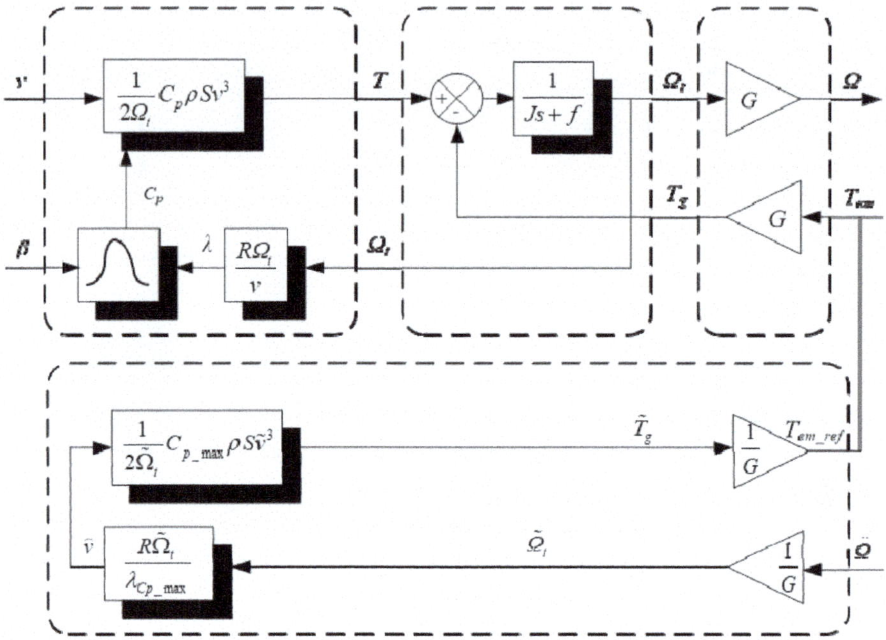

Fig. 2.28 Structure diagram of the turbine with the open-loop control

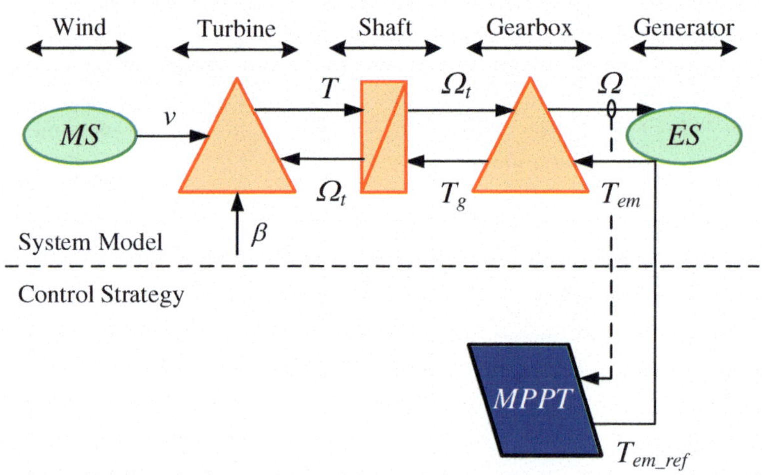

Fig. 2.29 EMR and simplified control diagram of the turbine and its drive shaft

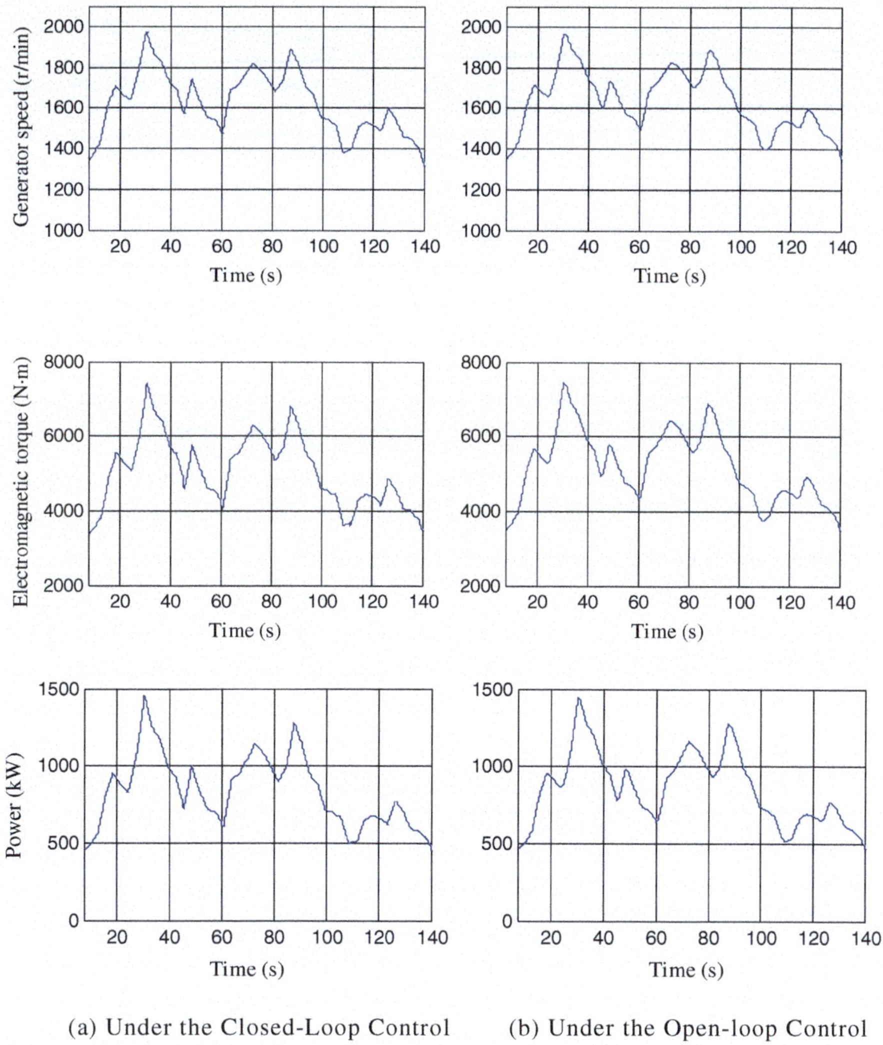

Fig. 2.30 Simulation results of the MPPT control of the wind turbine

References

1. Y. Ruan, B.S. Chen, *Electrical Drive and Automation System – Motion Control System*, 4th edn. (China Machine Press, Beijing, 2010)
2. G. Yang, Y.L. Luo, *Electrical Machine and Motion Control System*, 2nd edn. (Tsinghua University Press, Beijing, 2015)
3. R. Zanasi, Power-oriented graphs for modeling electrical machines, in *MELECOM'96* (1996), pp. 1211–1214
4. R. Schonfeld, G. Geitner, Power flow and information flow in motion control systems, in *EPE-PEMC 2004* (2004)

5. H. Piquet, H. Foch, F. Bernot, Design of electrical energy conversion systems with the aid of characterized elements. Int. J. Electr. Eng. **4**, 217–235 (2001)
6. J. Hautier, P. Barre, The causal ordering graph, a tool for system modelling and control law synthesis. Stud. Informat. Control **13**(4), 265–283 (2003)
7. P. Barre, A. Bouscayrol, P. Delarue et al., Inversion-based control of electromechanical systems using causal graphical descriptions. IECON **2006**, 5276–5281 (2006)
8. A. Bouscayrol, B. Davat, B. de Fornel et al., Multimachine multiconverter system: application for electromechanical drives. Eur. Phys. J. Appl. Phys. **10**(2), 131–147 (2000)
9. A. Bouscayrol, P. Delarue, Simplifications of the Maximum Control Structure of a wind conversion system with an induction generator. Int. J. Renew. Energy Eng. **4**(2), 479–485 (2002)
10. P. Ling, L. Yongdong, B. Francois, Modelling and control of the DFIG wind turbine system by using causal ordering graph during voltage dips, in *11th International Conference on Electrical Machines and Systems* (2018), pp. 2412–2417
11. P. Ling, Research on graphical modelling and low voltage ride-through control strategies of doubly fed induction wind generator system. Ph.D. Thesis, Tsinghua University, Beijing, China, 2010
12. P. Ling, L. Yongdong, C. Jianyun et al., Vector control a doubly fed induction generator for stand-alone shaft generator systems. J. Tsinghua Univ. (Sci. Technol.) **49**(7), 938–942 (2009)
13. E. Abdin, W. Xu, Control design and dynamic performance analysis of a wind turbine induction generator unit. IEEE Trans. Energy Convers. **15**(1), 91–96 (2000)

Chapter 3
Control Strategies for the DFIG-Based Wind Power System Under Ideal Grid Conditions

In this chapter, based on Chap. 2, various control algorithms[1] (including vector control, direct power control, and model predictive control) for the DFIG-based wind power system under ideal grid conditions are investigated, and related issues in various control algorithms (e.g., field orientation methods, magnetic chain observation, MPPT, etc.) are discussed.

3.1 Vector Control of the DFIG-Based Wind Power System

Vector control is a high-performance control method for electrical machines and their grid-connected converters by simplifying the dynamic mathematical models and decoupled the active and reactive currents through the field orientation technique, with the advantages of good dynamic and static characteristics and wide speed regulation range [1, 2], which is currently the most widely used high-performance control method for the DFIG-based wind power system.

3.1.1 Vector Control of the DFIG

The DFIG employs vector control for closed-loop control of the d-axis and q-axis components of the rotor current to achieve decoupled regulation of the active and reactive power of the generator. According to the selection of the orientation coordinate frame, the commonly used vector control can be divided into two types:

[1] Generally speaking, the control of variable speed and constant frequency wind power generation systems is divided into the generator control, the pitch control and the main control, and unless otherwise stated in this book, the wind power control refers specifically to its generator control.

H. Ma et al., *The Control Principle of Wind Power Generation System*,
https://doi.org/10.1007/978-981-99-6083-5_3

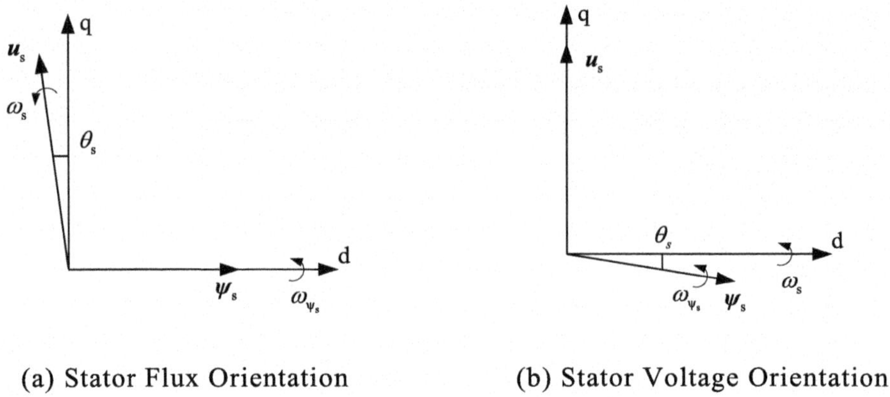

Fig. 3.1 Space vector diagrams of the DFIG

stator-flux-oriented vector control and stator-voltage-oriented vector control, whose space vector relations are shown in Fig. 3.1a and b, respectively.[2]

In the stator-flux-oriented coordinate frame, the d-axis coincides with the stator flux linkage vector, so that Eq. (2.20) can be simplified as follows.

$$\begin{cases} \psi_s^d = L_s i_s^d + L_M i_r^d = \psi_s \\ \psi_s^q = L_s i_s^q + L_M i_r^q = 0 \end{cases} \tag{3.1}$$

Substituting Eq. (3.1) into Eq. (2.19), it can be derived as follows.

$$\begin{cases} u_s^d = R_s i_s^d + p\psi_s^d \\ u_s^q = R_s i_s^q + \omega_{\psi_s}\psi_s^q \\ u_r^d = R_r i_r^d + p\psi_r^d - \left(\omega_{\psi_s} - \omega_r\right)\psi\omega_s \\ u_r^q = R_r i_r^q + p\psi_r^q + \left(\omega_{\psi_s} - \omega_r\right)\psi_r^q \end{cases} \tag{3.2}$$

[2] There are different understandings about the relation between flux orientation and voltage orientation, the main ones are: (1) the flux orientation means that the d-axis coincides with the flux linkage vector, and the voltage orientation means that the q-axis coincides with the voltage vector, which is based on the actual physical relation between the flux linkage vector and the voltage vector, so that a consistent mathematical description can be obtained for the same physical system in both orientation systems; (2) the flux linkage orientation means that the d-axis coincides with the flux linkage vector, and the voltage orientation means that the d-axis coincides with the voltage vector, which is based on the definition of the orientation itself and believes that a certain vector orientation means that the d-axis is coincident with the certain vector, so that the mathematical description for the same physical system is inconsistent (e.g., if the q-axis current is the active current under the flux orientation, then the d-axis current is the active current under the voltage orientation). The above two understandings are not right or wrong and do not affect the physical relations and physical behaviors of the oriented physical systems. To avoid ambiguity, the dq-axis voltage/flux distribution will be specified directly in the subsequent modeling and derivation of algorithms in this book.

Due to the stator flux orientation, the dq-frame rotates at the same angular speed as the stator flux linkage angular frequency. In the steady state, the frequency of the stator flux linkage is consistent with the grid voltage frequency, but in the transient process, the two will deviate. Setting the angle of the stator voltage vector leading the stator-flux-oriented q-axis as θ_s, the stator voltage equation can be obtained as

$$\begin{cases} u_s^d = -U_s \sin\theta_s \\ u_s^q = U_s \cos\theta_s \end{cases} \tag{3.3}$$

From the stator q-axis voltage in Eq. (3.2), the rotational angular speed of the oriented coordinate frame can be obtained as

$$\omega_c = \omega_{\psi_s} = \left(u_s^q - R_s i_s^q\right)/\psi_s^q \tag{3.4}$$

The motion equation of the rotating coordinate frame is

$$p\theta_s = \omega_s - \omega_{\psi_s} \tag{3.5}$$

Equations (3.1) to (3.5) form the fourth-order electromagnetic dynamic model of the DFIG in the stator-flux-oriented coordinate frame. Substituting Eq. (3.1) into the electromagnetic torque equation and the reactive power equation of the DFIG, the following can be obtained.

$$T_e = -n_p L_M \psi_s i_r^q / L_s \tag{3.6}$$

$$Q_s = \omega_{\psi_s}\psi_s\left(\psi_s - L_M i_s^d\right)/L_s \tag{3.7}$$

From Eqs. (3.6) and (3.7), if the stator flux linkage amplitude and angular frequency are kept constant, the q-axis component of the rotor current is linearly proportional to the electromagnetic torque, while the d-axis component of the rotor current is linearly proportional to the reactive power, and the decoupled control of the electromagnetic torque and the reactive power can be achieved by regulating the q-axis and d-axis components of the rotor current, respectively.

It can be seen that the use of stator flux orientation can achieve fully decoupled control of the DFIG, but this is predicated on accurate observation of the stator flux linkage amplitude and phase, for which a complex flux linkage observation algorithm needs to be designed. There are various methods for stator flux linkage observation, among which the one using a voltage model that neglects the stator resistance is the most commonly used, which will be discussed subsequently in this section.

Compared to the stator flux linkage, the stator voltage is more easily and accurately obtained in terms of phase information, so the stator voltage orientation can be used here by taking the q-axis of the rotating coordinate frame to coincide with the stator voltage vector, i.e.

$$\begin{cases} u_s^d = 0 \\ u_s^q = U_s \end{cases} \quad (3.8)$$

Since the stator voltage is identical to the grid voltage after grid connection, the angular speed of the stator-voltage-oriented rotating coordinate frame is constantly equal to the grid angular frequency during normal operation.

$$\omega_c = \omega_s = \omega_1 \quad (3.9)$$

Then the electromagnetic torque and the reactive power of the DFIG are

$$T_e = -n_p L_M \left(\psi_s^d i_r^q - \psi_s^q i_r^d\right)/L_s \quad (3.10)$$

$$Q_s = U_s i_s^d = U_s\left(\psi_s^d - L_M i_r^d\right)/L_s \quad (3.11)$$

where the stator flux linkage can be written as

$$\begin{aligned} \psi_s^d &= \left(U_s - R_s i_s^q + p\psi_s^q\right)/\omega_1 \\ \psi_s^q &= \left(R_s i_s^d + p\psi_s^d\right)/\omega_1 \end{aligned}$$

It can be seen that both electromagnetic torque and reactive power are related to the rotor d-axis and q-axis currents, making it impossible to decouple the DFIG control. However, considering that the grid voltage is stable during normal operation, the fluctuation of the stator flux linkage is small, and the voltage drop on the stator resistance can be neglected compared to the grid voltage, Eqs. (3.10) and (3.11) can thus be solved approximately as follows.

$$T_e \approx -n_p L_M \psi_s i_r^q / L_s \quad (3.12)$$

$$Q_s \approx U_s\left(\psi_s - L_M i_r^d\right)/L_s \quad (3.13)$$

In terms of decoupled control effect, the stator-flux-oriented vector control can achieve complete decoupled control of the DFIG, while the stator-voltage-oriented vector control can achieve decoupled control of the DFIG only after certain approximations have been made based on engineering experience, and will fail to decouple the electromagnetic torque and the reactive power when those assumptions cannot be met, such as the grid voltage fluctuation is large or the voltage drop across the stator resistance cannot be neglected.

In terms of the observation difficulty of the orientation angle, the stator voltage phase is much easier to observe than the stator flux linkage phase, with smaller calculations and simpler algorithms. However, the stator-voltage-oriented control system will fail when there is a sudden change in the grid voltage, especially for special conditions when the grid voltage is very low or even zero, whereas the stator flux linkage is current-dependent and does not change suddenly, so the stator-flux-oriented control system can still work.

In terms of the grid connection process, the DFIG needs to be excited to build up the stator voltage before it is connected to the grid and can only be connected to the grid after the stator voltage has been synchronised with the grid voltage. If the stator-flux-oriented vector control is used, the stator chain amplitude and phase control strategy needs to be designed in order to synchronise the stator voltage with the grid voltage, which is different from the power control after grid connection, resulting in a complex algorithm and the switching of the control strategies before and after grid connection. For the stator-voltage-oriented system, the synchronisation of the stator voltage before grid connection can be easily achieved by replacing the stator voltage with the grid voltage, and the control strategies before and after grid connection are the same and do not need to be switched.

Conclusively, stator voltage orientation is more suitable for the DFIG control under normal grid conditions, while stator flux orientation is more suitable for the analysis and control during grid anomalies.

The vector control of the DFIG controls the electromagnetic torque and the reactive power by means of a closed loop regulation of the rotor current.

Substituting Eq. (2.20) into Eq. (2.19) gives the voltage equation for the DFIG rotor winding as

$$\begin{cases} u_r^d = R_r i_r^d + \sigma L_r p i_r^d - \omega_{sl} \psi_r^q + \dfrac{L_M}{L_s} p \psi_s^d \\ u_r^q = R_r i_r^q + \sigma L_r p i_r^q + \omega_{sl} \psi_r^d + \dfrac{L_M}{L_s} p \psi_s^q \end{cases} \tag{3.14}$$

When the stator voltage orientation is used, $\omega_c = \omega_s = \omega_1$. When the stator flux orientation is used, $\omega_c = \omega_{\psi_s}$. When the grid voltage is stable, the differential term of the stator flux linkage can be considered as zero, and the simplified rotor winding voltage equation is

$$\begin{cases} u_r^d = R_r i_r^d + \sigma L_r p i_r^d + E_r^d \\ u_r^q = R_r i_r^q + \sigma L_r p i_r^q + E_r^d \end{cases} \tag{3.15}$$

where $E_r^d = -\omega_{sl} \psi_r^q$ and $E_r^q = \omega_{sl} \psi_r^d$ are the cross-coupling terms for the d-axis and q-axis rotor currents, respectively, which can be considered as external voltage disturbances to the rotor winding.

To reduce the disturbance to the closed-loop rotor current control by E_r^{dq}, a feed-forward compensation vector $\left(u_{rc}^{dq} = E_r^{dq}\right)$ can be added, giving the closed-loop rotor current controller output according to Eq. (3.15) as

$$\begin{cases} u_r^{d*} = C_{ri}^d \left(i_r^{d*} - i_r^d\right) + u_{rc}^d \\ u_r^{q*} = C_{ri}^q \left(i_r^{q*} - i_r^q\right) + u_{rc}^q \end{cases} \tag{3.16}$$

where C_{ri}^{d} and C_{ri}^{q} are the d-axis and q-axis PI controllers of the rotor current respectively, $i_{\text{r}}^{\text{d}*}$ and $i_{\text{r}}^{\text{q}*}$ are the d-axis and q-axis rotor current reference values respectively, $u_{\text{r}}^{\text{d}*}$ and $u_{\text{r}}^{\text{q}*}$ are the d-axis and q-axis control voltage reference values of the generator-side converter, respectively. Note: The superscript "*" indicates the reference value of the marked physical quantity, which is used in the rest of the book unless otherwise stated.

Because of the difficulty of accurately observing the electromagnetic torque, an open-loop control strategy is usually used in the DFIG. From Eq. (3.6), the q-axis rotor current is linearly proportional to the electromagnetic torque, which gives

$$i_{\text{r}}^{\text{q}*} = -L_{\text{s}}T_{\text{e}}^{*}/\left(n_{\text{p}}L_{\text{M}}\psi_{\text{s}}\right) \tag{3.17}$$

As a result of the motor reference system, T_{e}^{*} is less than zero and the corresponding $i_{\text{r}}^{\text{q}*}$ is greater than zero, when the DFIG is running as a generator.

The relation between the stator reactive power and the reference value of the d-axis rotor current can be obtained from Eq. (3.13) as

$$i_{\text{r}}^{\text{d}*} = \psi_{\text{s}}L_{\text{M}} - Q_{\text{s}}^{*}L_{\text{s}}/(U_{\text{s}}L_{\text{M}}) = i_{\text{ms}} - Q_{\text{s}}^{*}L_{\text{s}}/(U_{\text{s}}L_{\text{M}}) \tag{3.18}$$

When the DFIG operates at a unit power factor, the reactive power reference is assigned as 0. The reference value of the d-axis rotor current is obtained as

$$i_{\text{r}}^{\text{d}*} = i_{\text{ms}} \tag{3.19}$$

From the above analysis, the block diagram of the DFIG system using stator-voltage-oriented vector control can be obtained as shown in Fig. 3.2, where the regulators are PI controllers. The rotor voltage reference value obtained by the control algorithm is modulated by PWM to drive the rotor-side converter output to realise the control of the rotor current.

Figure 3.3 shows the block diagram of stator-flux-oriented vector control of the DFIG. The only difference with Fig. 3.2 is that the orientation angle and the angular frequency of the rotating coordinate frame are obtained differently: the stator-voltage-oriented system obtains the orientation information by observing the grid voltage, while the stator-flux-oriented system obtains the orientation information by observing the stator flux linkage. In order to achieve as accurate feedforward compensation terms as possible, the stator-voltage-orientated system also requires accurate observation of the d-axis and q-axis components of the stator flux linkage, but since the closed-loop control of the rotor current suppresses the inaccurately compensated disturbances, the stator-voltage-orientated system does not require a very high accuracy in the observation of the stator flux linkage.

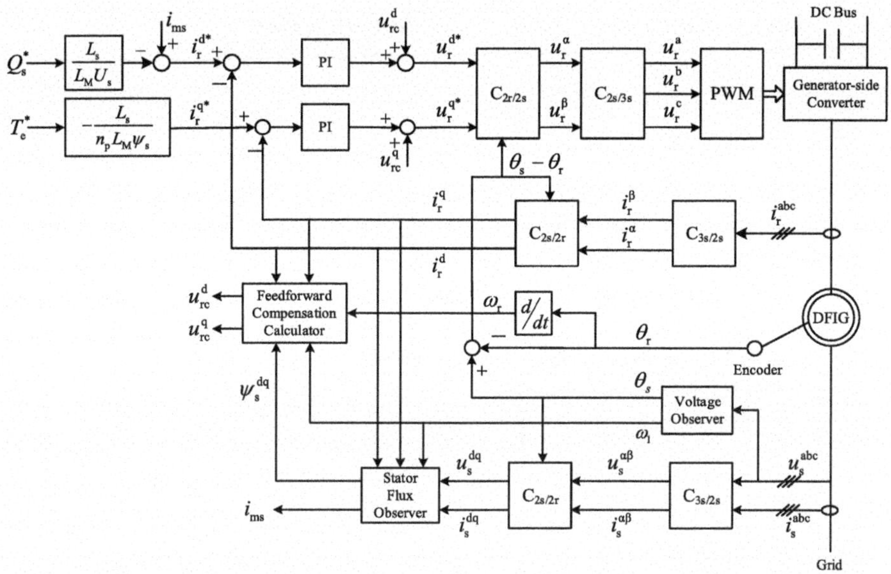

Fig. 3.2 Block diagram of stator-voltage-oriented vector control of the DFIG

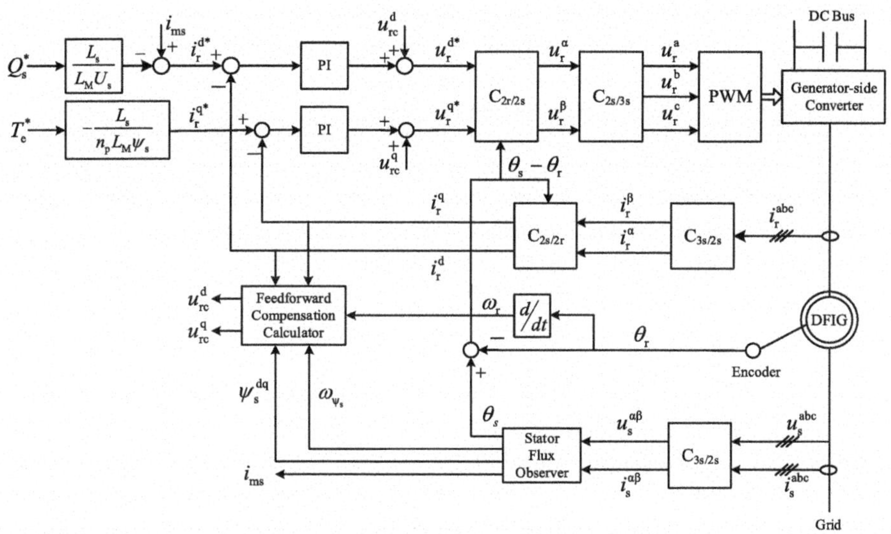

Fig. 3.3 Block diagram of stator-flux-oriented vector control of the DFIG

3.1.2 Vector Control of the Grid-Side Converter

There are two common control strategies for the grid-side converter: indirect current control and direct current control. This section focusses on the direct current control based on the grid-voltage-oriented vector control strategy.

In the grid-voltage-oriented rotating dq-frame, the grid voltage can be written as

$$\begin{cases} e_{\mathrm{g}}^{\mathrm{d}} = E_{\mathrm{g}} \\ e_{\mathrm{g}}^{\mathrm{q}} = 0 \end{cases} \tag{3.20}$$

Substituting Eq. (3.20) into Eq. (2.25), it gives

$$\begin{cases} P_{\mathrm{g}} = e_{\mathrm{g}}^{\mathrm{d}} i_{\mathrm{g}}^{\mathrm{d}} \\ Q_{\mathrm{g}} = -e_{\mathrm{g}}^{\mathrm{d}} i_{\mathrm{g}}^{\mathrm{q}} \end{cases} \tag{3.21}$$

According to Eq. (3.21), the d-axis component of the grid-side current is proportional to the active power of the converter and the q-axis component of the grid-side current is proportional to the reactive power of the converter, so the decoupled control of the active and reactive power of the grid-side converter can be achieved by controlling its d-axis and q-axis currents, respectively.

From Eq. (2.26) for the DC-bus dynamic behaviors, it can be seen that, when the rotor power is constant, increasing the active power input to the grid-side converter can charge the DC-bus capacitor and increase the DC-bus voltage, and decreasing the active power input to the grid-side converter can discharge the DC-bus and decrease the DC-bus voltage. Therefore, the DC-bus voltage can be controlled by regulating the active power of the grid-side converter.

In the grid-voltage-oriented dq-frame, Eq. (2.24) can be rewritten as

$$\begin{cases} u_{\mathrm{g}}^{\mathrm{d}} = R_{\mathrm{g}} i_{\mathrm{g}}^{\mathrm{d}} + L_{\mathrm{g}} p i_{\mathrm{g}}^{\mathrm{d}} + E_{\mathrm{g}}^{\mathrm{d}} \\ u_{\mathrm{g}}^{\mathrm{q}} = R_{\mathrm{g}} i_{\mathrm{g}}^{\mathrm{q}} + L_{\mathrm{g}} p i_{\mathrm{g}}^{\mathrm{q}} + E_{\mathrm{g}}^{\mathrm{q}} \end{cases} \tag{3.22}$$

where $E_{\mathrm{g}}^{\mathrm{d}} = -\omega_{\mathrm{e}} L_{\mathrm{g}} i_{\mathrm{g}}^{\mathrm{q}} + E_{\mathrm{g}}$ and $E_{\mathrm{g}}^{\mathrm{q}} = \omega_{\mathrm{e}} L_{\mathrm{g}} i_{\mathrm{g}}^{\mathrm{d}}$ are the cross-coupling terms for the d-axis and q-axis grid-side currents.

From Eq. (3.22) the output voltage of the grid-side current closed loop controller can be obtained as

$$\begin{cases} u_{\mathrm{g}}^{\mathrm{d}*} = C_{\mathrm{gi}}^{\mathrm{d}} \left(i_{\mathrm{g}}^{\mathrm{d}*} - i_{\mathrm{g}}^{\mathrm{d}} \right) + u_{\mathrm{gc}}^{\mathrm{d}} \\ u_{\mathrm{g}}^{\mathrm{q}*} = C_{\mathrm{gi}}^{\mathrm{q}} \left(i_{\mathrm{g}}^{\mathrm{q}*} - i_{\mathrm{g}}^{\mathrm{q}} \right) + u_{\mathrm{gc}}^{\mathrm{q}} \end{cases} \tag{3.23}$$

where $C_{\mathrm{gi}}^{\mathrm{d}}$ and $C_{\mathrm{gi}}^{\mathrm{q}}$ are the d-axis and q-axis PI controllers of the grid-side current respectively, $i_{\mathrm{g}}^{\mathrm{d}*}$ and $i_{\mathrm{g}}^{\mathrm{q}*}$ are the d-axis and q-axis grid-side current reference values

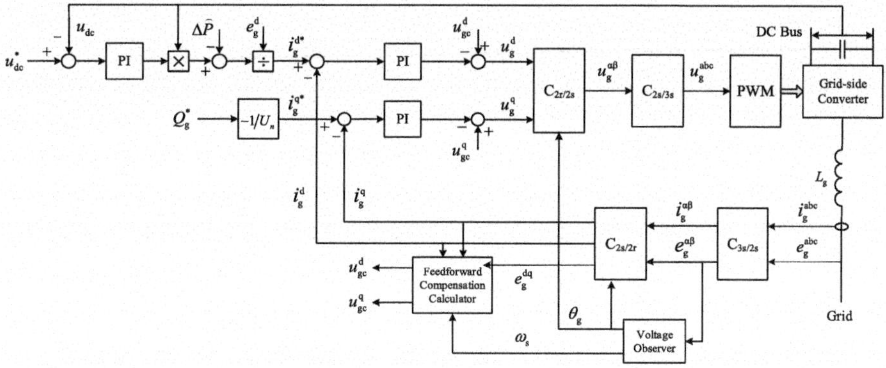

Fig. 3.4 Block diagram of grid-voltage-oriented vector control of the grid-side converter

respectively, $u_{\mathrm{g}}^{\mathrm{d}*}$ and $u_{\mathrm{g}}^{\mathrm{q}*}$ are the d-axis and q-axis control voltage reference values of the grid-side converter respectively, $u_{\mathrm{gc}}^{\mathrm{d}} = E_{\mathrm{g}}^{\mathrm{d}}$ and $u_{\mathrm{gc}}^{\mathrm{q}} = E_{\mathrm{g}}^{\mathrm{q}}$ are the d-axis and q-axis feedforward compensation values, respectively.

From Eq. (2.26) it follows that

$$P_{\mathrm{g}} = Cu_{\mathrm{dc}}pu_{\mathrm{dc}} = -u_{\mathrm{dc}}i_{\mathrm{dc}} - \Delta P \tag{3.24}$$

where ΔP can be estimated in the back-to-back system as, denoted $\Delta\widehat{P}$. Then the output of the DC-bus voltage loop controller is

$$P_{\mathrm{g}}^{*} = u_{\mathrm{dc}}C_{\mathrm{udc}}\left(u_{\mathrm{dc}}^{*} - u_{\mathrm{dc}}\right) - \Delta\widehat{P} \tag{3.25}$$

where C_{udc} is the PI controller of the DC-bus voltage loop, P_{g}^{*} is the grid-side active power reference and its relation with $i_{\mathrm{g}}^{\mathrm{d}*}$ is given by Eq. (3.21).

From Eq. (3.21), the relation between the reference value of the grid-side reactive power and the given value of the q-axis grid-side current is

$$i_{\mathrm{g}}^{\mathrm{q}*} = -Q_{\mathrm{g}}^{*}/E_{\mathrm{g}} \tag{3.26}$$

From the above analysis, the block diagram of the grid-voltage-oriented vector control based on PI controllers can be obtained as shown in Fig. 3.4.

3.1.3 MPPT Control

In the variable-speed and constant-frequency wind power system, MPPT control is often required to capture as much wind energy as possible, which can be achieved either by controlling the wind turbine system or by controlling the generator system

[3]. The wind turbine system is generally seen as the outer control loop of the generator control system, with a large response time, so in most cases the MPPT control is achieved through the generator control system. The specific implementation of MPPT is divided into the following main scenarios.

(1) If the power characteristic curve of the wind turbine is known in advance and the wind speed can be measured, the relation between wind speed—generator speed and optimal power P_{opt} can be stored in advance in the microcontroller chip as a data table. Then the difference between the actual generator speed ω_{r} and its reference ω_{r}^{*} from the table is processed by the regulator to obtain the torque current reference $i_{\text{r}}^{\text{q}*}$. Alternatively, the wind turbine can be controlled by giving the power reference or the torque current reference directly according to the optimal power curve of the wind turbine.
(2) If the power characteristic curve of the turbine is known in advance but the wind speed is difficult or impossible to measure, the wind speed can be estimated from the estimated output torque or power of the turbine and the power characteristic curve of the turbine.

In general, the load torque observer can be designed by the standard variable method according to the mechanical equation of the generator. It gives

$$p\begin{bmatrix}\tilde{\omega}_{\text{r}}\\ \tilde{T}_{\text{aux}}\end{bmatrix}=\begin{bmatrix}0 & 1/J\\ 0 & 0\end{bmatrix}\begin{bmatrix}\tilde{\omega}_{\text{r}}\\ \tilde{T}_{\text{aux}}\end{bmatrix}-\begin{bmatrix}p_{\text{n}}i_{\text{ms}}L_{\text{M}}^{2}/JL_{\text{s}}\\ 0\end{bmatrix}i_{\text{r}}^{\text{q}} \tag{3.27}$$

$$[\tilde{\omega}_{\text{r}}]=\begin{bmatrix}1 & 0\end{bmatrix}\begin{bmatrix}\tilde{\omega}_{\text{r}}\\ \tilde{T}_{\text{aux}}\end{bmatrix} \tag{3.28}$$

where $\tilde{\omega}_{\text{r}}$ is the predicted generator speed and $\tilde{T}_{\text{aux}}$ is the predicted auxiliary torque variable, the estimated speed $\widehat{\omega}_{\text{r}}$ and the estimated load torque $\widehat{T}_{m}$ can be given by

$$\widehat{\omega}_{\text{r}}=\tilde{\omega}_{\text{r}}+K_{1}(\omega_{\text{r}}-\tilde{\omega}_{\text{r}}) \tag{3.29}$$

$$\widehat{T}_{\text{aux}}=\tilde{T}_{\text{aux}}+K_{2}(\omega_{\text{r}}-\tilde{\omega}_{\text{r}}) \tag{3.30}$$

$$\widehat{T}_{\text{m}}=\widehat{T}_{\text{aux}}+B\omega_{\text{r}} \tag{3.31}$$

where K_1 and K_2 are coefficients determined according to the dynamic characteristics of the state observer. The method considers the generator as a wind speed instrument and obtains the generator power reference from the observed load torque.

The wind speed can also be estimated by observing the power output of the turbine. According to the measured total output power of the generator and the $P_{\text{T}}-I_{\text{c}}$ curve at the maximum efficiency point, the working current I_{c} can be found, then the corresponding efficiency η can also obtained from the $I_c - \eta$ curve of the generator. Therefore, the output power of the turbine can be calculated by $P_{wt} = P_T/\eta$. According to the following equation:

$$P_{wt} = \frac{1}{2}\pi\rho C_p(\lambda)R^5\frac{\Omega_t^3}{\lambda^3} \tag{3.32}$$

The wind speed can be estimated by finding the effective root of Eq. (3.22) and substitute it into the equation $\lambda = \Omega_t R/v$. This approach to estimating wind speed does not require strong robustness to machine parameters, but it requires iterative calculation of the tip-speed ratio, which is computationally intensive, and the accuracy of the estimated wind speed is influenced by the preset $P_T - I_c$ curve as well as the $I_c - \eta$ curve.

(3) If the wind power characteristic curve is not known in advance and the wind speed cannot be measured, MPPT control can be achieved by the method of peak power seeking [4], which does not require prior knowledge of the relation between wind speed—generator speed and optimal power, does not rely on the turbine parameters or the air density and can find the optimal power point quickly according to the point on the $P - \omega$ curve where $dP/d\omega = 0$ being the optimal power point. In order to improve the shortcomings of the fixed speed step algorithm, such as speed fluctuations and slow response to wind speed change, a variable step finding algorithm can be used, where the given speed step is determined by the following equation.

$$d\omega_g = K_{MPPT}dP_W/d\omega_g \tag{3.33}$$

where K_{MPPT} is the step coefficient of the given speed, which regulates the stability of the tracking algorithm. Due to the mechanical time of the generator, the change of the given speed cannot be too large and generally should be smaller than $\Delta\omega_{max} = T_{max}dT/J$, where T_{max} is the maximum mechanical torque of the generator, dT is the period of the speed loop and J is the rotational inertia.

3.1.4 Stator Flux Linkage Observation of the DFIG

In the stator-flux-oriented control system of the DFIG-based wind power system, the observation of the DFIG stator flux linkage plays a very important role, and accurate stator flux linkage observation will be directly related to control performance [5, 6].

In order to obtain accurate information of the flux linkage, earlier solutions of burying coils or laying magnetic sensitive elements inside the electrical machine were used. This direct measurement method is not only complex, but is also affected by the tooth grooves. Nowadays, practical systems mainly use the calculation method, which calculates the flux linkage in real time using the electrical machine equations based on the detected physical quantities such as voltage, current and rotor speed. The calculation method of the flux linkage can usually be divided into two types: one is the direct calculation through the electrical machine flux linkage equation, where errors in sampling and in the electrical machine parameters can directly affect the accuracy of the flux linkage estimation. The other one is closed-loop identification using the observer, which uses the state equations of the electrical machine to construct the state observer, can suppress various errors very well and can speed up the convergence of the observer by adjusting the gain matrix.

Currently, the widely used observation method for the stator flux linkage in the DFIG is open-loop.

The first method is the current-model-based approach. Since both the stator and rotor currents of the DFIG can be measured directly, it can be known from the flux linkage equation of the DFIG that

$$\boldsymbol{\psi}_{\mathrm{s}} = L_{\mathrm{s}}\boldsymbol{i}_{\mathrm{s}} + L_{\mathrm{m}}\boldsymbol{i}_{\mathrm{r}} \tag{3.34}$$

where $\boldsymbol{\psi}_{\mathrm{s}}$ is the stator flux linkage vector in stator stationary coordinate frame, $\boldsymbol{i}_{\mathrm{s}}$ and $\boldsymbol{i}_{\mathrm{r}}$ are the stator and rotor current vectors respectively. Therefore the stator flux linkage can be obtained simply by detecting the stator and rotor currents and calculating Eq. (3.34). However, this method is affected much by the accuracy of the stator self-inductance and mutual inductance parameters.

The second method is based on the voltage model with neglected stator resistance. Since the stator winding is directly connected to the grid, the voltage drop across the stator resistance is negligible compared to the grid voltage, and the following equation is true under steady state conditions.

$$\boldsymbol{u}_{\mathrm{s}} = j\omega_{\mathrm{s}}\boldsymbol{\psi}_{\mathrm{s}} \tag{3.35}$$

where $\boldsymbol{u}_{\mathrm{s}}$ is the stator voltage vector. Therefore, the stator flux linkage can be calculated by simply checking the stator voltage of the generator, which is currently the most widely used method, but does not allow for good observation of the stator flux linkage in the event of fluctuations in the grid voltage.

The third method is based on the complete voltage model, which calculates the stator flux linkage by integration of the stator electrical motive force as Eq. (3.36).

$$\boldsymbol{\psi}_{\mathrm{s}} = \int (\boldsymbol{u}_{\mathrm{s}} - R_{\mathrm{s}}\boldsymbol{i}_{\mathrm{s}})dt \tag{3.36}$$

This method is also relatively simple, uses fewer machine parameters and is highly robust to hardware parameters compared to the current-model-based method above. However, this method has its shortcomings: the absence of any correction mechanism results in a drift through the integration when there are detection or calculation errors in the voltage, current and stator resistance, especially when there is a sudden drop in the three-phase grid voltage, which will cause even more serious integration drift problems. If the integral drift can be eliminated, the stator flux linkage can be observed accurately. The simplest way to eliminate the drift (a direct flow) is to pass

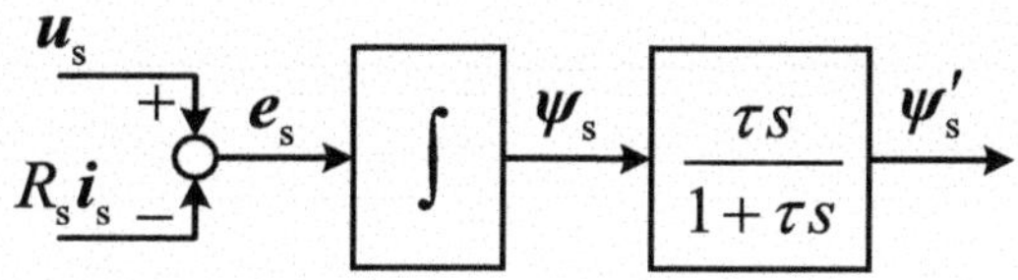

Fig. 3.5 Stator flux linkage observer based on the complete voltage model with a high-pass filter

the integral result through a high-pass filter, as shown in Fig. 3.5, where the stator electrical motive force vector is shown as $\boldsymbol{e}_{\rm s}$.

The high-pass filter can be represented by

$$\boldsymbol{\psi}_{\rm s} = \frac{\boldsymbol{u}_{\rm s} - R_{\rm s}\boldsymbol{i}_{\rm s}}{s}\frac{\tau s}{1+\tau s} \tag{3.37}$$

where $\omega_{\rm c} = 1/\tau$ and τ is the time constant of the filter.

After the high-pass filter, the integral drift is largely eliminated and the system can operate stably.

Further research found that although the use of the high-pass filter effectively eliminates integration drift, it introduces the new problem of a smaller amplitude and phase overshoot of the observed flux linkage compared to the actual one. In order to compensate for the amplitude degradation and the phase overshoot caused by the high-pass filter, an improved voltage-model-based method was proposed, in which the integral link and the low-pass filter are combined. Specifically,

$$y = \frac{1}{S+\omega_{\rm c}}x + \frac{\omega_{\rm c}}{S+\omega_{\rm c}}z \tag{3.38}$$

where y is the output of the integral part, x is the input of the integral part, z is the compensation signal and τ is the cut-off frequency. From the equation, it can be found that the modified integral part can be considered as a low-pass part and a pure integral part when $z = 0$ or $z = {\rm y}$ is taken respectively. Therefore, by appropriately tuning the amount of compensation, this improved integral part can act as an intermediate one between the pure integral part and the first order inertia part.

The structure diagram of the improved voltage-model-based stator flux linkage observer is illustrated as Fig. 3.6.

In Fig. 3.6, $e_{\rm s}^{\alpha}$ and $e_{\rm s}^{\beta}$ are the α-axis and β-axis components of the stator electrical motive force, respectively, as inputs to the stator flux observer. $\psi_{\rm s}^{\alpha}$ and $\psi_{\rm s}^{\beta}$ are the α-axis and β-axis components of the observed stator flux linkage, respectively, as outputs of the observer. The figure illustrates that the stator flux linkage components of $\boldsymbol{\psi}_{\rm sh}$ and $\boldsymbol{\psi}_{\rm sl}$ are first obtained by high-pass and low-pass filtering respectively, and then are summed to achieve the observed stator flux linkage $\widehat{\boldsymbol{\psi}}_{\rm s}$ in the current observation period as Eq. (3.39).

$$\boldsymbol{\psi}_{\rm s}(n) = \frac{\boldsymbol{u}_{\rm s} - R_{\rm s}\boldsymbol{i}_{\rm s}}{s+\omega_{\rm c}} + \frac{\omega_{\rm c}}{s+\omega_{\rm c}} \cdot \boldsymbol{\psi}_{\rm s}(n-1) \tag{3.39}$$

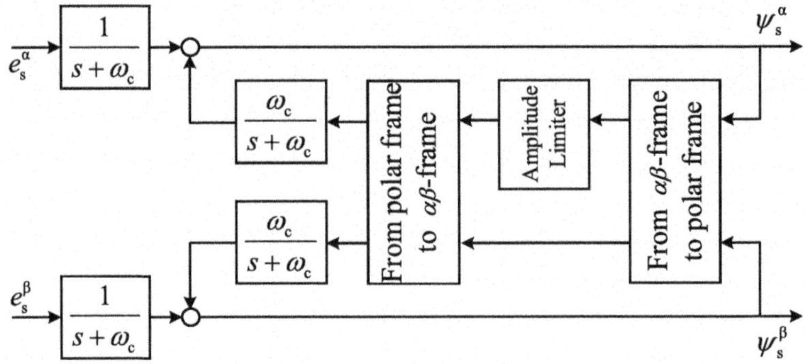

Fig. 3.6 Improved stator flux linkage observer based on the complete voltage model

Since the control period of the generator is very short, in this case it can be assumed that $\boldsymbol{\psi}_s(n) \approx \boldsymbol{\psi}_s(n-1)$, and therefore it gives

$$\boldsymbol{\psi}_s(n) = \frac{\boldsymbol{u}_s - R_s\boldsymbol{i}_s}{s} \cdot \frac{s}{s+\omega_c} + \frac{\omega_c}{s+\omega_c} \cdot \boldsymbol{\psi}_s(n-1) \approx \boldsymbol{\psi}_s(n) \cdot \frac{s+\omega_c}{s+\omega_c} = \boldsymbol{\psi}_s(n)$$

Therefore, the improved voltage-model-based stator flux linkage observer can theoretically obtain the same amplitude and phase as the true flux linkage. In the stator-flux-oriented DFIG wind power system, the stator flux linkage vector varies continuously with the stator voltage vector and $\boldsymbol{\psi}_s(n) \approx \boldsymbol{\psi}_s(n-1)$ is therefore established, so theoretically the improved voltage-model-based stator flux linkage observer yields an accurate stator flux linkage.

Figure 3.7 illustrates the vector synthesis in the improved voltage-model-based stator flux linkage observer, where $\boldsymbol{u}_s$ is the stator voltage vector, $\boldsymbol{e}_s$ is the stator electrical motive force vector and $\boldsymbol{i}_s$ is the stator current vector.

Figure 3.8 illustrates the comparison between the observed stator flux linkage, obtained by the improved voltage-model-based stator flux linkage observer and by the complete voltage-model-based stator flux linkage observer with a high-pass filter, and

Fig. 3.7 Vector relation in the improved stator flux linkage observer

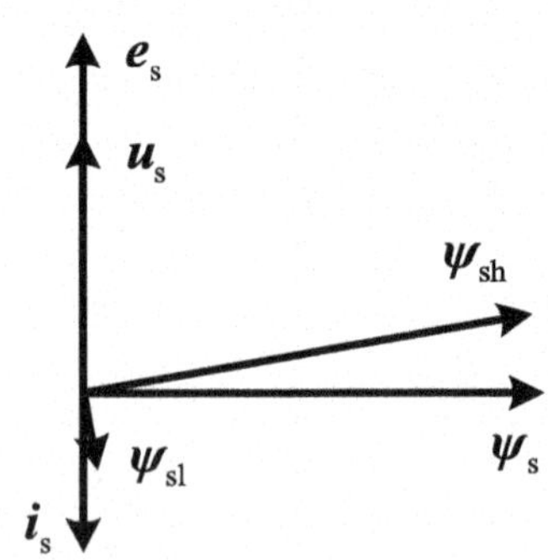

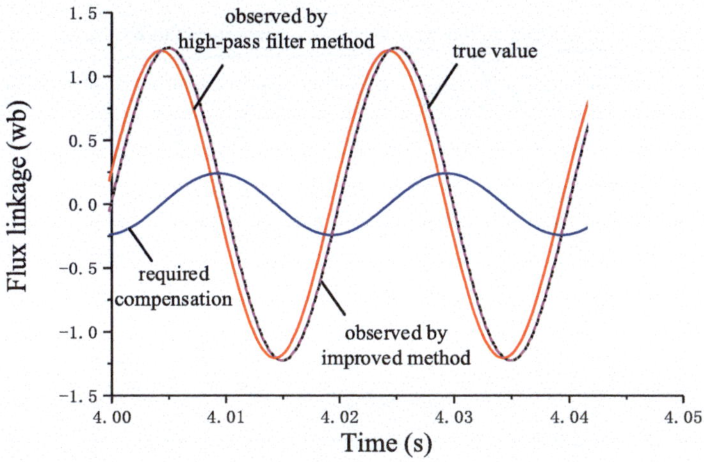

Fig. 3.8 Observed, true and compensating values of the stator flux linkage

the true stator flux linkage. It can be seen from the figure that the stator flux linkage, observed by the complete voltage-model-based stator flux linkage observer with a high-pass filter (marked as high-pass filter method), is closer to the true value, but its phase is somewhat ahead and its amplitude is smaller, while after the closed-loop feedback compensation, the stator flux linkage, observed by the improved voltage-model-based stator flux linkage observer (marked as improved method), and true values of the stator flux linkage agree in terms of phase and amplitude.

Figure 3.9 illustrates the comparison of the observed and true values of the two-phase stator flux linkage obtained by the improved voltage-model-based stator flux linkage observer. In the figure, there is a very good agreement between the true and observed values of the stator flux linkage.

The above simulation results show that the improved voltage-model-based stator flux linkage observer used in this section is feasible and can accurately observe the DFIG stator flux linkage.

3.1.5 Voltage Phase Detection

For the voltage-oriented vector control system, accurate observation of the phase of the oriented voltage vector is required to achieve good control performance. The hardware phase locked loop is highly accurate for the voltage phase detection, but is less reliable, more costly and more susceptible to external interference, so software algorithms are commonly used to detect the voltage phase. The two commonly used software algorithms are the direct calculation method and the software phase locked loop.

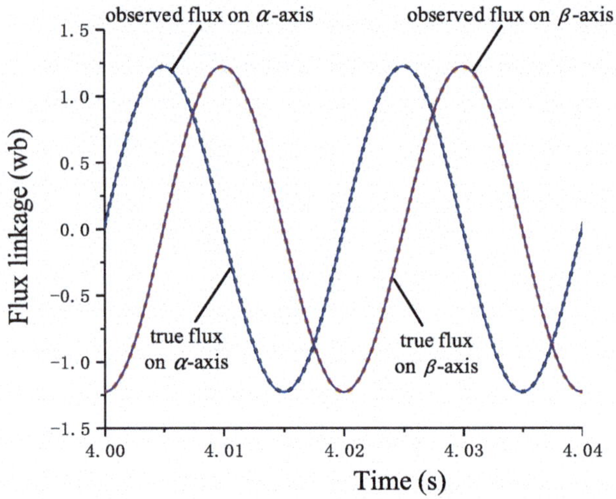

Fig. 3.9 Observed and true values of the stator flux linkage in the αβ-frame

3.1.5.1 Direct Calculation Method

The voltage vector in the αβ-frame is obtained by applying the $C_{3s/2s}$ transformation to the voltage vector in the ABC-frame, and is described in the polar coordinate system as

$$\begin{aligned} U &= \sqrt{(u^{\alpha})^2 + (u^{\beta})^2} \\ \angle \boldsymbol{u} &= \theta \end{aligned} \tag{3.40}$$

where the voltage phase can be calculated by

$$\begin{aligned} \sin\theta &= u^{\beta}/U \\ \cos\theta &= u^{\alpha}/U \end{aligned}$$

The direct calculation method is an open-loop observation method that can accurately obtain the phase of the instantaneous voltage vector with a small calculation and without the need to design a regulator, but the observation results are susceptible to the effects of voltage detection accuracy and harmonic interference.

3.1.5.2 Software Phase Lock Loop

In order to obtain accurate phase of the fundamental wave when the grid voltage is disturbed, a three-phase software Phase Lock Loop (PLL) can be used to detect the voltage phase, as shown in Fig. 3.10.

The relation between the dq-frame, with the PLL output phase angle as its orientation angle, and the voltage vector is shown in Fig. 3.11, where the voltage vector rotates at a fixed angular frequency and the dq-frame rotates at the output angular frequency of the phase lock loop. When $\theta^{\mathrm{PLL}} < 90°$, the input to the PI controller of PLL $\left(-u^{\mathrm{d}}\right)$ is negative, which will decrease ω^{PLL} and then result in a slower rotation of the dq-frame and an increasing θ^{PLL}; when $\theta^{\mathrm{PLL}} > 90°$, the input to the PI controller of PLL $\left(-u^{\mathrm{d}}\right)$ is positive, which will increase ω^{PLL} and then result in a faster rotation of the dq-frame and an decreasing θ^{PLL}. Eventually, the system will stabilise at $\theta^{\mathrm{PLL}} = 90°$, where the voltage vector coincides exactly with the q-axis and the output phase and frequency of the PLL coincides with the phase and frequency of the voltage vector.

Defining the intermediate variable x_{PLL}, the second-order dynamic model of the three-phase software PLL is

$$\begin{aligned} px^{\mathrm{PLL}} &= -u^{\mathrm{d}} \\ p\theta^{\mathrm{PLL}} &= \omega - \omega^{\mathrm{PLL}} \end{aligned} \tag{3.41}$$

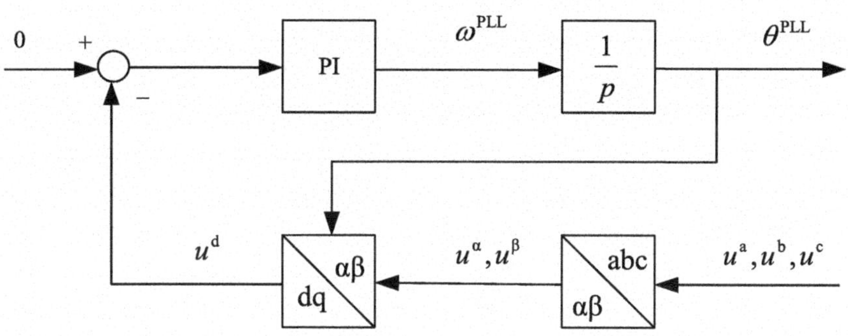

Fig. 3.10 Structure diagram of the 3-phase phase lock loop

Fig. 3.11 Voltage orientation based on PLL

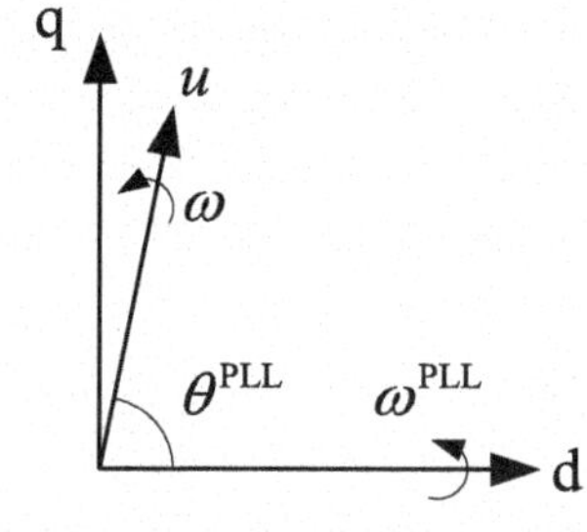

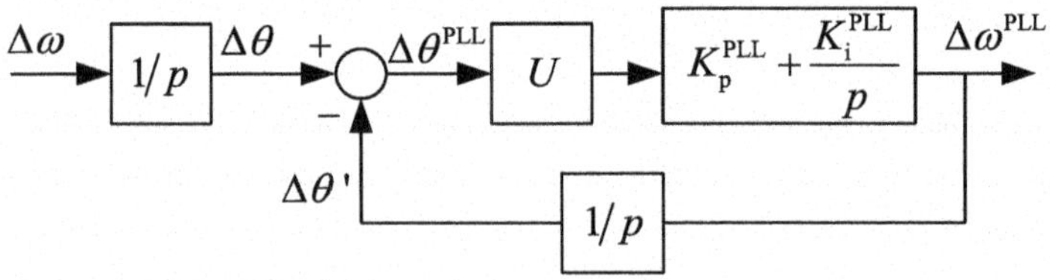

Fig. 3.12 Block diagram of the close-loop transfer function of the 3-phase software PLL

Where the angular frequency observed by the PLL is

$$\omega^{\mathrm{PLL}} = -K_{\mathrm{p}}^{\mathrm{PLL}} u^{\mathrm{d}} + K_{\mathrm{i}}^{\mathrm{PLL}} x^{\mathrm{PLL}}$$

When the PLL is operating normally, $\theta^{\mathrm{PLL}} = 90°$, and a local linearisation of the PLL dynamic model at this operating point can be done.

$$\begin{aligned} p\Delta x^{\mathrm{PLL}} &= U\Delta\theta^{\mathrm{PLL}} \\ p\Delta\theta^{\mathrm{PLL}} &= \Delta\omega - \Delta\omega^{\mathrm{PLL}} \end{aligned} \tag{3.42}$$

From Eq. (3.42) the block diagram of the transfer function of the three-phase PLL small signal model can be obtained as shown in Fig. 3.12, where θ' is the voltage phase observed by the PLL and θ is the actual voltage phase.

The close-loop transfer function of the PLL is

$$G^{\mathrm{PLL}} = \frac{\Delta\theta'}{\Delta\theta} = \frac{U\left(K_{\mathrm{p}}^{\mathrm{PLL}} p + K_{\mathrm{i}}^{\mathrm{PLL}}\right)}{p^2 + U\left(K_{\mathrm{p}}^{\mathrm{PLL}} p + K_{\mathrm{i}}^{\mathrm{PLL}}\right)} \tag{3.43}$$

The relation between the parameters of the PLL PI controller and the dynamic characteristics of the phase observation can be written as

$$K_{\mathrm{p}}^{\mathrm{PLL}} = 2\delta/U \tag{3.44}$$

$$K_{\mathrm{I}}^{\mathrm{PLL}} = U_{\mathrm{I}}\left(K_{\mathrm{p}}^{\mathrm{PLL}}\right)^2/(4\varsigma^2) \tag{3.45}$$

The direct calculation method and the software PLL were used to observe the phase of the same disturbed grid voltage, respectively, and the experimental results are shown in Fig. 3.13. Compared with the direct calculation method, the software PLL is almost unaffected by the disturbance and can obtain the accurate voltage phase.

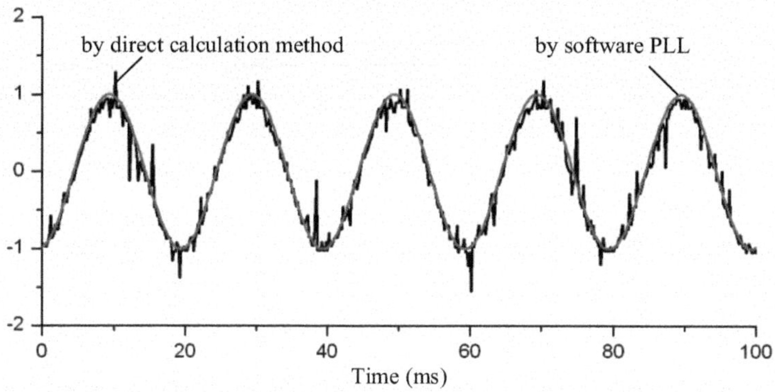

Fig. 3.13 Experimental results of the two voltage phase detection methods under the identity disturbed grid voltage

3.1.6 Smooth Grid Connection

The DFIG needs to control the stator voltage to be identical to the grid voltage in terms of amplitude, frequency, phase and phase sequence before it can be connected to the grid. The voltage-oriented vector control strategy shown in Fig. 3.2 makes it easy to connect the DFIG to the grid, but this method is an open-loop algorithm that can only achieve smooth grid connection when the generator parameters and rotor position are detected accurately and has the potential to fail when the generator mutual inductance and rotor position are detected with errors. To automatically correct for the errors that may exist, the closed-loop control of the amplitude and phase of the stator voltage is required.

Before grid connection, the stator current is zero, so the dynamic voltage equation of the stator voltage in the grid-voltage-oriented DFIG wind power system can be written as

$$u_{\mathrm{s}}^{\mathrm{d}} = L_{\mathrm{M}} p i_{\mathrm{r}}^{\mathrm{d}} - \omega_{\mathrm{n}} L_{\mathrm{M}} i_{\mathrm{r}}^{\mathrm{q}} \tag{3.46}$$

$$u_{\mathrm{s}}^{\mathrm{q}} = L_{\mathrm{M}} p i_{\mathrm{r}}^{\mathrm{q}} + \omega_{\mathrm{n}} L_{\mathrm{M}} i_{\mathrm{r}}^{\mathrm{d}} \tag{3.47}$$

In the steady state, if $u_{\mathrm{s}}^{\mathrm{q}} = E_{\mathrm{g}}$ and $u_{\mathrm{s}}^{\mathrm{d}} = 0$, the stator voltage is synchronised with the grid voltage and the DFIG meets the grid connection conditions, when the steady state values of the rotor currents satisfy $i_{\mathrm{r}}^{\mathrm{q}} = 0$ and $i_{\mathrm{r}}^{\mathrm{d}} = E_{\mathrm{g}}/(\omega_1 L_{\mathrm{M}})$. It can be seen that the DFIG can be controlled to meet the grid connection conditions by simply making $i_{\mathrm{r}}^{\mathrm{q}*} = 0$ and $i_{\mathrm{r}}^{\mathrm{d}*} = i_{\mathrm{ms}}$ in its grid-voltage-oriented vector control algorithm before grid connection. However, as the motor mutual inductance is used in the calculation of i_{ms}, if this parameter is inaccurate, the stator voltage amplitude before grid connection will not be consistent with the grid voltage, which will cause a large current shock if it is connected to the grid forcibly at this time.

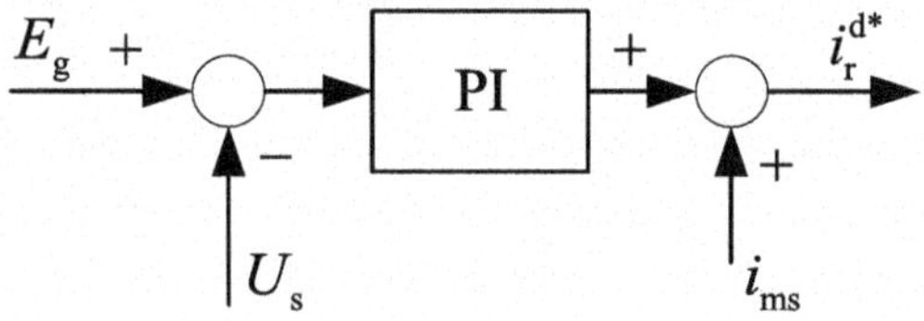

Fig. 3.14 Control diagram of the close-loop control of the stator voltage amplitude

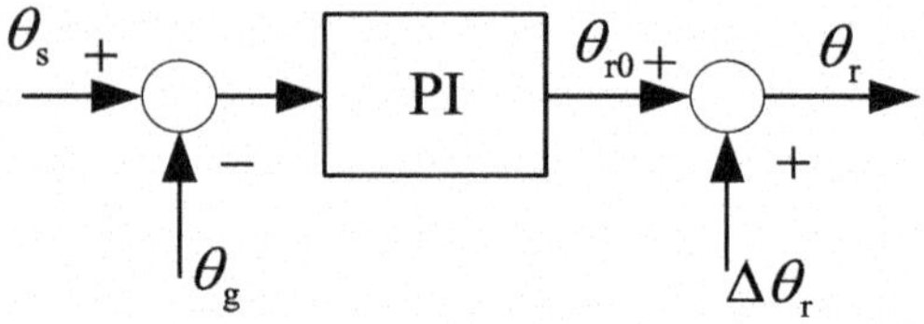

Fig. 3.15 Control diagram of the close-loop control of the stator voltage phase

From Eq. (3.47), the amplitude of the stator voltage in steady state is linearly related to i_r^{d*}. Therefore, the compensation of i_r^{d*} can eliminate the stator voltage amplitude control error caused by the inaccuracy of parameter L_M. Accordingly, the block diagram of the closed-loop control of the stator voltage can be designed as shown in Fig. 3.14.

The phase control of the stator voltage depends on the accuracy of the rotor position detection. The absolute rotor position θ_r is equal to the superposition of the initial encoder position θ_{r0} and the position increment $\Delta\theta_r$ obtained from the incremental encoder, but as the initial encoder position is not easily available, resulting in the possibility that the grid connection conditions for the stator voltage phase will not be met. For this reason, the rotor position compensator shown in Fig. 3.15 can be designed to use the phase difference between the grid voltage and the stator voltage as the feed-back signal for compensating the error in the initial rotor position. The compensated result is summed with the incremental rotor position $\Delta\theta_r$ obtained from the incremental encoder to provide the true absolute rotor position.

The closed-loop control algorithm of the stator voltage phase requires simultaneous detection of the stator voltage phase and the grid voltage phase. However, before grid connection, the harmonics of the generator-side converter output PWM voltage is divided between the motor mutual inductance and the rotor leakage inductance, resulting in large harmonic components in the stator voltage shown as Fig. 3.16. It can be seen that the stator voltage here contains harmonic components at the switching frequency, and the only way to obtain its accurate phase information is to use the software PLL algorithm.

After the closed-loop control of the stator voltage amplitude and phase has reached steady state, the DFIG can be connected to the grid. After grid connection: the closed-loop controller of the stator phase will no longer work, and its output will be used as a constant to compensate for rotor position detection; the closed-loop control of the stator voltage amplitude can continue to work to regulate the voltage amplitude at

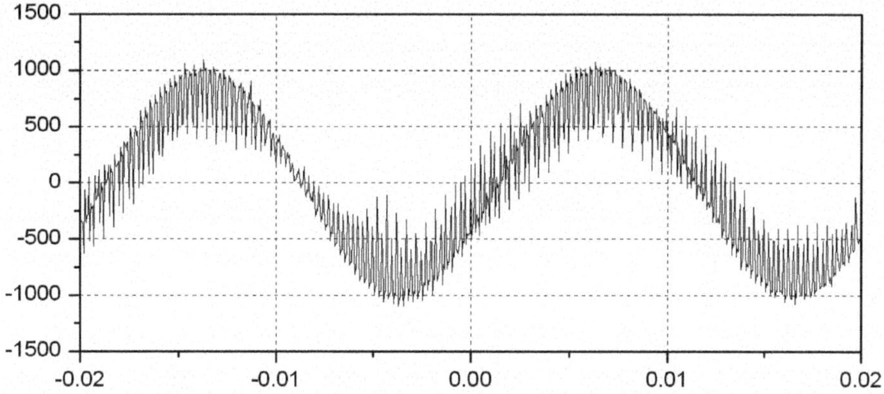

Fig. 3.16 Actual waveform of the stator voltage before grid connection

the stator terminal by controlling the reactive power output from the DFIG to achieve reactive support for the grid voltage.

3.2 Direct Power Control of the DFIG

In this section, the Direct Power Control (DPC) method of the DFIG will be discussed. This method eliminates the need for an inner current loop in the traditional vector control method and directly controls the active and reactive power on the stator side of the DFIG through a single close-loop structure, which is characterised by a small controller number, a simple control structure and a fast dynamic response [7, 8].

3.2.1 Direct Stator Power Control Based on PI-SVM

In the stator-voltage-oriented synchronous rotating coordinate system, and neglecting the stator resistance, there is

$$\begin{cases} \psi_{\mathrm{s}}^{\mathrm{d}} = 0 \\ \psi_{\mathrm{s}}^{\mathrm{q}} = -u_{\mathrm{s}}^{\mathrm{d}}/\omega_1 = -e_{\mathrm{g}}^{\mathrm{d}}/\omega_1 \end{cases} \tag{3.48}$$

Substituting Eqs. (3.8) and (3.9) into Eqs. (2.9), (2.20) and (2.21), it gives

$$\begin{cases} u_{\mathrm{r}}^{\mathrm{d}} = -\dfrac{L_{\mathrm{s}}(R_{\mathrm{r}} + \sigma L_{\mathrm{r}} p)}{L_{\mathrm{M}}} i_{\mathrm{s}}^{\mathrm{d}} + \dfrac{\omega_{\mathrm{sl}} \sigma L_{\mathrm{s}} L_{\mathrm{r}}}{L_{\mathrm{M}}} i_{\mathrm{s}}^{\mathrm{q}} - \dfrac{\omega_{\mathrm{sl}} L_{\mathrm{r}}}{\omega_1 L_{\mathrm{M}}} e_{\mathrm{g}}^{\mathrm{d}} \\ u_{\mathrm{r}}^{\mathrm{q}} = -\dfrac{\omega_{\mathrm{sl}} \sigma L_{\mathrm{s}} L_{\mathrm{r}}}{L_{\mathrm{M}}} i_{\mathrm{s}}^{\mathrm{d}} - \dfrac{L_{\mathrm{s}}(R_{\mathrm{r}} + \sigma L_{\mathrm{r}} p)}{L_{\mathrm{M}}} i_{\mathrm{s}}^{\mathrm{q}} + \dfrac{R_{\mathrm{r}}}{\omega_1 L_{\mathrm{M}}} e_{\mathrm{g}}^{\mathrm{d}} \end{cases} \tag{3.49}$$

Then stator power of the DFIG can be written as

$$\begin{cases} P_s = u_s^d i_s^d + u_s^q i_s^q = e_g^d i_s^d \\ Q_s = u_s^q i_s^d - u_s^d i_s^q = -e_g^d i_s^q \end{cases} \tag{3.50}$$

Substituting Eq. (3.50) into Eq. (3.49), it gives

$$\begin{cases} u_r^d = m(R_r + \sigma L_r p)P_s + m\omega_{sl}\sigma L_r Q_s - \dfrac{\omega_{sl} L_r}{\omega_1 L_M} e_g^d \\ u_r^q = m\omega_{sl}\sigma L_r P_s - m(R_r + \sigma L_r p)Q_s + \dfrac{R_r}{\omega_1 L_M} e_g^d \end{cases} \tag{3.51}$$

where $m = -L_s u_g / L_M$.

According to Eq. (3.51), the output voltage reference value of the generator-side converter can be obtained as

$$\begin{cases} u_r^{d*} = \left(K_{p\_sp}^d + \dfrac{K_{i\_sp}^d}{p}\right)(P_s^* - P_s) + m\omega_{sl}\sigma L_r Q_s + u_{rc}^d \\ u_r^{q*} = \left(K_{p\_sp}^q + \dfrac{K_{i\_sp}^q}{p}\right)(Q_s^* - Q_s) + m\omega_{sl}\sigma L_r P_s + u_{rc}^q \end{cases} \tag{3.52}$$

where $K_{p\_sp}^d$ and $K_{i\_sp}^d$ are the proportional and integral parameters of the PI controller for the active power loop respectively, $K_{p\_sp}^q$ and $K_{i\_sp}^q$ are the proportional and integral parameters of the PI controller for the reactive power loop respectively, $u_{rc}^d = -\omega_{sl} L_r e_g^d / \omega_1 L_M$ and $u_{rc}^q = R_r e_g^d / \omega_1 L_M$ are the feedforward compensation of the active and reactive power controllers respectively.

From Eq. (3.52), the control diagram of the direct stator power control of the DFIG is shown in Fig. 3.17. Thanks to the use of PI controllers and PWM (by space vector modulation—SVM) techniques, this direct power control method has the advantage of a fixed switching frequency.

3.2.2 Direct Stator Power Control Based on HC-SST

The equivalent circuit of the DFIG is the rotor 2-phase stationary coordinate system (marked as $\alpha_r - \beta_r$) is as Fig. 3.18.

From Fig. 3.18, the electromagnetic relation of the DFIG in the $\alpha_r - \beta_r$-frame can be expressed as

$$\boldsymbol{u}_s^{\alpha\beta r} = R_s \boldsymbol{i}_s^{\alpha\beta r} + p\boldsymbol{\psi}_s^{\alpha\beta r} + j\omega_r \boldsymbol{\psi}_s^{\alpha\beta r} \tag{3.53}$$

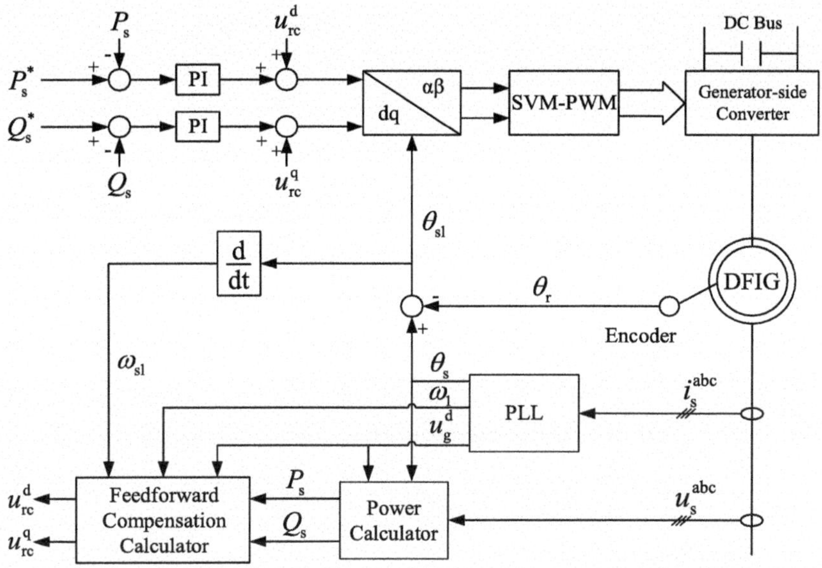

Fig. 3.17 Control diagram of the direct stator power control based on PI-SVM

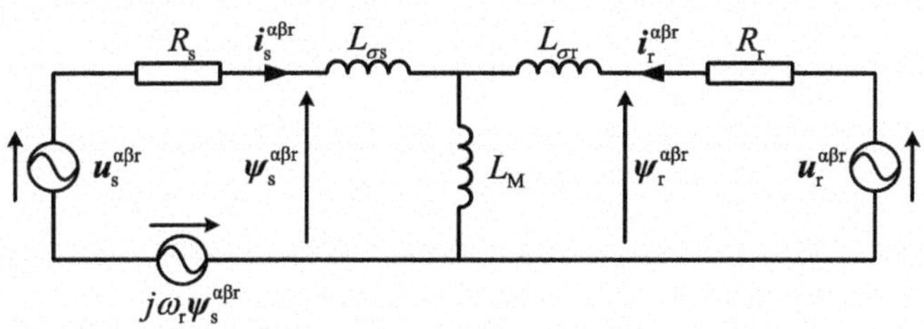

Fig. 3.18 Equivalent circuit of the DFIG in the rotor 2-phase stationary coordinate system

$$\boldsymbol{\psi}_{\mathrm{s}}^{\alpha\beta\mathrm{r}} = L_{\mathrm{s}}\boldsymbol{i}_{\mathrm{s}}^{\alpha\beta\mathrm{r}} + L_{\mathrm{M}}\boldsymbol{i}_{\mathrm{r}}^{\alpha\beta\mathrm{r}} \tag{3.54}$$

$$\boldsymbol{\psi}_{\mathrm{r}}^{\alpha\beta\mathrm{r}} = L_{\mathrm{r}}\boldsymbol{i}_{\mathrm{r}}^{\alpha\beta\mathrm{r}} + L_{\mathrm{M}}\boldsymbol{i}_{\mathrm{s}}^{\alpha\beta\mathrm{r}} \tag{3.55}$$

$$S_{\mathrm{s}} = P_{\mathrm{s}} + jQ_{\mathrm{s}} = \boldsymbol{u}_{\mathrm{s}}^{\alpha\beta\mathrm{r}}\left(\boldsymbol{i}_{\mathrm{s}}^{\alpha\beta\mathrm{r}}\right)^{*} \tag{3.56}$$

For the grid connected DFIG, the stator voltage is the grid voltage. Providing the grid voltage is constant and neglecting the stator resistance, the Eqs. (3.53)–(3.56) gives

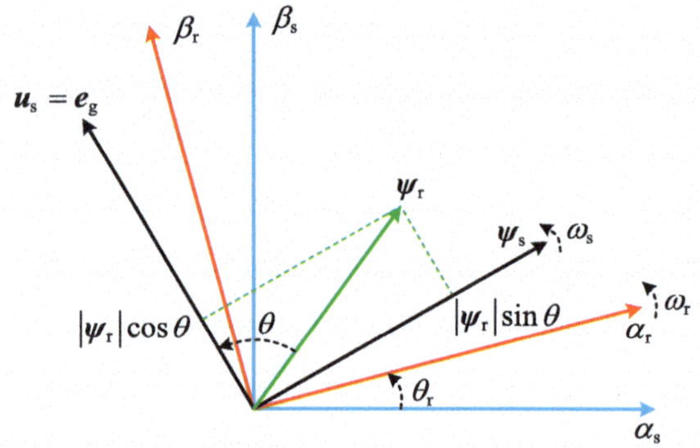

Fig. 3.19 Stator voltage vector, stator flux linkage and rotor flux linkage

$$\begin{cases} P_s = -\frac{L_M}{\sigma L_s L_r}\omega_s\left|\boldsymbol{\psi}_s^{\alpha\beta r}\right|\left|\boldsymbol{\psi}_r^{\alpha\beta r}\right|\cos\theta \\ Q_s = \frac{L_M}{\sigma L_s L_r}\omega_s\left|\boldsymbol{\psi}_s^{\alpha\beta r}\right|\left(\frac{L_r}{L_M}\left|\boldsymbol{\psi}_s^{\alpha\beta r}\right| - \left|\boldsymbol{\psi}_r^{\alpha\beta r}\right|\sin\theta\right) \end{cases} \tag{3.57}$$

where θ is the angle between the rotor flux linkage vector and the grid voltage vector shown as Fig. 3.19.

Taking the derivative of both sides of Eq. (3.57) with respect to time, it gives

$$\begin{cases} \frac{dP_s}{dt} = -\frac{L_M}{\sigma L_s L_r}\omega_s\left|\boldsymbol{\psi}_s^{\alpha\beta r}\right|\frac{d}{dt}\left(\left|\boldsymbol{\psi}_r^{\alpha\beta r}\right|\cos\theta\right) \\ \frac{dQ_s}{dt} = -\frac{L_M}{\sigma L_s L_r}\omega_s\left|\boldsymbol{\psi}_s^{\alpha\beta r}\right|\frac{d}{dt}\left(\left|\boldsymbol{\psi}_r^{\alpha\beta r}\right|\sin\theta\right) \end{cases} \tag{3.58}$$

From Eq. (3.58), it can be seen that the amount of stator active and reactive power change during each control period can be controlled by the two quantities $\left|\boldsymbol{\psi}_r^{\alpha\beta r}\right|\cos\theta$ and $\left|\boldsymbol{\psi}_r^{\alpha\beta r}\right|\sin\theta$ respectively.

From Fig. 3.19, neglecting the rotor resistance, the change of the rotor flux linkage can be expressed as

$$\Delta\boldsymbol{\psi}_r^{\alpha\beta r} \approx \boldsymbol{u}_s^{\alpha\beta r}T_s \tag{3.59}$$

The above equation shows that in each control period, the DPC algorithm can control the amount of change in the rotor flux linkage vector and thus the stator power following its reference value by selecting the appropriate voltage vector in the switching state table (SST, as shown in Table 3.1).

The DTC algorithm usually uses a hysteresis controller (HC). As a result, the main problem with this DTC is that the switching frequency is not constant, which leads to problems such as difficult estimation of power losses and increased difficulty in filter design. For this reason, various improved DPC algorithms with a constant switching frequency have been proposed. Compared to vector control, DPC does not require current inner loops, rotating coordinate transformations and special modulation methods, which significantly reduces the complexity of the algorithm

Table 3.1 Optimised switching state table of DPC [9]

Output of HC		Sector where the stator flux locates					
ΔP_s	ΔQ_s	I	II	III	IV	V	VI
1	1	v5	v4	v6	v2	v3	v1
1	0	v4	v6	v2	v3	v1	v5
1	−1	v6	v2	v3	v1	v5	v4
0	1	v1	v5	v4	v6	v2	v3
0	0	v7/v0	v7/v0	v7/v0	v7/v0	v7/v0	v7/v0
0	−1	v2	v3	v1	v5	v4	v6
−1	1	v1	v5	v4	v6	v2	v3
−1	0	v3	v1	v5	v4	v6	v2
−1	−1	v2	v3	v1	v5	v4	v6

and its dependence on system parameters. It has been shown that DPC also has the advantages of a good steady-state characteristic and a fast dynamic response [9, 10].

3.2.3 *Direct Imaginary Power Control*

Traditional DPC algorithms cannot be used for grid connection control of the DFIG because there is always no power output on the stator side before grid connection, which in turn renders the DPC algorithm useless as it cannot form the closed loop of the stator power [11, 12]. A commonly adopted solution is to add another vector-control-based method especially for controlling the stator voltage synchronised with the grid voltage before the DFIG is connected to the grid, and to switch the algorithm back to the DPC method after the DFIG is connected to the grid. This scheme deviates from the original purpose of the DPC algorithm due to the re-introduction of vector control, increases the complexity of the algorithm, and requires switching between two completely different algorithms, and a smooth transition is also a problem that needs to be considered. At present, there is little literature dealing with the unified use of DPC for the control of the DFIG before and after grid connection, and the controller structure needs to produce large changes before and after grid connection.

In this section, the definition of power is extended by introducing the concept of Imaginary Power, which allows the traditional DPC to still work properly before grid connection, using a unified DPC approach to control the DFIG before and after grid connection. This method, known as Direct Imaginary Power Control (DIPC) for grid connection, has the following characteristics.

(1) Simple controller structure without the need for PI controllers and rotating coordinate transformations.
(2) Sharing the same switch table as the traditional DPC for the DFIG, the algorithm is minimally modified and only the definition of the power feedback needs to

be changed before grid connection, which can be achieved with the traditional DPC method.

(3) The main structure of the controller remains unchanged before and after grid connection.

3.2.3.1 Definition of Imaginary Power

In order for the traditional DPC algorithm to still establish an effective close loop before grid connection, a quantity needs to be constructed to replace the real power which is constant zero on the stator side before grid connection. This constructed quantity is the imaginary power defined in this section, which is defined and used only before grid connection and which satisfies the following four settings adequately but not necessarily.

(1) The imaginary power has the same dimension as the real power. Since the imaginary power is constructed to replace the real power feedback on the stator side in the control structure, it is then considered to have the same dimension as the real power, and so, similar to Eq. (3.56), the imaginary power can be defined as

$$S_{\text{sv}} = P_{\text{sv}} + jQ_{\text{sv}} = \boldsymbol{u}_{\text{sv}}^{\alpha\beta\text{r}}\left(\boldsymbol{i}_{\text{sv}}^{\alpha\beta\text{r}}\right)^{*} \tag{3.60}$$

(2) The imaginary power contains information about the grid voltage vector. As the imaginary power is constructed for the synchronous control of the stator voltage with the grid voltage, this quantity should then contain information on the amplitude and phase of the grid voltage vector and can be considered as

$$\boldsymbol{u}_{\text{sv}}^{\alpha\beta\text{r}} = M\boldsymbol{e}_{\text{g}}^{\alpha\beta\text{r}}(M \in F \wedge M \neq 0) \tag{3.61}$$

where mathematically F denotes the domain of complex numbers and M is any non-zero complex number in the domain of complex numbers.

(3) The imaginary power is a controllable quantity of the rotor voltage. In the DFIG, the quantity that can be directly controlled by the algorithm is the output voltage of the generator-side converter. Any quantity can be controlled by the algorithm of the generator-side converter only if it is a controllable quantity of the rotor voltage. Under the conditions of settings (1) and (2), the imaginary current is required to be a controllable quantity of the rotor voltage in order to satisfy setting (3). Before the DFIG is connected to the grid, the only controllable quantity of the rotor voltage with a current dimension is the rotor current, so it is considered that there is

$$\boldsymbol{I}_{\text{sv}}^{\alpha\beta\text{r}} = N\boldsymbol{I}_{\text{r}}^{\alpha\beta\text{r}}(N \in F \wedge M \neq 0) \tag{3.62}$$

where N is any non-zero complex number in the complex domain F.

(4) In one control period, the same amount of change in the rotor flux linkage causes the same amount of change in power for both imaginary power and real power. As can be seen from the previous analysis, the essence of the traditional DPC is that in each control period, a suitable switching state is selected in the switching state table to control the rotor output voltage and, based on the volt-second product principle, the amount of change in the rotor flux linkage, and hence the amount of change in power. Therefore, this setting ensures that the traditional DPC algorithm based on the same switching state table has the same control capability and control effect on the imaginary power and the real power.

Substituting Eqs. (3.61) and (3.62) into Eq. (3.60), and with the rotor flux linkage being only generated by the rotor excitation current before grid connection, it gives

$$\begin{cases} P_{\mathrm{sv}} = \frac{MN}{L_r}\left|\boldsymbol{e}_{\mathrm{g}}^{\alpha\beta\mathrm{r}}\right|\left|\boldsymbol{\psi}_{\mathrm{r}}^{\alpha\beta\mathrm{r}}\right|\cos\theta \\ Q_{\mathrm{sv}} = \frac{MN}{L_r}\left|\boldsymbol{e}_{\mathrm{g}}^{\alpha\beta\mathrm{r}}\right|\left|\boldsymbol{\psi}_{\mathrm{r}}^{\alpha\beta\mathrm{r}}\right|\sin\theta \end{cases} \tag{3.63}$$

Taking the derivative of both sides of Eq. (3.63), it gives

$$\begin{cases} \frac{dP_{\mathrm{sv}}}{dt} = \frac{MN}{L_r}\left|\boldsymbol{e}_{\mathrm{g}}^{\alpha\beta\mathrm{r}}\right|\frac{d}{dt}\left(\left|\boldsymbol{\psi}_{\mathrm{r}}^{\alpha\beta\mathrm{r}}\right|\cos\theta\right) \\ \frac{dQ_{\mathrm{sv}}}{dt} = \frac{MN}{L_r}\left|\boldsymbol{e}_{\mathrm{g}}^{\alpha\beta\mathrm{r}}\right|\frac{d}{dt}\left(\left|\boldsymbol{\psi}_{\mathrm{r}}^{\alpha\beta\mathrm{r}}\right|\sin\theta\right) \end{cases} \tag{3.64}$$

In order to satisfy setting (3.64), it requires

$$\frac{dP_{\mathrm{sv}}}{dt} = \frac{dP_{\mathrm{s}}}{dt}, \frac{dQ_{\mathrm{sv}}}{dt} = \frac{dQ_{\mathrm{s}}}{dt} \tag{3.65}$$

From Eqs. (3.58), (3.64) and (3.65), it is known as

$$MN = -\frac{L_{\mathrm{M}}}{\sigma L_{\mathrm{s}}} \tag{3.66}$$

In Eq. (3.66), only the product of M and N is defined, not the respective values of M and N. From Eq. (3.63), it can be seen that the respective values of M and N do not affect the expression of the definition of imaginary power,[3] and considering the convenience and simplicity, it may be taken as

$$M = 1, N = -\frac{L_{\mathrm{M}}}{\sigma L_{\mathrm{s}}} \tag{3.67}$$

As can be seen from the above definition and derivation of imaginary power, this section gives a general idea and process for constructing imaginary power, a method of construction that is sufficient but not necessary, and the formulation of imaginary

[3] This implies that M and N can be defined in a variety of ways, only one of which is discussed in this paper.

power as an artificially constructed mathematical quantity is not unique. This section gives a concise and effective formulation of it, defined by Eqs. (3.60), (3.61), (3.62) and (3.67).

3.2.3.2 Grid Connection Control Based on Direct Imaginary Power Control

Before the DFIG is connected to the grid, the stator voltage and the grid voltage need to be controlled to match in phase and amplitude. Whereas before the DFIG is connected to the grid, the vector relation in the DFIG is shown in Fig. 3.20 as the stator side is open circuit in Fig. 3.18 and therefore the rotor flux linkage is entirely generated by the rotor current excitation.

Before grid connection, the stator flux linkage is also generated entirely by rotor current excitation, so there is

$$\boldsymbol{u}_{\mathrm{s}}^{\alpha\beta\mathrm{r}} = j\omega_{\mathrm{s}}\boldsymbol{\psi}_{\mathrm{s}}^{\alpha\beta\mathrm{r}} = j\omega_{\mathrm{s}}L_{\mathrm{M}}\boldsymbol{I}_{\mathrm{r}}^{\alpha\beta\mathrm{r}} = -j\omega_{\mathrm{s}}\sigma L_{\mathrm{s}}\boldsymbol{I}_{\mathrm{sv}}^{\alpha\beta\mathrm{r}} \tag{3.68}$$

From the above equation, the stator voltage lags the stator imaginary current a vector angle of $\pi/2$. In order to make the stator voltage and the grid voltage the same phase, the stator imaginary current is required to be ahead of the grid voltage a vector angle of $\pi/2$ (i.e. in Fig. 3.20 $\theta = \pi/2$), then it is known from Eq. (3.63) the stator-side imaginary active power is zero at this time. In the direct imaginary power control (DIPC) algorithm, the imaginary active power can be controlled by setting its reference value, i.e. there is

$$P_{\mathrm{sv}}^{*} = 0 \tag{3.69}$$

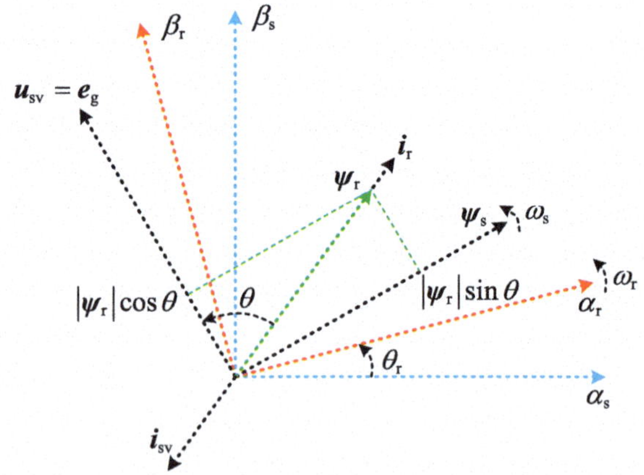

Fig. 3.20 Relation of electromagnetic vectors before grid connection in the DFIG

In order to make the stator voltage and the grid voltage have the same amplitude, it is necessary to control the imaginary reactive power to an appropriate value. In Eq. (3.68), let the stator voltage equal to the grid voltage, find the stator imaginary current and substitute it into the definition of imaginary reactive power, then the imaginary reactive power at this time is

$$Q_{\mathrm{sv}} = -\frac{1}{\sigma \omega_{\mathrm{s}} L_{\mathrm{s}}} \left| e_{\mathrm{g}}^{\alpha\beta\mathrm{r}} \right|^2 \tag{3.70}$$

The above equation is the stator-side imaginary reactive power required to synchronise the stator voltage with the grid voltage amplitude.

From the above analysis, it can be found that in the process of grid connection control, the imaginary active power mainly affects the phase synchronisation and the imaginary reactive power mainly affects the amplitude synchronisation. The interrelationship is: when the imaginary active power is greater than 0, the stator voltage phase lags the grid voltage and vice versa; when the imaginary reactive power is greater than the value given in Eq. (3.70), the stator voltage amplitude is smaller than the grid voltage amplitude and vice versa.

To avoid additional disturbances to the original DPC algorithm caused by different power feedback calculations before and after grid connection, here it uses the power observer shown in Fig. 3.21, where a compensation amount is added to the feedback path of the imaginary reactive power so that Eq. (3.70) can be satisfied by setting the reactive power reference a value of zero, i.e.

$$Q_{\mathrm{sv}}^{*} = 0 \tag{3.71}$$

In summary, the DIPC strategy is shown in Fig. 3.22. As can be seen, the DIPC for grid connection uses the same switching table as the traditional DPC, and the main

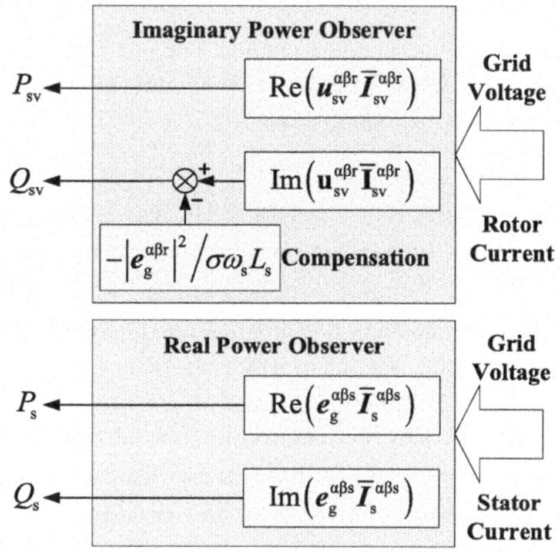

Fig. 3.21 Extended power observer

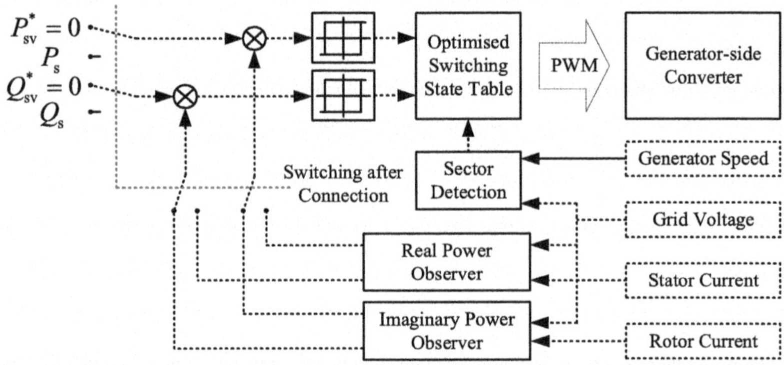

Fig. 3.22 Control diagram of the DIPC grid connection method

structure of the controller remains the same before and after grid connection, only the power feedback and power reference need to be switched. However, this switching does not affect the performance of the control system, as both the imaginary and real power feedbacks and their reference values are zero before and after the switching.

From the introduction of the concept of imaginary power and the derivation process of the DIPC, it is clear that the DIPC proposed in this section is widely applicable to existing DPC methods on the stator side of the DFIG (e.g. PI-SVM-based direct stator power control in Sect. 3.2.1).

In summary, this section presents the methods related to the DPC of the DFIG. It is worth pointing out that similar DPC methods in which the rotor current inner loop is eliminated have a more limited ability to control the rotor current, which to some extent limits their application in situations where high quality rotor current is required, and care should be taken to deal with possible rotor overcurrent problems when using them.

3.3 Model Predictive Control of the DFIG-Based Wind Power System

Model Predictive Control (MPC) is a control method that uses a system model to determine the current control action by predicting the future state of the system in a finite time domain, which is a non-linear, optimal control method and an important branch of predictive control with the advantages of good dynamic and static control performance, strong parameter robustness, easy tuning and the ability to achieve uniform optimisation of multiple control objectives.

A typical MPC system consists of the main parts of the predictive model, rolling optimisation and feedback correction, and its control diagram is shown in Fig. 3.23.

At each current instant, the prediction model predicts the future output of the controlled quantity in a certain time domain, which should be a function of the

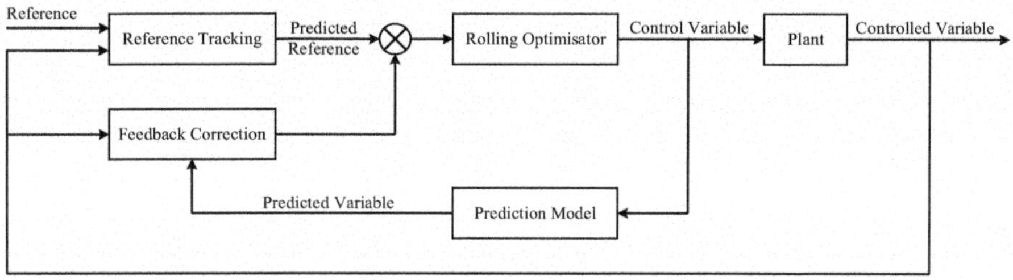

Fig. 3.23 Control diagram of the model predictive control

current sampled quantity, the current control quantity and the current controlled quantity, and the prediction sequence, with the controlled quantity and its predicted reference from reference trajectory, is sent to the feedback correction block; in the feedback correction block, the prediction sequence from the prediction model is corrected according to the latest measured data of the system, and the corrected result is sent to the rolling optimisation block; in the rolling optimisation block, the system will determine the values of the current and future control quantities according to a performance evaluation function, which will enable the prediction sequence to reach its reference in an optimal way according to the limits of the performance evaluation function, and among the control quantities obtained, only the current control quantity will be output and really act on the controlled object. The above is the basic idea of model predictive control.

This section introduces the MPC method to the DFIG-based wind power system and combines it with space vector pulse width modulation (SVPWM) to discuss the MPC control strategy for the DFIG-based wind power system under ideal grid conditions and to investigate its grid connection method.

3.3.1 *Model Predictive Control of the Grid-Side Converter*

The control objective of the grid-side converter is to maintain DC-bus voltage stability and grid-side unit power factor operation. Here the traditional control structure of DC-bus voltage outer loop and grid-side current inner loop is still used and MPC algorithms are derived for the grid-side current loop and the DC-bus voltage loop respectively [13].

3.3.1.1 Basic Principles of the Grid-Side Converter Control

The q-axis component of the grid voltage in the grid-voltage-oriented synchronous coordinate system can be written as

$$e_{\mathrm{g}}^{\mathrm{q}} = 0 \tag{3.72}$$

Substituting Eq. (3.72) into Eq. (2.25), it gives

$$\begin{cases} P_g = e_g^q i_g^q \\ Q_g = -e_g^d i_g^q \end{cases} \tag{3.73}$$

Equation (2.24) shows that the control of the d-axis and q-axis components of the grid-side current of the grid-side converter can be achieved by regulating the d-axis and q-axis components of the output voltage of the grid-side converter, while Eq. (3.73) shows that the decoupled control of the active and reactive power injected into the grid by the grid-side converter can be achieved by controlling the d-axis and q-axis components of the grid-side current separately, which is the basic principle of the grid-side current loop of the grid-side converter.

Equation (2.26) shows that by controlling the active power injected into the grid by the grid-side converter, the DC bus can be discharged or charged, thus enabling control of the DC-bus voltage, which is the basic principle of the DC-bus voltage loop of the grid-side converter.

3.3.1.2 Grid-Side Current Loop

Setting the switching period of the grid-side converter to Ts (the switching period, sampling period and current loop control period are considered to be the same here), the discrete expression of the grid-side current of the grid-side converter can be obtained by discretizing Eq. (2.24) using the forward Euler method[4] as

$$\begin{cases} i_g^d(k+1|k) = a_g i_g^d(k) + b_g i_g^q(k) - c_g e_g^d(k) + c_g u_g^d(k) \\ i_g^q(k+1|k) = a_g i_g^q(k) - b_g i_g^d(k) - c_g e_g^q(k) + c_g u_g^q(k) \end{cases} \tag{3.74}$$

where $k \in N$ indicates the instantaneous value of the corresponding physical quantity at the moment of kT_s, $a_g = 1 - R_g T_s / L_g$, $b_g = \omega_s T_s$, $c_g = T_s / L_g$, $i_g^d(k)$ and $i_g^q(k)$ indicate the d-axis and q-axis components of the instantaneous grid-side current at the moment of kT_s respectively, $e_g^d(k)$ and $e_g^q(k)$ indicate the d-axis and q-axis components of the instantaneous grid voltage at the moment of kT_s respectively, $u_g^d(k)$ and $u_g^q(k)$ indicate the d-axis and q-axis components of the instantaneous grid-side converter output voltage at the moment of kT_s respectively, $i_g^d(k+1|k)$ and $i_g^q(k+1|k)$ indicate the d-axis and q-axis components of the grid-side current predicted at the moment of kT_s for the moment of $(k+1)T_s$. The above equation shows that at any sampling moment, the grid-side current at the moment of $(k+1)T_s$ can be predicted as long as the grid-side current, the grid voltage and the grid-side converter output voltage are known at the current sampling moment.

---

[4] Other methods can also be used here to obtain the discretized expression, such as the backward Euler method and the trapezoidal method. Here, for simplicity of derivation, the forward Euler method is used to discretize the formula.

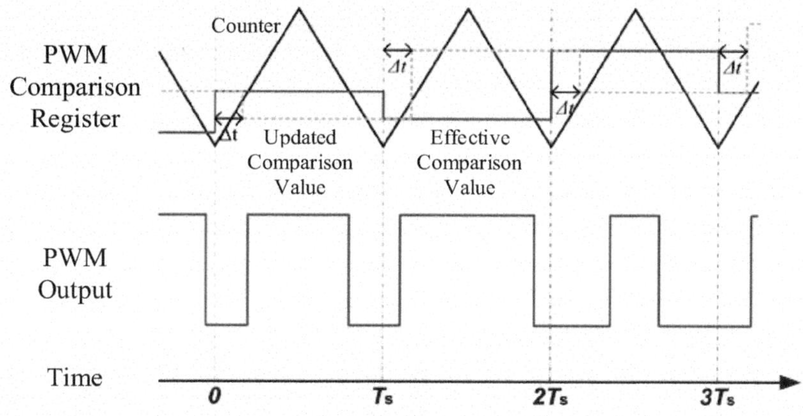

Fig. 3.24 Timing chart of the PWM module in DSP

At present, the DFIG-based wind power system commonly uses various digital processor-based fully digital control systems, which introduce the control delay at the sampling period level in the control process. Taking the digital signal processor (DSP) as an example, the timing relation between the value update of the comparison register in the PWM module and the duty cycle update of the actual PWM output is shown in Fig. 3.24.

In Fig. 3.24, at the moment of T_s, the digital chip obtains the sampling value required by the system, then gets the control voltage of the converter output according to the control algorithm, and then gets the value of the comparison register of the PWM module through the modulation algorithm, and then updates the value of the comparison register of the PWM module, and completes the above process at the moment of $T_s + \Delta t$; however, the duty cycle of the PWM output waveform does not change immediately until the next sampling moment $2T_s$, when the duty cycle of the PWM output waveform is updated according to the value of the comparison register updated at that moment. Obviously, there is a delay of about one sampling period between the moment when the control algorithm generates the required control voltage and the moment when this control voltage is actually applied to the controlled object, which is determined by the operating principle of the digital control chip. As this delay is closer to the step length of the current prediction in Eq. (3.74), it should not simply be ignored, but needs to be taken into account in the prediction process.

The impact of digital control system delay on control performance can be reduced by using methods such as oversampling or changing the computational timing, but this increases the system sampling frequency, reduces the available running time of the control algorithms in the digital chip, puts pressure on the implementation of large-scale systems, and this approach cannot theoretically completely eliminate the impact of digital chip delay on the system. In this section, a two-step prediction method is used that can theoretically completely eliminate the effect of digital control system delay on the prediction result without changing the system sampling frequency, for which a correction to Eq. (3.74) is required.

At any sampling moment kT_s, taking into account the above digital control system delay effect, the converter control voltage that really acts on the controlled object should be $u_g^{dq}(k-1)$, so that Eq. (3.74) can be modified as

$$\begin{cases} i_g^d(k+1|k) = a_g i_g^d(k) + b_g i_g^q(k) - c_g e_g^d(k) + c_g u_g^d(k-1) \\ i_g^q(k+1|k) = a_g i_g^q(k) - b_g i_g^d(k) - c_g e_g^q(k) + c_g u_g^q(k-1) \end{cases} \quad (3.75)$$

Using the current prediction result of Eq. (3.75) as a basis, further prediction of the current at the moment of $(k+2)T_s$ can be obtained as

$$\begin{cases} i_g^d(k+2|k) = a_g i_g^d(k+1|k) + b_g i_g^q(k+1|k) - c_g e_g^d(k+1) + c_g u_g^d(k) \\ i_g^q(k+2|k) = a_g i_g^q(k+1|k) - b_g i_g^d(k+1|k) - c_g e_g^q(k+1) + c_g u_g^q(k) \end{cases} \quad (3.76)$$

where $e_g^d(k+1)$ and $e_g^q(k+1)$ indicate the d-axis and q-axis components of the grid voltage at the moment of $(k+1)T_s$ respectively, and on ideal grid conditions it can be considered that $e_g^d(k+1) = e_g^d(k)$ and $e_g^q(k+1) = e_g^q(k)$. Equation (3.76) shows that the control voltage given by the control algorithm at the moment of kT_s will have an effect on the current prediction result at the moment of $(k+2)T_s$. With the two-step prediction of Eqs. (3.75) and (3.76), the prediction process of the grid-side current is consistent with the actual operating process of the control system, completely offsetting the effect caused by the digital control chip delay. For the convenience of the subsequent derivation, Eq. (3.76) is further written in the form of the converter output voltage increment, as shown in Eq. (3.77).

$$\begin{cases} i_g^d(k+2|k) = a_g i_g^d(k+1|k) + b_g i_g^q(k+1|k) \\ \qquad -c_g e_g^d(k+1) + c_g u_g^d(k-1) + c_g \Delta u_g^d(k) \\ i_g^q(k+2|k) = a_g i_g^q(k+1|k) - b_g i_g^d(k+1|k) \\ \qquad -c_g e_g^q(k+1) + c_g u_g^q(k-1) + c_g \Delta u_g^q(k) \end{cases} \quad (3.77)$$

At this point, the prediction model for the grid-side current can be obtained as shown in Eqs. (3.75) and (3.77). Obviously, this current prediction model is an open-loop model and its prediction accuracy will be affected by many factors such as the accuracy of system parameters, time variability, spurious parameters, etc. In order to reduce the influence of these non-ideal factors on the current prediction model, it is necessary to introduce a feedback correction term into this prediction model to obtain a closed-loop prediction model for the grid-side current, as shown in Eqs. (3.77) and (3.79).

$$\begin{cases} i_{gm}^d(k+1|k) = a_g i_g^d(k) + b_g i_g^q(k) - c_g e_g^d(k) + c_g u_g^d(k-1) + x_g^d(k) \\ i_{gm}^q(k+1|k) = a_g i_g^q(k) - b_g i_g^d(k) - c_g e_g^q(k) + c_g u_g^q(k-1) + x_g^q(k) \end{cases} \quad (3.78)$$

$$\begin{cases} i_{\mathrm{gm}}^{\mathrm{d}}(k+2|k) = a_{\mathrm{g}} i_{\mathrm{gm}}^{\mathrm{d}}(k+1|k) + b_{\mathrm{g}} i_{\mathrm{gm}}^{\mathrm{q}}(k+1|k) - c_{\mathrm{g}} e_{\mathrm{g}}^{\mathrm{d}}(k+1) \\ \qquad + c_{\mathrm{g}} u_{\mathrm{g}}^{\mathrm{d}}(k-1) + c_{\mathrm{g}} \Delta u_{\mathrm{g}}^{\mathrm{d}}(k) + x_{\mathrm{g}}^{\mathrm{d}}(k) \\ i_{\mathrm{gm}}^{\mathrm{q}}(k+2|k) = a_{\mathrm{g}} i_{\mathrm{gm}}^{\mathrm{q}}(k+1|k) - b_{\mathrm{g}} i_{\mathrm{gm}}^{\mathrm{d}}(k+1|k) - c_{\mathrm{g}} e_{\mathrm{g}}^{\mathrm{q}}(k+1) \\ \qquad + c_{\mathrm{g}} u_{\mathrm{g}}^{\mathrm{q}}(k-1) + c_{\mathrm{g}} \Delta u_{\mathrm{g}}^{\mathrm{q}}(k) + x_{\mathrm{g}}^{\mathrm{q}}(k) \end{cases} \tag{3.79}$$

where $x_{\mathrm{g}}^{\mathrm{d}}(k)$ and $x_{\mathrm{g}}^{\mathrm{q}}(k)$ are the d-axis and q-axis components of the feedback correction term of the grid-side current at the moment of kT_{s}, respectively, which include the error information between the current predicted at the moment of $(k-1)T_{\mathrm{s}}$ for the moment of kT_{s} and the current sampled at kT_{s} the moment of and can be written as

$$\begin{cases} x_{\mathrm{g}}^{\mathrm{d}}(k) = f_{\mathrm{g}}^{\mathrm{d}} \cdot \left[i_{\mathrm{g}}^{\mathrm{d}}(k) - i_{\mathrm{gm}}^{\mathrm{d}}(k|k-1) \right] \\ x_{\mathrm{g}}^{\mathrm{q}}(k) = f_{\mathrm{g}}^{\mathrm{q}} \cdot \left[i_{\mathrm{g}}^{\mathrm{q}}(k) - i_{\mathrm{gm}}^{\mathrm{q}}(k|k-1) \right] \end{cases} \tag{3.80}$$

where $f_{\mathrm{g}}^{\mathrm{d}}$ and $f_{\mathrm{g}}^{\mathrm{q}}$ are the feedback correction factors of the d-axis and q-axis components of the grid-side current, respectively.

To achieve optimal control of the grid-side current, an evaluation function characterising its performance index is required. Here, the control objective of the grid-side current loop is that the current value follows its reference value and that the variation of the converter output voltage is not too large. For this reason, the evaluation function of the grid-side current loop can be given as

$$\begin{aligned} J_{\mathrm{g}}(k) = {} & \lambda_{\mathrm{g}}^{\mathrm{d}} \left[i_{\mathrm{g}}^{\mathrm{d}*}(k) - i_{\mathrm{gm}}^{\mathrm{d}}(k+2|k) \right]^2 + \lambda_{\mathrm{g}}^{\mathrm{q}} \left[i_{\mathrm{g}}^{\mathrm{q}*}(k) - i_{\mathrm{gm}}^{\mathrm{q}}(k+2|k) \right]^2 \\ & + \varepsilon_{\mathrm{g}}^{\mathrm{d}} \left[\Delta u_{\mathrm{g}}^{\mathrm{d}}(k) \right]^2 + \varepsilon_{\mathrm{g}}^{\mathrm{q}} [\Delta u_{\mathrm{g}}^{\mathrm{q}}(k)]^2 \end{aligned} \tag{3.81}$$

where $i_{\mathrm{g}}^{\mathrm{d}*}(k)$ and $i_{\mathrm{g}}^{\mathrm{q}*}(k)$ indicate the reference values for the d-axis and q-axis components of the grid-side current, respectively, $\lambda_{\mathrm{g}}^{\mathrm{d}}(k)$ and $\lambda_{\mathrm{g}}^{\mathrm{q}}(k)$ indicate the weight factors for the errors in the d-axis and q-axis components of the grid-side current, respectively, and $\varepsilon_{\mathrm{g}}^{\mathrm{d}}(k)$ and $\varepsilon_{\mathrm{g}}^{\mathrm{q}}(k)$ indicate the weight factors for the d-axis and q-axis components of the control voltage increment, respectively.

From the evaluation function as Eq. (3.81), it can be seen that the grid-side current follows its reference value as fast as possible without considering the amount of the converter output voltage increment and its output voltage capability, etc., but there is still a delay of 2 sampling periods between the current value and the reference value, which, as can be seen from the previous analysis, is introduced by the digital control chip. At any sampling moment of kT_{s}, in order to obtain the optimal converter control voltage increment, it is only necessary to obtain the control voltage increment that minimises the evaluation function $J_{\mathrm{g}}(k)$. By substituting Eq. (3.79) into Eq. (3.81), the optimal grid-side converter output voltage increment at the moment of kT_{s} can be obtained as shown in Eq. (3.82) through optimal solution, where $\Delta u_{\mathrm{g}}^{\mathrm{d}*}(k)$ and

$\Delta u_{\mathrm{g}}^{\mathrm{q}^*}(k)$ represent the optimal increment of the control voltage given by the control algorithm at the sampling moment of kT_{s}. Then at any sampling moment of kT_{s}, the control algorithm can obtain the optimal increment of the control voltage at the current moment according to Eqs. (3.78), (3.80) and (3.82), then the d-axis and q-axis components of the optimal control voltage, $u_{\mathrm{g}}^{\mathrm{d}^*}(k)$ and $u_{\mathrm{g}}^{\mathrm{q}^*}(k)$, are obtained as $u_{\mathrm{g}}^{\mathrm{d}^*}(k) = u_{\mathrm{g}}^{\mathrm{d}}(k-1) + \Delta u_{\mathrm{g}}^{\mathrm{d}^*}(k)$ and $u_{\mathrm{g}}^{\mathrm{q}^*}(k) = u_{\mathrm{g}}^{\mathrm{q}}(k-1) + \Delta u_{\mathrm{g}}^{\mathrm{q}^*}(k)$, respectively.

$$\begin{cases} \Delta u_{\mathrm{g}}^{\mathrm{d}^*}(k) = \frac{c_{\mathrm{g}}\lambda_{\mathrm{g}}^{\mathrm{d}}}{c_{\mathrm{g}}^2\lambda_{\mathrm{g}}^{\mathrm{d}}+\varepsilon_{\mathrm{g}}^{\mathrm{d}}}\left[i_{\mathrm{g}}^{\mathrm{d}^*}(k) - a_{\mathrm{g}} i_{\mathrm{gm}}^{\mathrm{d}}(k+1|k) - b_{\mathrm{g}} i_{\mathrm{gm}}^{\mathrm{q}}(k+1|k)\right. \\ \quad \left. + c_{\mathrm{g}} e_{\mathrm{g}}^{\mathrm{d}}(k+1) - c_{\mathrm{g}} u_{\mathrm{g}}^{\mathrm{d}}(k-1) - x_{\mathrm{g}}^{\mathrm{d}}(k)\right] \\ \Delta u_{\mathrm{g}}^{\mathrm{q}^*}(k) = \frac{c_{\mathrm{g}}\lambda_{\mathrm{g}}^{\mathrm{q}}}{c_{\mathrm{g}}^2\lambda_{\mathrm{g}}^{\mathrm{q}}+\varepsilon_{\mathrm{g}}^{\mathrm{q}}}\left[i_{\mathrm{g}}^{\mathrm{q}^*}(k) - a_{\mathrm{g}} i_{\mathrm{gm}}^{\mathrm{q}}(k+1|k) + b_{\mathrm{g}} i_{\mathrm{gm}}^{\mathrm{q}}(k+1|k)\right. \\ \quad \left. + c_{\mathrm{g}} e_{\mathrm{g}}^{\mathrm{q}}(k+1) - c_{\mathrm{g}} u_{\mathrm{g}}^{\mathrm{q}}(k-1) - x_{\mathrm{g}}^{\mathrm{q}}(k)\right] \end{cases} \tag{3.82}$$

It should be noted here that, in order to simplify the optimisation process, there is no restriction on the control voltage increment in the above optimisation process for the Evaluation Function (3.81). However, in a practical system the control voltage outputted by the grid-side converter should be in a certain range which is governed by the factors such as the DC-bus voltage and the modulation method. When the control voltage given by the algorithm exceeds the range, the system will overshoot, so a limiter needs to be added to limit the output voltage of the converter. In summary, the control diagram of the grid-side current loop of the grid-side converter with a limiter can be obtained as shown in Fig. 3.25.

3.3.1.3 DC-Bus Voltage Loop

In a double close-loop control structure, the outer-loop bandwidth is usually required to be lower than the inner-loop bandwidth, so that the bandwidth of the DC-bus voltage loop cannot be higher than that of the grid-side current loop. As can be seen from Sect. 3.3.1.2, the control period of the grid-side current loop is T_{s} and the control period of the DC-bus voltage loop is T_{dc} ($T_{\mathrm{dc}} = rT_{\mathrm{s}}$, r is usually taken as a natural number between 5 and 15), at which point the effect of the digital control chip delay effect on the DC-bus voltage loop can be neglected.

In Eq. (2.26), neglecting the effect of external disturbances ΔP, and again discretizing it using the forward Euler method, the open-loop prediction model for the DC-bus voltage can be obtained as

$$u_{\mathrm{dc}}^2(m+1|m) = u_{\mathrm{dc}}^2(m) - h\overline{P_{\mathrm{g}}}(m-1) - h\Delta\overline{P_{\mathrm{g}}}(m) \tag{3.83}$$

where $m = k/r$, $m \in N$ and $h = 2T_{\mathrm{dc}}/C$, $u_{\mathrm{dc}}(m+1|m)$ indicates the DC-bus voltage predicted at the moment of mT_{dc} for the moment of $(m+1)T_{\mathrm{dc}}$, $u_{\mathrm{dc}}(m)$ indicates the DC-bus voltage sampled at the moment of mT_{dc}, $\overline{P_{\mathrm{g}}}(m-1)$ indicates the average active power injected into the grid by the grid-side converter from the moment of $(m-1)T_{\mathrm{dc}}$ to the moment of mT_{dc}, and $\Delta\overline{P_{\mathrm{g}}}(m)$ indicates the increment

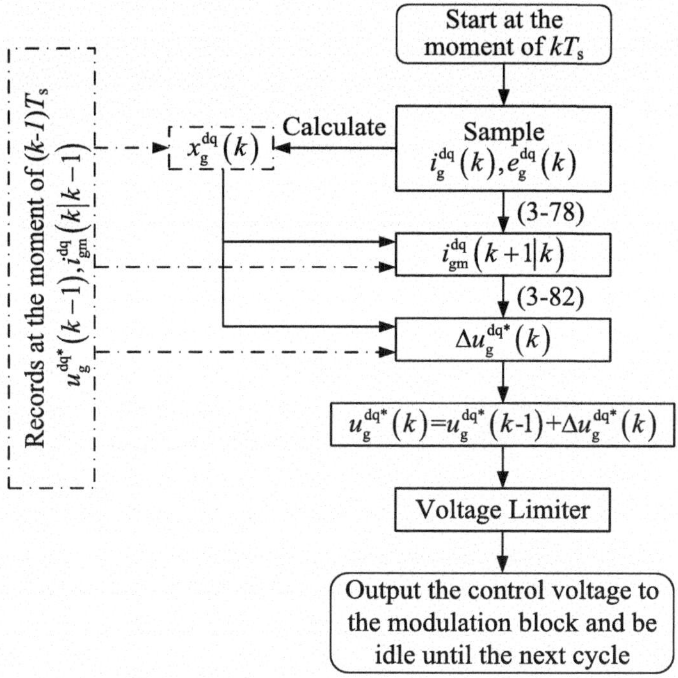

Fig. 3.25 Control diagram of the current loop of the grid-side converter based on MPC

of the average active power injected into the grid by the grid-side converter from the moment of mT_{dc} to the moment of $(m+1)T_{dc}$.

Replacing the grid-side current in Eq. (3.73) by the average current during a current-loop control period, the active power injected into the grid by the converter during current-loop control period T_s can be obtained as

$$P_g(k) = e_g^d(k)\overline{i_g^d}(k) = e_g^d(k) \cdot \pm\frac{i_g^d(k) + i_g^d(k+1)}{2} \tag{3.84}$$

According to the definition of m in Eq. (3.83), the timing relation of the control period between the DC-bus voltage loop and the grid-side current loop can be known as shown in Fig. 3.26.

From the above figure, it can be seen that between moments of $(m-1)T_{dc}$ and mT_{dc}, the DC-bus voltage loop undergoes one control period, while the grid-side current loop undergoes $r = T_{dc}/T_s$ control periods. Therefore, in the calculation of the voltage loop, the average active power $\overline{P_g}(m-1)$ injected into the grid by the grid-side converter is taken as the average of the active power injected into the grid between moments of $(k-r)T_s$, $(k-r+1)T_s$ … and $(k-1)T_s$, which is obtained from Eq. (3.84) as

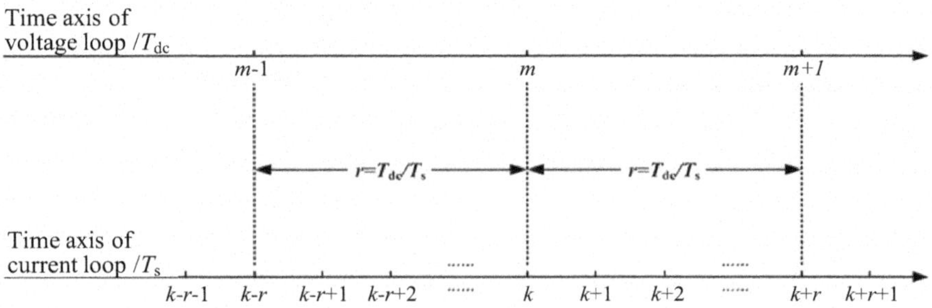

Fig. 3.26 Timing relation between the DC-bus voltage loop and the grid-side current loop

$$\overline{P_g}(m-1) = \frac{1}{r}\sum_{j=k-r}^{k-1} P_g(j) \tag{3.85}$$

To reduce the influence of non-ideal factors in the control system on the voltage prediction results, a feedback correction term is introduced into the open-loop prediction model, and the closed-loop prediction model of the DC-bus voltage is obtained as

$$u_{dcm}^2(m+1|m) = u_{dc}^2(m) - h\overline{P_g}(m-1) - h\Delta\overline{P_g}(m) + y_{dc}(m) \tag{3.86}$$

where $y_{dc}(m)$ is the feedback correction term of the DC-bus voltage at the moment of mT_{dc}, which includes the error between the DC-bus voltage predicted at the moment of $(m-1)T_{dc}$ for the moment of mT_{dc} and the DC-bus voltage sampled at the moment of mT_{dc} and can be written mathematically as

$$y_{dc}(m) = l_{dc}\cdot\left[u_{dc}^2(m) - u_{dcm}^2(m|m-1)\right]\overrightarrow{AB} \tag{3.87}$$

where l_{dc} is the feedback correction factor of the DC-bus voltage.

To achieve optimal control of the DC-bus voltage, an evaluation function characterising its performance index is required. Here, the control objective of the DC-bus voltage loop is that the DC-bus voltage value follows its reference value and that the variation of the active power injected into the grid is not too large. For this reason, the evaluation function of the DC-bus voltage loop can be given as

$$W_g(m) = \lambda_{dc}\cdot\left[u_{dc}^{*2}(m) - u_{dcm}^2(m+1|m)\right] + \varepsilon_{dc}\left[\Delta\overline{P_g}(m)\right] \tag{3.88}$$

where $u_{dc}^*(m)$ indicates the DC-bus voltage reference value, λ_{dc} indicates the weight factor for the error of the square of the DC-bus voltage, and ε_{dc} indicates the weight factor for the increment of active power injected into the grid given by the algorithm.

At any sampling moment mT_{dc}, in order to obtain the optimal increment of the average active power injected into the grid by the converter, it is only necessary to obtain the average active power increment that minimizes the evaluation function $W_g(m)$. Substituting Eq. (3.86) into Eq. (3.88), the optimal increment of the average active power injected into the grid by the grid-side converter between the moment of mT_{dc} and the moment of $(m+1)T_{dc}$ can be obtained through optimal solution as

$$\Delta\overline{P_g^*}(m) = \frac{h\lambda_{dc}}{h^2\lambda_{dc}+\varepsilon_{dc}}\left[u_{dc}^{*2}(m) - u_{dc}^2(m) + h\overline{P}_g(m-1) - y_{dc}(m)\right] \tag{3.89}$$

where $\Delta\overline{P_g^*}(m)$ indicates the optimal increment of the average active power injected into the grid by the converter given by the algorithm at moment of mT_{dc}. Then at any sampling moment of mT_{dc}, the control algorithm can obtain the optimal increment of the average active power injected into the grid by the converter at the current moment according to Eqs. (3.87) and (3.89), and through the accumulation of the optimal increment, the optimal average active power $\overline{P_g^*}(m)$ injected into the grid by the converter at the current moment can be obtained. Considering that the active power reference value given by the outer voltage loop is stable in the future period of T_{dc}, according to the timing relation between the current loop and the voltage loop in Fig. 3.26, the active power reference values obtained by the current loop at the moments of kT_s,$(k+1)T_s$, ..., $(k+r-1)T_s$ are all $\overline{P_g^*}(m)$, i.e., there is

$$P_g^*(k) = \overline{P_g^*}(m), \quad k \in [rm, rm+r-1) \tag{3.90}$$

From Eq. (3.73), the power reference value can be converted to the current reference value in order to be used as the input to the grid-side current inner loop, and the conversion interface function is obtained as

$$\begin{cases} i_g^{d*}(k) = \frac{P_g^*(k)}{e_g^d(k)} \\ i_g^{q*}(k) = \frac{-Q_g^*(k)}{e_g^d(k)} \end{cases} \tag{3.91}$$

To simplify the optimization process, the above-mentioned Evaluation Function (3.89) is optimised without any limitation on the incremental power injected into the grid by the converter. In a practical system, the active power injected into the grid by the converter is constrained by the capacity of the grid-side converter, which has a certain range. To ensure that the system is within the safe working range, it is necessary to add a limiter to limit the reference value of the current loop, so as to obtain the control diagram of the DC-bus voltage loop as shown in Fig. 3.27.

As can be seen from Sects. 3.3.1.1 and 3.3.1.2, in the control strategy of the grid-side converter, the control periods of the DC-bus voltage outer loop and the grid-side current inner loop are not the same: the DC-bus voltage loop performs a control action at every interval of $T_{dc} = rT_s$ to update the grid-side active power reference value and, after the conversion interface function, obtains the grid-side d-axis current reference value, which is locked and stored until the next DC-bus voltage loop operation; while the grid-side current loop performs a control action at each interval of T_s to obtain the current reference value and complete the related operation

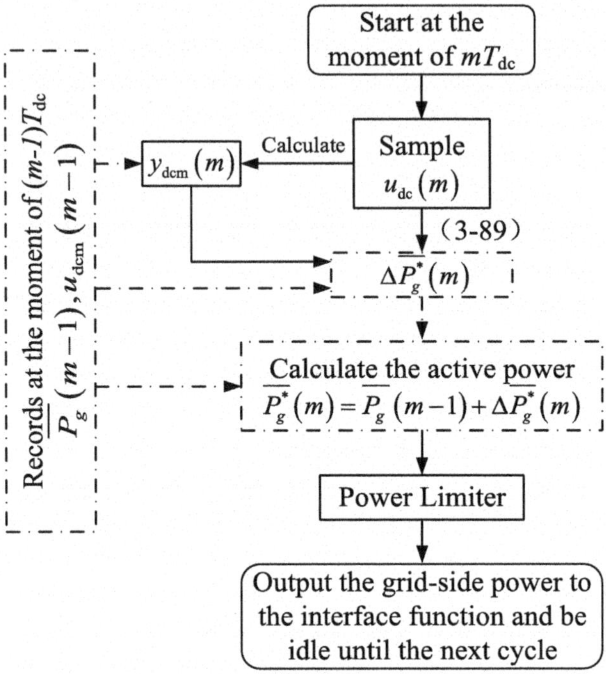

Fig. 3.27 Control diagram of the DC-bus voltage loop based on MPC

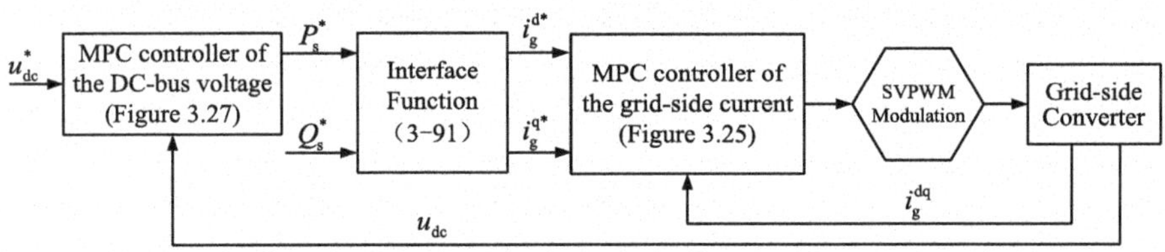

Fig. 3.28 Control diagram of the grid-side converter with the double close-loop structure based on MPC

to update the PWM duty cycle. In summary, the complete double close-loop control diagram of the grid-side converter is shown in Fig. 3.28.

3.3.2 Model Predictive Control of the DFIG

In grid-connected operation of the DFIG, the control objective of the generator-side converter is to achieve decoupled control of the active and reactive power on the stator side of the DFIG. Here, the stator power outer loop & rotor current inner loop control structure is still used, and MPC algorithms are derived for the rotor current loop and stator power loop, respectively.

3.3.2.1 Rotor Current Loop

Similar to the derivation of the grid-side current loop of the grid-side converter, the forward Euler method is used to discretize Eq. (2.19), and the delay effect of the digital control chip is taken into account to obtain a two-step prediction model for the rotor current with a feedback correction term as

$$\begin{cases} i_{\mathrm{rm}}^{\mathrm{d}}(k+1|k) = a_{\mathrm{r}} i_{\mathrm{r}}^{\mathrm{d}}(k) + b_{\mathrm{r}} i_{\mathrm{r}}^{\mathrm{q}}(k) - c_{\mathrm{r}} e_{\mathrm{g}}^{\mathrm{d}}(k) + \mathrm{d}_{\mathrm{r}} u_{\mathrm{r}}^{\mathrm{d}}(k-1) + x_{\mathrm{r}}^{\mathrm{d}}(k) \\ i_{\mathrm{rm}}^{\mathrm{q}}(k+1|k) = a_{\mathrm{r}} i_{\mathrm{r}}^{\mathrm{q}}(k) - b_{\mathrm{r}} i_{\mathrm{r}}^{\mathrm{d}}(k) + \mathrm{d}_{\mathrm{r}} u_{\mathrm{r}}^{\mathrm{q}}(k-1) + x_{\mathrm{r}}^{\mathrm{q}}(k) \end{cases} \tag{3.92}$$

$$\begin{cases} i_{\mathrm{rm}}^{\mathrm{d}}(k+2|k) = a_{\mathrm{r}} i_{\mathrm{rm}}^{\mathrm{d}}(k) + b_{\mathrm{r}} i_{\mathrm{rm}}^{\mathrm{q}}(k) - c_{\mathrm{r}} e_{\mathrm{g}}^{\mathrm{d}}(k) \\ \qquad + d_{\mathrm{r}} u_{\mathrm{r}}^{\mathrm{d}}(k-1) + d_{\mathrm{r}} \Delta u_{\mathrm{r}}^{\mathrm{d}}(k) + x_{\mathrm{r}}^{\mathrm{d}}(k) \\ i_{\mathrm{rm}}^{\mathrm{q}}(k+2|k) = a_{\mathrm{r}} i_{\mathrm{rm}}^{\mathrm{q}}(k) - b_{\mathrm{r}} i_{\mathrm{rm}}^{\mathrm{d}}(k) \\ \qquad + d_{\mathrm{r}} u_{\mathrm{r}}^{\mathrm{q}}(k-1) + d_{\mathrm{r}} \Delta u_{\mathrm{r}}^{\mathrm{q}}(k) + x_{\mathrm{r}}^{\mathrm{q}}(k) \end{cases} \tag{3.93}$$

where $k \in N$, $a_{\mathrm{r}} = 1 - R_{\mathrm{r}} T_{\mathrm{s}} / L_{\mathrm{r}}$, $b_{\mathrm{r}} = \omega_{\mathrm{sl}} T_{\mathrm{s}}$, $c_{\mathrm{r}} = \omega_{\mathrm{sl}} L_m T_{\mathrm{s}} / (\omega_{\mathrm{s}} \sigma L_{\mathrm{s}} L_{\mathrm{r}})$, $d_{\mathrm{r}} = T_{\mathrm{s}} / L_{\mathrm{r}}$, $i_{\mathrm{rm}}^{\mathrm{d}}(k+1|k)$ and $i_{\mathrm{rm}}^{\mathrm{q}}(k+1|k)$ indicate the d-axis and q-axis components of the instantaneous rotor current predicted at the moment of kT_{s} for the moment of $(k+1)T_{\mathrm{s}}$ respectively, $i_{\mathrm{rm}}^{\mathrm{d}}(k+2|k)$ and $i_{\mathrm{rm}}^{\mathrm{q}}(k+2|k)$ indicate the d-axis and q-axis components of the instantaneous rotor current predicted at the moment of kT_{s} for the moment of $(k+2)T_{\mathrm{s}}$ respectively, $i_{\mathrm{r}}^{\mathrm{d}}(k)$ and $i_{\mathrm{r}}^{\mathrm{q}}(k)$ indicate the d-axis and q-axis components of the instantaneous rotor current sampled at the moment of kT_{s} respectively, $u_{\mathrm{r}}^{\mathrm{d}}(k-1)$ and $u_{\mathrm{r}}^{\mathrm{q}}(k-1)$ indicate the d-axis and q-axis components of the control voltage given by the rotor current loop at the moment of $(k-1)T_{\mathrm{s}}$ respectively, $\Delta u_{\mathrm{r}}^{\mathrm{d}}(k)$ and $\Delta u_{\mathrm{r}}^{\mathrm{q}}(k)$ indicate the d-axis and q-axis components of the control voltage increment at the moment of kT_{s} respectively, $x_{\mathrm{r}}^{\mathrm{d}}(k)$ and $x_{\mathrm{r}}^{\mathrm{q}}(k)$ indicate the d-axis and q-axis components of the feedback correction term at the moment kT_{s} respectively, which include the error between the rotor current predicted at the moment of $(k-1)T_{\mathrm{s}}$ for the moment of kT_{s} and the rotor current sampled at the moment of kT_{s} and can be mathematically written as

$$\begin{cases} x_{\mathrm{r}}^{\mathrm{d}}(k) = f_{\mathrm{r}}^{\mathrm{d}} \cdot \left[i_{\mathrm{r}}^{\mathrm{d}}(k) - i_{\mathrm{rm}}^{\mathrm{d}}(k|k-1) \right] \\ x_{\mathrm{r}}^{\mathrm{q}}(k) = f_{\mathrm{r}}^{\mathrm{q}} \cdot \left[i_{\mathrm{r}}^{\mathrm{q}}(k) - i_{\mathrm{rm}}^{\mathrm{q}}(k|k-1) \right] \end{cases} \tag{3.94}$$

where $f_{\mathrm{r}}^{\mathrm{d}}$ and $f_{\mathrm{r}}^{\mathrm{q}}$ are the feedback correction factors of the d-axis and q-axis components of the rotor current, respectively.

To achieve optimal control of the rotor current, an evaluation function characterising its performance index is required. Here, the control objective of the rotor current loop is that the rotor current value follows its reference value and that the variation of the generator-side converter output voltage is not too large. For this reason, the evaluation function of the rotor current loop can be given as

$$\begin{aligned} J_r(k) &= \lambda_r^d\left[i_r^{d*}(k) - i_{rm}^d(k+2|k)\right]^2 + \lambda_r^q\left[i_r^{q*}(k) - i_{rm}^q(k+2|k)\right]^2 \\ &+\varepsilon_r^d\left[\Delta u_r^d(k)\right]^2 + \varepsilon_r^q\left[\Delta u_r^q(k)\right]^2 \end{aligned} \tag{3.95}$$

where $i_r^{d*}(k)$ and $i_r^{q*}(k)$ indicate the reference values for the d-axis and q-axis components of the rotor current, respectively, λ_r^d and λ_r^q indicate the weight factors for the errors in the d-axis and q-axis components of the rotor current, respectively, and ε_r^d and ε_r^q indicate the weight factors for the d-axis and q-axis components of the generator-side control voltage increment, respectively.

At any sampling moment of kT_s, in order to obtain the optimal converter output voltage increment, it is only necessary to obtain the value of the voltage increment that minimizes the evaluation function $J_r(k)$, substituting Eq. (3.93) into Eq. (3.95), the optimal output voltage increment of the generator-side converter at the moment of kT_s can be obtained by the optimal solution method as

$$\begin{cases} \Delta u_r^{d*}(k) = \frac{d_r\lambda_r^d}{d_r^2\lambda_r^d+\varepsilon_r^d}\left[i_r^{d*}(k) - a_r i_{rm}^d(k+1|k) - b_r i_{rm}^q(k+1|k)\right. \\ \qquad \left. + c_r e_g^d(k) - d_r u_r^d(k-1) - x_r^d(k)\right] \\ \Delta u_r^{q*}(k) = \frac{d_r\lambda_r^q}{d_r^2\lambda_r^q+\varepsilon_r^q}\left[i_r^{q*}(k) - a_r i_{rm}^q(k+1|k) + b_r i_{rm}^d(k+1|k)\right. \\ \qquad \left. - d_r u_r^q(k-1) - x_r^q(k)\right] \end{cases} \tag{3.96}$$

where $\Delta u_r^{d*}(k)$ and $\Delta u_r^{q*}(k)$ indicate the d-axis and q-axis components of the optimal control voltage increment at the moment of kT_s given by the rotor current loop, respectively. Then at any sampling moment of kT_s, the optimal control voltage increment at the current moment can be obtained, according to Eqs. (3.92), (3.94) and (3.96), and the optimal control voltage at the current moment can be obtained by the equations of $u_r^{d*}(k) = u_r^d(k-1) + \Delta u_r^{d*}(k)$ and $u_r^{q*}(k) = u_r^q(k-1) + \Delta u_r^{q*}(k)$.

In a practical system, to avoid overshooting the generator-side converter, it is also necessary to limit the given output voltage. The generator-side converter rotor current control diagram with a limiter is shown in Fig. 3.29.

3.3.2.2 Stator Power Loop

From Eq. (2.22), it can be seen that, under definite grid voltage conditions, the stator power is entirely determined by the stator current, and therefore the close-loop control of the stator current is the close-loop control of the stator power. For ease of derivation, this section first derives the close-loop method of the stator current, on the basis of which the stator power loop is implemented.

The stator current loop is used as the outer loop of the rotor current loop, and its control period is taken to be T_p ($T_p = zT_s$, z is usually taken to be a natural number

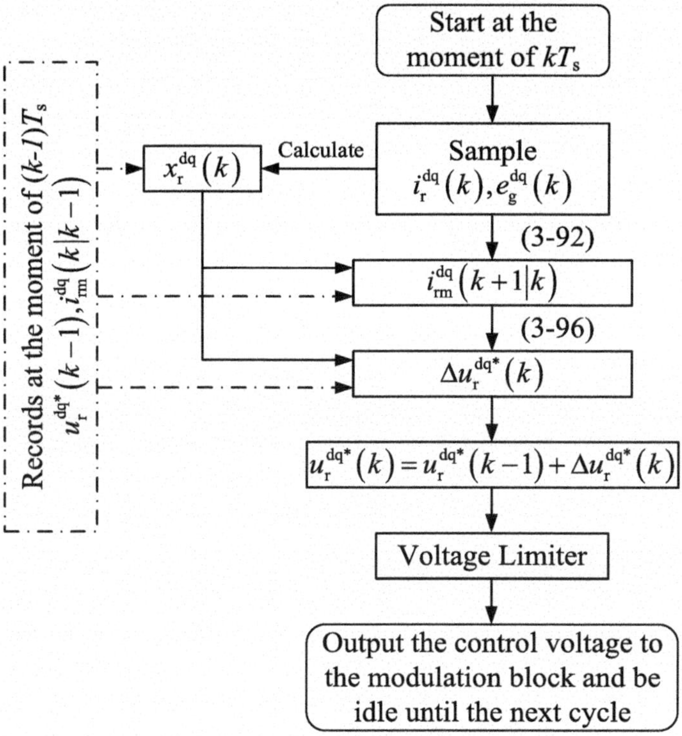

Fig. 3.29 Control diagram of the rotor current loop based on MPC

between 5 and 15), at which point the delay effect of the digital control chip can be neglected.

In grid-connected operation of DFIG, the stator is directly connected to the grid and the stator voltage is considered to be the grid voltage. When the stator voltage orientation technology is used, it gives

$$\begin{cases} u_s^d = e_g^d \\ u_s^q = e_g^q = 0 \end{cases} \tag{3.97}$$

Substituting Eq. (2.30) into Eq. (2.19) and neglecting the stator resistance, under the steady state conditions, there is

$$\begin{cases} \varphi_s^d = 0 \\ \varphi_s^q = -\frac{e_g^d}{\omega_s} = \varphi_m \end{cases} \tag{3.98}$$

Substituting Eq. (3.98) into Eq. (2.20), the stator current can be written as

$$\begin{cases} i_{\mathrm{s}}^{\mathrm{d}} = -\frac{L_{\mathrm{m}}}{L_{\mathrm{s}}} i_{\mathrm{r}}^{\mathrm{d}} \\ i_{\mathrm{s}}^{\mathrm{q}} = \frac{\varphi_{\mathrm{m}} - L_{\mathrm{m}} i_{\mathrm{r}}^{\mathrm{q}}}{L_{\mathrm{s}}} \end{cases} \tag{3.99}$$

By discretizing Eq. (3.99) and adding a feedback correction term, the prediction model for the stator current can be obtained as

$$\begin{cases} i_{\mathrm{sm}}^{\mathrm{d}}(n+1|n) = i_{\mathrm{s}}^{\mathrm{d}}(n) - \dfrac{L_{\mathrm{m}}}{L_{\mathrm{s}}} \Delta i_{\mathrm{r}}^{\mathrm{d}}(n) + y_{\mathrm{s}}^{\mathrm{d}}(n) \\ i_{\mathrm{sm}}^{\mathrm{q}}(n+1|n) = i_{\mathrm{s}}^{\mathrm{q}}(n) - \dfrac{L_{\mathrm{m}}}{L_{\mathrm{s}}} \Delta i_{\mathrm{r}}^{\mathrm{q}}(n) + y_{\mathrm{s}}^{\mathrm{q}}(n) \end{cases} \tag{3.100}$$

where $n = k/z$ and $n \in N$, $i_{\mathrm{sm}}^{\mathrm{d}}(n+1|n)$ and $i_{\mathrm{sm}}^{\mathrm{q}}(n+1|n)$ indicate the d-axis and q-axis components of the stator current predicted at the moment of nT_{p} for the moment of $(n+1)T_{\mathrm{p}}$ respectively, $i_{\mathrm{s}}^{\mathrm{d}}(n)$ and $i_{\mathrm{s}}^{\mathrm{q}}(n)$ indicate the stator current sampled at the moment of nT_{p} respectively, $\Delta i_{\mathrm{r}}^{\mathrm{d}}(n)$ and $\Delta i_{\mathrm{r}}^{\mathrm{q}}(n)$ indicate the d-axis and q-axis components of the rotor current increment at the moment of nT_{p} respectively, $y_{\mathrm{s}}^{\mathrm{d}}(n)$ and $y_{\mathrm{s}}^{\mathrm{q}}(n)$ indicate the d-axis and q-axis components of the feedback correction term at the moment of nT_{p} respectively, which includes the error between the stator current predicted at the moment of $(n-1)T_{\mathrm{p}}$ for the moment of nT_{p} and the stator current sampled at the moment of nT_{p} respectively and can be mathematically written as

$$\begin{cases} y_{\mathrm{s}}^{\mathrm{d}}(n) = l_{\mathrm{s}}^{\mathrm{d}} \cdot \left[i_{\mathrm{s}}^{\mathrm{d}}(n) - i_{\mathrm{sm}}^{\mathrm{d}}(n|n-1) \right] \\ y_{\mathrm{s}}^{\mathrm{q}}(n) = l_{\mathrm{s}}^{\mathrm{q}} \cdot \left[i_{\mathrm{s}}^{\mathrm{q}}(n) - i_{\mathrm{sm}}^{\mathrm{q}}(n|n-1) \right] \end{cases} \tag{3.101}$$

where $l_{\mathrm{s}}^{\mathrm{d}}$ and $l_{\mathrm{s}}^{\mathrm{q}}$ are the feedback correction factors of the d-axis and q-axis components of the stator current, respectively.

The control objective of the stator current loop is that the stator current value follows its reference value and that the rotor current does not vary too much, for which the evaluation function of the stator current loop can be given as

$$\begin{aligned} W_{\mathrm{s}}(n) = {} & \lambda_{\mathrm{s}}^{\mathrm{d}} \left[i_{\mathrm{s}}^{\mathrm{d}*}(n) - i_{\mathrm{sm}}^{\mathrm{d}}(n+1|n) \right]^2 + \lambda_{\mathrm{s}}^{\mathrm{q}} \left[i_{\mathrm{s}}^{\mathrm{q}*}(n) - i_{\mathrm{sm}}^{\mathrm{q}}(n+1|n) \right]^2 \\ & + \varepsilon_{\mathrm{s}}^{\mathrm{d}} \left[\Delta i_{\mathrm{r}}^{\mathrm{d}}(n) \right]^2 + \varepsilon_{\mathrm{s}}^{\mathrm{q}} \left[\Delta i_{\mathrm{r}}^{\mathrm{q}}(n) \right]^2 \end{aligned} \tag{3.102}$$

where $\lambda_{\mathrm{s}}^{\mathrm{d}}$ and $\lambda_{\mathrm{s}}^{\mathrm{q}}$ indicate the weight factors of the d-axis and q-axis components of the stator current respectively, $\varepsilon_{\mathrm{s}}^{\mathrm{d}}$ and $\varepsilon_{\mathrm{s}}^{\mathrm{q}}$ indicate the weight factors of the d-axis and q-axis components of the rotor current respectively.

At any sampling moment of nT_{p}, in order to obtain the optimal increment of the rotor current, it is only necessary to obtain the increment of the rotor current that minimises the evaluation function $W_{\mathrm{s}}(n)$. Substituting Eq. (3.100) into Eq. (3.102) and solving optimally, the optimal increment of the rotor current at the moment of nT_{p} is obtained as

$$
\begin{cases}
\Delta i_{\mathrm{r}}^{\mathrm{d}*}(n) = \dfrac{-L_{\mathrm{m}}L_{\mathrm{s}}\lambda_{\mathrm{s}}^{\mathrm{d}}}{L_{\mathrm{m}}^{2}\lambda_{\mathrm{s}}^{\mathrm{d}} + L_{\mathrm{s}}^{2}\varepsilon_{\mathrm{s}}^{\mathrm{d}}}\left[i_{\mathrm{s}}^{\mathrm{d}*}(n) - i_{\mathrm{s}}^{\mathrm{d}}(n) - y_{\mathrm{s}}^{\mathrm{d}}(n)\right] \\
\Delta i_{\mathrm{r}}^{\mathrm{q}*}(n) = \dfrac{-L_{\mathrm{m}}L_{\mathrm{s}}\lambda_{\mathrm{s}}^{\mathrm{d}}}{L_{\mathrm{m}}^{2}\lambda_{\mathrm{s}}^{\mathrm{d}} + L_{\mathrm{s}}^{2}\varepsilon_{\mathrm{s}}^{\mathrm{d}}}\left[i_{\mathrm{s}}^{\mathrm{q}*}(n) - i_{\mathrm{s}}^{\mathrm{q}}(n) - y_{\mathrm{s}}^{\mathrm{q}}(n)\right]
\end{cases}
\tag{3.103}
$$

where $\Delta i_{\mathrm{r}}^{\mathrm{d}*}(n)$ and $\Delta i_{\mathrm{r}}^{\mathrm{q}*}(n)$ indicate the d-axis and q-axis components of the optimal rotor current increment given by the outer stator current loop at the moment of nT_{p} respectively. $i_{\mathrm{r}}^{\mathrm{d}*}(n)$ and $i_{\mathrm{r}}^{\mathrm{q}*}(n)$ indicate the d-axis and q-axis components of the optimal rotor current obtained by accumulating the optimal increment respectively. The rotor current reference given by the outer stator current loop is considered to be stable during the interval of T_{p} and according to the timing relation between the stator current loop and the rotor current loop, the rotor current reference obtained by the rotor current loop at the moments of kT_{s}, $(k+1)T_{\mathrm{s}}$, …, $(k+z-1)T_{\mathrm{s}}$ are all $i_{\mathrm{r}}^{\mathrm{d}*}(n)$ and $i_{\mathrm{r}}^{\mathrm{q}*}(n)$, i.e. there is

$$
\begin{cases}
i_{\mathrm{r}}^{\mathrm{d}*}(k) = i_{\mathrm{r}}^{\mathrm{d}*}(n) \\
i_{\mathrm{r}}^{\mathrm{q}*}(k) = i_{\mathrm{r}}^{\mathrm{q}*}(n)
\end{cases}
\quad k \in [zn, zn + z - 1)
\tag{3.104}
$$

From Eq. (2.22), the stator power reference can be converted into the stator current reference and the conversion interface function can be written as

$$
\begin{cases}
i_{\mathrm{s}}^{\mathrm{d}*} = \dfrac{P_{\mathrm{s}}^{*}}{e_{\mathrm{g}}^{\mathrm{d}}} \\
i_{\mathrm{s}}^{\mathrm{q}*} = -\dfrac{Q_{\mathrm{s}}^{*}}{e_{\mathrm{g}}^{\mathrm{d}}}
\end{cases}
\tag{3.105}
$$

To simplify the optimization process, the above-mentioned Evaluation Function (3.102) is optimised without any limitation on the incremental rotor current. In a practical system, the rotor current is constrained by the capacity of the generator-side converter, which has a certain range. To ensure that the system is within the safe working range, it is necessary to add a limiter to limit the reference value of the current loop, so as to obtain the control diagram of the stator power loop as shown in Fig. 3.30.

As can be seen from Sects. 3.3.2.1 and 3.3.2.2, in the control strategy of the generator-side converter, the control periods of the stator power outer loop and the rotor current inner loop are not the same: the stator power loop performs a control action at every interval of $T_{\mathrm{p}} = zT_{\mathrm{s}}$ to update the rotor current reference, which is locked and stored until the next stator power loop operation; while the rotor current loop performs a control action at each interval of T_{s} to obtain the current reference and complete the related operation to update the PWM duty cycle. In summary, the complete double close-loop control diagram of the generator-side converter is shown in Fig. 3.31.

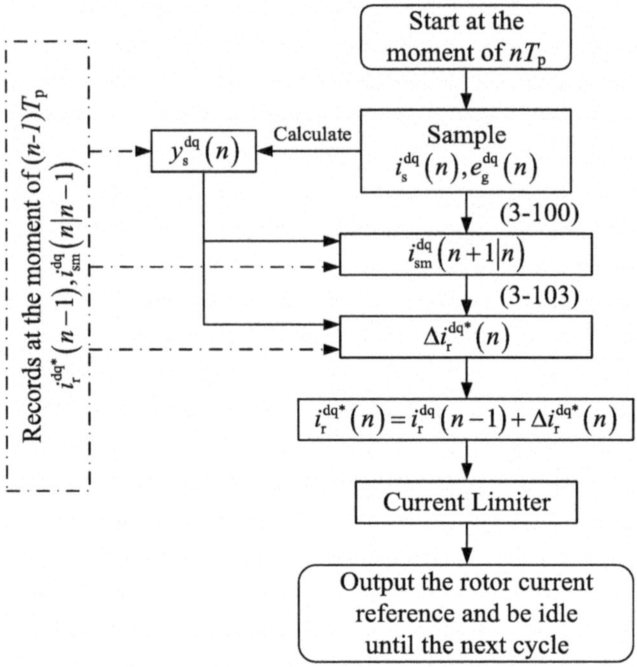

Fig. 3.30 Control diagram of the stator power loop based on MPC

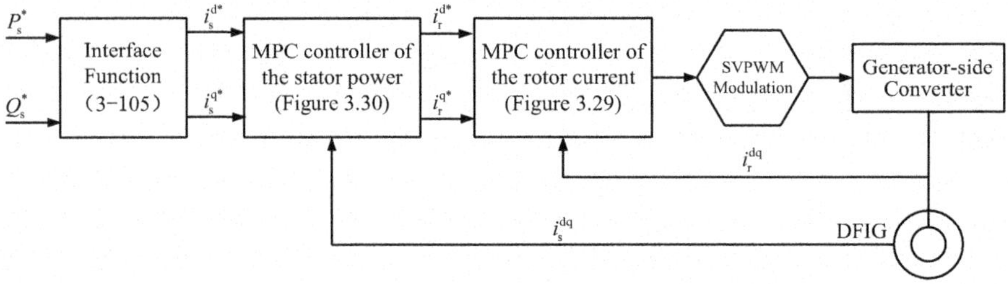

Fig. 3.31 Control diagram of the generator-side converter with the double close-loop structure based on MPC

3.3.3 *Smooth Grid Connection Based on MPC*

Before the DFIG system is connected to the grid, its stator voltage needs to be controlled to be consistent with the grid voltage in terms of magnitude, phase and frequency to reduce the current impact at the moment of stator-side switch closure and to achieve smooth grid connection.

3.3.3.1 Basic Principles of the Stator Voltage Control

Before the stator-side grid connection switch is closed, the DFIG stator-side is open-circuited and its current is constant at 0, i.e.

$$i_{\mathrm{s}}^{\mathrm{d}} = i_{\mathrm{s}}^{\mathrm{q}} = 0 \tag{3.106}$$

Substituting Eq. (3.106) into Eq. (2.20), it gives

$$\begin{cases} \psi_{\mathrm{s}}^{\mathrm{d}} = L_{\mathrm{M}} i_{\mathrm{r}}^{\mathrm{d}} \\ \psi_{\mathrm{s}}^{\mathrm{q}} = L_{\mathrm{M}} i_{\mathrm{r}}^{\mathrm{q}} \\ \psi_{\mathrm{r}}^{\mathrm{d}} = L_{\mathrm{r}} i_{\mathrm{r}}^{\mathrm{d}} \\ \psi_{\mathrm{r}}^{\mathrm{q}} = L_{\mathrm{r}} i_{\mathrm{r}}^{\mathrm{q}} \end{cases} \tag{3.107}$$

Substituting the rotor flux linkage formulas in Eq. (3.107) into the rotor voltage formulas in Eq. (2.19), it gives

$$\begin{cases} u_{\mathrm{r}}^{\mathrm{d}} = R_{\mathrm{r}} i_{\mathrm{r}}^{\mathrm{d}} + L_{\mathrm{r}} p i_{\mathrm{r}}^{\mathrm{d}} - \omega_{\mathrm{sl}} L_{\mathrm{r}} i_{\mathrm{r}}^{\mathrm{q}} \\ u_{\mathrm{r}}^{\mathrm{q}} = R_{\mathrm{r}} i_{\mathrm{r}}^{\mathrm{q}} + L_{\mathrm{r}} p i_{\mathrm{r}}^{\mathrm{q}} + \omega_{\mathrm{sl}} L_{\mathrm{r}} i_{\mathrm{r}}^{\mathrm{d}} \end{cases} \tag{3.108}$$

Substituting the stator flux linkage formulas in Eq. (3.107) into the stator voltage formulas in Eq. (2.19) and neglecting the transient change of the stator flux linkage, it gives

$$\begin{cases} u_{\mathrm{s}}^{\mathrm{d}} = -\omega_{\mathrm{s}} L_{\mathrm{M}} i_{\mathrm{r}}^{\mathrm{q}} \\ u_{\mathrm{s}}^{\mathrm{q}} = \omega_{\mathrm{s}} L_{\mathrm{M}} i_{\mathrm{r}}^{\mathrm{d}} \end{cases} \tag{3.109}$$

Equation (3.109) indicates that the DFIG stator voltage can be controlled by the rotor current before grid connection, while Eq. (3.108) indicates that the DFIG rotor current can be controlled by the generator-side converter output voltage, which is the basic principle of the stator voltage outer loop and rotor current inner loop structure in stator voltage synchronous control.

3.3.3.2 Rotor Current Loop

Similar to the derivation of the grid-side current loop of the grid-side converter, a two-step prediction model of the rotor current with a feedback correction term is obtained, by utilizing the forward Euler method to discretize Eq. (3.108) with consideration of the delay effect caused by the digital control chip, as

$$\begin{cases} i_{\mathrm{rm0}}^{\mathrm{d}}(k+1|k) = a_{\mathrm{r0}} i_{\mathrm{r}}^{\mathrm{d}}(k) + b_{\mathrm{r0}} i_{\mathrm{r}}^{\mathrm{q}}(k) + c_{\mathrm{r0}} u_{\mathrm{r}}^{\mathrm{d}}(k-1) + x_{\mathrm{r0}}^{\mathrm{d}}(k) \\ i_{\mathrm{rm0}}^{\mathrm{q}}(k+1|k) = a_{\mathrm{r0}} i_{\mathrm{r}}^{\mathrm{q}}(k) - b_{\mathrm{r0}} i_{\mathrm{r}}^{\mathrm{d}}(k) + c_{\mathrm{r0}} u_{\mathrm{r}}^{\mathrm{q}}(k-1) + x_{\mathrm{r0}}^{\mathrm{q}}(k) \end{cases} \tag{3.110}$$

$$\begin{cases} i_{\mathrm{rm0}}^{\mathrm{d}}(k+2|k) = a_{\mathrm{r0}} i_{\mathrm{rm0}}^{\mathrm{d}}(k+1|k) + b_{\mathrm{r0}} i_{\mathrm{rm0}}^{\mathrm{q}}(k+1|k) \\ \qquad + c_{\mathrm{r0}} u_{\mathrm{r}}^{\mathrm{d}}(k-1) + c_{\mathrm{r0}} \Delta u_{\mathrm{r}}^{\mathrm{d}}(k) + x_{\mathrm{r0}}^{\mathrm{d}}(k) \\ i_{\mathrm{rm0}}^{\mathrm{q}}(k+2|k) = a_{\mathrm{r0}} i_{\mathrm{rm0}}^{\mathrm{q}}(k+1|k) - b_{\mathrm{r0}} i_{\mathrm{rm0}}^{\mathrm{d}}(k+1|k) \\ \qquad + c_{\mathrm{r0}} u_{\mathrm{r}}^{\mathrm{q}}(k-1) + c_{\mathrm{r0}} \Delta u_{\mathrm{r}}^{\mathrm{q}}(k) + x_{\mathrm{r0}}^{\mathrm{q}}(k) \end{cases} \tag{3.111}$$

where $k \in N$, $a_{\mathrm{r0}} = 1 - R_{\mathrm{r}} T_{\mathrm{s}}/L_{\mathrm{r}}$, $b_{\mathrm{r0}} = \omega_{\mathrm{sl}} T_{\mathrm{s}}$, $c_{\mathrm{r0}} = T_{\mathrm{s}}/L_{\mathrm{r}}$, $i_{\mathrm{rm0}}^{\mathrm{d}}(k+1|k)$ and $i_{\mathrm{rm0}}^{\mathrm{q}}(k+1|k)$ indicate the d-axis and q-axis components of the rotor current predicted at the moment of kT_{s} for the moment of $(k+1)T_{\mathrm{s}}$, respectively, $i_{\mathrm{r}}^{\mathrm{d}}(k)$ and $i_{\mathrm{r}}^{\mathrm{q}}(k)$ indicate the d-axis and q-axis components of the rotor current sampled at the moment of kT_{s}, respectively, $u_{\mathrm{r}}^{\mathrm{d}}(k-1)$ and $u_{\mathrm{r}}^{\mathrm{q}}(k-1)$ indicate the d-axis and q-axis components of the control voltage outputted by the generator controller at the moment of $(k-1)T_{\mathrm{s}}$, respectively, $\Delta u_{\mathrm{r}}^{\mathrm{d}}(k)$ and $\Delta u_{\mathrm{r}}^{\mathrm{q}}(k)$ indicate the d-axis and q-axis components of control voltage increment outputted by the generator controller at the moment of kT_{s}, respectively, $x_{\mathrm{r0}}^{\mathrm{d}}(k)$ and $x_{\mathrm{r0}}^{\mathrm{q}}(k)$ indicate the d-axis and q-axis components of the feedback correction term at the moment of kT_{s}, respectively, which includes the error between the rotor current predicted at the moment of $(k-1)T_{\mathrm{s}}$ for the moment of kT_{s} and the rotor current sampled at the moment of kT_{s} and can be mathematically written as

$$\begin{cases} x_{\mathrm{r0}}^{\mathrm{d}}(k) = f_{\mathrm{r0}}^{\mathrm{d}} \cdot \left[i_{\mathrm{r}}^{\mathrm{d}}(k) - i_{\mathrm{rm0}}^{\mathrm{d}}(k|k-1) \right] \\ x_{\mathrm{r0}}^{\mathrm{q}}(k) = f_{\mathrm{r0}}^{\mathrm{q}} \cdot \left[i_{\mathrm{r}}^{\mathrm{q}}(k) - i_{\mathrm{rm0}}^{\mathrm{q}}(k|k-1) \right] \end{cases} \tag{3.112}$$

where $f_{\mathrm{r0}}^{\mathrm{d}}$ and $f_{\mathrm{r0}}^{\mathrm{q}}$ are the feedback correction factors of the d-axis and q-axis components of the rotor current, respectively.

The control objective of the rotor current loop is that the rotor current value follows its reference value and that the control voltage does not vary too much, for which the evaluation function of the rotor current loop can be given as

$$\begin{aligned} J_{\mathrm{r0}}(k) = {} & \lambda_{\mathrm{r0}}^{\mathrm{d}} \left[i_{\mathrm{r}}^{\mathrm{d}*}(k) - i_{\mathrm{rm0}}^{\mathrm{d}}(k+2|k) \right]^2 + \lambda_{\mathrm{r0}}^{\mathrm{q}} \left[i_{\mathrm{r}}^{\mathrm{q}*}(k) - i_{\mathrm{rm0}}^{\mathrm{q}}(k+2|k) \right]^2 \\ & + \varepsilon_{\mathrm{r0}}^{\mathrm{d}} \left[\Delta u_{\mathrm{r}}^{\mathrm{d}}(k) \right]^2 + \varepsilon_{\mathrm{r0}}^{\mathrm{q}} \left[\Delta u_{\mathrm{r}}^{\mathrm{q}}(k) \right]^2 \end{aligned} \tag{3.113}$$

where $i_{\mathrm{r}}^{\mathrm{d}*}(k)$ and $i_{\mathrm{r}}^{\mathrm{q}*}(k)$ indicate the d-axis and q-axis components of the rotor current reference, respectively, $\lambda_{\mathrm{r0}}^{\mathrm{d}}$ and $\lambda_{\mathrm{r0}}^{\mathrm{q}}$ indicate the weight factors of the d-axis and q-axis components of the rotor current, respectively, $\varepsilon_{\mathrm{r0}}^{\mathrm{d}}$ and $\varepsilon_{\mathrm{r0}}^{\mathrm{q}}$ indicate the weight factors of the d-axis and q-axis components of the control voltage increment outputted by the generator-side converter, respectively.

At any sampling moment of kT_{s}, in order to obtain the optimal increment of the generator-side converter output voltage, it is only necessary to obtain the increment of the generator-side converter output voltage that minimises the evaluation function $J_{\mathrm{r0}}(k)$. Substituting Eq. (3.111) into Eq. (3.113) and solving optimally, the optimal increment of the generator-side converter output voltage at the moment of kT_{s} is obtained as

$$\begin{cases} \Delta u_{\rm r}^{\rm d*}(k) = \frac{c_{\rm r0}\lambda_{\rm r0}^{\rm d}}{c_{\rm r0}^2\lambda_{\rm r0}^{\rm d}+\varepsilon_{\rm r0}^{\rm d}}\left[i_{\rm r}^{\rm d*}(k) - a_{\rm r0}i_{\rm rm0}^{\rm d}(k+1|k) - b_{\rm r0}i_{\rm rm0}^{\rm q}(k+1|k)\right. \\ \qquad \left. - c_{\rm r0}u_{\rm r}^{\rm d}(k-1) - x_{\rm r0}^{\rm d}(k)\right] \\ \Delta u_{\rm r}^{\rm q*}(k) = \frac{c_{\rm r0}\lambda_{\rm r0}^{\rm q}}{c_{\rm r0}^2\lambda_{\rm r0}^{\rm q}+\varepsilon_{\rm r0}^{\rm q}}\left[i_{\rm r}^{\rm q*}(k) - a_{\rm r0}i_{\rm rm0}^{\rm q}(k+1|k) + b_{\rm r0}i_{\rm rm0}^{\rm d}(k+1|k)\right. \\ \qquad \left. - c_{\rm r0}u_{\rm r}^{\rm q}(k-1) - x_{\rm r0}^{\rm q}(k)\right] \end{cases} \tag{3.114}$$

where $\Delta u_{\rm r}^{\rm d*}(k)$ and $\Delta u_{\rm r}^{\rm q*}(k)$ indicate the d-axis and q-axis components of the optimal increment of the generator-side converter output voltage assigned by the control algorithm at the moment of $kT_{\rm s}$, respectively. Therefore the optimal increment of the generator-side converter output voltage at any sampling moment can be obtained by the control algorithm according to Eqs. (3.110), (3.112) and (3.114). By accumulating the series optimal increments, the optimal output voltage of the generator-side converter can be found as $u_{\rm r}^{\rm d*}(k)$ and $u_{\rm r}^{\rm q*}(k)$.

In a practical system, the output voltage of the generator-side converter also needs to be limited in order to avoid overshooting. The control diagram of the rotor current loop of the generator-side converter with a limiter is shown in Fig. 3.32.

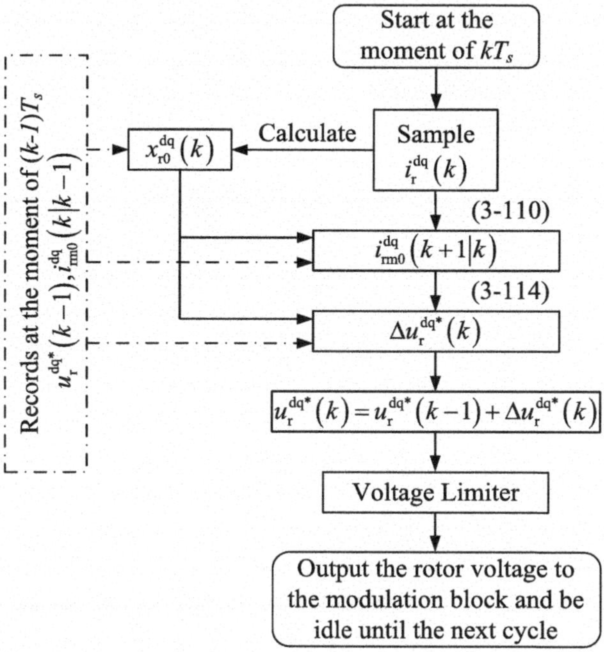

Fig. 3.32 Control diagram of the MPC-based current loop of the generator-side converter for smooth grid connection

3.3.3.3 Stator Voltage Loop

The stator voltage loop is used as the outer loop of the rotor current loop, and its control period is taken to be T_{st} ($T_{st} = z_0 T_s$, z_0 is usually taken to be a natural number between 5 and 15), at which point the delay effect of the digital control chip can be neglected.

By discretizing Eq. (3.109) and adding a feedback correction term, the prediction model for the stator voltage is obtained as

$$\begin{cases} u_{sm}^{d}(j+1|j) = -\omega_s L_m i_r^q(j) - \omega_s L_m \Delta i_r^q(j) + y_{st}^{d}(j) \\ u_{sm}^{q}(j+1|j) = \omega_s L_m i_r^d(j) + \omega_s L_m \Delta i_r^d(j) + y_{st}^{q}(j) \end{cases} \tag{3.115}$$

where $j = k/z_0$ and $j \in N$, $u_{sm}^{d}(j+1|j)$ and $u_{sm}^{q}(j+1|j)$ indicate the d-axis and q-axis components of the stator voltage predicted at the moment of jT_{st} for the moment of $(j+1)T_{st}$, respectively, $i_r^d(j)$ and $i_r^q(j)$ indicate the d-axis and q-axis components of the rotor current sampled at the moment of jT_{st}, respectively, $\Delta i_r^d(j)$ and $\Delta i_r^q(j)$ respectively indicate the d-axis and q-axis components of the rotor current increment at the moment of jT_{st}, respectively, $y_{st}^{d}(j)$ and $y_{st}^{q}(j)$ indicate the d-axis and q-axis components of the feedback correction term of the rotor current, respectively, which includes the error between the stator voltage predicted at the moment of $(j-1)T_{st}$ for the moment of jT_{st} and the stator voltage sampled at the moment of jT_{st} and can be mathematically written as

$$\begin{cases} y_{st}^{d}(j) = l_{st}^{d} \cdot \left[u_s^d(j) - u_{sm}^{d}(j|j-1)\right] \\ y_{st}^{q}(j) = l_{st}^{q} \cdot \left[u_s^q(j) - u_{sm}^{q}(j|j-1)\right] \end{cases} \tag{3.116}$$

where l_{st}^{d} and l_{st}^{q} are the feedback correction factors of the d-axis and q-axis components of the stator voltage, respectively.

The control objective of the stator voltage loop is that the stator voltage value follows its reference value and that the rotor current does not vary too much, for which the evaluation function of the stator voltage loop can be given as

$$\begin{aligned} W_{st}(j) = {} & \lambda_{st}^{d}\left[e_g^d(j) - u_{sm}^{d}(j+1|j)\right]^2 + \lambda_{st}^{q}\left[e_g^q(j) - u_{sm}^{q}(j+1|j)\right]^2 \\ & + \varepsilon_{st}^{d}\left[\Delta i_r^d(j)\right]^2 + \varepsilon_{st}^{q}\left[\Delta i_r^q(j)\right]^2 \end{aligned} \tag{3.117}$$

where λ_{st}^{d} and λ_{st}^{q} indicate the weight factors of the d-axis and q-axis components of the stator voltage, respectively, ε_{st}^{d} and ε_{st}^{q} indicate the weight factors of the d-axis and q-axis components of the rotor current, respectively.

At any sampling moment of jT_{st}, in order to obtain the optimal increment of the rotor current, it is only necessary to obtain the increment of the rotor current that minimises the evaluation function $W_{st}(j)$. In this section, the grid voltage orientation is used, i.e. $e_g^q(j) = 0$. Accordingly, substituting Eq. (3.115) into Eq. (3.117) and solving optimally, the optimal increment of the rotor current at the moment of jT_{st} is obtained as

$$\begin{cases} \Delta i_{\mathrm{r}}^{\mathrm{d}*}(j) = \frac{-\omega_{\mathrm{s}} L_{\mathrm{m}} \lambda_{\mathrm{st}}^{\mathrm{q}}}{\omega_{\mathrm{s}}^2 L_{\mathrm{m}}^2 \lambda_{\mathrm{st}}^{\mathrm{q}} + \varepsilon_{\mathrm{st}}^{\mathrm{d}}} \left[\omega_{\mathrm{s}} L_{\mathrm{m}} i_{\mathrm{r}}^{\mathrm{d}}(j) + y_{\mathrm{st}}^{\mathrm{q}}(j)\right] \\ \Delta i_{\mathrm{r}}^{\mathrm{q}*}(j) = \frac{-\omega_{\mathrm{s}} L_{\mathrm{m}} \lambda_{\mathrm{st}}^{\mathrm{d}}}{\omega_{\mathrm{s}}^2 L_{\mathrm{m}}^2 \lambda_{\mathrm{st}}^{\mathrm{d}} + \varepsilon_{\mathrm{st}}^{\mathrm{q}}} \left[e_{\mathrm{g}}^{\mathrm{d}}(j) + \omega_{\mathrm{s}} L_{\mathrm{m}} i_{\mathrm{r}}^{\mathrm{q}}(j) - y_{\mathrm{st}}^{\mathrm{d}}(j)\right] \end{cases} \tag{3.118}$$

where $\Delta i_{\mathrm{r}}^{\mathrm{d}*}(j)$ and $\Delta i_{\mathrm{r}}^{\mathrm{q}*}(j)$ indicate the d-axis and q-axis components of the optimal rotor current increment given by the outer stator voltage loop at the moment of jT_{st}, respectively. $i_{\mathrm{r}}^{\mathrm{d}*}(j)$ and $i_{\mathrm{r}}^{\mathrm{q}*}(j)$ indicate the d-axis and q-axis components of the optimal rotor current obtained by accumulating the optimal increment, respectively. The rotor current reference given by the outer stator voltage loop is considered to be stable during the interval of T_{st} and according to the timing relation between the stator voltage loop and the rotor current loop, the rotor current reference obtained by the rotor current loop at the moments of kT_{st}, $(k+1)T_{\mathrm{st}}$, …, $(k+z_0-1)T_{\mathrm{st}}$ are all $i_{\mathrm{r}}^{\mathrm{d}*}(j)$ and $i_{\mathrm{r}}^{\mathrm{q}*}(j)$, i.e. there is

$$\begin{cases} i_{\mathrm{r}}^{\mathrm{d}*}(k) = i_{\mathrm{r}}^{\mathrm{d}*}(j) \\ i_{\mathrm{r}}^{\mathrm{q}*}(k) = i_{\mathrm{r}}^{\mathrm{d}*}(j) \end{cases} \quad k \in \left[z_0 j, z_0 j + z_0 - 1\right) \tag{3.119}$$

In summary, the control diagrams of the stator voltage outer loop and the whole smooth grid connection algorithm can be illustrated in Fig. 3.33 and 3.34 respectively.

In stator voltage synchronisation control, the voltage synchronisation effect depends on the accuracy of the rotor position detection. Generally, the rotor position of the DFIG is obtained by an incremental photoelectric encoder. The absolute rotor position θ_{r} is obtained by adding the initial position θ_{r0} of the encoder and the position increment $\Delta\theta_{\mathrm{r}}$ of the encoder. In a practical system, the initial position angle θ_{r0} of the photoelectric encoder is not easily obtained due to problems such as shaft misalignment. For this reason, this section provides a method for automatically detecting the initial position angle of the photoelectric encoder. The method first assumes that the position increment of the encoder is the absolute rotor position, inputs 0 and $i_{\mathrm{r\_test}}^{\mathrm{q}*}$ as the d-axis and q-axis components of the rotor current reference, respectively, into the rotor current loop in Fig. 3.34, and opens the loop of its stator voltage, where $i_{\mathrm{r\_test}}^{\mathrm{q}*}$ is a certain current value less than the rated rotor current of the DFIG, and then detects the DFIG stator voltage $u_{\mathrm{s}}^{\mathrm{dq}}$. At this point, the vector relation between the rotor current and the stator voltage in the DFIG is shown in Fig. 3.35.

It is known that the initial rotor position angle at this point can be calculated from the d-axis and q-axis components of the stator voltage as follow.

$$\theta_{\mathrm{r0}} = \begin{cases} \arctan\left(u_{\mathrm{s}}^{\mathrm{q}}//u_{\mathrm{s}}^{\mathrm{d}}\right) & u_{\mathrm{s}}^{\mathrm{d}} > 0 \\ \pi + \arctan\left(u_{\mathrm{s}}^{\mathrm{q}}/u_{\mathrm{s}}^{\mathrm{d}}\right) & u_{\mathrm{s}}^{\mathrm{d}} < 0 \\ -\pi/2 & u_{\mathrm{s}}^{\mathrm{d}} = 0,\, u_{\mathrm{s}}^{\mathrm{q}} < 0 \\ \pi/2 & u_{\mathrm{s}}^{\mathrm{d}} = 0,\, u_{\mathrm{s}}^{\mathrm{q}} > 0 \end{cases} \tag{3.120}$$

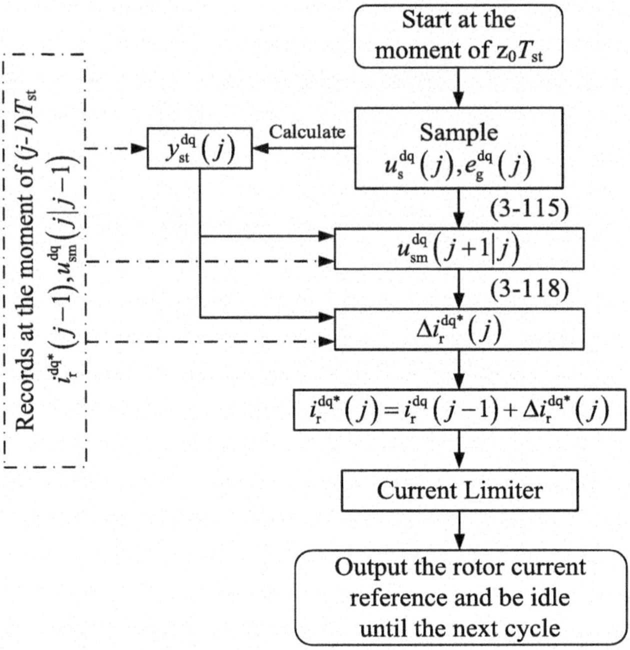

Fig. 3.33 Control diagram of the MPC-based stator voltage loop of the generator-side converter for smooth grid connection

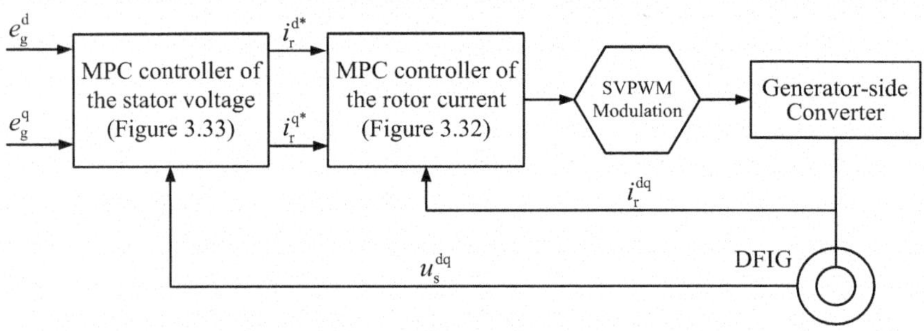

Fig. 3.34 Control diagram of the generator-side converter for smooth grid connection with the double close-loop structure based on MPC

When the detected change in θ_{r0} over a continuous period of time is less than a certain threshold, the detection is considered complete, the detection algorithm no longer works and the detected θ_{r0} value is saved for stator voltage synchronisation and grid-connected operation control.

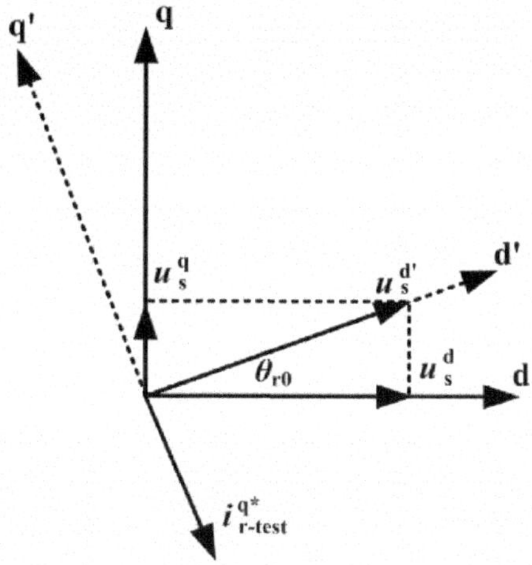

Fig. 3.35 Vector relation between rotor current and stator voltage

References

1. S. Bhowmik, R. Spee, J.H.R. Enslin, Performance optimization for doubly fed wind power generation systems. IEEE Trans. Ind. Appl. **35**(4), 949–958 (1999)
2. Y. Zheng, Research on operating characteristic and control strategy of grid-connected doubly fed induction wind generation system. Ph.D. Thesis, Tsinghua University, Beijing, China, 2009
3. R. Pena, J.C. Clare, G.M. Asher, Doubly fed induction generator using back to back PWM converter and its application to variable speed wind energy generation. IEE Proc. Electrical Power Appl. **143**(3), 231–241 (1996)
4. J. Yaoqin, Y. Zhongqing, G. Binggang, A new maximum power point tracking control scheme for wind generation, in *International Conference on Power System Technology. Proceedings, PowerCon'2002* (2002), pp. 144–148
5. B. Hopfensperger, D.J. Atkinson, R.A. Lakin, Stator flux oriented control of a doubly fed induction machine with and without position encoder. IEE Proc. Electric Power Appl. **147**(4), 241–250 (2000)
6. H. Ju, New integration algorithms for estimating motor flux over a wide speed range, in *Power Electronics Specialists Conference, PESC'97* (1997), pp. 1075–1081
7. M. Kayikc, J.V. Milanovic, Reactive power control strategies for DFIG-based plants. IEEE Trans. Energy Convers. **22**(2), 389–396 (2007)
8. C.X. Wang, Y. Zhang, *Wind Power Generation* (China Electric Power Press, Beijing, 2003)
9. L. Xu, P. Cartwright, Direct active and reactive power control of DFIG for wind energy generation. IEEE Trans. Energy Convers. **21**(3), 750–758 (2006)
10. M. Tazil, V. Kumar, R.C. Bansal et al., Three-phase double fed induction generators: an overview. IET Electric Power Appl. **4**(2), 75–89 (2010)
11. G. Tapia, G. Santamaria, M. Telleria et al., Methodology for smooth connection of doubly fed induction generators to the grid. IEEE Trans. Energy Convers. **24**(4), 959–971 (2009)
12. J.B.G. Manel, A. Jihen, S.B. Ilhem, A novel approach of direct active and reactive power control allowing the connection of the DFIG to the grid, in *Power Electronics and Applications 13th European Conference*, Barcelona, USA (2009)
13. J. Wang, Y.D. Li, PWM rectifier using a time sequencing method in the current control loop. J. Tsinghua Univ. (Sci. Technol.) **49**(10), 1–4 (2009)

Chapter 4
Operational Characteristics of the DFIG-Based Wind Power System Under Non-ideal Grid Conditions

Wind power systems are often built in areas with rich wind resources, which are mostly remote, far from the main grid, with a fragile grid structure and a complex surrounding electricity environment, where grid voltage jitters, dips and asymmetries often occur, while short circuits and grounding faults in the grid will also cause various grid voltage faults [1, 2], thus having a serious impact on the wind power generator and even the whole wind power system [3, 4].

In the DFIG-based wind power system, the stator of the DFIG is directly connected to the grid via a grid connection switch, making the system very sensitive to grid voltage faults. When grid voltage faults occur, intense electromagnetic processes are introduced in the DFIG-based system, which in turn cause a range of problems such as overcurrent, overvoltage, power pulsations, mechanical pulsations and uneven heating, thus reducing the system lifetime and even causing permanent damage to the system hardware.

This chapter will first discuss the classification of grid fault types, then analyse the operational characteristics of the DFIG and the grid-side converter under symmetrical and asymmetrical grid fault conditions, respectively, and on this basis further discuss the impact of grid faults on the operational characteristics of the DFIG-based wind power system, and finally introduce the grid connection specifications for modern wind power systems.

H. Ma et al., *The Control Principle of Wind Power Generation System*,
https://doi.org/10.1007/978-981-99-6083-5_4

4.1 Classification of Grid Faults

The main grid fault faced by wind power systems is the grid voltage dip,[1] including both the symmetrical and the asymmetrical dips.[2]

For the symmetrical grid voltage dip, it can usually be described by two parameters, i.e. the voltage dip depth and the voltage dip duration. The voltage dip depth is the ratio between the voltage amplitude of the faulty grid at the time of the voltage dip and the normal grid voltage amplitude before the voltage dip, and is used to represent the severity of the voltage dip; the voltage dip duration is the time elapsed from the moment when the grid voltage amplitude drops to 90% of its normal value to the moment when the grid voltage amplitude recovers to 90% of its normal value again.

For the asymmetrical grid voltage dip, the voltage dip depth varies from phase to phase and may be accompanied by phase changes during the fault, and even the same type of voltage dip generated at the same fault point may change its type when it is transmitted via different transformer connections in the transmission line to the common coupling point of the wind farm or to the generator machine end. Therefore, the asymmetrical grid voltage dip is even more complex and difficult to express in terms of only two parameters, the voltage dip depth and the voltage dip duration. For this reason, a voltage space vector representation can be used to classify the grid voltage fault. Under the assumption that the positive sequence component, negative sequence component and zero-sequence component of the transmission line impedance are equal, the grid voltage fault is classified into seven types as shown in Table 4.1 according to the system fault type and the transformer winding connection, where $\boldsymbol{E}$ indicates the normal grid phase voltage amplitude before the dip and $\boldsymbol{U}$ indicates the faulty grid phase voltage amplitude after the dip.

This method reflects very intuitively the changes in magnitude and phase during the grid voltage fault. It classifies the grid voltage fault of all kinds into seven fault types in a complete and clear way, which facilitates further analysis and discussion of the grid fault.

In particular, a parameter called asymmetry degree is now also commonly used in the literature to quantify the degree of asymmetry of the asymmetrical grid voltage and is defined as the amplitude ratio of the negative sequence voltage component to the positive sequence voltage component when the asymmetrical voltage fault occurs. This parameter will also be used in subsequent discussions in this book.

---

[1] A voltage dip is a sudden drop in grid voltage amplitude to 10–90% of the rated value and lasts for 0.5 power cycles to a few seconds.

[2] About 60% of the voltage dip comes from lightning strikes caused by insulator flashover and line to earth discharge, and a portion of the voltage dip mainly from transmission line faults, including three-phase short circuits, single-phase grounding, two-phase grounding and phase-to-phase short circuits, where the voltage dip caused by three-phase short circuits is symmetrical, and the voltage dip caused by the other three line faults is asymmetrical, while the latter three occupy a major part of the actual system faults.

Table 4.1 Seven types of the grid voltage fault

Fault type	Voltage space and vector diagram	Voltage space vector	Fault description
A	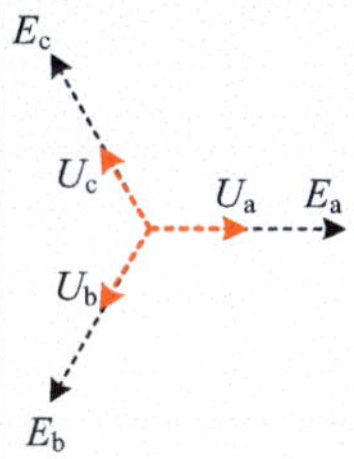	$U_a = U$ $U_b = -\frac{1}{2}U - j\frac{\sqrt{3}}{2}U$ $U_c = -\frac{1}{2}U + j\frac{\sqrt{3}}{2}U$	Three-phase short-circuit fault: symmetrical dip in three-phase voltage
B	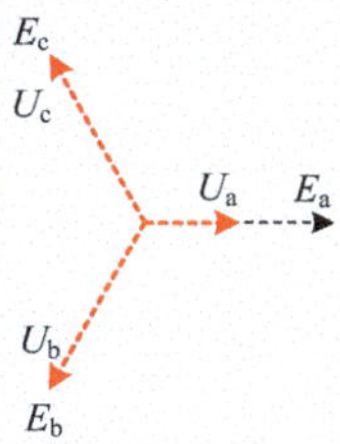	$U_a = U$ $U_b = -\frac{1}{2}E - j\frac{\sqrt{3}}{2}E$ $U_c = -\frac{1}{2}E + j\frac{\sqrt{3}}{2}E$	Single-phase grounding fault: the grounded phase voltage dips, and the other two phase voltages are unchanged
C	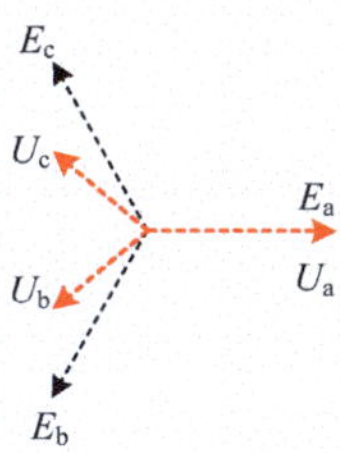	$U_a = E$ $U_b = -\frac{1}{2}E - j\frac{\sqrt{3}}{2}U$ $U_c = -\frac{1}{2}E + j\frac{\sqrt{3}}{2}U$	Inter-phase short-circuit fault: the two short-circuited phase voltages dip in amplitude and jump in phase, the other phase voltage remains unchanged A B-type fault via a DY-type transformer
D	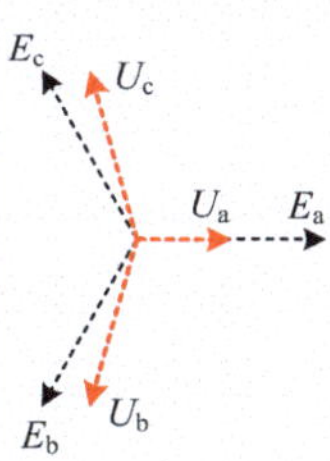	$U_a = U$ $U_b = -\frac{1}{2}U - j\frac{\sqrt{3}}{2}E$ $U_c = -\frac{1}{2}U + j\frac{\sqrt{3}}{2}E$	A C-type fault via a DY-type transformer A B-type fault via two Dy-type transformers

(continued)

Table 4.1 (continued)

Fault type	Voltage space and vector diagram	Voltage space vector	Fault description
E	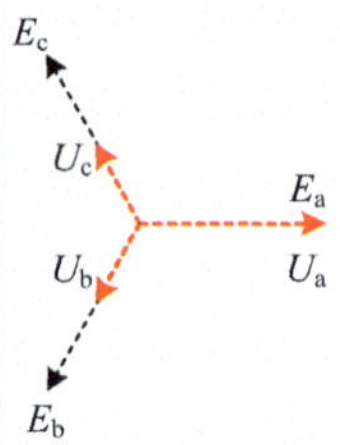	$U_a = E$ $U_b = -\frac{1}{2}E - j\frac{\sqrt{3}}{2}U$ $U_c = -\frac{1}{2}E + j\frac{\sqrt{3}}{2}U$	Two-phase grounding fault: the two grounded phase voltages dip, and the other phase voltage remains unchanged
F	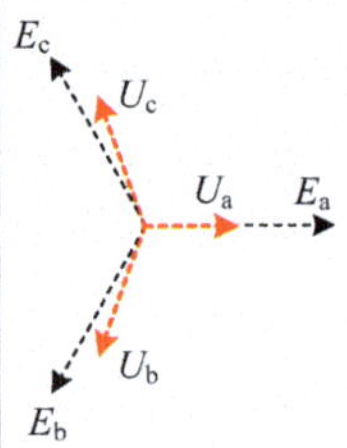	$U_a = U$ $U_b = -\frac{1}{2}U - j\frac{\sqrt{3}}{3}E - j\frac{\sqrt{3}}{6}U$ $U_c = -\frac{1}{2}U + j\frac{\sqrt{3}}{2}E + j\frac{\sqrt{3}}{6}U$	A E-type fault via a DY-type transformer
G	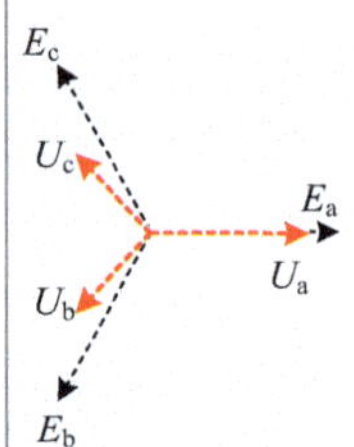	$U_a = \frac{2}{3}E + \frac{1}{3}U$ $U_b = -\frac{1}{3}E - \frac{1}{6}U - j\frac{\sqrt{3}}{2}U$ $U_c = -\frac{1}{3}E - \frac{1}{6}U + j\frac{\sqrt{3}}{2}U$	A E-type fault via two DY-type transformers

4.2 Operational Characteristics of the DFIG Under Grid Faults

4.2.1 Operational Characteristics of the DFIG under the Symmetrical Grid Voltage Dip

In the simultaneous rotation coordinate frame, taking the d-axis and q-axis as the real and imaginary axes respectively, an arbitrary electromagnetic vector can be expressed as

$$X = X^{\mathrm{d}} + jX^{\mathrm{q}} \tag{4.1}$$

where $\boldsymbol{X}$ indicates the arbitrary electromagnetic vector, $\boldsymbol{X}^{\mathrm{d}}$ and $\boldsymbol{X}^{\mathrm{q}}$ indicate the d-axis and q-axis components of the arbitrary vector respectively. In the same expression way, Eqs. (2.19) and (2.20) can be rewritten as

$$\begin{cases} \boldsymbol{u}_{\mathrm{r}} = R_{\mathrm{r}}\boldsymbol{i}_{\mathrm{r}} + p\boldsymbol{\psi}_{\mathrm{r}} + j\omega_{\mathrm{sl}}\boldsymbol{\psi}_{\mathrm{r}} \\ \boldsymbol{u}_{\mathrm{s}} = R_{\mathrm{s}}\boldsymbol{i}_{\mathrm{s}} + p\boldsymbol{\psi}_{\mathrm{s}} + j\omega_{1}\boldsymbol{\psi}_{\mathrm{s}} \end{cases} \tag{4.2}$$

$$\begin{cases} \boldsymbol{\psi}_{\mathrm{r}} = L_{\mathrm{r}}\boldsymbol{i}_{\mathrm{r}} + L_{\mathrm{M}}\boldsymbol{i}_{\mathrm{s}} \\ \boldsymbol{\psi}_{\mathrm{s}} = L_{\mathrm{M}}\boldsymbol{i}_{\mathrm{r}} + L_{\mathrm{s}}\boldsymbol{i}_{\mathrm{s}} \end{cases} \tag{4.3}$$

In the process of a three-phase symmetrical voltage dip fault in the grid, the dip and recovery are considered to be in the form of a step [5], then the DFIG stator voltage can be expressed as

$$\boldsymbol{u}_{\mathrm{s}} = \begin{cases} \boldsymbol{u}_{\mathrm{norm1}} = u_{\mathrm{norm}} e^{j\omega_1 t + \phi_{\mathrm{norm}}} & t < t_{\mathrm{sag}} \\ \boldsymbol{u}_{\mathrm{sag}} = u_{\mathrm{sag}} e^{j\omega_1 t + \phi_{\mathrm{sag}}} & t_{\mathrm{sag}} \le t < t_{\mathrm{rec}} \\ \boldsymbol{u}_{\mathrm{norm2}} = u_{\mathrm{norm}} e^{j\omega_1 t + \phi_{\mathrm{rec}}} & t \ge t_{\mathrm{rec}} \end{cases} \tag{4.4}$$

where $\boldsymbol{u}_{\mathrm{norm1}}$, u_{norm} and ϕ_{norm1} indicate the vector, amplitude and initial phase of the normal grid voltage before the grid dip fault, respectively, $\boldsymbol{u}_{\mathrm{sag}}$, u_{sag} and ϕ_{sag} indicate the vector, amplitude and initial phase of the faulty grid voltage after the grid dip fault, respectively, $\boldsymbol{u}_{\mathrm{norm2}}$ and ϕ_{rec} indicate the vector and initial phase of the recovered normal grid voltage, respectively, t_{sag} and t_{rec} indicate the moments when the grid voltage dips and recovers, respectively.

For the convenience of discussion, the stator voltage is considered to be determined by the grid voltage and, due to the slow regulation of the generator speed, the generator speed can be considered to remain constant during the grid fault. The operational characteristics of the DFIG under symmetrical the grid voltage dip are not only governed by the grid condition shown in Eq. (4.4), but also by the control voltage applied by the generator-side converter. The operational characteristics of the DFIG under the symmetrical grid voltage dip are analysed below according to the different control cases of the generator-side converter, respectively.

4.2.1.1 Fault Operation Characteristics of the DFIG with a Constant Rotor Current Vector

Substituting the stator voltage formulas in Eq. (4.2) into the stator flux linkage formulas in Eq. (4.3), it gives

$$p\boldsymbol{\psi}_{\mathrm{s}} = \boldsymbol{u}_{s} - \left(\frac{R_{\mathrm{s}}}{L_{\mathrm{s}}} + j\omega_{1}\right)\boldsymbol{\psi}_{\mathrm{s}} + \frac{R_{\mathrm{s}}L_{\mathrm{M}}}{L_{\mathrm{s}}}\boldsymbol{i}_{\mathrm{r}} \tag{4.5}$$

It can be seen that the stator flux linkage will be constrained by both $\boldsymbol{u}_s$ and $\boldsymbol{i}_s$. To obtain an analytical expression for the stator flux linkage, it is assumed that the generator-side converter has sufficient bandwidth and control capability to maintain the rotor current vector constant throughout the whole low-voltage process [6]. Then Eq. (4.5) is a first order differential equation for the stator flux linkage which, when the stator voltage is described by Eq. (4.4), can be expressed as a superposition of the forced component $\boldsymbol{\psi}_{\mathrm{s\_force}}$ and the natural component $\boldsymbol{\psi}_{\mathrm{s\_natural}}$ as

$$\boldsymbol{\psi}_{\mathrm{s}} = \boldsymbol{\psi}_{\mathrm{s\_force}} + \boldsymbol{\psi}_{\mathrm{s\_natural}} \tag{4.6}$$

where the expressions of $\boldsymbol{\psi}_{\mathrm{s\_force}}$ and $\boldsymbol{\psi}_{\mathrm{s\_natural}}$ for the grid before, during and after recovery from the fault are as follows

(i) when $t < t_{\mathrm{sag}}$, $\boldsymbol{\psi}_{\mathrm{s\_force}}$ is the steady-state value of the stator flux linkage under the normal grid voltage and $\boldsymbol{\psi}_{\mathrm{s\_natural}}$ is 0, i.e.

$$\begin{cases} \boldsymbol{\psi}_{\mathrm{s\_force}} = \dfrac{L_{\mathrm{s}}\boldsymbol{u}_{\mathrm{norm}} + R_{\mathrm{s}}L_{\mathrm{M}}\boldsymbol{i}_{\mathrm{r}}}{R_{\mathrm{s}} + j\omega_{1}L_{\mathrm{s}}} \\ \boldsymbol{\psi}_{\mathrm{s\_natural}} = 0 \end{cases} \tag{4.7}$$

(ii) when $t_{\mathrm{sag}} \leq t < t_{\mathrm{rec}}$, $\boldsymbol{\psi}_{\mathrm{s\_force}}$ is the steady-state value of the stator flux linkage under the faulty grid voltage and $\boldsymbol{\psi}_{\mathrm{s\_natural}}$ is determined by both the instantaneous value of the stator flux linkage immediately before the fault and the steady-state value of the stator flux linkage after the fault, i.e.

$$\begin{cases} \boldsymbol{\psi}_{\mathrm{s\_force}} = \dfrac{L_{\mathrm{s}}\boldsymbol{u}_{\mathrm{sag}} + R_{\mathrm{s}}L_{\mathrm{M}}\boldsymbol{i}_{\mathrm{r}}}{R_{\mathrm{s}} + j\omega_{1}L_{\mathrm{s}}} \\ \boldsymbol{\psi}_{\mathrm{s\_natural}} = \dfrac{L_{\mathrm{s}}\left(\boldsymbol{u}_{\mathrm{norm}} - \boldsymbol{u}_{\mathrm{sag}}\right)}{R_{\mathrm{s}} + j\omega_{1}L_{\mathrm{s}}} e^{-j\omega_{1}\left(t - t_{\mathrm{sag}}\right)} e^{-\left(t - t_{\mathrm{sag}}\right)R_{\mathrm{s}}/L_{\mathrm{s}}} \end{cases} \tag{4.8}$$

(iii) when $t \geq t_{\mathrm{rec}}$, $\boldsymbol{\psi}_{\mathrm{s\_force}}$ is the steady-state value of the stator flux linkage under the normal grid voltage and $\boldsymbol{\psi}_{\mathrm{s\_natural}}$ is determined by both the instantaneous and steady-state values of the stator flux linkage at the end of $t_{sag} \leq t < t_{rec}$, i.e.

$$\begin{cases} \boldsymbol{\psi}_{s\_force} = \dfrac{L_s \boldsymbol{u}_{norm} + R_s L_M \boldsymbol{i}_r}{R_s + j\omega_1 L_s} \\ \boldsymbol{\psi}_{s\_natural} = \dfrac{L_s(\boldsymbol{u}_{sag} - \boldsymbol{u}_{norm})\left[1 - e^{-j\omega_1(t_{rec}-t_{sag})} e^{\frac{-(t_{rec}-t_{sag})R_s}{L_s}}\right]}{R_s + j\omega_1 L_s} e^{-j\omega_1(t-t_{rec})} e^{\frac{-(t-t_{rec})R_s}{L_s}} \end{cases} \tag{4.9}$$

As can be seen from Eqs. (4.6) to (4.9), during the symmetrical dip and recovery of the grid voltage, there will be a forced component in the DFIG stator flux linkage that remains constant in the synchronous coordinate frame, which depends only on the grid voltage vector and has no relation to the dip or recovery time; at the same time, there will be a natural component that pulsates at a synchronous frequency and gradually decays in the synchronous coordinate frame, which is not only related to the grid voltage vector, but also to the dip and recovery time.

Figure 4.1 illustrates the trajectory of the stator flux linkage vector in vector space during the symmetrical grid fault. When the grid voltage is normal, the stator flux linkage trajectory is located at Point a (the effect of the stator resistance is ignored here); when the symmetrical voltage dip occurs, the stator flux linkage trajectory will rotate clockwise at a synchronous angular speed of ω_1 along Circle 1 centred on Point b, where Point b is the location of the forced component of the stator flux linkage after the voltage dip, and the vector pointing from Point b to the current location of the stator flux linkage on Circle 1 is the natural component of the stator flux linkage, and consequently the radius of Circle 1 will decay with a time constant of L_s/R_s until it reaches 0, whereupon the stator flux linkage vector will again stabilise at Point b; When the grid voltage is recovered, the stator flux linkage vector starts to rotate clockwise at a synchronous angular speed of ω_1 from its current position along a circular trajectory centred on a, shown as Circle 2 or Circle 3, and the radius of the circular trajectory also decays with a time constant of L_s/R_s until the flux linkage vector returns to Point a again, at which point the transient process ends and the system returns to its normal operating state. Obviously, the farther the flux linkage trajectory is from Point a, the greater the amplitude of the natural component is introduced after the recovery, resulting in a more intense transient process, e.g. the flux linkage process is significantly more intense when the grid is recovered at Point f (illustrated as Circle 3) than when the grid is recovered at Point c or e (illustrated as Circle 2).

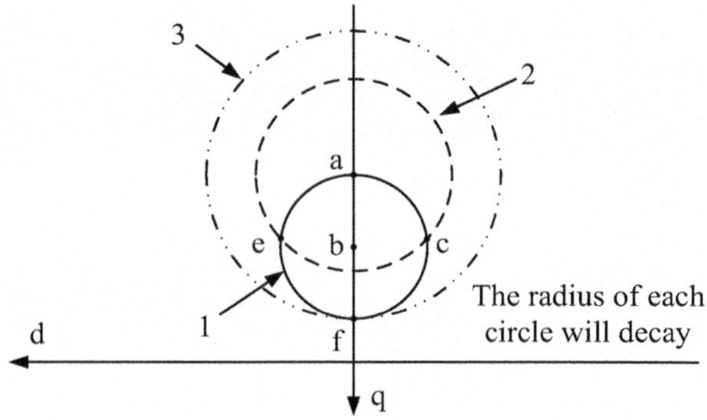

Fig. 4.1 Trajectories of the stator flux linkage in vector space during the symmetrical grid fault (with constant rotor current)

Substituting Eqs. (4.6) to (4.9) into Eq. (4.3), the stator current can be found as [6]

$$\boldsymbol{i}_s = \boldsymbol{i}_{s\_force} + \boldsymbol{i}_{s\_natural} \tag{4.10}$$

where

$$t < t_{sag}: \begin{cases} \boldsymbol{i}_{s\_force} = \dfrac{\boldsymbol{u}_{norm} - j\omega_1 L_M \boldsymbol{i}_r}{R_s + j\omega_1 L_s} \\ \boldsymbol{i}_{s\_natural} = 0 \end{cases} \tag{4.11}$$

$$t_{sag} \le t < t_{rec}: \begin{cases} \boldsymbol{i}_{s\_force} = \dfrac{\boldsymbol{u}_{sag} - j\omega_1 L_M \boldsymbol{i}_r}{R_s + j\omega_1 L_s} \\ \boldsymbol{i}_{s\_natural} = \dfrac{\boldsymbol{u}_{norm} - \boldsymbol{u}_{sag}}{R_s + j\omega_1 L_s} e^{-j\omega_1 (t - t_{sag})} e^{-(t - t_{sag}) R_s / L_s} \end{cases} \tag{4.12}$$

$$t \ge t_{rec}: \begin{cases} \boldsymbol{i}_{s\_force} = \dfrac{\boldsymbol{u}_{norm} - j\omega_1 L_M \boldsymbol{i}_r}{R_s + j\omega_1 L_s} \\ \boldsymbol{i}_{s\_natural} = \dfrac{(\boldsymbol{u}_{sag} - \boldsymbol{u}_{norm}) \left[1 - e^{-j\omega_1 (t_{rec} - t_{sag})} e^{\frac{-(t_{rec} - t_{sag}) R_s}{L_s}} \right] e^{-j\omega_1 (t - t_{rec})} e^{\frac{-(t - t_{rec}) R_s}{L_s}}}{R_s + j\omega_1 L_s} \end{cases} \tag{4.13}$$

From Eqs. (4.10) to (4.13), it can be seen that during the symmetrical dip and recovery of the grid voltage, the dynamic process of the DFIG stator flux linkage will trigger a similar dynamic process of the stator current in the generator, so that there will also be a forced component in the stator current that remains constant in the synchronous coordinate frame and a natural component that pulsates at a synchronous frequency and gradually decays in the synchronous coordinate frame, where the natural component of the stator current will lead to a surge and pulsation of the stator current.

From the above analysis, it can be seen that if the rotor current vector is controlled to be constant, during the symmetrical grid fault, the stator flux linkage will pulsate at the synchronous frequency of the grid and the pulsation will gradually decay until the stator flux reaches its steady state; the pulsation of the stator flux linkage will cause the stator current to pulsate at the same frequency, causing a certain impact on the generator stator, and through the electromagnetic coupling between the stator and rotor, it will also cause the rotor side voltage to rise with a possible overvoltage fault in the generator-side converter.

4.2.1.2 Fault Operation Characteristics of the DFIG with a Constant Sotor Current Vector

From the stator voltage formulas in Eq. (4.2), it gives

$$p\boldsymbol{\psi}_{\mathrm{s}} = j\omega_1\boldsymbol{\psi}_{\mathrm{s}} - \boldsymbol{u}_s + R_{\mathrm{s}}\boldsymbol{i}_{\mathrm{s}} \tag{4.14}$$

It can be seen that the stator flux linkage will be constrained by both u_s and i_s. To obtain an analytical expression for the stator flux linkage, it is assumed that the generator-side converter has sufficient bandwidth and control capability to maintain the stator current vector constant throughout the whole low-voltage process. Then Eq. (4.14) is a first order differential equation for the stator flux linkage which, when the stator voltage is described by Eq. (4.4), can be expressed as a superposition of the forced component $\boldsymbol{\psi}_{\mathrm{s\_force}}$ and the natural component $\boldsymbol{\psi}_{\mathrm{s\_natural}}$ as

$$\boldsymbol{\psi}_{\mathrm{s}} = \boldsymbol{\psi}_{\mathrm{s\_force}} + \boldsymbol{\psi}_{\mathrm{s\_natural}} \tag{4.15}$$

where the expressions of $\boldsymbol{\psi}_{\mathrm{s\_force}}$ and $\boldsymbol{\psi}_{\mathrm{s\_natural}}$ for the grid before, during and after recovery from the fault are as follows

(i) when $t < t_{\mathrm{sag}}$, $\boldsymbol{\psi}_{\mathrm{s\_force}}$ is the steady-state value of the stator flux linkage under the normal grid voltage and $\boldsymbol{\psi}_{\mathrm{s\_natural}}$ is 0, i.e.

$$\begin{cases} \boldsymbol{\psi}_{\mathrm{s\_force}} = \dfrac{\boldsymbol{u}_{\mathrm{norm}} - R_{\mathrm{s}}\boldsymbol{i}_{\mathrm{s}}}{j\omega_1} \\ \boldsymbol{\psi}_{\mathrm{s\_natural}} = 0 \end{cases} \tag{4.16}$$

(ii) when $t_{\mathrm{sag}} \le t < t_{\mathrm{rec}}$, $\boldsymbol{\psi}_{\mathrm{s\_force}}$ is the steady-state value of the stator flux linkage under the faulty grid voltage and $\boldsymbol{\psi}_{\mathrm{s\_natural}}$ is determined by both the instantaneous value of the stator flux linkage immediately before the fault and the steady-state value of the stator flux linkage after the fault, i.e.

$$\begin{cases} \boldsymbol{\psi}_{\mathrm{s\_force}} = \dfrac{\boldsymbol{u}_{\mathrm{sag}} - R_{\mathrm{s}}\boldsymbol{i}_{\mathrm{s}}}{j\omega_1} \\ \boldsymbol{\psi}_{\mathrm{s\_natural}} = \dfrac{\boldsymbol{u}_{\mathrm{norm}} - \boldsymbol{u}_{\mathrm{sag}}}{j\omega_1} e^{-j\omega_1(t-t_{\mathrm{sag}})} \end{cases} \tag{4.17}$$

(iii) when $t \geq t_{rec}$, $\boldsymbol{\psi}_{s\_force}$ is the steady-state value of the stator flux linkage under the normal grid voltage and $\boldsymbol{\psi}_{s\_natural}$ is determined by both the instantaneous and steady-state values of the stator flux linkage at the end of $t_{sag} \leq t < t_{rec}$, i.e.

$$\begin{cases} \boldsymbol{\psi}_{s\_force} = \dfrac{\boldsymbol{u}_{norm} - R_s \boldsymbol{i}_s}{j\omega_1} \\ \boldsymbol{\psi}_{s\_natural} = \dfrac{\left(\boldsymbol{u}_{sag} - \boldsymbol{u}_{norm}\right)\left[1 - e^{-j\omega_1 (t_{rec} - t_{sag})}\right]}{j\omega_1} e^{-j\omega_1 (t - t_{rec})} \end{cases} \tag{4.18}$$

As can be seen from Eqs. (4.15) to (4.18), during the symmetrical dip and recovery of the grid voltage, there will be a forced component in the DFIG stator flux linkage that remains constant in the synchronous coordinate frame, which depends only on the grid voltage vector and has no relation to the dip or recovery time; at the same time, there will be a natural component that pulsates at a synchronous frequency in the synchronous coordinate frame, which is not only related to the grid voltage vector, but also to the dip and recovery time. It is worth noting that, unlike the case in Sect. 4.2.1.1, here the natural component of the stator flux linkage will no longer decay, but will continue to pulsate with a constant amplitude.

Figure 4.2 illustrates the trajectory of the stator flux linkage vector in vector space during the symmetrical grid fault. In Fig. 4.2, the vector space trajectory of the stator flux linkage is basically the same as that depicted in Fig. 4.1, except that the radii of Circles 1, 2, and 3 therein will remain constant and do not decay, i.e., once the voltage dips, the natural component of the stator flux linkage will persist, and the system will not be able to return to the initial normal operating state point (Point a) even after the fault is cleared. The above results indicate that during symmetric grid fault ride-through, controlling the stator current vector forcefully to be constant will be detrimental to the recovery of the generator flux linkage. At the same time, the intensity of the flux linkage transients triggered by the recovery process is still related to the moment of fault recovery, i.e., the recovery process at Point f in Fig. 4.2 is more intense than that at Point c or Point e.

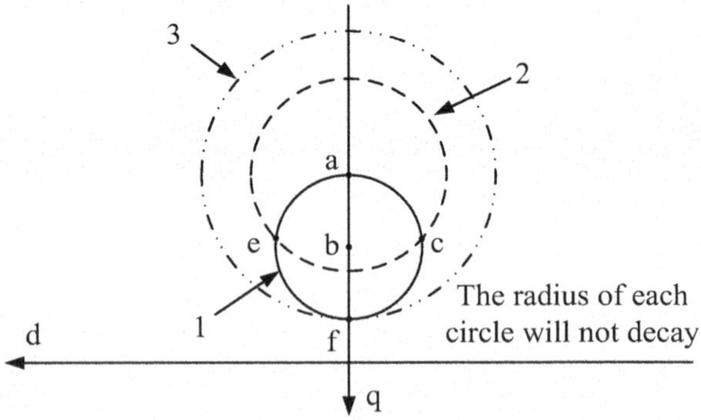

Fig. 4.2 Trajectories of the stator flux linkage in vector space during the symmetrical grid fault (with constant stator current)

Substituting Eqs. (4.15) to (4.18) into Eq. (4.3), the resulting rotor current can be found as

$$\boldsymbol{i}_{\mathrm{r}} = \boldsymbol{i}_{\mathrm{r\_force}} + \boldsymbol{i}_{\mathrm{r\_natural}} \tag{4.19}$$

where

$$t < t_{\mathrm{sag}}: \begin{cases} \boldsymbol{i}_{\mathrm{r\_force}} = \dfrac{\boldsymbol{u}_{\mathrm{norm}} - (R_{\mathrm{s}} + j\omega_1 L_{\mathrm{s}})\boldsymbol{i}_{\mathrm{s}}}{j\omega_1 L_{\mathrm{M}}} \\ \boldsymbol{i}_{\mathrm{r\_natural}} = 0 \end{cases} \tag{4.20}$$

$$t_{\mathrm{sag}} \le t < t_{\mathrm{rec}}: \begin{cases} \boldsymbol{i}_{\mathrm{r\_force}} = \dfrac{\boldsymbol{u}_{\mathrm{sag}} - (R_{\mathrm{s}} + j\omega_1 L_{\mathrm{s}})\boldsymbol{i}_{\mathrm{s}}}{j\omega_1 L_{\mathrm{M}}} \\ \boldsymbol{i}_{\mathrm{r\_natural}} = \dfrac{\boldsymbol{u}_{\mathrm{norm}} - \boldsymbol{u}_{\mathrm{sag}}}{j\omega_1 L_{\mathrm{M}}} e^{-j\omega_1 (t - t_{\mathrm{sag}})} \end{cases} \tag{4.21}$$

$$t \ge t_{\mathrm{rec}}: \begin{cases} \boldsymbol{i}_{\mathrm{r\_force}} = \dfrac{\boldsymbol{u}_{\mathrm{norm}} - (R_{\mathrm{s}} + j\omega_{\mathrm{s}} L_{\mathrm{s}})\boldsymbol{i}_{\mathrm{s}}}{j\omega_{\mathrm{s}} L_{\mathrm{M}}} \\ \boldsymbol{i}_{\mathrm{r\_natural}} = \dfrac{(\boldsymbol{u}_{\mathrm{norm}} - \boldsymbol{u}_{\mathrm{sag}})\left[1 - e^{-j\omega_{\mathrm{s}}(t_{\mathrm{rec}} - t_{\mathrm{sag}})}\right]}{j\omega_{\mathrm{s}} L_{\mathrm{M}}} e^{-j\omega_{\mathrm{s}}(t - t_{\mathrm{rec}})} \end{cases} \tag{4.22}$$

From Eqs. (4.19) to (4.22), it can be seen that during the symmetrical dip and recovery of the grid voltage, the dynamic process of the DFIG stator flux linkage will trigger a similar dynamic process of the rotor current in the generator, so that there will also be a forced component in the rotor current that remains constant in the synchronous coordinate frame and a natural component that pulsates at a synchronous frequency and doesn't decay in the synchronous coordinate frame, where the natural component of the rotor current will lead to a surge and pulsation of the rotor current.

From the above analysis, it can be seen that if the stator current vector is controlled to be constant, during the symmetrical grid fault, the stator flux linkage will pulsate at the grid synchronous frequency, and the pulsation will not decay, even after the grid fault is cleared, which will exist all the time, making the system unable to return to the normal working condition before the fault. The pulsation of the stator flux linkage will cause the rotor current to pulsate at the same frequency, which will easily trigger the rotor overcurrent fault and also increase the control voltage that the generator-side converter needs to output, thus reducing the control range of the generator-side converter, which is not good for the system control.

4.2.1.3 Faulty Operational Characteristics of the DFIG in Other Conditions

Constant rotor current vector and constant stator current vector are the control objectives that many DFIG control algorithms are presently concerned with, therefore

Sects. 4.2.1.1 and 4.2.1.2 analyse the operational characteristics of the DFIG during the symmetrical grid fault based on these two assumptions, respectively. Although not all the DFIG-based wind power systems fully satisfy the above two assumptions during the grid fault, an analytical description of the DFIG characteristics during the grid fault can be obtained based on these two assumptions, which is beneficial for understanding the transient behaviour of the generator during the grid fault.

The operational characteristics of the DFIG under the symmetrical grid fault are further complicated when the operating conditions of the DFIG do not satisfy the assumptions in Sects. 4.2.1.1 and 4.2.1.2. When the stator voltage changes abruptly, the stator flux linkage is not able to change abruptly due to the integration effect and a natural component $\boldsymbol{\psi}_{\mathrm{s\_natural}}$ will be generated which will trigger a series of violent electromagnetic processes in the generator. In the absence of effective control of both stator and rotor currents, the natural component of the stator flux linkage will produce pulsating components simultaneously in the stator and rotor currents at synchronous grid frequency, resulting in overcurrent on both the stator and rotor sides; through the electromagnetic coupling between the stator and rotor, a natural component $\boldsymbol{\psi}_{\mathrm{r\_natural}}$ in rotor flux linkage will also be produced corresponding to the stator flux linkage, which is expressed as a DC component in the abc stationary coordinate frame and, due to the high speed of the DFIG in normal operation, will be cut by the rotor at a high relative speed, resulting in rotor overvoltage.

From the above analysis, it is clear that in order to achieve uninterrupted operation of the DFIG-based wind power system under the symmetrical grid fault, corresponding measures need to be taken to control and protect the system.

4.2.2 Operational Characteristics of the DFIG Under the Asymmetrical Grid Voltage Dip

4.2.2.1 Symmetrical Component Method

The symmetrical component method is an effective method for studying the characteristics of asymmetrical grids, stating that any asymmetrical three-phase sinusoidal power system can be represented as a superposition of three symmetrical systems—a positive sequence component system, a negative sequence component system and a zero-sequence component system.

As the DFIG-based wind power system is mostly connected to the grid via a star/delta transformer, the zero-sequence component is usually not taken into account, at which point a visual representation of the symmetrical component method is shown in Fig. 4.3.

Where $\boldsymbol{F}$ is an arbitrary asymmetrical electromagnetic vector in the αβ coordinate frame, which can be equated to a vector synthesis of a positive sequence component $\boldsymbol{F}_+$ rotating positively at angular speed ω_1 and a negative sequence component $\boldsymbol{F}_-$ rotating negatively at angular speed ω_1. The positive sequence component $\boldsymbol{F}_+$ is

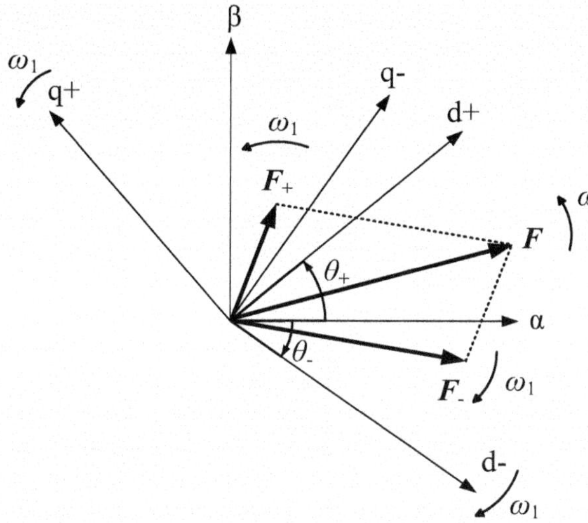

Fig. 4.3 Symmetrical component representation of an arbitrary electromagnetic vector

represented as a constant vector in the positive synchronous rotation coordinate frame dq+, while the negative sequence component $\boldsymbol{F}_-$ is also represented as a constant vector in the negative synchronous rotation coordinate frame dq−, denoted as $\boldsymbol{F}_+^{\mathrm{dq}+}$ and $\boldsymbol{F}_-^{\mathrm{dq}-}$ respectively, where the superscript + and − indicate that the current vector is in the positive and negative synchronous coordinate frames respectively, and the subscript + and − indicate the positive and negative sequence components respectively, and the following similar indications have the same meaning.

To facilitate the derivation, it is worth taking the angle between the dq+ and αβ coordinate frames as $\theta_+ = \omega_1 t$, the angle between the dq− and αβ coordinate frames as $\theta_- = \omega_1 t$, then the angle between the dq+ and dq− coordinate frames as $\theta = \theta_+ + \theta_- = 2\omega_1 t$. From this, the conversion relation between the electromagnetic vectors in the αβ, dq+ and dq− coordinate frames is

$$\begin{cases} \boldsymbol{F}^{\alpha\beta} = \boldsymbol{F}^{\mathrm{dq}+} e^{j\omega_1 t}, \boldsymbol{F}^{\mathrm{dq}+} = \boldsymbol{F}^{\alpha\beta} e^{-j\omega_1 t} \\ \boldsymbol{F}^{\alpha\beta} = \boldsymbol{F}^{\mathrm{dq}-} e^{-j\omega_1 t}, \boldsymbol{F}^{\mathrm{dq}-} = \boldsymbol{F}^{\alpha\beta} e^{j\omega_1 t} \\ \boldsymbol{F}^{\mathrm{dq}+} = \boldsymbol{F}^{\mathrm{dq}-} e^{-j2\omega_1 t}, \boldsymbol{F}^{\mathrm{dq}-} = \boldsymbol{F}^{\mathrm{dq}+} e^{j2\omega_1 t} \end{cases} \quad (4.23)$$

Then an arbitrary electromagnetic vector in a positive synchronous coordinate frame can be expressed as

$$\boldsymbol{F}^{\mathrm{dq}+} = \boldsymbol{F}_+^{\mathrm{dq}+} + \boldsymbol{F}_-^{\mathrm{dq}+} = \boldsymbol{F}_+^{\mathrm{dq}+} + \boldsymbol{F}_-^{\mathrm{dq}-} e^{-j2\omega_1 t} \quad (4.24)$$

4.2.2.2 Faulty Operational Characteristics of the DFIG with the Asymmetrical Grid

The following analysis on the operational characteristics of the DFIG under the asymmetrical grid fault scenario is carried out using the symmetric component method.

When the grid voltage has an asymmetrical fault, in the positive synchronous coordinate frame, rewrite Eqs. (2.19), (2.20), (2.21) and (2.22) to obtain the mathematical model of the DFIG as

$$\begin{cases} u_{\mathrm{s}}^{\mathrm{d}+} = R_{\mathrm{s}} i_{\mathrm{s}}^{\mathrm{d}+} + p\psi_{\mathrm{s}}^{\mathrm{d}+} - \omega_1 \psi_{\mathrm{s}}^{\mathrm{q}+} \\ u_{\mathrm{s}}^{\mathrm{q}+} = R_{\mathrm{s}} i_{\mathrm{s}}^{\mathrm{q}+} + p\psi_{\mathrm{s}}^{\mathrm{q}+} + \omega_1 \psi_{\mathrm{s}}^{\mathrm{d}+} \\ u_{\mathrm{r}}^{\mathrm{d}+} = R_{\mathrm{r}} i_{\mathrm{r}}^{\mathrm{d}+} + p\psi_{\mathrm{r}}^{\mathrm{d}+} - \omega_{\mathrm{sl}} \psi_{\mathrm{r}}^{\mathrm{q}+} \\ u_{\mathrm{r}}^{\mathrm{q}+} = R_{\mathrm{r}} i_{\mathrm{r}}^{\mathrm{q}+} + p\psi_{\mathrm{r}}^{\mathrm{q}+} + \omega_{\mathrm{sl}} \psi_{\mathrm{r}}^{\mathrm{d}+} \end{cases} \tag{4.25}$$

$$\begin{cases} \psi_{\mathrm{s}}^{\mathrm{d}+} = L_{\mathrm{s}} i_{\mathrm{s}}^{\mathrm{d}+} + L_{\mathrm{M}} i_{\mathrm{r}}^{\mathrm{d}+} \\ \psi_{\mathrm{s}}^{\mathrm{q}+} = L_{\mathrm{s}} i_{\mathrm{s}}^{\mathrm{q}+} + L_{\mathrm{M}} i_{\mathrm{r}}^{\mathrm{q}+} \\ \psi_{\mathrm{r}}^{\mathrm{d}+} = L_{\mathrm{r}} i_{\mathrm{r}}^{\mathrm{d}+} + L_{\mathrm{M}} i_{\mathrm{s}}^{\mathrm{d}+} \\ \psi_{\mathrm{r}}^{\mathrm{q}+} = L_{\mathrm{r}} i_{\mathrm{r}}^{\mathrm{q}+} + L_{\mathrm{M}} i_{\mathrm{s}}^{\mathrm{q}+} \end{cases} \tag{4.26}$$

$$T_{\mathrm{e}} = n_{\mathrm{p}} L_{\mathrm{M}} \left(i_{\mathrm{s}}^{\mathrm{q}+} i_{\mathrm{r}}^{\mathrm{d}+} - i_{\mathrm{s}}^{\mathrm{d}+} i_{\mathrm{r}}^{\mathrm{q}+} \right) \tag{4.27}$$

$$\begin{cases} P_{\mathrm{s}} = u_{\mathrm{s}}^{\mathrm{d}+} i_{\mathrm{s}}^{\mathrm{d}+} + u_{s}^{\mathrm{q}+} i_{s}^{\mathrm{q}+} \\ Q_{\mathrm{s}} = u_{s}^{\mathrm{q}+} i_{\mathrm{s}}^{\mathrm{d}+} - u_{\mathrm{s}}^{\mathrm{d}+} i_{\mathrm{s}}^{\mathrm{q}+} \end{cases} \tag{4.28}$$

Equations (4.25) to (4.28) are the mathematical models of the DFIG under asymmetrical grid conditions. When an asymmetrical grid voltage fault occurs, the negative sequence component of the grid voltage will excite the negative sequence component in the stator flux linkage, according to Eq. (4.24), the stator voltage and the stator flux linkage can be expressed as

$$\begin{cases} u_{\mathrm{s}}^{\mathrm{d}+} = u_{\mathrm{s}+}^{\mathrm{d}+} + u_{\mathrm{s}-}^{\mathrm{d}-} \cos(2\omega_1 t) + u_{\mathrm{s}-}^{\mathrm{d}-} \sin(2\omega_1 t) \\ u_{\mathrm{s}}^{\mathrm{q}+} = u_{\mathrm{s}+}^{\mathrm{q}+} + u_{\mathrm{s}-}^{\mathrm{q}-} \cos(2\omega_1 t) + u_{\mathrm{s}-}^{\mathrm{q}-} \sin(2\omega_1 t) \end{cases} \tag{4.29}$$

$$\begin{cases} \psi_{\mathrm{s}}^{\mathrm{d}+} = \psi_{\mathrm{s}+}^{\mathrm{d}+} + \psi_{\mathrm{s}-}^{\mathrm{d}-} \cos(2\omega_1 t) + \psi_{\mathrm{s}-}^{\mathrm{d}-} \sin(2\omega_1 t) \\ \psi_{\mathrm{s}}^{\mathrm{q}+} = \psi_{\mathrm{s}+}^{\mathrm{q}+} + \psi_{\mathrm{s}-}^{\mathrm{q}-} \cos(2\omega_1 t) + \psi_{\mathrm{s}-}^{\mathrm{q}-} \sin(2\omega_1 t) \end{cases} \tag{4.30}$$

Figure 4.4 gives the trajectories of the grid voltage and stator flux linkage in vector space under several typical asymmetrical grid faults. As can be seen from the figure, the spatial trajectories of the grid voltage and stator flux linkage under different grid faults are different, but they all change from circular under the normal grid condition to elliptical under the asymmetrical fault, which is caused by the presence of the

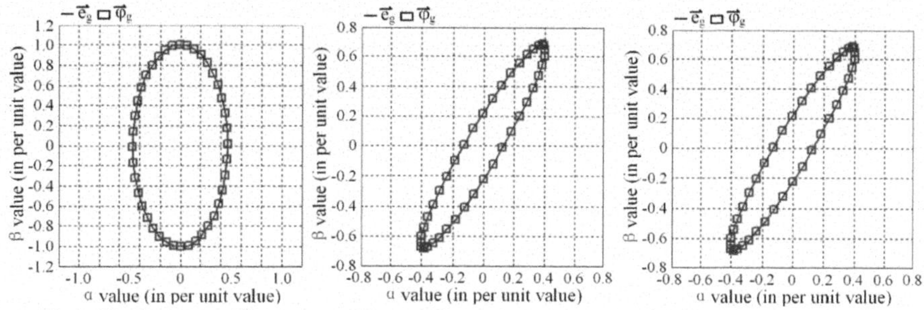

(a) single phase grounding (b) two phase grounding (c) interphase short circuit

Fig. 4.4 Trajectories of grid voltage and stator flux linkage under typical asymmetrical grid faults **a** single phase grounding, **b** two phase grounding, **c** interphase short circuit

negative sequence components, and this distortion of the voltage and flux linkage makes the operating state of the system different from the normal state.

The asymmetrical grid voltage and the stator flux linkage cause a negative sequence component in the stator current and, through the electromagnetic coupling between the stator and rotor, cause a negative sequence component in the rotor current as well, thus, the stator and rotor currents can be expressed as

$$\begin{cases} i_{\mathrm{s}}^{\mathrm{d+}} = i_{\mathrm{s+}}^{\mathrm{d+}} + i_{\mathrm{s-}}^{\mathrm{d-}}\cos(2\omega_1 t) + i_{\mathrm{s-}}^{\mathrm{d-}}\sin(2\omega_1 t) \\ i_{\mathrm{s}}^{\mathrm{q+}} = i_{\mathrm{s+}}^{\mathrm{q+}} + i_{\mathrm{s-}}^{\mathrm{q-}}\cos(2\omega_1 t) + i_{\mathrm{s-}}^{\mathrm{q-}}\sin(2\omega_1 t) \end{cases} \tag{4.31}$$

$$\begin{cases} i_{\mathrm{r}}^{\mathrm{d+}} = i_{\mathrm{r+}}^{\mathrm{d+d+}} + i_{\mathrm{r-}}^{\mathrm{d-}}\cos(2\omega_1 t) + i_{\mathrm{r-}}^{\mathrm{d-}}\sin(2\omega_1 t) \\ i_{\mathrm{r}}^{\mathrm{q+}} = i_{\mathrm{r+}}^{\mathrm{q+}} + i_{\mathrm{r-}}^{\mathrm{q-}}\cos(2\omega_1 t) + i_{\mathrm{r-}}^{\mathrm{q-}}\sin(2\omega_1 t) \end{cases} \tag{4.32}$$

The above equations show that there are negative sequence components of twice the grid frequency[3] in the stator and rotor currents at this time. Substituting Eqs. (4.29) and (4.41) into Eq. (4.28), it gives

$$\begin{cases} P_{\mathrm{s}} = P_{\mathrm{sconst}} + P_{\mathrm{s}2\omega_1} \\ Q_{\mathrm{s}} = Q_{\mathrm{sconst}} + Q_{\mathrm{s}2\omega_1} \end{cases} \tag{4.33}$$

where P_{sconst} and Q_{sconst} are the stable components of the stator active and reactive power, respectively, $P_{\mathrm{s}2\omega_1}$ and $Q_{\mathrm{s}2\omega_1}$ are the components of the stator active and reactive power pulsating at twice the synchronous frequency, respectively, which take the following specific forms.

[3] For the DFIG in grid-connected operation, it is considered that $\omega_1 = \omega_{\mathrm{e}}$ in its stator-voltage-oriented model, which works for the subsequent discussion unless otherwise specified.

$$\begin{cases} P_{\text{sconst}} = u_{s+}^{d+} i_{s+}^{d+} + u_{s+}^{q+} i_{s+}^{q+} + u_{s-}^{d-} i_{s-}^{d-} + u_{s-}^{q-} i_{s-}^{q-} \\ P_{s2\omega_1} = P_{s2\omega_1 \cdot \cos} \cos(2\omega_1 t) + P_{s2\omega_1 \cdot \sin} \sin(2\omega_1 t) \\ \quad = \left(u_{s+}^{d+} i_{s-}^{d-} + u_{s+}^{q+} i_{s-}^{q-} + u_{s-}^{d-} i_{s+}^{d+} + u_{s-}^{q-} i_{s+}^{q+} \right) \cos(2\omega_1 t) \\ \quad + \left(u_{s+}^{d+} i_{s-}^{q-} - u_{s+}^{q+} i_{s-}^{d-} - u_{s-}^{d-} i_{s+}^{q+} + u_{s-}^{q-} i_{s+}^{d+} \right) \sin(2\omega_1 t) \end{cases} \tag{4.33a}$$

$$\begin{cases} Q_{\text{sconst}} = u_{s+}^{q+} i_{s+}^{d+} - u_{s+}^{d+} i_{s+}^{q+} + u_{s-}^{q-} i_{s-}^{d-} - u_{s-}^{d-} i_{s-}^{q-} \\ Q_{s2\omega_1} = Q_{s2\omega_1 \cdot \cos} \cos(2\omega_1 t) + Q_{s2\omega_1 \cdot \sin} \sin(2\omega_1 t) \\ \quad = \left(-u_{s+}^{d+} i_{s-}^{q-} + u_{s+}^{q+} i_{s-}^{d-} - u_{s-}^{d-} i_{s+}^{q+} + u_{s-}^{q-} i_{s+}^{d+} \right) \cos(2\omega_1 t) \\ \quad + \left(u_{s+}^{d+} i_{s-}^{d-} + u_{s+}^{q+} i_{s-}^{q-} - u_{s-}^{d-} i_{s+}^{d+} - u_{s-}^{q-} i_{s+}^{q+} \right) \sin(2\omega_1 t) \end{cases} \tag{4.33b}$$

Substituting Eqs. (4.31) and (4.32) into Eq. (4.27), the electromagnetic torque of the generator can be found as

$$T_e = T_{\text{econst}} + T_{e2\omega_1} \tag{4.34}$$

where T_{econst} is the stable component of the electromagnetic torque, $T_{e2\omega_1}$ is the component of the electromagnetic torque pulsating at twice the synchronous frequency, which take the following specific forms.

$$\begin{cases} T_{\text{econst}} = \frac{n_p}{\omega_s} \left(u_{s-}^{d-} i_{s-}^{d-} + u_{s-}^{q-} i_{s-}^{q-} - u_{s+}^{d+} i_{s+}^{d+} - u_{s+}^{q+} i_{s+}^{q+} \right) \\ T_{e2\omega_1} = T_{e2\omega_1 \cdot \cos} \cos(2\omega_1 t) + T_{e2\omega_1 \cdot \sin} \sin(2\omega_1 t) \\ \quad = \frac{n_p}{\omega_s} \Big[\left(u_{s-}^{d-} i_{s+}^{d+} + u_{s-}^{q-} i_{s+}^{q+} - u_{s+}^{d+} i_{s-}^{d-} - u_{s+}^{q+} i_{s-}^{q-} \right) \cos(2\omega_1 t) \\ \quad + \left(u_{s+}^{q+} i_{s-}^{d-} + u_{s-}^{q-} i_{s+}^{d+} - u_{s+}^{d+} i_{s-}^{q-} - u_{s-}^{d-} i_{s+}^{q+} \right) \sin(2\omega_s t) \Big] \end{cases} \tag{4.34a}$$

As can be seen from the analysis in this section, the impact of an asymmetrical grid fault on the DFIG-based wind power system is mainly due to the presence of the negative sequence component of the grid voltage, which excites negative sequence components in the various electromagnetic quantities of the DFIG system. These negative sequence components will persist until the grid voltage returns to normal, thus causing pulsations in the current, flux linkage, power and torque at twice the synchronous frequency, which may result in a system failure shutdown or even permanent hardware damage. For this reason, the effective suppression of the negative sequence voltage component in asymmetric grid is the key to the uninterrupted operation of the DFIG-based wind power system with asymmetrical grid faults.

4.3 Operational Characteristics of the Grid-Side Converter Under Grid Faults

The symmetrical grid fault has a low impact on the grid-side converter. For the grid-side converter with a traditional control structure of an inner current loop and an outer DC-bus voltage loop, the symmetrical grid fault mainly generates a dynamic disturbance in the phase-locked loop and thus triggers a transient process in the current and voltage loops, therefore, the dynamic characteristics of the phase-locked loop have a major influence on this transient process. The grid-side converter with sufficiently good dynamic characteristics can complete this transient process (which may be accompanied by transient current spikes and DC-bus voltage fluctuations) within a very short time and enter a new stable state again, realising uninterrupted operation with symmetrical faults, where the relevant control problems have been solved relatively well.

The asymmetrical grid fault has a high impact on the grid-side converter. This is due to the fact that the asymmetric grid voltage may persist and thus may have a relatively continuous effect on the grid-side converter, while conventional vector control methods for the grid-side converter are modelled and derived on the basis of ideal grid conditions and are therefore theoretically not capable of overcoming asymmetrical disturbances, which will further increase the effect of the asymmetrical grid fault on the grid-side converter. In the following section, the operational characteristics of the grid-side converter under the asymmetrical grid voltage are further analysed with the help of the symmetrical component method in Sect. 4.2.2.

When the asymmetric fault occurs in the grid voltage, Eqs. (2.24) and (2.25) are rewritten in the positive synchronous coordinate frame to obtain the mathematical model of the grid-side converter at this time as

$$\begin{cases} u_{\mathrm{g}}^{\mathrm{d}+} = R_{\mathrm{g}} i_{\mathrm{g}}^{\mathrm{d}+} + L_{\mathrm{g}} p i_{\mathrm{g}}^{\mathrm{d}+} - \omega_1 L_{\mathrm{g}} i_{\mathrm{g}}^{\mathrm{q}+} + e_{\mathrm{g}}^{\mathrm{d}+} \\ u_{\mathrm{g}}^{\mathrm{q}+} = R_{\mathrm{g}} i_{\mathrm{g}}^{\mathrm{q}+} + L_{\mathrm{g}} p i_{\mathrm{g}}^{\mathrm{q}+} + \omega_1 L_{\mathrm{g}} i_{\mathrm{g}}^{\mathrm{d}+} + e_{\mathrm{g}}^{\mathrm{q}} \end{cases} \tag{4.35}$$

$$\begin{cases} P_{\mathrm{g}} = e_{\mathrm{g}}^{\mathrm{d}+} i_{\mathrm{g}}^{\mathrm{d}+} + e_{\mathrm{g}}^{\mathrm{q}+} i_{\mathrm{g}}^{\mathrm{q}+} \\ Q_{\mathrm{g}} = e_{\mathrm{g}}^{\mathrm{q}+} i_{\mathrm{g}}^{\mathrm{d}+} - e_{\mathrm{g}}^{\mathrm{d}+} i_{\mathrm{g}}^{\mathrm{q}+} \end{cases} \tag{4.36}$$

Equations (2.26), (4.35) and (4.36) represent the mathematical model of the grid-side converter under the asymmetrical grid condition, and according to Eq. (4.24), the grid voltage and the grid-side current can be expressed as

$$\begin{cases} e_{\mathrm{g}}^{\mathrm{d}+} = e_{\mathrm{g}+}^{\mathrm{d}+} + e_{\mathrm{g}-}^{\mathrm{d}-} \cos(2\omega_1 t) + e_{\mathrm{g}-}^{\mathrm{d}-} \sin(2\omega_1 t) \\ e_{\mathrm{g}}^{\mathrm{q}+} = e_{\mathrm{g}+}^{\mathrm{q}+} + e_{\mathrm{g}-}^{\mathrm{q}-} \cos(2\omega_1 t) + e_{\mathrm{g}-}^{\mathrm{q}-} \sin(2\omega_1 t) \end{cases} \tag{4.37}$$

$$\begin{cases} i_{\mathrm{g}}^{\mathrm{d}+} = i_{\mathrm{g}+}^{\mathrm{d}+} + i_{\mathrm{g}-}^{\mathrm{d}-} \cos(2\omega_1 t) + i_{\mathrm{g}-}^{\mathrm{d}-} \sin(2\omega_1 t) \\ i_{\mathrm{g}}^{\mathrm{q}+} = i_{\mathrm{g}+}^{\mathrm{q}+} + i_{\mathrm{g}-}^{\mathrm{q}-} \cos(2\omega_1 t) + i_{\mathrm{g}-}^{\mathrm{q}-} \sin(2\omega_1 t) \end{cases} \tag{4.38}$$

The above equations show that there is a negative sequence component of twice the grid frequency in the grid-side current at this time. Substituting Eqs. (4.37) and (4.38) into Eq. (4.36), it gives

$$\begin{cases} P_{\mathrm{g}} = P_{\mathrm{gconst}} + P_{\mathrm{g}2\omega_1} \\ Q_{\mathrm{g}} = Q_{\mathrm{gconst}} + Q_{\mathrm{g}2\omega_1} \end{cases} \tag{4.39}$$

where P_{gconst} and Q_{gconst} are the stable components of the grid-side active and reactive power, respectively, $P_{\mathrm{g}2\omega_1}$ and $Q_{\mathrm{g}2\omega_1}$ are the components of the grid-side active and reactive power pulsating at twice the synchronous frequency, respectively, which take the following specific forms.

$$\begin{cases} P_{\mathrm{gconst}} = e_{\mathrm{g+}}^{\mathrm{d+}} i_{\mathrm{g+}}^{\mathrm{d+}} + e_{\mathrm{g+}}^{\mathrm{q+}} i_{\mathrm{g+}}^{\mathrm{q+}} + e_{\mathrm{g-}}^{\mathrm{d-}} i_{\mathrm{g-}}^{\mathrm{d-}} + e_{\mathrm{g-}}^{\mathrm{q-}} i_{\mathrm{g-}}^{\mathrm{q-}} \\ P_{\mathrm{g}2\omega_1} = P_{\mathrm{g}2\omega_1\cdot\cos} \cos(2\omega_1 t) + P_{\mathrm{g}2\omega_1\cdot\sin} \sin(2\omega_1 t) \\ \quad = \left(e_{\mathrm{g+}}^{\mathrm{d+}} i_{\mathrm{g-}}^{\mathrm{d-}} + e_{\mathrm{g+}}^{\mathrm{q+}} i_{\mathrm{g-}}^{\mathrm{q-}} + e_{\mathrm{g-}}^{\mathrm{d-}} i_{\mathrm{g+}}^{\mathrm{d+}} + e_{\mathrm{g-}}^{\mathrm{q-}} i_{\mathrm{g+}}^{\mathrm{q+}}\right) \cos(2\omega_1 t) \\ \quad + \left(e_{\mathrm{g+}}^{\mathrm{d+}} i_{\mathrm{g-}}^{\mathrm{q-}} - e_{\mathrm{g+}}^{\mathrm{q+}} i_{\mathrm{g-}}^{\mathrm{d-}} - e_{\mathrm{g-}}^{\mathrm{d-}} i_{\mathrm{g+}}^{\mathrm{q+}} + e_{\mathrm{g-}}^{\mathrm{q-}} i_{\mathrm{g+}}^{\mathrm{d+}}\right) \sin(2\omega_1 t) \end{cases} \tag{4.39a}$$

$$\begin{cases} Q_{\mathrm{gconst}} = e_{\mathrm{g+}}^{\mathrm{q+}} i_{\mathrm{g+}}^{\mathrm{d+}} - e_{\mathrm{g+}}^{\mathrm{d+}} i_{\mathrm{g+}}^{\mathrm{q+}} + e_{\mathrm{g-}}^{\mathrm{q-}} i_{\mathrm{g-}}^{\mathrm{d-}} - e_{\mathrm{g-}}^{\mathrm{d-}} i_{\mathrm{g-}}^{\mathrm{q-}} \\ Q_{\mathrm{g}2\omega_1} = Q_{\mathrm{g}2\omega_1\cdot\cos} \cos(2\omega_1 t) + Q_{\mathrm{g}2\omega_1\cdot\sin} \sin(2\omega_1 t) \\ \quad = \left(-e_{\mathrm{g+}}^{\mathrm{d+}} i_{\mathrm{g-}}^{\mathrm{q-}} + e_{\mathrm{g+}}^{\mathrm{q+}} i_{\mathrm{g-}}^{\mathrm{d-}} - e_{\mathrm{g-}}^{\mathrm{d-}} i_{\mathrm{g+}}^{\mathrm{q+}} + e_{\mathrm{g-}}^{\mathrm{q-}} i_{\mathrm{g+}}^{\mathrm{d+}}\right) \cos(2\omega_1 t) \\ \quad + \left(e_{\mathrm{g+}}^{\mathrm{d+}} i_{\mathrm{g-}}^{\mathrm{d-}} + e_{\mathrm{g+}}^{\mathrm{q+}} i_{\mathrm{g-}}^{\mathrm{q-}} - e_{\mathrm{g-}}^{\mathrm{d-}} i_{\mathrm{g+}}^{\mathrm{d+}} - e_{\mathrm{g-}}^{\mathrm{q-}} i_{\mathrm{g+}}^{\mathrm{q+}}\right) \sin(2\omega_1 t) \end{cases} \tag{4.39b}$$

Substituting the active power expression in Eq. (4.38) into Eq. (2.36), it gives

$$u_{\mathrm{dc}} i_{\mathrm{dc}} = C u_{\mathrm{dc}} p u_{\mathrm{dc}} = -P_{\mathrm{gconst}} - P_{\mathrm{g}2\omega_1} - \Delta P \tag{4.40}$$

At this point, the disturbance to the DC-bus voltage is not only from the active power on the generator side, but also from the grid side with a pulsation at twice the synchronous frequency of the grid, which in turn leads to a pulsation in the DC-bus voltage at twice the synchronous frequency.

From the above analysis, it can be seen that when the grid voltage asymmetry fault occurs, due to the existence of the negative sequence component of the grid voltage, it will lead to the appearance of the negative sequence component of the grid-side current, which is manifested by the pulsation of the grid-side current at twice the grid synchronous frequency, a surge in harmonic content and deterioration of the current quality; at the same time, the grid voltage containing positive and negative sequence components interacts with the grid-side current, resulting in a pulsation of twice the grid synchronous frequency in the active and reactive power exchanged between the grid-side converter and the grid, where the pulsating active power surges into the DC bus, causing a pulsation of twice the grid synchronous frequency in the DC-bus voltage and reducing the lifetime of the DC-bus capacitors..

4.4 Impact of Grid Faults on the DFIG-Based Wind Power System

As can be seen from the analysis in Sects. 4.2 and 4.3, the impact of grid voltage dips on the DFIG-based wind power system will be very significant due to its own topological feature.

The voltage dip at the generator end causes a DC component in the stator flux linkage, which in turn causes a rapid increase in the generator stator current. Through the magnetic coupling between the rotor and the stator, the rapidly increasing stator current causes the rotor current to rise sharply as well. At the same time, as the generator generally runs at a high speed during normal operation, the DC component in the stator flux linkage will have a high slip ratio relative to the rotor, which can cause overvoltage in the rotor winding. In addition, due to the slow speed regulation of the wind turbine, the wind energy absorbed by the wind turbine will not be significantly reduced during a grid fault, but the voltage at this time drops so that the system delivers less power to the grid, which makes part of the wind energy cannot be fed into the grid in time so that it stays inside the wind power system, either causing the DC-bus voltage to rise or causing the generator speed to surge beyond its effective working range. At the same time, the natural components in the stator and rotor flux linkages and currents can also cause pulsations in the electromagnetic torque of the generator and increase the mechanical stress on the drive shaft, gearbox and impeller. If no measures are taken, the converter for the generator and the drive shaft system for the turbine will most likely be damaged.

In particular, when the grid voltage is asymmetrical, the presence of the negative sequence component of the grid voltage will lead to a negative sequence component in the DFIG stator flux linkage, which in turn will lead to a negative sequence component in the stator current and, through the electromagnetic coupling between the stator and rotor, to a negative sequence component in the rotor current, resulting in an increase in the harmonics of the stator and rotor currents and a deterioration of the current quality, causing the generator to heat up. The power exchanged between the motor and the grid, and between the grid-side converter and the grid, also has the pulsating component of twice the grid synchronous frequency. The power pulsating component shocks the grid on the stator side and the grid side, and shocks the DC-bus capacitor on the rotor side, which is detrimental to the long-term reliable operation of the system. At the same time, the electromagnetic torque of the generator also has the pulsating component of twice the grid synchronous frequency, which increases the mechanical shock and stress on the drive shaft, gearbox, impeller, etc., and increases the probability of system damage.

From the above analysis it can be seen that in order to achieve uninterrupted operation in the event of a grid fault, the DFIG-based wind power system must meet the following basic requirements:

(1) In the event of a grid fault, damage to the converter caused by rotor overcurrent and DC-bus overvoltage can be avoided.

(2) The impact of torque pulsations on the gearbox and drive shaft during a grid fault can be minimised to prevent mechanical damage to the gearbox and wind turbine.
(3) It can quickly return to its pre-fault normal operating state once the fault is cleared.
(4) It can meet the LVRT operation specifications of the grid

In order to achieve good control and proper protection of the DFIG-based wind power system during grid faults, there are generally two ways: one is to enhance the system's ability to control the zero-sequence and negative sequence components by improving the control strategy; the other is to add the hardware protection circuit with a corresponding fault control strategy in the event of a grid fault. As the former approach is physically limited by the output voltage capability of the generator-side converter (which is usually not high), it is mostly only applicable in cases where the voltage dip is not very severe, while for a deep voltage dip, the hardware protection circuit must be used to achieve uninterrupted operation under grid faults.

4.5 Grid Connection Code for Modern Wind Power Systems

As wind power accounts for an increasing proportion of the grid, in order to ensure safe and reliable operation of the power system, countries have made clear requirements for the performance index of grid-connected wind power systems, and the relevant standards are available on the official websites of the relevant organisations in the countries.

The requirements of China's national standard for low-voltage ride-through of wind farms are shown in Fig. 4.5. When the voltage at the grid-connection point of the wind farm drops to 20% of the nominal voltage, the wind power units in the wind farm are required to ensure uninterrupted operation without going off-grid for 625 ms. when the voltage at the grid-connection point of the wind farm recovers to 90% of the nominal voltage within 2 s after a drop, the wind power units in the wind farm are required to ensure uninterrupted operation without going off-grid. When different types of faults occur in the power system, if the voltage at the grid-connection point of the wind farm (e.g. Table 4.2) is all within the area at and above the voltage contour line shown in Fig. 4.5, the wind power units are required to ensure uninterrupted operation without going off-grid; for the wind farm that does not cut off during a power system fault, its active power is expected to recover quickly after the fault is cleared, with a power increasing rate of at least 10% of the nominal power per second recovering to the value before the fault. It is worth pointing out that not all the fault types given in Sect. 4.1 are fully covered in this standardised Fault Ride-through Criteria.

For the wind farm with a total installed capacity of GW and more, it requires the dynamic reactive power support capability during a low-voltage ride-through when

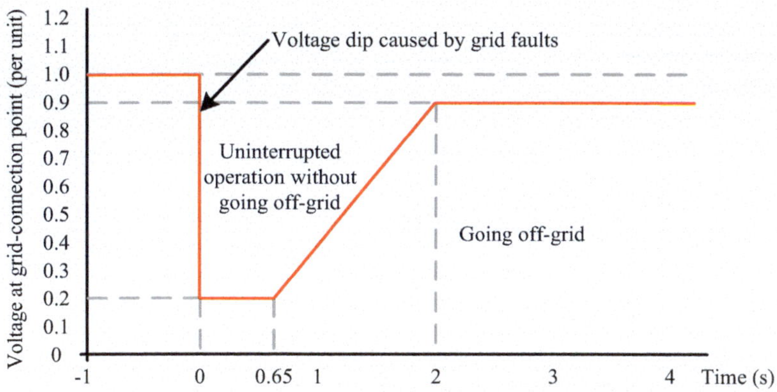

Fig. 4.5 China national standard for low-voltage ride-through of wind farms

Table 4.2 Detected voltage types for low-voltage ride-through of wind farms

Fault type	Detected voltage type
Three-phase short circuit	Line voltage at the grid-connection point
Two-phase short circuit	Line voltage at the grid-connection point
Single-phase short circuit to earth	Phase voltage at the grid-connection point

the voltage dip is triggered by a three-phase short-circuit fault in the grid: when the voltage at the grid-connection point is in the range of 20–90% of the nominal voltage, the wind farm is required to support the voltage recovery by injecting reactive current; from the moment when the voltage dip occurs at the grid-connection point, the response time of the dynamic reactive current control is no more than 75 ms and the duration is no less than 550 ms; the dynamic reactive current I_T injected into the grid by the wind farm satisfies the Formula (4.40).

$$I_{\mathrm{T}} \geq 1.5 \times (0.9 - U_{\mathrm{T}})I_{\mathrm{N}} \tag{4.41}$$

where U_{T} indicates the per unit value of the voltage at the grid-connection point of the wind farm, I_{N} indicates the rated current of the wind farm.

The national standard makes the following requirements for the operating environment of wind farms: when the voltage at the grid-connection point of the wind farm is between 90 and 110% of the nominal voltage, its wind power units are required to be able to operate normally; when the voltage fluctuations and current harmonics at the grid-connection point of the wind farm meet the requirements of Tables 4.3 and 4.4 respectively, the wind power units in the wind farm are required to be able to operate normally; when the steady-state asymmetry of the voltage at the grid-connection point of the wind farm is no more than 2%, and the transient asymmetry is no more than 4%, the wind power units in the wind farm are required to be able to operate normally.

Table 4.3 Limit values of voltage fluctuation at the grid-connection point

r (times/hour)	d (%)	
	No more than 35 kV	35–220 kV
$r \leq 1$	4	3
$1 < r \leq 10$	3	2.5
$10 < r \leq 100$	2	1.5
$100 < r \leq 1000$	1.25	1

Table 4.4 Limit values of current harmonic at the grid-connection point

Standard voltage (kV)	Baseline short-circuit capacity (MVA)	Number of harmonics and RMS value of permissible harmonic currents (A)								
		2	3	4	5	6	7	8	9	10
0.38	10	78	62	39	62	26	44	19	21	16
6	100	43	34	21	34	14	24	11	11	8.5
10	100	26	20	13	20	8.5	15	6.4	6.8	5.1
35	250	15	12	7.7	12	5.1	8.8	3.8	4.1	3.1
66	500	16	13	8.1	13	5.4	9.3	4.1	4.3	3.3
110	750	12	9.6	6.0	9.6	4.0	6.8	3.0	3.2	2.4

The above is a reference to China National Standards GB/T 19963–2011, GB/T 12326–2008, GB/T 14549–1993 and GB/T 15543–1995 on key parameters for safe and stable operation of wind power systems and grid environmental requirements, for the reference of relevant R&D and technical personnel, and more specific standards are available for free in China's Full Text Public System of National Standards.

References

1. Y. Wang, D.L. Zhao, B. Zhao et al., A review of research status on LVRT technology in doubly-fed wind turbine generator system, in *International Conference on Electrical and Control Engineering* (2010), pp. 4948–4953
2. M. Bollen, Characterization of voltage sags experienced by three phase adjustable-speed drives. IEEE Trans. Power Deliv.Deliv. **12**, 1666–1671 (1997)
3. M. Bollen, R. Graaff, Behavior of AC and DC drives during voltage sags with phase-angle jump and three-phase unbalance. IEEE Power Eng. Soc. Win. Meeting 1225–1230 (1999)
4. M. Bollen, G. Olguin, M. Martins, Voltage dips at the terminals of wind power installations. Wind Energy **8**, 307–318 (2005)
5. A. Rolan, F. Corcoles, J. Pedra, Doubly fed induction generator subject to symmetrical voltage sags. IEEE Trans. Energy Convers. **26**(4), 1219–1229 (2011)
6. J. Jopez, P. Sanchis, X. Roboam et al., Dynamic behavior of the doubly fed induction generator during three-phase voltage dips. IEEE Trans. Energy Convers. **22**(3), 709–717 (2007)

Chapter 5
Low-Voltage Ride-Through Technologies for the DFIG-Based Wind Power System

As can be seen in Chap. 4, grid voltage dips can trigger violent electromagnetic processes in the DFIG-based wind power system, which can have a series of adverse effects on the control performance, electrical components and even mechanical structures, and even damage the system hardware. In order to meet the requirements of national standards for the low-voltage ride-through (LVRT) function of the DFIG-based wind power system, improvements need to be made to both the software and hardware aspects of the system.

This chapter first gives an overview of the research on LVRT in the DFIG-based wind power system, then introduces three LVRT control methods based on improved vector control and model predictive control, respectively, and analyses their control domains, and finally discusses in detail the design methods and control strategies of several typical hardware protection circuits for LVRT processes.

5.1 Overview of LVRT Technologies

To achieve LVRT in the DFIG-based wind power system, there are generally two kinds of ways: one is to improve the control strategy to enhance the system's ability to control each transient variable; the other is to access the hardware protection circuit in the event of a grid fault to provide a channel for releasing excess energy within the system.

H. Ma et al., *The Control Principle of Wind Power Generation System*,
https://doi.org/10.1007/978-981-99-6083-5_5

5.1.1 LVRT Methods Based on Improved Control Strategies

In order to increase costs as little as possible, many scholars have researched low-voltage ride-through methods for the DFIG-based wind power system from the viewpoint of not adding hardware circuits but only improving control strategies, and most of these methods are based on various improved algorithms of the traditional vector control, which is because the traditional vector control has the advantages of high control accuracy, good steady-state characteristics, constant switching frequency and the ability to achieve decoupling control, and is currently the more widely used control strategy in the DFIG system. A simple improvement is to moderately increase the proportional factor of the current-loop proportional-integral (PI) controller in the generator-side converter of a conventional DFIG system [1], which essentially increases the response bandwidth of the controller to achieve uninterrupted operation under symmetrical grid faults within a certain operating range. However, this method will still cause rotor side overcurrent and overvoltage problems when a grid fault causes a serious drop in the DC-bus voltage, and increasing the proportional factor of the PI controller may have an impact on the operating characteristics of the system under normal grid conditions. Another improvement is to add a decoupling compensation term or dynamic compensation term to the rotor-current loop of conventional vector control to counteract the effect of the stator flux linkage change on the rotor current during grid faults and to improve the system control performance, at the same time, an "active resistor" is introduced to increase the stator resistance, thus increasing the decay rate of the DC component of the stator flux linkage and helping the generator to recover from the transient state as soon as possible [2], and furthermore, by applying an additional excitation voltage [3], a rotor current space vector opposite to the transient DC component of the stator flux linkage can be created in the rotor to counteract the influence of the DC component of the stator flux linkage on the system and to suppress rotor overcurrent and overvoltage. However, under grid fault conditions, the rotor current is usually already high and the use of the above-mentioned methods to counteract the effect of the stator flux linkage will potentially increase the rotor current further and thus trigger rotor-side overcurrent protection. The use of the dynamic compensation term [4] places high demands on dynamic flux linkage observation and is highly dependent on the system parameters.

Considering the relatively complex structure of vector control and its high dependence on parameters, some researchers have explored and investigated other control methods such as direct power control (DPC) and flux magnitude and angle control (FMAC). By using the direct control idea in DPC [5], the rotor flux linkage, electromagnetic torque and rotor current of the DFIG can be separately controlled in a direct closed-loop manner, so as to accelerate the transient process, reduce the mechanical shock and suppress the rotor overcurrent during the low-voltage process, respectively, of which the first two direct control methods are more dependent on the system parameters, while the third direct control method is less dependent on the parameters. Compared to traditional vector control, the above methods are simple in structure and good in dynamic characteristics. In contrast, the FMAC method [6] controls the machine terminal voltage by controlling the rotor flux linkage amplitude and the active power output from the generator by controlling the rotor flux linkage phase angle. Compared to vector control, it has better decoupling characteristics, and especially in the case of small dips in the grid, it shows better control performance, but its control process requires rotor flux linkage amplitude and phase information, which places high demands on rotor flux linkage observation.

In addition, various improved non-linear algorithms, such as robust control and fuzzy control, have been applied to the symmetrical grid fault ride-through process for DFIG-based systems, but their practicality needs further research due to the complexity of their design process and the fact that their controllable range has not yet been described in the literature.

5.1.2 LVRT Methods Based on Hardware Protection Circuits

As can be seen from Sect. 5.1.1, the various improvement algorithms essentially respond to the zero-sequence transient component introduced by a grid fault by increasing the control voltage from the generator-side converter during a grid fault, thus achieving uninterrupted operation of the DFIG system under grid fault conditions. However, as the DFIG-based wind power system uses a partial-scale converter, its ability to control the system under abnormal grid conditions is very limited. Therefore, when a severe voltage fault occurs in the grid, it is no longer possible to achieve fault ride-through by relying exclusively on software control strategy improvements, and hardware protection units have to be added [7].

Currently, the most commonly used hardware protection circuit in the DFIG-based wind power system is an energy consumption unit connected to the rotor, which is named as crowbar and is composed of an unloading resistor and a power electronic switch. When the rotor current exceeds its threshold under grid fault conditions, the unloading resistor is shorted to the rotor winding, thereby bypassing and protecting the wind power converter and, at the same time, providing an additional energy path for the rotor current to accelerate the motor transient process and suppress rotor-side overcurrent. There are various configurations of the crowbar suitable for the DFIG-based wind power system [8], and several conventional hardware configurations of the crowbar are shown in Fig. 5.1.

The conventional crowbar uses the thyristor-based structure and does not have a self-shutoff function. As a result, a system based on this crowbar structure cannot automatically return to its normal state after a grid fault is recovered and needs to be reconnected to the grid, so this crowbar is called the passive crowbar. Due to the difficulty of adapting the passive crowbar to the new grid code requirements, the active crowbar structure with a self-shutting function is becoming widely used. The active crowbar can be connected to the rotor to protect the system in the event of a grid fault causing rotor overcurrent, and can be disconnected from the rotor when the rotor current is recovered, allowing the system to return to normal operation quickly and flexibly.

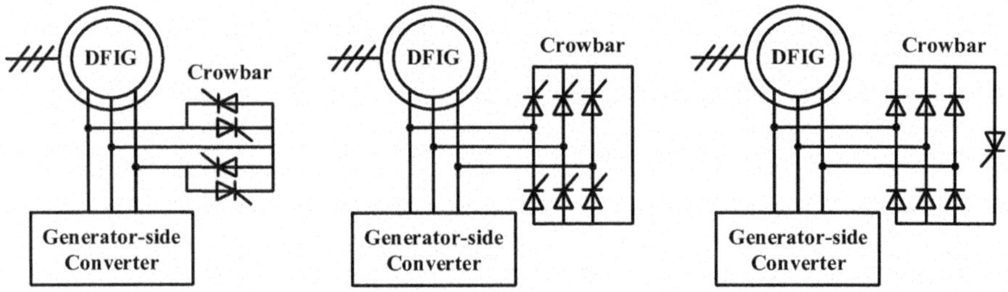

Fig. 5.1 Crowbars with different hardware configurations

In order to accelerate the decay of the rotor transient current, the active crowbar usually contains an unloading resistor. The resistance value of this resistor is related to the generator parameters and has an important influence on the operational performance of the wind power system during fault ride-through: when its value is small, it does not effectively limit rotor overcurrent and electromagnetic torque shocks, but the rotor voltage is not high at this time; while a larger value reduces rotor current and electromagnetic torque shocks, it causes excessive voltage to be applied to the generator-side converter and may damage the power electronic devices. Therefore, a suitable unloading resistor must be selected for the crowbar circuit. At present, the selection of a crowbar unloading resistor is mostly based on numerical simulations, where the control effects of different crowbar unloading resistors are tried and tested.

For the rightmost crowbar structure in Fig. 5.2, the analytical analysis allows the unloading resistance to be limited to [9]

$$\begin{cases} R_{\min} = \dfrac{\omega_r}{I_{\text{safe}}} \sqrt{\left(\dfrac{U_s}{\omega_s}\right)^2 - (L_\sigma I_{\text{safe}})^2} \\ R_{\max} = \dfrac{U_{\text{dc}} \omega_r L_\sigma}{\sqrt{3(U_s \omega_r / \omega_s)^2 - U_{\text{dc}}^2}} \end{cases} \tag{5.1}$$

where $R_{\min}$ and $R_{\max}$ are the lower and upper limits of the crowbar unloading resistance respectively, U_s and U_{dc} are the stator and DC-bus voltages respectively, ω_r and

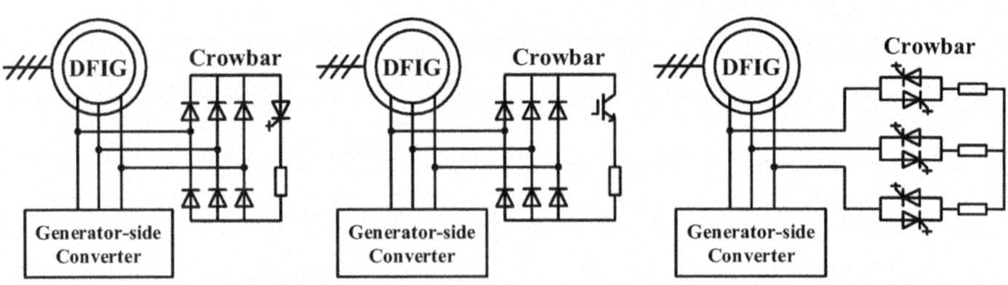

Fig. 5.2 Several active crowbars

ω_s are the rotor angular frequency and grid synchronous angular frequency respectively, L_σ is the sum of the stator and rotor leakage inductance and I_{safe} is the rotor safety current. After obtaining the range of the crowbar unloading resistance from Eq. (5.1), the simulation method is then used to continuously try to find the unloading resistance within this range that meets the performance requirements. This method provides guidance and reference for subsequent simulations by determining the resistance range through an analytical method, reducing the complexity and workload of the simulation and is an important step in the theoretical and analytical solution of the crowbar resistance.

The crowbar performance function can also be used to guide the selection of its unloading resistance. For both the current control and flux linkage control methods, the crowbar resistance performance function that measures the effect of the dynamic process on the system is [10]

$$Index = \alpha_1 \int_{t_0}^{t_1} I_r^2 dt + \alpha_2 \int_{t_0}^{t_1} |Q| dt + \alpha_3 \int_{t_0}^{t_1} |\Delta T_e| dt \tag{5.2}$$

where *Index* is the value of the performance function, I_r is the rotor current amplitude, Q is the absorbed reactive power of the system, ΔT_e is the value of the electromagnetic torque pulsation, α_1, α_2 and α_3 are the weighting factors of the corresponding terms in the performance function, t_0 and t_1 are the moments at which the performance function calculation is started and completed, respectively. This method gives a quantitative measure for the performance of the crowbar resistance, but it still relies entirely on simulation to find the optimal value of the crowbar resistance that minimises Eq. (5.2), which is a large amount of work and has low operability, and it is difficult to determine the moment of t_1 in the performance function during the practical calculation.

In summary, the selection of the crowbar unloading resistor is an important issue in the design of hardware protection circuits, and there is still a lack of relevant literature, most of which relies entirely on simulation methods for qualitative discussions, with a large workload and a complex selection process.

Another commonly used hardware protection circuit for the DFIG-based wind power system is the DC-bus protection circuit, which is used to protect the DC-bus voltage within a safe range. One of the commonly used structures is the DC-bus chopper circuit [11], which is shown in Fig. 5.3. It consists of a power electronic switch and an unloading resistor, connected in parallel to the two terminals of the DC-bus capacitor [12]. In the process of a heavy grid fault, due to the capacity limitation of the grid-side converter, it may occur that it cannot effectively complete the energy exchange between the DC bus and the grid, thus causing energy to accumulate on the DC-bus capacitor. At this time, the switching device in the DC-bus chopper circuit is switched on and the unloading resistor provides a temporary path for the energy in the DC-bus capacitor, thus preventing the DC-bus from being over-voltage. After the DC-bus voltage is recovered, the DC-bus chopper is cut out from the system and the system resumes normal operation. Such a DC bus chopper is simple in structure and easy to control, but it will be accompanied by energy consumption and heat

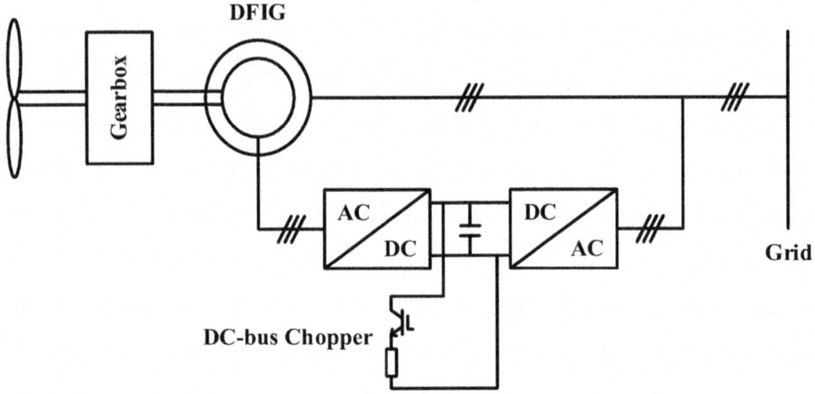

Fig. 5.3 The DFIG-based wind power system with A DC-bus chopper

generation, reducing the system efficiency. For this reason, energy storage devices such as the battery, supercapacitor and auxiliary converter can be used instead of the DC-bus chopper connected in parallel at both terminals of the DC-bus capacitor. When a grid fault causes the DC-bus voltage to rise, the excess energy surging into the DC bus will be stored in the energy storage device and fed into the grid after the grid fault is eliminated, improving the system efficiency. The problem with this method is that it will cause an increase in the cost and size of the system, creating a certain barrier to its practical application.

It is to be noted that the crowbar/dc-bus chopper helps the system to recover as quickly as possible by providing an additional, temporary energy path for the transient energy in the wind power system. When an asymmetrical fault occurs in the grid, the negative sequence component of the grid voltage is not a transient presence but persists until the grid recovers from the fault. Therefore, for the heavy asymmetrical grid fault, the crowbar/dc-bus chopper can help protect the system, but cannot help the system to recover control, or it can be considered that the system is beyond the control capability of the converter, and for this reason, other more complex hardware protection circuits have been introduced to help the system to achieve uninterrupted operation under the heavy the asymmetrical grid fault.

Figure 5.4 illustrates a hardware protection circuit with a three-phase auxiliary inductor [13]. When the grid voltage is asymmetrical, the grid-side converter will regulate the currents flowing through the inductors L_c and L_g and make them asymmetrical. In this way, the stator voltage can be controlled in a relatively more symmetrical manner even in the case of a large grid voltage asymmetry, reducing the impact of grid voltage asymmetry on the DFIG system to some extent.

Figure 5.5 illustrates a hardware protection circuit with an auxiliary converter and a series transformer on the stator side for asymmetrical low-voltage ride-through [14]. When the grid voltage is asymmetrical, the auxiliary converter is controlled to operate and compensate the generator stator voltage via the stator series transformer, so that

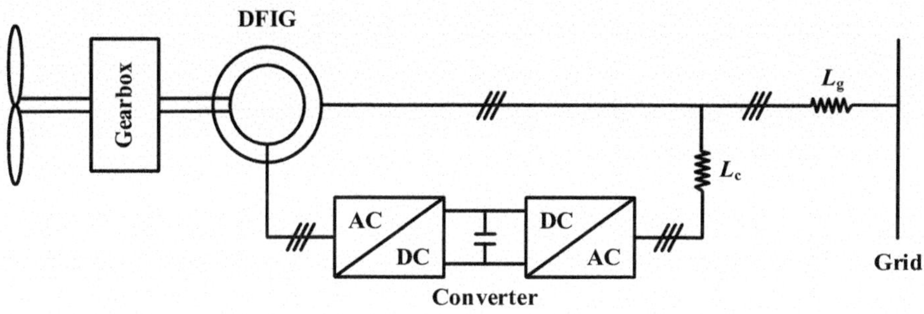

Fig. 5.4 The DFIG-based wind power system with a three-phase auxiliary inductor

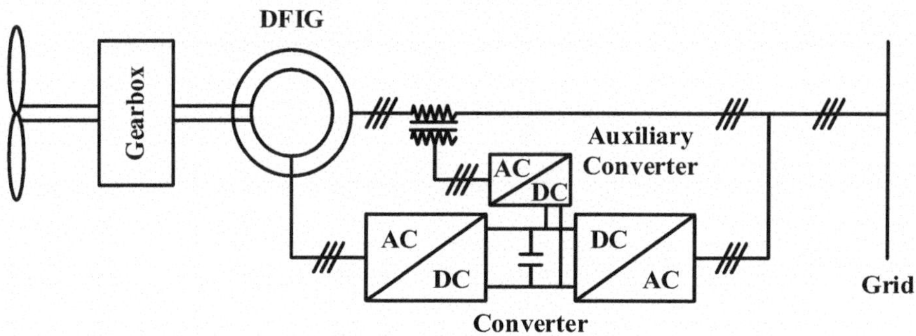

Fig. 5.5 The DFIG-based wind power system with an auxiliary converter and a series transformer

the stator voltage remains at the level of the normal grid voltage and the effects of the asymmetrical grid voltage on the generator are avoided.

Hardware protection circuits for asymmetrical grid voltage compensation are generally connected in series with the DFIG stator or even the grid in order to compensate for the effect of the asymmetrical grid voltage on the stator voltage. As a result, the capacity of the auxiliary hardware protection circuits is generally very large, which increases the size, weight and cost of the wind power system and is of low practicality. Therefore, such methods are hardly used in practice and are not the focus of this book.

5.2 Improved Control Strategies for LVRT

5.2.1 Vector Control Based on Dynamic Feedforward Compensation

In the traditional vector control method based on the stator flux orientation, in order to simplify the design of the rotor current regulator, the stator flux linkage is usually considered to be constant and always placed on the d-axis of the synchronous rotation coordinate frame. However, when the grid voltage dips, the stator flux linkage decays and a DC component is created in the stator flux linkage, so it is not possible to ensure that the stator flux linkage vector is accurately oriented on the d-axis, i.e. $\varphi_{sq} \neq 0$. Also the differential term of the stator flux linkage is no longer equal to zero.

Then the voltage equation for the DFIG in the dq-frame can be written as

$$u_{s}^{d} = R_{s}i_{s}^{d} + p\psi_{s}^{d} - \omega_{1}\psi_{s}^{q} \tag{5.3}$$

$$u_{s}^{q} = R_{s}i_{s}^{q} + p\psi_{s}^{q} + \omega_{1}\psi_{s}^{d} \tag{5.4}$$

$$u_{r}^{d} = R_{r}i_{r}^{d} + p\psi_{r}^{d} - \omega_{sl}\psi_{r}^{q} \tag{5.5}$$

$$u_{r}^{q} = R_{r}i_{r}^{q} + p\psi_{r}^{q} + \omega_{sl}\psi_{r}^{d} \tag{5.6}$$

The stator flux linkage of the DFIG can be written as

$$\psi_{s}^{d} = L_{s}i_{s}^{d} + L_{M}i_{r}^{d} \tag{5.7}$$

$$\psi_{s}^{q} = L_{s}i_{s}^{q} + L_{M}i_{r}^{q} \tag{5.8}$$

The stator current can be found, according to Eqs. (5.7) and (5.8), as

$$i_{s}^{d} = \frac{\psi_{s}^{d} - L_{M}i_{r}^{d}}{L_{s}} \tag{5.9}$$

$$i_{s}^{q} = \frac{\psi_{s}^{q} - L_{M}i_{r}^{q}}{L_{s}} \tag{5.10}$$

The rotor flux linkage can be expressed as

$$\psi_{r}^{d} = L_{M}i_{s}^{d} + L_{r}i_{r}^{d} \tag{5.11}$$

$$\psi_{r}^{q} = L_{M}i_{s}^{q} + L_{r}i_{r}^{q} \tag{5.12}$$

Substituting Eqs. (5.9) and (5.10) into Eqs. (5.11) and (5.12) respectively, it gives

$$\psi_{\mathrm{r}}^{\mathrm{d}} = \sigma L_{\mathrm{r}} i_{\mathrm{r}}^{\mathrm{d}} + \frac{L_{\mathrm{M}}}{L_{\mathrm{s}}} \psi_{\mathrm{s}}^{\mathrm{d}} \tag{5.13}$$

$$\psi_{\mathrm{r}}^{\mathrm{q}} = \sigma L_{\mathrm{r}} i_{\mathrm{r}}^{\mathrm{q}} + \frac{L_{\mathrm{M}}}{L_{\mathrm{s}}} \psi_{\mathrm{s}}^{\mathrm{q}} \tag{5.14}$$

Substituting Eqs. (5.9) and (5.10) into Eqs. (5.3) and (5.4) respectively, the stator voltage can be rewritten as

$$u_{\mathrm{s}}^{\mathrm{d}} = \frac{R_{\mathrm{s}}}{L_{\mathrm{s}}} \psi_{\mathrm{s}}^{\mathrm{d}} - \frac{R_{\mathrm{s}}}{L_{\mathrm{s}}} i_{\mathrm{r}}^{\mathrm{d}} + p\psi_{\mathrm{s}}^{\mathrm{d}} \tag{5.15}$$

$$u_{\mathrm{s}}^{\mathrm{d}} = \frac{R_{\mathrm{s}}}{L_{\mathrm{s}}} \psi_{\mathrm{s}}^{\mathrm{q}} - \frac{R_{\mathrm{s}}}{L_{\mathrm{s}}} i_{\mathrm{r}}^{\mathrm{q}} + p\psi_{\mathrm{s}}^{\mathrm{q}} \tag{5.16}$$

Similarly, substituting Eqs. (5.13) and (5.14) into Eqs. (5.5) and (5.6) respectively, the rotor voltage can be rewritten as

$$u_{\mathrm{r}}^{\mathrm{d}} = R_{\mathrm{r}} i_{\mathrm{r}}^{\mathrm{d}} + \sigma L_{\mathrm{r}} p i_{\mathrm{r}}^{\mathrm{d}} - \omega_{\mathrm{sl}} \sigma L_{\mathrm{r}} i_{\mathrm{r}}^{\mathrm{q}} - \omega_{\mathrm{sl}} \frac{L_{\mathrm{M}}}{L_{\mathrm{s}}} \psi_{\mathrm{s}}^{\mathrm{q}} + \frac{L_{\mathrm{M}}}{L_{\mathrm{s}}} p\psi_{\mathrm{s}}^{\mathrm{d}} \tag{5.17}$$

$$u_{\mathrm{r}}^{\mathrm{q}} = R_{\mathrm{r}} i_{\mathrm{r}}^{\mathrm{q}} + \sigma L_{\mathrm{r}} p i_{\mathrm{r}}^{\mathrm{q}} + \omega_{\mathrm{sl}} \sigma L_{\mathrm{r}} i_{\mathrm{r}}^{\mathrm{d}} + \omega_{\mathrm{sl}} \frac{L_{\mathrm{M}}}{L_{\mathrm{s}}} \psi_{\mathrm{s}}^{\mathrm{d}} + \frac{L_{\mathrm{M}}}{L_{\mathrm{s}}} p\psi_{\mathrm{s}}^{\mathrm{q}} \tag{5.18}$$

The complete mathematical model of the DFIG considering the transient flux linkage is obtained from the above, compared to the mathematical model under ideal grid conditions, it can be seen that to achieve high performance control of the rotor current in the complete mathematical model, the rotor voltage needs to overcome not only the equivalent electrical motive force generated by the original cross-axis coupling term of current, but also the equivalent electrical motive force caused by the transient stator flux linkage, which can be achieved in the practical control algorithm by superimposing such transient electrical motive forces in the form of feedforward compensation components to the rotor output voltage, as shown in the control block diagram in Fig. 5.6.

Compared with the traditional vector control, the improved control algorithm adds $e_{\varphi}^{\mathrm{d}*}$ and $e_{\varphi}^{\mathrm{q}*}$ to the feedforward compensation term to take into account the dynamic change of the stator flux linkage, which is used to suppress the electromagnetic transition process caused by the fall of the stator flux linkage when the voltage dips. When the grid voltage is stable, the feedforward compensation term degenerates to e_{φ}^{*}. Therefore, the improved vector control method can obtain the same control effect as the traditional method in the normal state of the grid.

Figures 5.7, 5.8 and 5.9 illustrate the simulation results of the LVRT process for a 1.5 MW DFIG-based wind power system when the voltage at the common coupling

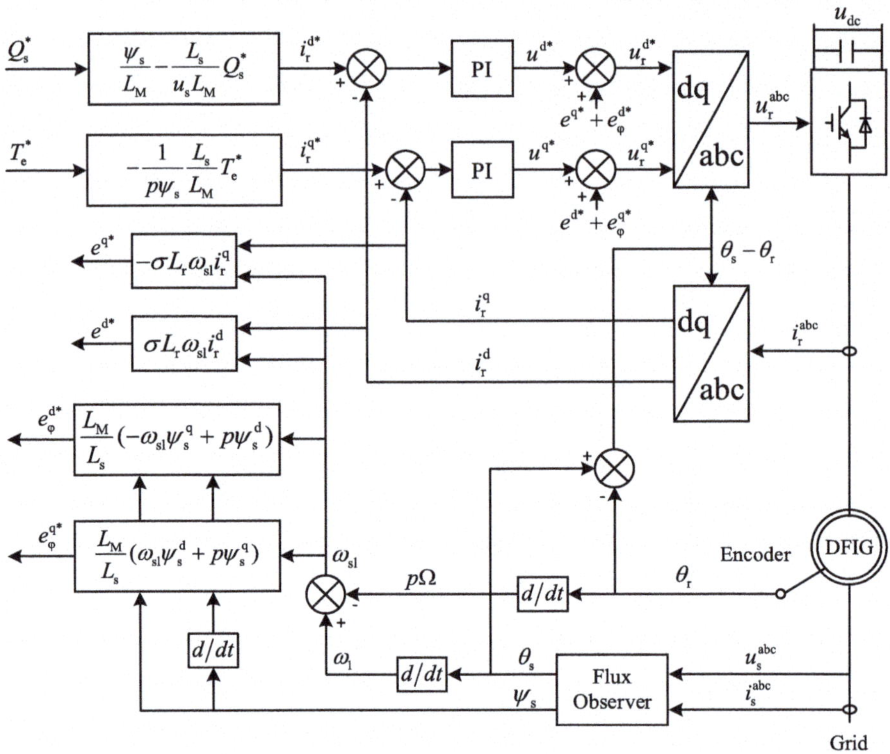

Fig. 5.6 Control diagram of vector control for the DFIG with feedforward compensation

point of the grid dips to 15% of the nominal value, thus causing the voltage at the stator terminal of the generator to dip to 67% of the nominal value.

Figure 5.7 illustrates the simulation results of the rotor current in the dq-frame. It can be seen from the figure that the inclusion of the feedforward compensation term

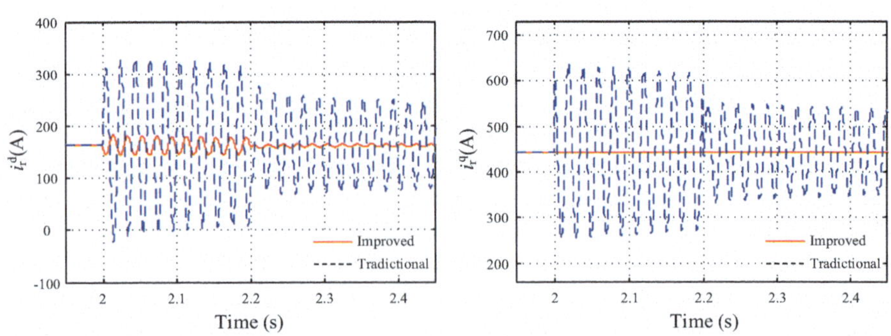

Fig. 5.7 Simulation waveforms of rotor current in synchronous rotating coordinate frame (Stator voltage dips to 67% of the nominal value)

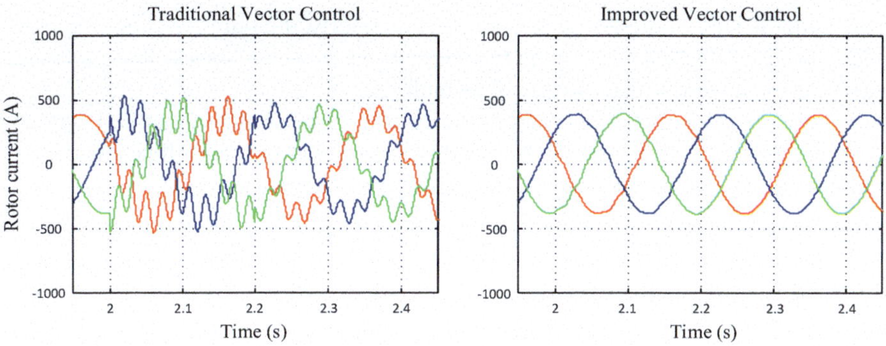

Fig. 5.8 Simulation waveforms of rotor current in ABC-frame (Stator voltage dips to 67% of the nominal value)

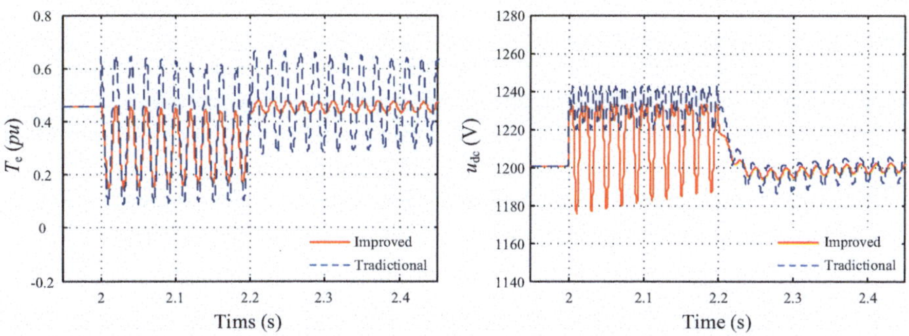

Fig. 5.9 Simulation waveforms of electromagnetic torque and DC-bus voltage (Stator voltage dips to 67% of the nominal value)

for the dynamic process of the stator flux linkage results in a good suppression of the oscillations in the rotor current, which effectively reduces the overcurrent in the rotor winding under the grid voltage dip.

Figure 5.8 compares the simulation results of the generator three-phase rotor and stator currents for the two control methods, traditional vector control and improved vector control. As can be seen from the figure, the improved vector control method is able to maintain effective control of the rotor current under faults and also greatly improves the waveforms of the rotor and stator currents, reducing their harmonic content.

Figure 5.9 compares the effect of the two control methods on the generator electromagnetic torque and DC-bus voltage. As can be seen from Fig. 5.9a, the voltage dip at the generator terminal causes the electromagnetic torque to oscillate, and the use of the improved vector control method effectively reduces the magnitude of the oscillation, thereby reducing the mechanical shock to the drive shaft and gearbox. From Fig. 5.9b, the power surge from the rotor side of the generator causes the DC-bus voltage to rise and oscillate, which can be gradually controlled to a stable value with the help of the grid-side converter. Similarly, the improved vector control method is also effective in reducing the fluctuations in the DC-bus voltage, so that a stable value can be reached quickly after fault recovery.

It is worth noting that the control effectiveness of the improved method depends on the accurate compensation of the dynamic changes in the stator flux linkage and therefore requires a high accuracy of the generator parameters.

5.2.2 Vector Control Based on PIR Controllers

From the analysis of the DFIG operational characteristics during the grid fault in Sect. 4.2, it can be seen that during the symmetrical dip and recovery of the grid voltage, the DFIG stator flux linkage is not able to change abruptly due to the integral effect and produces a natural component $\boldsymbol{\psi}_{s\_natural}$ which can be considered as a disturbance at the grid synchronous frequency introduced into the DFIG-based wind power system under the traditional vector control. The PI controller used in traditional vector control is unable to respond effectively to the synchronous frequency disturbance, thus making the entire control system unable to effectively suppress a series of effects caused by the natural component of the stator flux linkage.

The proportional integral resonant (PIR) controller is made by connecting a resonant unit in parallel to a conventional PI controller, so that the system has a high open-loop gain near its resonant point and thus responds effectively to signals near the resonant frequency, which means that this closed-loop system based on the PIR controller is able to follow the reference value of the signal at the frequency near the resonant point and is also able to suppress disturbances in the same frequency range (Fig. 5.10).

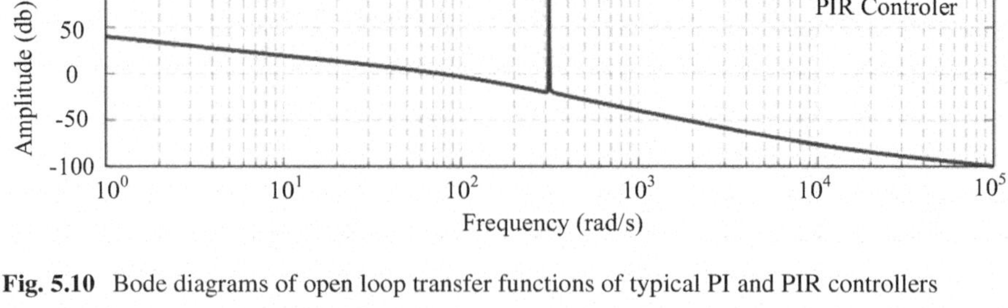

Fig. 5.10 Bode diagrams of open loop transfer functions of typical PI and PIR controllers

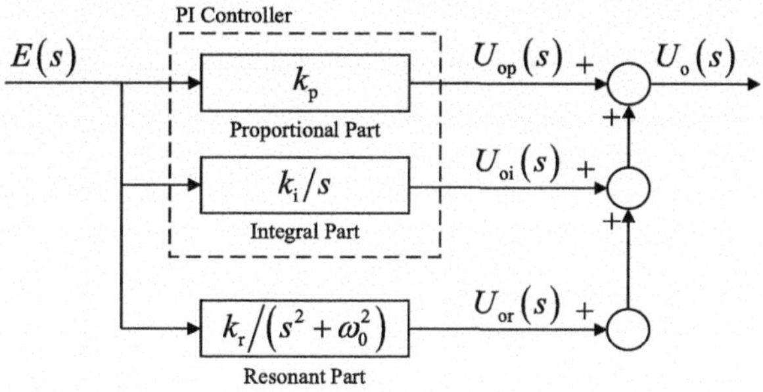

Fig. 5.11 Block diagram of the PIR controller

The transfer function of the PIR controller and its block diagram are shown in Eq. (5.19) and Fig. 5.11 respectively, where a is the resonance frequency.

$$G_{\mathrm{PIR}}(s) = k_{\mathrm{p}} + \frac{k_{\mathrm{i}}}{s} + \frac{k_{\mathrm{r}}s}{s^2 + \omega_0^2} \tag{5.19}$$

Therefore, in the traditional vector control method, the PI controller is replaced by the PIR controller and the resonance point frequency is set at the grid synchronous frequency, which can be used to suppress the grid synchronous frequency disturbance in the DFIG system during the grid voltage dip, thus realising the low-voltage ride-through of the DFIG-based wind power system, the control diagram of which is shown in Fig. 5.12.

As can be seen, this PIR controller-based method for dynamic compensation of the stator flux linkage relies entirely on a closed-loop structure rather than on feedforward term, so the dependence on generator parameters is reduced, but the complexity of the controller parameter tuning process is increased compared to the traditional vector control method.

It is worth pointing out that the implementation of the PIR controller in the practical system is somewhat tricky. As can be seen from Fig. 5.11:

$$U_{\mathrm{o}}(s) = U_{\mathrm{op}}(s) + U_{\mathrm{oi}}(s) + U_{\mathrm{or}}(s) \tag{5.20}$$

where

$$U_{\mathrm{op}}(s) = k_{\mathrm{p}}E(s) \tag{5.21}$$

$$U_{\mathrm{oi}}(s) = \frac{k_{\mathrm{i}}E(s)}{s} \tag{5.22}$$

$$U_{\mathrm{or}}(s) = \frac{k_{\mathrm{r}}sE(s)}{s^2 + \omega_0^2} \tag{5.23}$$

It can be known from Eq. (5.23) that

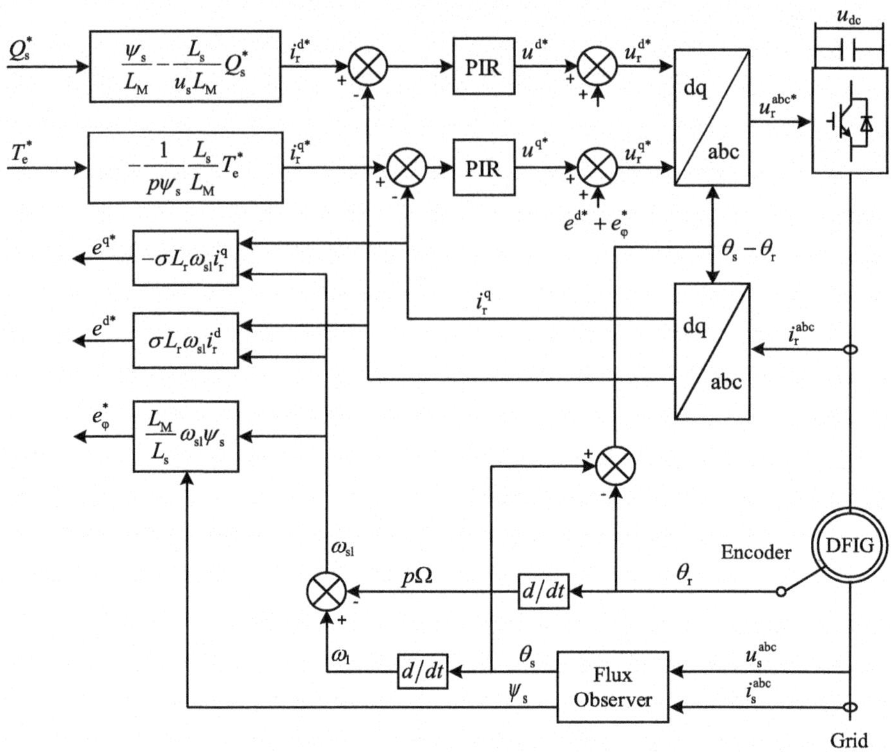

Fig. 5.12 Control diagram of vector control based on PIR controllers

$$U_{\text{or}}(s) = \frac{k_r E(s) - \frac{\omega_0^2}{s} U_{\text{or}}(s)}{s} \tag{5.24}$$

Therefore, the block diagram of $U_o(s)$ is illustrated as

The above results show that the resonant part in the PIR controller can be achieved by the integral calculation, which is a simple solution to implement. When an analogue circuit is used, the PIR controller circuit diagram is shown in Fig. 5.14, where each proportional as well as integral calculation is implemented using operational amplifier circuits.

When a digital control system is used, only the integral part $1/s$ needs to be discretized. The output of the PIR controller can be obtained by simple four arithmetic calculations, and the meaning of each variable is shown in Fig. 5.13. Taking the control period as T_s, the output of the controller at the moment of $(k+1)T_s$ is

$$\begin{cases} U_{\text{op}}(k+1) = k_{\text{p}}E(k+1) \\ U_{\text{oi}}(k+1) = U_{\text{oi}}(k) + k_{\text{i}}E(k+1)T_{\text{s}} \\ X(k+1) = X(k) + U_{\text{or}}(k)T_{\text{s}} \\ U_{\text{or}}(k+1) = U_{\text{or}}(k) + \left[k_{\text{r}}E(k+1) - \omega_0^2 X(k+1)\right]T_{\text{s}} \\ U_{\text{o}}(k+1) = U_{\text{op}}(k+1) + U_{\text{oi}}(k+1) + U_{\text{or}}(k+1) \end{cases} \tag{5.25}$$

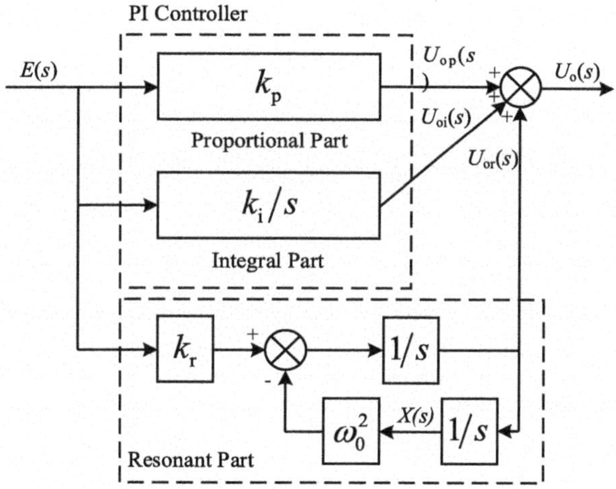

Fig. 5.13 Block diagram of the PIR controller

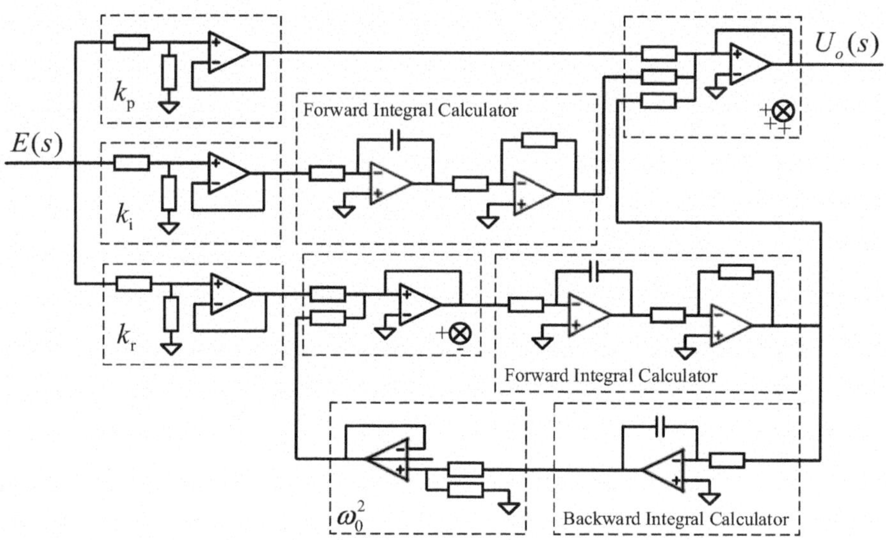

Fig. 5.14 Analog circuit implementation of the PIR controller

From Eq. (5.25), it can be seen that each calculation only needs the history value of each variable last control period and only a small number of four arithmetic calculations, so the PIR controller implemented in the digital control system based on this process has the advantages of simple scheme, small calculation, easy implementation, etc.

Some of the experimental waveforms for the above control method are shown in Fig. 5.15.

5.2.3 *Model Predictive Control with Consideration of the Transient Characteristics of the Flux Linkage*

The MPC method discussed in Chap. 3 is based on the grid-voltage-oriented method, i.e. the grid voltage is oriented on the d-axis of the synchronous coordinate frame, at which point the generator stator flux linkage is considered to satisfy the relation given in Eq. (3.98) for convenience of derivation. From the analysis in Sect. 4.2, it can be seen that during the symmetrical grid fault, the stator flux linkage does not fully satisfy the limits in Eq. (3.98), i.e. both the d-axis and q-axis components of the stator flux linkage suffer from pulsations introduced by their natural components, thus causing a reduction in the control effect of the algorithm, or even a complete failure. In order to achieve uninterrupted operation of the DFIG-based wind power system under the symmetrical grid fault, the transient characteristics of the stator flux linkage need to be fully taken into account in the control algorithm.

5.2.3.1 MPC Strategy for the Rotor Current Loop

The grid voltage orientation is still used to obtain the position relation between the current synchronous coordinate frame and the stationary coordinate frame, but the stator flux linkage is no longer used in the simplified form of Eq. (3.98), but should be used directly in its complete form. Substituting Eq. (2.20) into Eq. (2.19), it gives

$$\begin{cases} u_{\mathrm{r}}^{\mathrm{d}} = R_{\mathrm{r}} i_{\mathrm{r}}^{\mathrm{d}} + \sigma L_{\mathrm{r}} p i_{\mathrm{r}}^{\mathrm{d}} + \dfrac{\omega_1 L_{\mathrm{M}}^2 - \omega_{\mathrm{sl}} L_{\mathrm{s}} L_{\mathrm{r}}}{L_{\mathrm{s}}} i_{\mathrm{r}}^{\mathrm{q}} - \dfrac{L_{\mathrm{M}} R_{\mathrm{s}}}{L_{\mathrm{s}}} i_{\mathrm{s}}^{\mathrm{d}} + \omega_{\mathrm{r}} L_{\mathrm{M}} i_{\mathrm{s}}^{\mathrm{q}} + \dfrac{L_{\mathrm{M}}}{L_{\mathrm{s}}} e_{\mathrm{g}}^{\mathrm{d}} \\ u_{\mathrm{r}}^{\mathrm{q}} = R_{\mathrm{r}} i_{\mathrm{r}}^{\mathrm{q}} + \sigma L_{\mathrm{r}} p i_{\mathrm{r}}^{\mathrm{q}} - \dfrac{\omega_1 L_{\mathrm{M}}^2 - \omega_{\mathrm{sl}} L_{\mathrm{s}} L_{\mathrm{r}}}{L_{\mathrm{s}}} i_{\mathrm{r}}^{\mathrm{d}} - \dfrac{L_{\mathrm{M}} R_{\mathrm{s}}}{L_{\mathrm{s}}} i_{\mathrm{s}}^{\mathrm{q}} - \omega_{\mathrm{r}} L_{\mathrm{M}} i_{\mathrm{s}}^{\mathrm{d}} + \dfrac{L_{\mathrm{M}}}{L_{\mathrm{s}}} e_{\mathrm{g}}^{\mathrm{q}} \\ e_{\mathrm{g}}^{\mathrm{d}} = R_{\mathrm{s}} i_{\mathrm{s}}^{\mathrm{d}} + \sigma L_{\mathrm{s}} p i_{\mathrm{s}}^{\mathrm{d}} + \dfrac{\omega_{\mathrm{sl}} L_{\mathrm{M}}^2 - \omega_1 L_{\mathrm{s}} L_{\mathrm{r}}}{L_{\mathrm{r}}} i_{\mathrm{s}}^{\mathrm{q}} - \dfrac{L_{\mathrm{M}} R_{\mathrm{r}}}{L_{\mathrm{r}}} i_{\mathrm{r}}^{\mathrm{d}} - \omega_{\mathrm{r}} L_{\mathrm{M}} i_{\mathrm{r}}^{\mathrm{q}} + \dfrac{L_{\mathrm{M}}}{L_{\mathrm{r}}} u_{\mathrm{r}}^{\mathrm{d}} \\ e_{\mathrm{g}}^{\mathrm{q}} = R_{\mathrm{s}} i_{\mathrm{s}}^{\mathrm{q}} + \sigma L_{\mathrm{s}} p i_{\mathrm{s}}^{\mathrm{q}} - \dfrac{\omega_{\mathrm{sl}} L_{\mathrm{M}}^2 - \omega_1 L_{\mathrm{s}} L_{\mathrm{r}}}{L_{\mathrm{r}}} i_{\mathrm{s}}^{\mathrm{d}} - \dfrac{L_{\mathrm{M}} R_{\mathrm{r}}}{L_{\mathrm{r}}} i_{\mathrm{r}}^{\mathrm{q}} + \omega_{\mathrm{r}} L_{\mathrm{M}} i_{\mathrm{r}}^{\mathrm{d}} + \dfrac{L_{\mathrm{M}}}{L_{\mathrm{r}}} u_{\mathrm{r}}^{\mathrm{q}} \end{cases} \tag{5.26}$$

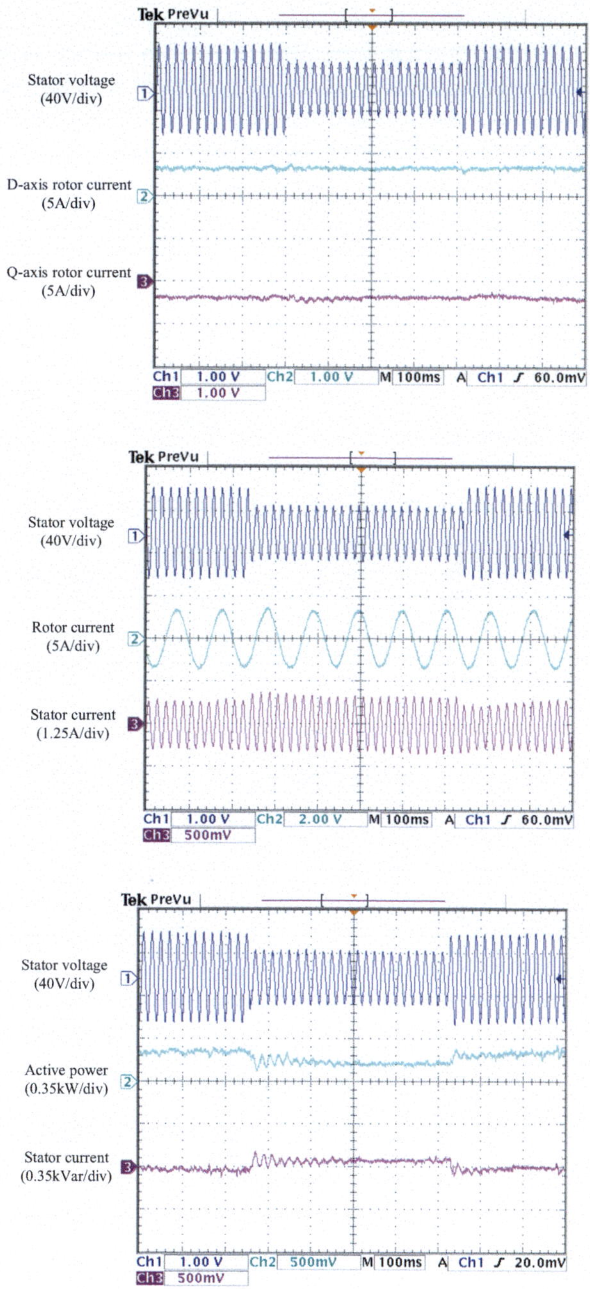

Fig. 5.15 Experimental waveforms of the DFIG LVRT with the PIR-based vector control

Equation (5.26) is the stator voltage equation and rotor voltage equation considering the transient process of the stator flux linkage. At this time, not only the d-axis and q-axis components of the stator current are coupled with each other and as well as the d-axis and q-axis components of the rotor current, but also the stator and rotor currents are coupled with each other, i.e. the d-axis stator current, the q-axis stator current, the d-axis rotor current and the q-axis rotor current are coupled with each other, which is quite complicated.

Similar to the rotor current prediction model under ideal grid conditions, a two-step prediction of the rotor current is performed using the forward Eulerian discretization of Eq. (5.26), taking into account the delay effect of the digital control chip. It is important to note that the stator current will also affect the rotor current, so a step for predicting the stator current is added between the two prediction steps for the rotor current. The two-step rotor current prediction model with a feedback correction term can be obtained as

$$\begin{cases} i_{\mathrm{rm}}^{\mathrm{d}}(k+1|k) = a_{\mathrm{r}} i_{\mathrm{r}}^{\mathrm{d}}(k) + b_{\mathrm{r}} i_{\mathrm{r}}^{\mathrm{q}}(k) + c_{\mathrm{r}} e_{\mathrm{g}}^{\mathrm{d}}(k) + d_{\mathrm{r}} u_{\mathrm{r}}^{\mathrm{d}}(k-1) + g_{\mathrm{r}} i_{\mathrm{s}}^{\mathrm{d}}(k) + h_{\mathrm{r}} i_{\mathrm{s}}^{\mathrm{q}}(k) + x_{\mathrm{r}}^{\mathrm{d}}(k) \\ i_{\mathrm{rm}}^{q}(k+1|k) = a_{\mathrm{r}} i_{\mathrm{r}}^{\mathrm{q}}(k) - b_{\mathrm{r}} i_{\mathrm{r}}^{\mathrm{d}}(k) + c_{\mathrm{r}} e_{\mathrm{g}}^{\mathrm{q}}(k) + d_{\mathrm{r}} u_{\mathrm{r}}^{\mathrm{q}}(k-1) + g_{\mathrm{r}} i_{\mathrm{s}}^{\mathrm{q}}(k) - h_{\mathrm{r}} i_{\mathrm{s}}^{\mathrm{d}}(k) + x_{\mathrm{r}}^{\mathrm{q}}(k) \end{cases} \tag{5.27}$$

$$\begin{cases} i_{\mathrm{sm}}^{\mathrm{d}}(k+1|k) = a_{\mathrm{s}} i_{\mathrm{s}}^{\mathrm{d}}(k) + b_{\mathrm{s}} i_{\mathrm{s}}^{\mathrm{q}}(k) + c_{\mathrm{s}} e_{\mathrm{g}}^{\mathrm{d}}(k) + d_{\mathrm{s}} u_{\mathrm{r}}^{\mathrm{d}}(k-1) + g_{s} i_{\mathrm{r}}^{\mathrm{d}}(k) + h_{\mathrm{s}} i_{\mathrm{r}}^{\mathrm{q}}(k) + x_{\mathrm{s}}^{\mathrm{d}}(k) \\ i_{\mathrm{sm}}^{\mathrm{q}}(k+1|k) = a_{\mathrm{s}} i_{\mathrm{s}}^{\mathrm{q}}(k) - b_{\mathrm{s}} i_{\mathrm{s}}^{\mathrm{d}}(k) + c_{\mathrm{s}} e_{\mathrm{g}}^{\mathrm{q}}(k) + d_{\mathrm{s}} u_{\mathrm{r}}^{\mathrm{q}}(k-1) + g_{s} i_{\mathrm{r}}^{\mathrm{q}}(k) - h_{\mathrm{s}} i_{\mathrm{r}}^{\mathrm{d}}(k) + x_{\mathrm{s}}^{\mathrm{q}}(k) \end{cases} \tag{5.28}$$

$$\begin{cases} i_{\mathrm{rm}}^{\mathrm{d}}(k+2|k) = a_{\mathrm{r}} i_{\mathrm{rm}}^{\mathrm{d}}(k+1|k) + b_{\mathrm{r}} i_{\mathrm{rm}}^{\mathrm{q}}(k+1|k) + c_{\mathrm{r}} e_{\mathrm{g}}^{\mathrm{d}}(k+1) + d_{\mathrm{r}} u_{\mathrm{r}}^{\mathrm{d}}(k-1) \\ \qquad + d_{\mathrm{r}} \Delta u_{\mathrm{r}}^{\mathrm{d}}(k) + g_{\mathrm{r}} i_{\mathrm{sm}}^{\mathrm{d}}(k+1|k) + h_{\mathrm{r}} i_{\mathrm{sm}}^{\mathrm{q}}(k+1|k) + x_{\mathrm{r}}^{\mathrm{d}}(k) \\ i_{\mathrm{rm}}^{\mathrm{q}}(k+2|k) = a_{\mathrm{r}} i_{\mathrm{rm}}^{\mathrm{q}}(k+1|k) - b_{\mathrm{r}} i_{\mathrm{rm}}^{\mathrm{d}}(k+1|k) + c_{\mathrm{r}} e_{\mathrm{g}}^{\mathrm{q}}(k+1) + d_{\mathrm{r}} u_{\mathrm{r}}^{\mathrm{q}}(k-1) \\ \qquad + d_{\mathrm{r}} \Delta u_{\mathrm{r}}^{\mathrm{q}}(k) + g_{\mathrm{r}} i_{\mathrm{sm}}^{\mathrm{q}}(k+1|k) - h_{\mathrm{r}} i_{\mathrm{sm}}^{\mathrm{d}}(k+1|k) + x_{\mathrm{r}}^{\mathrm{q}}(k) \end{cases} \tag{5.29}$$

where $k \in N$, $i_{\mathrm{rm}}^{\mathrm{d}}(k+1|k)$ and $i_{\mathrm{rm}}^{\mathrm{q}}(k+1|k)$ indicate the d-axis and q-axis components of the rotor current predicted at the moment of kT_{s} for the moment of $(k+1)T_{\mathrm{s}}$, respectively, $i_{\mathrm{sm}}^{\mathrm{d}}(k+1|k)$ and $i_{\mathrm{sm}}^{\mathrm{q}}(k+1|k)$ indicate the d-axis and q-axis components of the stator current predicted at the moment of kT_{s} for the moment of $(k+1)T_{\mathrm{s}}$, respectively, $i_{\mathrm{rm}}^{\mathrm{d}}(k+2|k)$ and $i_{\mathrm{rm}}^{\mathrm{q}}(k+2|k)$ indicate the d-axis and q-axis components of the rotor current predicted at the moment of kT_{s} for the moment of $(k+2)T_{\mathrm{s}}$, respectively, $i_{\mathrm{r}}^{\mathrm{d}}(k)$ and $i_{\mathrm{r}}^{\mathrm{q}}(k)$ indicate the d-axis and q-axis components of the rotor current sampled at the moment of kT_{s}, respectively, $i_{\mathrm{s}}^{\mathrm{d}}(k)$ and $i_{\mathrm{s}}^{\mathrm{q}}(k)$ indicate the d-axis and q-axis components of the stator current sampled at the moment of kT_{s}, respectively, $e_{\mathrm{g}}^{\mathrm{d}}(k)$ and $e_{\mathrm{g}}^{\mathrm{q}}(k)$ indicate the d-axis and q-axis components of the grid voltage sampled at the moment of kT_{s}, respectively, $e_{\mathrm{g}}^{\mathrm{d}}(k+1)$ and $e_{\mathrm{g}}^{\mathrm{q}}(k+1)$ indicate the d-axis and q-axis components of the grid voltage sampled at the moment of $(k+1)T_{\mathrm{s}}$,[1]

[1] Under symmetrical grid conditions, neglecting the process of voltage jumps, it can be considered that $e_{\mathrm{g}}^{\mathrm{d}}(k+1) = e_{\mathrm{g}}^{\mathrm{d}}(k)$ and $e_{\mathrm{g}}^{\mathrm{q}}(k+1) = e_{\mathrm{g}}^{\mathrm{q}}(k)$.

respectively, $u_r^d(k-1)$ and $u_r^q(k-1)$ indicate the d-axis and q-axis components of the control voltage, respectively, given by the algorithm at the moment of $(k-1)T_s$, $\Delta u_r^d(k)$ and $\Delta u_r^q(k)$ indicate the d-axis and q-axis components of the control voltage increment, respectively, given by the algorithm at the moment of kT_s, $x_r^d(k)$ and $x_r^q(k)$ indicate the feedback correction factors of the d-axis and q-axis components of the rotor current, respectively, which includes the error between the rotor current predicted at the moment of $(k-1)T_s$ for the moment of kT_s and the rotor current sampled at the moment of kT_s, $x_s^d(k)$ and $x_s^q(k)$ indicate the feedback correction factors of the d-axis and q-axis components of the stator current, respectively, which includes the error between the stator current predicted at the moment of $(k-1)T_s$ for the moment of kT_s and the stator current sampled at the moment of kT_s.

The coefficients in Eqs. (5.27)–(5.29) and the expressions for the stator-current and rotor-current feedback correction terms are

Coefficients in rotor-current expression
$$\begin{cases} a_r = 1 - R_r T_s/(\sigma L_r) \\ b_r = \left(\omega_{sl} L_s L_r - \omega_1 L_M^2\right) T_s \big/ (\sigma L_s L_r) \\ c_r = -L_M T_s/(\sigma L_s L_r) \\ d_r = T_s/(\sigma L_r) \\ g_r = R_s L_M T_s/(\sigma L_s L_r) \\ h_r = -\omega_r L_M T_s/(\sigma L_r) \end{cases}$$

Coefficients in stator-current expression
$$\begin{cases} a_s = 1 - R_s T_s/(\sigma L_s) \\ b_s = \left(\omega_1 L_s L_r - \omega_{sl} L_M^2\right) T_s \big/ (\sigma L_s L_r) \\ c_s = T_s/(\sigma L_s) \\ d_s = -L_M T_s/(\sigma L_s L_r) \\ g_s = R_r L_M T_s/(\sigma L_s L_r) \\ h_s = \omega_r L_M T_s/(\sigma L_s) \end{cases}$$

Feedback correction terms of rotor current
$$\begin{cases} x_r^d(k) = f_r^d \cdot \left[i_r^d(k) - i_{rm}^d(k|k-1)\right] \\ x_r^q(k) = f_r^q \cdot \left[i_r^q(k) - i_{rm}^q(k|k-1)\right] \end{cases}$$

Feedback correction terms of stator current
$$\begin{cases} x_s^d(k) = f_s^d \cdot \left[i_s^d(k) - i_{sm}^d(k|k-1)\right] \\ x_s^q(k) = f_s^q \cdot \left[i_s^q(k) - i_{sm}^q(k|k-1)\right] \end{cases}$$

where f_r^d, f_r^q, f_s^d and f_s^q are the coefficients in the feedback correction terms of the stator and rotor currents, respectively.

At this point, the evaluation function of the rotor current loop is still as shown in Eq. (3.97). By substituting Eq. (5.29) into Eq. (3.97), the optimal increment of rotor voltage that minimizes the evaluation function can be found as

$$
\begin{cases}
\Delta u_{\mathrm{r}}^{\mathrm{d}*}(k) = \dfrac{d_{\mathrm{r}}\varepsilon_{\mathrm{r}}^{\mathrm{d}}}{d_{\mathrm{r}}^{2}\varepsilon_{\mathrm{r}}^{\mathrm{d}} + \lambda_{\mathrm{r}}^{\mathrm{d}}}\Big[i_{\mathrm{r}}^{\mathrm{d}*}(k) - a_{\mathrm{r}} i_{\mathrm{rm}}^{\mathrm{d}}(k+1|k) - b_{\mathrm{r}} i_{\mathrm{rm}}^{\mathrm{q}}(k+1|k) - c_{\mathrm{r}} e_{\mathrm{g}}^{\mathrm{d}}(k+1) \\
\qquad - d_{\mathrm{r}} u_{\mathrm{r}}^{\mathrm{d}}(k-1) - g_{\mathrm{r}} i_{\mathrm{sm}}^{\mathrm{d}}(k+1|k) - h_{\mathrm{r}} i_{\mathrm{sm}}^{\mathrm{q}}(k+1|k) - x_{\mathrm{r}}^{\mathrm{d}}(k)\Big] \\
\Delta u_{\mathrm{r}}^{\mathrm{q}*}(k) = \dfrac{d_{\mathrm{r}}\varepsilon_{\mathrm{r}}^{\mathrm{q}}}{d_{\mathrm{r}}^{2}\varepsilon_{\mathrm{r}}^{\mathrm{q}} + \lambda_{\mathrm{r}}^{\mathrm{q}}}\Big[i_{\mathrm{r}}^{\mathrm{q}*}(k) - a_{\mathrm{r}} i_{\mathrm{rm}}^{\mathrm{q}}(k+1|k) + b_{\mathrm{r}} i_{\mathrm{rm}}^{\mathrm{d}}(k+1|k) - c_{\mathrm{r}} e_{\mathrm{g}}^{\mathrm{q}}(k+1) \\
\qquad - d_{\mathrm{r}} u_{\mathrm{r}}^{\mathrm{q}}(k-1) - g_{\mathrm{r}} i_{\mathrm{sm}}^{\mathrm{q}}(k+1|k) + h_{\mathrm{r}} i_{\mathrm{sm}}^{\mathrm{d}}(k+1|k) - x_{\mathrm{r}}^{\mathrm{q}}(k)\Big]
\end{cases}
\tag{5.30}
$$

where $\Delta u_{\mathrm{r}}^{\mathrm{d}*}(k)$ and $\Delta u_{\mathrm{r}}^{\mathrm{q}*}(k)$ indicate the d-axis and q-axis components of the optimal increment of the converter output voltage given by the control algorithm at sample moment of kT_{s}, respectively. Then, at any sampling moment of kT_{s}, the control algorithm can obtain the optimal increment of the converter output voltage at the current moment according to Eqs. (5.27), (5.28) and (5.29). By accumulating the optimal increment, the current converter optimal output voltage, consisting of the d-axis and q-axis components as $u_{\mathrm{r}}^{\mathrm{d}*}(k)$ and $u_{\mathrm{r}}^{\mathrm{q}*}(k)$ respectively, can be obtained.

In summary, the algorithm diagram of the rotor current loop based on MPC considering the dynamic process of the flux linkage can be obtained as Fig. 5.16.

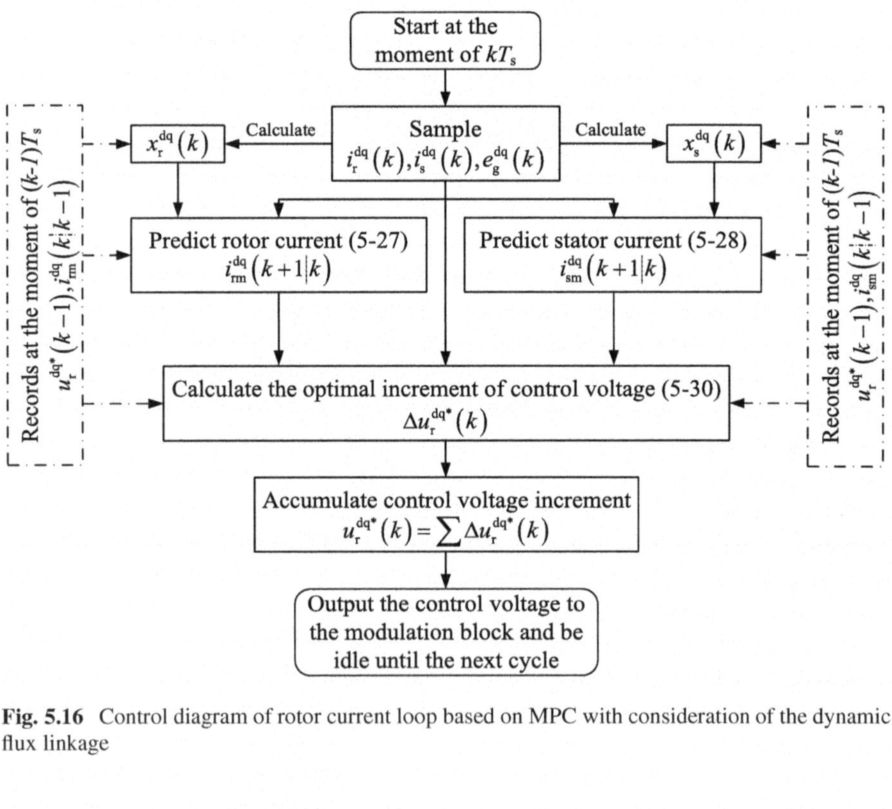

Fig. 5.16 Control diagram of rotor current loop based on MPC with consideration of the dynamic flux linkage

5.2.3.2 MPC Strategy for the Stator Power

From the analysis in Sect. 4.2.1, it is clear that when the symmetrical grid fault occurs, controlling the stator current forcefully to be constant before the stator flux linkage returns to its steady state is not beneficial to the decay of the natural component of the stator flux linkage, for this reason, an open-loop stator power control strategy will be used here for the DFIG under the symmetrical grid fault. The stator current reference can be obtained from the stator power reference as shown in Eq. (3.104). Substituting Eq. (3.104) into Eq. (3.99), the rotor current reference is obtained as

$$\begin{cases} i_{\mathrm{r}}^{\mathrm{d}*}(k) = -\dfrac{L_{\mathrm{s}}P_{\mathrm{s}}^{*}}{L_{\mathrm{m}}e_{\mathrm{g}}^{\mathrm{d}}} \\ i_{\mathrm{r}}^{\mathrm{q}*}(k) = \dfrac{1}{L_{\mathrm{m}}}\left(\dfrac{e_{\mathrm{g}}^{\mathrm{d}}}{\omega_{\mathrm{s}}} + \dfrac{L_{\mathrm{s}}Q_{\mathrm{s}}^{*}}{e_{\mathrm{g}}^{\mathrm{d}}}\right) \end{cases} \tag{5.31}$$

As can be seen from Eq. (5.31), the rotor current reference value is calculated from the stator-rotor current relation when the stator flux linkage is in steady state, which can reduce the frequent changes of the rotor current reference value during the control process and is conducive to the stability of the stator flux linkage.

In summary, the control diagram of the generator-side converter for the DFIG-based wind power system under the symmetrical grid fault is shown in Fig. 5.17.

Compared to the various improved vector control methods in Sects. 5.2.1, the MPC-based method discussed in this section is less sensitive to generator parameters and shows better parameter robustness. Some of its simulation results are shown in Figs. 5.18 and 5.19.

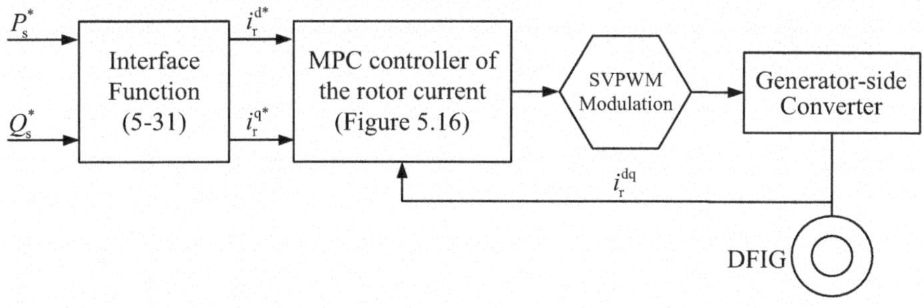

Fig. 5.17 Control diagram of the generator-side converter under symmetrical grid faults

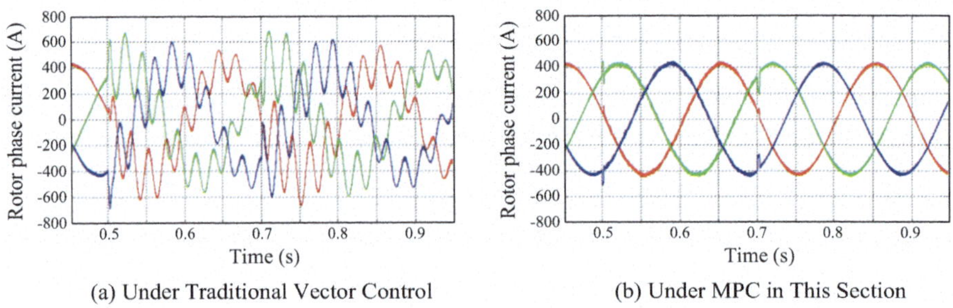

Fig. 5.18 Simulation results of rotor currents under the symmetrical grid fault. **a** Under traditional vector control. **b** Under MPC in this section

5.3 Hardware Protection Strategies

The most commonly used hardware protection circuit in the DFIG-based wind power system for LVRT is the rotor energy unit, named as crowbar, which is available in a variety of topologies, whereas the most widely used is the active crowbar consisting of an IGBT and an unloading resistor. In the event of a heavy symmetrical voltage fault in the grid, the crowbar is able to provide an additional, temporary energy pathway for transient energy in the DFIG-based wind power system to protect the generator-side converter from damage and help recover the control of the DFIG-based system.

5.3.1 Design and Control of the Crowbar

The crowbar topology discussed in this section is shown in Fig. 5.20. It consists of a three-phase bi-directional switch in series with an unloading resistor R_{crow}. The crowbar is connected to the rotor winding and paralleled to the generator-side converter. When a heavy voltage fault occurs in the grid and causes a surge in rotor current, the bi-directional switch in the crowbar will be switched on and bypass the generator-side converter, and the unloading resistor is directly connected to the rotor, helping the system to release energy and accelerate the transient process of the flux linkage, preventing the generator-side converter from overcurrent, thus protecting the system hardware, and at an appropriate moment the crowbar will be cut out and the generator-side converter will be controlled to resume operation, thus achieving grid fault ride-through of the system.

In the crowbar unit, the selection of the unloading resistance is an important issue in relation to the operational characteristics of the system. The smaller the unloading resistance, the faster the recovery of the system transient process triggered by grid faults, but a too small unloading resistance can cause problems such as excessive rotor current and electromagnetic torque shock; the larger the unloading resistance, the lower the rotor current and the smaller the electromagnetic torque shock, but

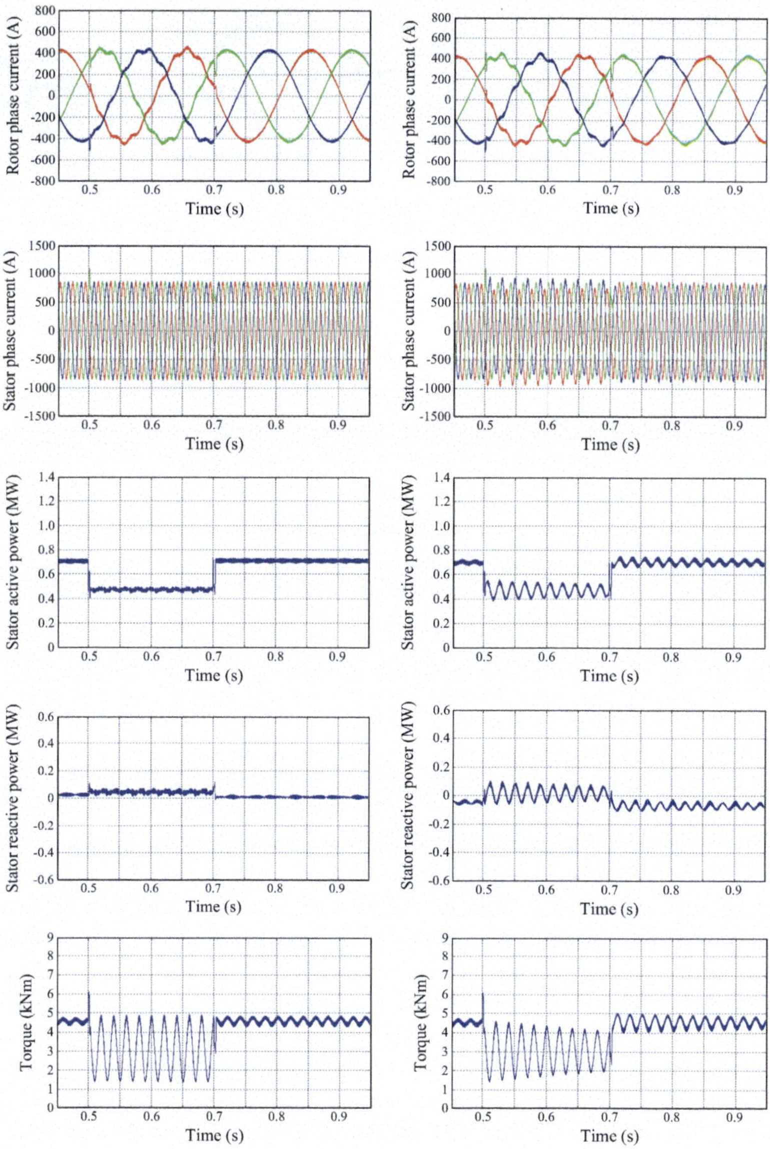

Fig. 5.19 Simulation results of LVRT with DFIG parameter errors[2] (Mutual inductance errors of +30% and −30% for left and right columns respectively)

[2] Since the (symmetrical) LVRT control of the DFIG-based wind power system is mainly related to the generator-side converter and has little influence on the grid-side converter, therefore the grid-side converter in the simulated system here is still use the MPC strategy described in Sect. 3.3.1.

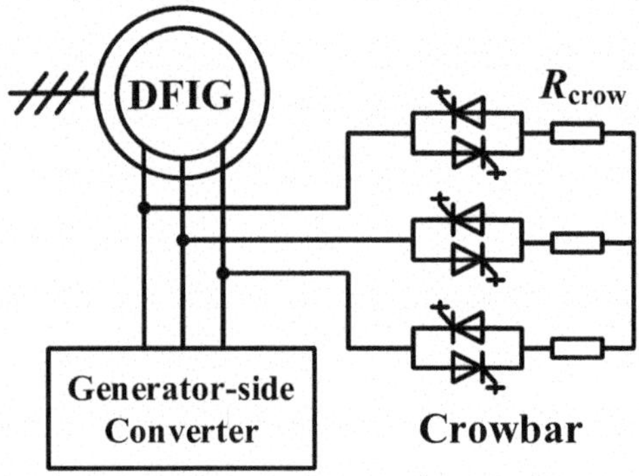

Fig. 5.20 DFIG and its crowbar

a too large unloading resistance can cause problems such as slow system transient process and DC bus overvoltage. As the transient characteristics of the DFIG in the crowbar access state are not yet well described analytically, the selection of the crowbar resistance is mostly based on a range estimation (e.g. Eq. (5.1)) combined with simulation.

Under heavy grid fault conditions, the system can be protected by the crowbar cut-in to achieve fault ride-through. However, once the crowbar cuts into the system, the generator-side converter will be bypassed and the DFIG will operate in the mode of a wound asynchronous generator, where the excitation current will no longer be supplied from the rotor side but from the stator side, thus absorbing a large amount of reactive power from the grid, which is detrimental to the recovery from grid faults. Therefore, the control strategy should minimise the effective cut-in period of the crowbar while protecting the power generation system. This section introduces a rotor-current hysteresis-loop control strategy, the control diagram of which is shown in Fig. 5.21. Wherein, the control strategy samples the rotor phase current on in real time, get to its current amplitude through Clarke transformation, and input it to the current hysteresis controller, where it is compared with the current threshold: When the rotor current amplitude is greater than the on-state threshold $i_{r\_crowbaron}$, the controller will output turn on the three-phase bi-directional switching device in the crowbar and bypass the generator-side converter, at which time the crowbar unit will cut into the system, suppress the rotor overcurrent and accelerate the system transient process; when the rotor current amplitude is lower than the off-state threshold $i_{r\_crowbaroff}$, the controller will turn off the three-phase bi-directional switching device in the crowbar unit and enable the generator-side converter, at which time the crowbar unit will cut out of the system and the generator-side converter will resume the control of the system and complete the control of the generator according to the preset control target. The above crowbar control strategy can monitor the rotor current change in real time and dynamically regulate the cut-in and cut-out actions of the crowbar according to the current amplitude change, reducing the effective cut-in period of the crowbar while protecting the system, reducing the absorption of reactive power by the power generation system during grid faults and facilitating the recovery from grid faults.

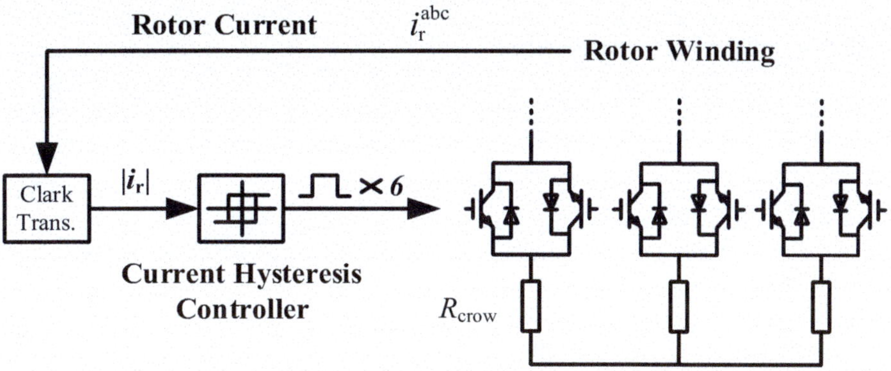

Fig. 5.21 Current hysteresis control of crowbar

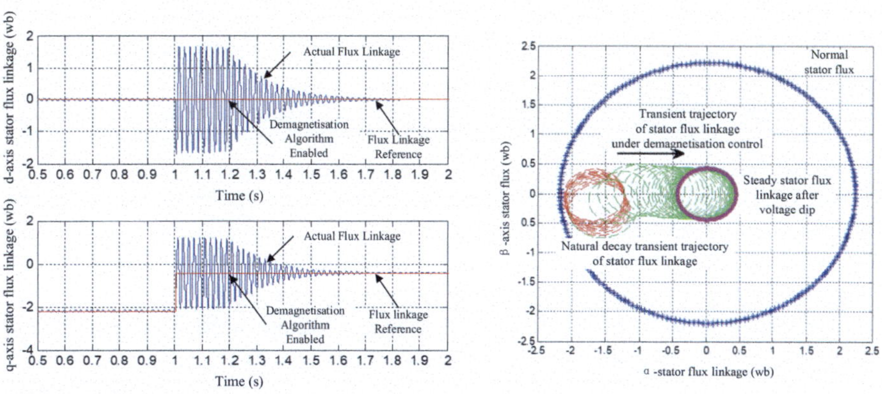

Fig. 5.22 Dynamic process of stator flux linkage of DFIG under the active demagnetisation algorithm (Symmetrical dip in grid voltage at 0.1 s)

5.3.2 Active Demagnetisation Control

In order to accelerate the decay rate of the electromagnetic transient process of the DFIG under deep voltage dips in the grid and to provide the necessary reactive support to the grid during faults, appropriate control of the DFIG and the grid-side converter is required.

The analysis in Sect. 4.2.1 shows that the main reason for the violent electromagnetic transition process in the DFIG under voltage dips is the generation of the DC component in the stator flux linkage. For this reason, the generator can be actively demagnetised by controlling the rotor current to counteract the effect of the DC component in the stator flux linkage on the rotor side. This is done by closed-loop control of the stator flux linkage so that its d-axis and q-axis components transition to their steady-state values as soon as possible after the voltage dip. Based on the relations among the generator electromagnetic variables, this closed-loop controller of the flux linkage can be designed as

$$i_{\mathrm{r}}^{\mathrm{d}*} = C_{\mathrm{PI}}^{\mathrm{d}}[\psi_{\mathrm{s}}^{\mathrm{d}*} - \psi_{\mathrm{s}}^{\mathrm{d}}] \tag{5.32}$$

$$i_{\mathrm{r}}^{\mathrm{q}*} = C_{\mathrm{PI}}^{\mathrm{d}}[\psi_{\mathrm{s}}^{\mathrm{q}*} - \psi_{\mathrm{s}}^{\mathrm{q}}] \tag{5.33}$$

where $C_{\mathrm{PI}}^{\mathrm{d}}$ and $C_{\mathrm{PI}}^{\mathrm{q}}$ are the PI controllers for the d-axis and q-axis components of the stator flux linkage, respectively, $\psi_{\mathrm{s}}^{\mathrm{d}}$ and $\psi_{\mathrm{s}}^{\mathrm{q}}$ are the observed values of the d-axis and q-axis components of the stator flux linkage, respectively, and the reference value of the stator flux linkage is its forced component under the faulty grid voltage, which can be written as

$$\psi_{\mathrm{s}}^{\mathrm{d}*} = \frac{U_{\mathrm{s}}}{\omega_1} \tag{5.34}$$

$$\psi_{\mathrm{s}}^{\mathrm{q}*} = 0 \tag{5.35}$$

It should be noted that the demagnetisation control can only take effect after the crowbar has been cut out, as the generator-side converter is bypassed when the crowbar is in operation. After the grid fault has been cleared, the recovery of the grid voltage will cause the stator flux linkage of the DFIG to oscillate again, so it is also necessary to use the crowbar protection and the demagnetisation control together to reduce the electromagnetic transition time of the generator (Fig. 5.22).

Comparing the results in Fig. 5.23, it can be seen that the active demagnetisation method in this section stabilises the generator stator flux linkage faster and has a better demagnetisation effect than the method that relies on crowbar cut-in, but this is achieved relying on the effective control of the rotor current by the generator-side converter, and the generator-side converter may experience high current throughout the process, which results in a greater potential to cause overcurrent in the generator-side converter under heavy grid voltage dip conditions. Therefore, a combination of crowbar and active demagnetisation algorithms can be used to demagnetise the generator for a period of time after the grid voltage dip and recovery, and then the system can be protected using the crowbar unit, which prevents overcurrent faults in the generator-side converter and enables the generator stator flux linkage to be restored to stability as soon as possible, thus resuming the control of the generator by the generator-side converter.

If the voltage dip lasts for such a long time that the grid voltage is still in a dip after the stator flux linkage has decayed to its steady value under the combination of the active crowbar method and the demagnetisation control, it is also necessary to control the generator to output as much reactive current as possible to the grid in order to achieve maximum reactive power compensation for the grid voltage.

Compared to the generator-side converter, the grid-side converter is less affected by grid voltage dips. As the DFIG needs to absorb reactive power from the grid for

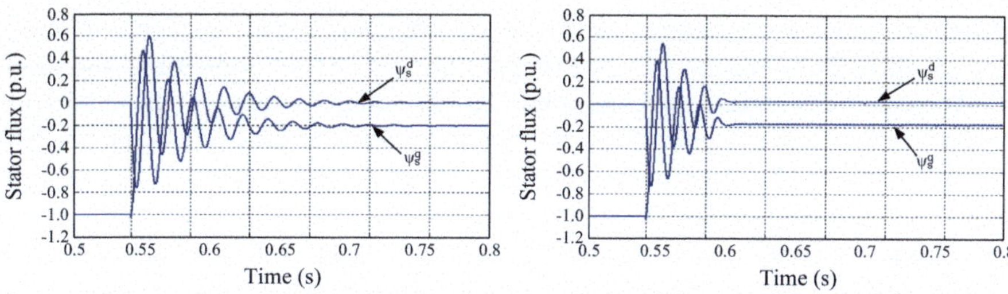

(a) Stator Flux After Crowbar Cut-in (b) Stator Flux under Demagnetisation Control

Fig. 5.23 Performance comparison between crowbar-based method and demagnetisation method. **a** Stator flux after crowbar cut-in. **b** Stator flux under demagnetisation control

excitation after the crowbar has been switched on, the grid-side converter is required to output as much reactive power as possible to the grid during a grid voltage dip, in addition to controlling the DC-bus voltage recovery to the reference value, to compensate for the reactive power absorbed by the generator on the one hand and to provide reactive power support for grid voltage recovery on the other hand.

5.3.3 Crowbar-Based LVRT for Heavy Grid Faults

Combining the rotor current hysteresis control of the active crowbar, the active demagnetisation control of the stator flux linkage, the improved vector control method for LVRT and the reactive power compensation scheme for the whole wind power system, it is possible to obtain the LVRT strategy for heavy grid faults using the crowbar hardware protection unit in the following steps:

Step 1: After a voltage dip at the stator terminal caused by a grid fault, the active crowbar is controlled to protect the rotor from overcurrent by the rotor current hysteresis loop based on the rotor current measurement, providing a pathway for the rotor transient overcurrent. As soon as the active crowbar cuts out of the system, the generator-side converter immediately starts to demagnetise the generator, further accelerating the decay of the generator stator transient flux linkage. Due to the output voltage capability of the converter, the de-magnetisation control may trigger the active crowbar protection again, but the rotor current will return to a safe range again due to the unloading resistor. In this way, the electromagnetic transient process of the DFIG quickly transitions to a steady state due to the combined effect of the active-crowbar rotor-current hysteresis control and the generator demagnetisation control. During the electromagnetic transition of the DFIG, the main task of the grid-side converter is to maintain the stability of the DC-bus voltage and to prevent overvoltage on the DC side.

Step 2: Once the DFIG has transitioned to a steady state under grid voltage dips, the generator's control strategy switches from demagnetisation control to normal control. At this point the generator-side converter controls the generator to inject as much reactive current as possible into the grid. At the same time, the grid-side converter is required to control the DC-bus voltage and output the maximum reactive power to the grid. Thus, the generator and the grid-side converter can together provide reactive power support to the grid during faults.

Step 3: When the grid voltage is recovered after the fault is cleared, similar to the first step, the DFIG is again controlled by the combination of the active-crowbar rotor-current hysteresis method and the generator demagnetisation control to transition the DFIG to a steady state operation as soon as possible. During this process, it is also necessary to control the grid-side converter to feed the transient energy flowing from the rotor to the DC bus to the grid in order to maintain the stability of the DC-bus voltage.

Step 4: After the DFIG has returned to a steady state, the generator-side converter resumes normal operation and controls the generator to resume active power output as soon as possible and return to maximum power point tracking mode. At the same time the grid-side converter controls the DC-bus voltage to its reference value and recovers to unit power factor operation.

Figure 5.24 shows the LVRT waveform of the DFIG-based wind power system in the simulation system. At the moment of 3 s the grid voltage dips to 15% of its rated value and the improved vector control strategy of the DFIG still loses effective control of the rotor current due to the capacity limitation of the generator-side converter, so overcurrent occurs in the rotor winding. With the protection of the active crowbar circuit, the rotor overcurrent does not flow into the generator-side converter and therefore effectively protects the converter from damage. With the hysteresis protection of the crowbar and the demagnetisation control, the stator flux linkage decays to a stable value within 100 ms. Since the grid voltage has not yet recovered, the generator-side converter starts to control the generator to output reactive current to the grid, while the grid-side converter also starts to inject the maximum reactive current into the grid, so that the voltage at the generator terminal rises. As can be seen from the trigger signal of the active crowbar during the voltage dip, the protection circuit only works for a very short time, which means that the DFIG is controllable for most of the time during the voltage dip and only absorbs a small amount of reactive power from the grid. at the moment of 3.5 s, the grid voltage is restored, which also triggers the protection circuit of the crowbar, and after about 300 ms the system is able to return to its pre-fault normal operation and deliver the rated power to the grid.

The simulation results show that with the help of the LVRT technology, the DFIG-based wind power system can still meet the requirements of the grid during deep voltage dips, remain connected to the grid during grid faults and also provide some reactive power support to the grid to help recover the grid voltage; after the grid fault is cleared the system can quickly return to its pre-fault normal operation to maintain the stability of the grid.

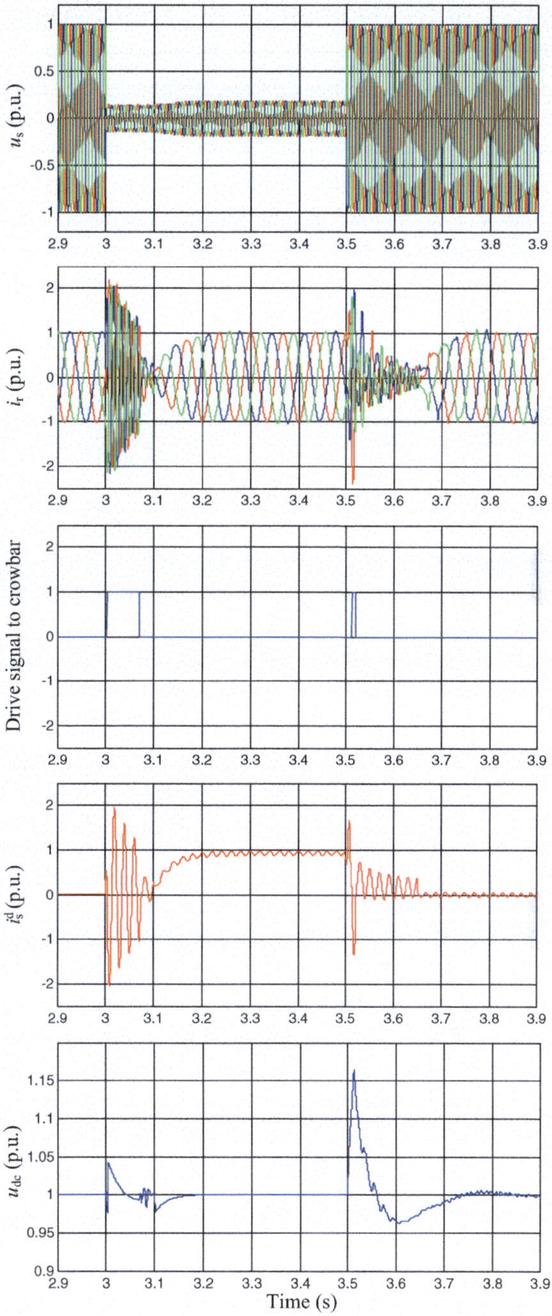

Fig. 5.24 LVRT simulation under crowbar-based strategy for heavy grid faults ($R_{crow} = 0.1|Z_s|$, where Z_s is the stator impedance)

5.4 Control Domain of the LVRT Algorithm

As can be seen from the previous analysis, when the DFIG-based wind power system undergoes LVRT, the rotor current is controlled by increasing the output voltage of the generator-side converter to suppress rotor overcurrent; at the same time, the grid-side converter maintains the stability of the DC-bus voltage by transferring the pulsating slip power to the grid. As the DFIG-based wind power system uses the partial-scale power converter, the capacity of the generator-side converter to output voltage and the capacity of the grid-side converter to transfer the slip power are both limited by the capacity of the hardware itself.

When the rotor current vector is constant, substitute Eq. (4.3) into Eq. (4.2) to obtain the control voltage vector that the generator-side converter needs to output as

$$\boldsymbol{u}_{\mathrm{r}} = (R_{\mathrm{r}} + j\omega_{\mathrm{sl}}\sigma L_{\mathrm{r}})\boldsymbol{i}_{\mathrm{r}} + (p + j\omega_{\mathrm{sl}})\frac{L_{\mathrm{M}}}{L_{\mathrm{s}}}\boldsymbol{\Psi}_{\mathrm{s}} \tag{5.36}$$

The above equation shows that the control voltage $\boldsymbol{u}_{\mathrm{r}}$ required from the generator-side converter is related to the current rotor current, the rotor slip and the stator flux linkage. For a wind power system operating under normal grid conditions, the rotor current in steady state can be considered to be a slip-related quantity, i.e. $\boldsymbol{u}_{\mathrm{r}}$ can be considered to be related to the current system slip and the state of the stator flux linkage. The slip depends on the current wind speed and the command of the main control system, while the stator flux linkage is related to the voltage dip depth d_{sag} and the voltage recovery moment t_{rec}, as can be seen from the analysis in Sect. 4.2.1. The required control voltage of the rotor is therefore noted as a function of the slip, the grid voltage dip depth, and the grid voltage recovery moment:

$$\boldsymbol{u}_{\mathrm{r}} = \mathrm{function}\left(\omega_{\mathrm{sl}}, d_{\mathrm{sag}}, t_{\mathrm{rec}}\right) \tag{5.37}$$

For a generator-side converter with SVPWM modulation strategy, the maximum voltage vector output amplitude is approximately $0.866u_{\mathrm{dc}}$. Therefore, the generator-side converter is controllable provided that the conditions shown in Eq. (5.38) are satisfied. For the case where Eq. (5.38) is not satisfied, the rotor current vector will not be controlled to be constant and the rotor current will increase and its harmonics will appear, but the system can still be considered to achieve the uninterrupted operation under grid faults only so long as it has not reached the upper limit of the rotor converter current resistance.

$$|\boldsymbol{u}_{\mathrm{r}}| = \left|\mathrm{function}\left(\omega_{\mathrm{sl}}, d_{\mathrm{sag}}, t_{\mathrm{rec}}\right)\right| \le 0.866u_{\mathrm{dc}} \tag{5.38}$$

The slip power from the rotor side of the generator is the main disturbance to the DC-bus voltage loop of the grid-side converter during grid fault ride-through, requiring the slip power not to exceed the maximum power transfer capability of the grid-side converter, i.e.

$$|P_{\rm r}| = \left|{\rm Re}\left(\boldsymbol{u}_{\rm r}\boldsymbol{i}_{\rm r}^{*}\right)\right| \leq 0.866u_{\rm dc} \tag{5.39}$$

where $P_{\rm r}$ is the slip power, determined by the rotor voltage shown in Eq. (5.38) and the rotor current. It can be seen that the slip power during grid fault ride-through is also mainly related to the slip, voltage dip depth and voltage recovery time; $\left|P_{\rm g}\right|_{\rm max}$ indicates the maximum power that can be exchanged with the grid under the hardware conditions of the grid-side converter, which is mainly determined by the maximum current of the grid-side converter and the grid conditions.

From the above analysis, it can be seen that the controllable range of the DFIG-based system during the symmetrical grid fault ride-through process is closely related to its operating state, the grid fault state and the selection of system hardware parameters, especially when the rotor current vector fails to be controlled as a constant but the converter is not yet overcurrent, the electromagnetic processes of the system are complex and difficult to be described analytically, so the control domain of the system will be investigated here by means of simulation.

In a 1.5 MW DFIG-based wind power simulation system, according to the system operation state and design requirements, the IGBT of the grid-side converter is FF700R17K (1700 V, 700 A) and the IGBT of the generator-side converter is FF1200R17K (1700 V, 1200 A). The control domain of the system during symmetric fault ride-through is shown in Fig. 5.25, where the points marked in black are the state points for simulation. Wherein Fig. 5.25a shows the control effect of the control strategy at different operating areas of the system when the grid voltage dips, and Fig. 5.25b shows the control effect of the control strategy at different operating areas of the system when the grid voltage is recovered. According to the analysis in Sect. 4.2.1, it can be considered that the electromagnetic process of the system is the most violent and the operating conditions are the most demanding when the grid voltage is recovered half an power cycle after a dip has occurred. Therefore, the results in Fig. 5.25b are obtained under such conditions. Figure 5.25 shows the distribution of the control domain of the DFIG-based system, which is limited by the hardware capability of the partial-scale power converter, and will be reduced in practice due to the control strategy, control delay, etc.

From the above analysis it is clear that for less severe symmetrical grid faults the DFIG-based wind power system can realise uninterrupted operation by means of pure software algorithms; for some severe symmetrical grid faults the DFIG-based wind power system will experience overcurrent and overvoltage faults due to the limited control capability of the partial-scale power converter, in which case the system must be protected by means of hardware auxiliary units as necessary in order to improve the system ability to ride through symmetrical faults.

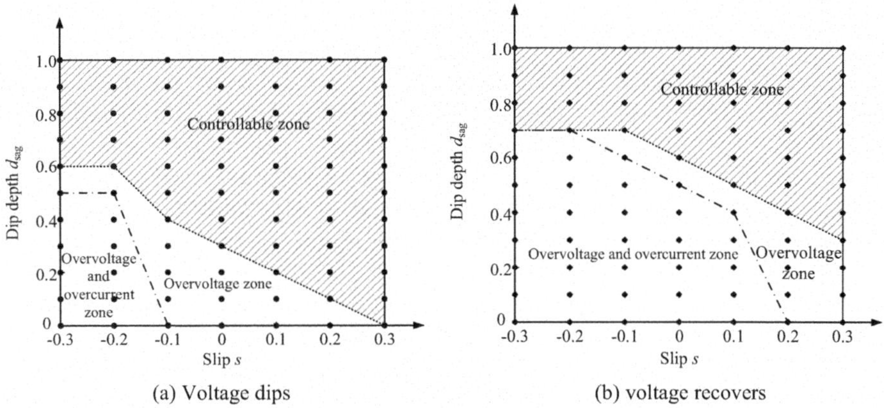

Fig. 5.25 Control domain of the LVRT strategy for symmetrical grid faults. **a** Voltage dips, **b** voltage recovers

References

1. J. Ekanayake, L. Holdsworth, X. Wu et al., Dynamic modeling of doubly fed induction generator wind turbines. IEEE Trans. Power Syst. **18**(2), 803–809 (2003)
2. A. Petersson, L. Harnefors, T. Thiringer, Evaluation of current control methods for wind turbines using doubly-fed induction machines. IEEE Trans. Power Electron. **20**(1), 227–235 (2005)
3. D. Xiang, L. Ran, P. Tavner et al., Control of a doubly fed induction generator in a wind turbine during grid fault ride-through. IEEE Trans. Energy Convers. **21**(3), 652–662 (2006)
4. Y. He, J. Hu, R. Zhao, Modeling and control of wind-turbine used DFIG under network fault conditions, in *International Conference on Electrical Machines and Systems*, vol. 2 (2005), pp. 986–991
5. G. Xiaoming, H. Yikang, H. Benteng, Direct power control of doubly fed induction generator wind turbines with a constant switching frequency. Autom. Electr. Power Syst. **32**(1), 61–65 (2008)
6. G. Changchun, Y. Xingjia, J. Yanjun et al., Characteristics of wind turbines under weak grid. Sun Energy **3**, 41–43 (2008)
7. H. Shuju, L. Jianlin, X. Honghua, Analysis on protection circuits suitable for VSCF-WECS to cope with grid faults. Convert. Technol. Electr. Traction **1**, 45–55 (2008)
8. J. Niiranen, Simulation of doubly fed induction generator wind turbine with an active crowbar. EPE-PEMC 2004 (2004)
9. W. Zhang, P. Zhou, Y.K. He, Analysis of the by-pass resistance of an active crowbar for doubly-fed induction generator based wind turbines under grid faults, in *International Conference on Electrical Machines and Systems* (2008), pp. 2316–2321
10. O. Anaya-Lara, Z.F. Liu, G. Quinonez-Varela et al., Optimal DFIG crowbar resistance design under different controllers during grid faults, in *3rd International Conference on Electric Utility Deregulation and Restructuring and Power Technologies* (2008), pp. 2580–2585
11. I. Erlick, J. Kretschmann, J. Fortmann et al., Modeling of wind turbines based on doubly-fed induction generators for power system stability studies. IEEE Trans. Power Syst. **22**(3), 909–919 (2007)
12. C. Abbey, G. Joos, Supercapacitor energy storage for wind energy applications. IEEE Trans. Ind. Appl. **43**(3), 769–776 (2007)

13. M.R. Rathi, P.P. Jose, N. Mohan, A Novel H∞ based controller for wind turbine applications operating under unbalanced voltage conditions, in *Proceedings of the 13th International Conference on Intelligent Systems Application to Power Systems* (2005)
14. S.F. Patrick, V. Giri, Unbalanced voltage sag ride-through of a doubly fed induction generator wind turbine with series grid-side converter. IEEE Trans. Ind. Appl. **45**(5), 1879–1887 (2009)

Chapter 6
Control Strategies for the DFIG-Based Wind Power System Under Asymmetrical Grid Conditions

Under asymmetric grid conditions, the persistence of the negative sequence grid voltage introduces more complex and violent and non-spontaneously decaying electromagnetic processes in DFIG-base wind power system, causing a series of problems such as stator overcurrent, rotor overcurrent, rotor overvoltage, DC bus overvoltage, electromagnetic torque pulsation, power pulsation and DC-bus voltage pulsation. The grid requires the wind power system to have the ability of uninterrupted operation under asymmetrical grid faults, e.g. the "Emergency Grid Operation Regulations" requires the power equipment to be able to operate stably for a long time in a grid environment with a steady-state asymmetry of 2% and a transient asymmetry of no more than 5% [1], while the uninterrupted operation of the DFIG-based wind power system under more severe asymmetrical grid conditions and the improvement of its operational performance is also a hot issue for current research.

This chapter first provides an overview of the operational control for the DFIG-based wind power system under asymmetrical grid conditions, then details the separation of positive and negative sequence components and vector orientation methods in the three-phase asymmetrical system, which is the basis for the operational control of the DFIG-based wind power system under asymmetrical grid conditions. On this basis, several typical control methods are described in detail, including vector control based on PI controllers in positive and negative sequence synchronous coordinate frames, vector control based on PIR/MFPIR controllers in the positive sequence synchronous coordinate frame and model predictive control under asymmetrical conditions.

H. Ma et al., *The Control Principle of Wind Power Generation System*,
https://doi.org/10.1007/978-981-99-6083-5_6

6.1 Overview of Operation Control for the DFIG-Based Wind Power System Under Asymmetrical Grid Conditions

As far as pure software algorithms without hardware protection circuits concerned, the control strategies for DFIG-based systems under asymmetrical grid conditions are also mostly based on various improvements of vector control methods. For example, the typical vector control method based on positive and negative sequence separation firstly separates the positive and negative sequence components of the electromagnetic quantities in the system from each other, then designs the positive and negative sequence vector controllers for the positive and negative sequence components respectively, and finally combines the outputs of the positive and negative sequence controllers to generate the control signal for the whole system. At present, this kind of vector control based on the separation of positive and negative sequence is the most common control strategy used in the asymmetrical operation control of DFIG-based systems, which is easy to understand and has a clear structure, but it requires several rotational coordinate transformations and the separation and synthesis of positive and negative sequence vectors, which is a large number of operations and a complex algorithm, and there are many controller parameters and a high dependency on the system parameters, which bring some difficulties to the practical system tuning.

In order to reduce the computational effort, a number of vector control methods with feedforward compensation terms are beginning to emerge. By adding a feedforward compensation term to the rotor current controller, this method counteracts the influence of the negative sequence component on the current loop of a traditional vector control, thus enabling asymmetrical operation of the DFIG system. Compared to vector control based on positive and negative sequence separation, this method does not require the decoupling of the positive and negative sequence components in the controlled object, which simplifies the control structure and the amount of operations, but the accuracy of the feedforward compensation term is affected by the accuracy of the generator parameters and the system operating conditions.

An alternative improvement to traditional vector control is the use of a Multi-Frequency Proportional Integral Resonant (MFPIR) controller. By setting the resonant frequency of the controller at 1x, 2x and 3x the grid frequency, it is possible to control both the positive and negative sequence components just in the positive synchronous coordinate frame, while improving the current waveform. Compared to vector control based on positive and negative sequence separation, this method reduces the number of rotational transformations and the number of controllers by almost half, but the number of parameters to be tuned for the controller is equally high, which brings inconvenience to the practical system tuning, and the dynamic characteristics are still significantly influenced by the system parameters.

In addition to various improvements to vector control, researchers have also investigated and explored other control algorithms, such as the Vector-Based Hysteresis Current Regulator (VBHCR) method and the improved Direct Power Control Plus (DPC+) method. The former can respond to both the positive and negative sequence components of the rotor current in the positive synchronous coordinate frame with the help of the improved hysteresis controller, and the system based on hysteresis controller is simple in structure, easy to implement and has good dynamic characteristics, but is difficult to apply in high power applications due to the variable switching frequency of its power devices. The latter, with the power hysteresis controller and optimised switching table, enables direct control of the stator power of the DFIG under asymmetrical grid conditions in the stationary coordinate frame and achieves four control modes: the mode for stator current maintains the stator current sinusoidal and symmetrical and improves the current quality; the mode based for stator power maintains the stability of stator power and reduces electrical shocks; the mode for electromagnetic torque and stator reactive power maintains the stability of electromagnetic torque and reactive power and reduces mechanical shocks; and the fourth mode achieves a compromise among the control performance of the above three modes. Due to the use of hysteresis controllers to realise power closed loop, the problem of variable switching frequency in the above process still exists, making it unsuitable for high power applications. Although constant switching frequency can be obtained by introducing methods such as power prediction and vector modulation, this method eliminates the rotor current internal loop, which is detrimental to the control of rotor current during grid faults, and does not take into account the dynamic process of the flux linkage during voltage dips.

In addition, various improved non-linear algorithms, such as H∞ control and sliding mode control, have been applied to the uninterrupted operation of DFIG systems with asymmetrical grid faults, but as their controller design and implementation are more complex, the related methods are still under further refinement and research.

6.2 Separation of Positive and Negative Sequence Components and Phase Lock Loop

6.2.1 Separation Methods of Positive and Negative Sequence Components

In Chap. 4, Sect. 4.2.2, the symmetrical component method for the study of asymmetrical three-phase systems is introduced, while in practical control it is necessary to obtain the positive sequence component, the negative sequence component and the zero-sequence component (if it exists) from the time-sequence waveform. Positive and negative sequence separation is the basis for field orientation and the calculation of certain important physical quantities in asymmetrical systems. There are two common methods of positive and negative sequence separation [2–5]: the filter/trap-based separation method and the delay-based separation method.

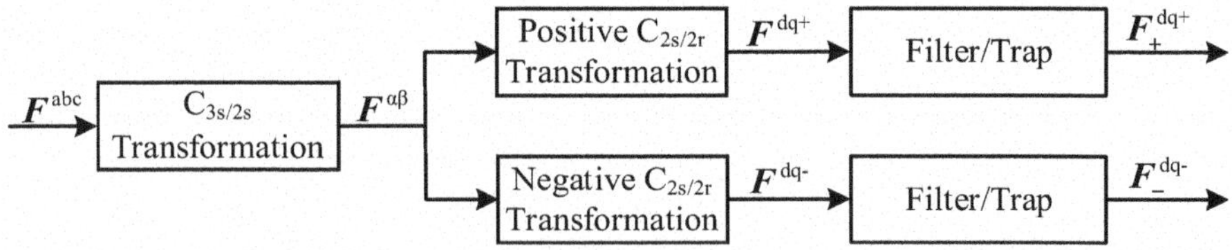

Fig. 6.1 Flow chart of the filter/trap-based separation method for positive and negative sequences

(1) The filter/trap-based separation method for positive and negative sequences

From Fig. 4.3 and Eq. (4.23), it can be seen that the positive sequence component $\boldsymbol{F}_+$ of the asymmetrical system is represented as a DC quantity in the positive synchronous coordinate frame dq+ and as an AC quantity of 2 times the synchronous frequency in the negative synchronous coordinate frame dq−. The negative sequence component $\boldsymbol{F}_-$ of the asymmetrical system is represented as a DC quantity in the negative synchronous coordinate frame dq− and as an AC quantity of 2 times the synchronous frequency in the positive synchronous coordinate frame dq+. Therefore, by projecting the asymmetrical vector into the positive and negative synchronous rotation coordinate frames respectively, and filtering the result through a filter or trap to remove the AC signal of two times the synchronous frequency, the positive and negative sequence components of the current vector can be obtained in real time, thus realising the separation of positive and negative sequences (Fig. 6.1).

(2) The delay-based separation method for positive and negative sequences

From Fig. 4.3 and Eq. (4.33), it can be seen that, in the αβ frame, an asymmetrical electromagnetic vector $\boldsymbol{F}$ can be written as

$$\begin{cases} F^{\alpha}(t) = F_+^{\alpha}(t) + F_-^{\alpha}(t) = F_+\cos(\omega_1 t + \phi_+) + F_-\cos(-\omega_1 t + \phi_-) \\ F^{\beta}(t) = F_+^{\beta}(t) + F_-^{\beta}(t) = F_+\sin(\omega_1 t + \phi_+) + F_-\sin(-\omega_1 t + \phi_-) \end{cases} \tag{6.1}$$

where F_+ and F_- are the positive and negative sequence component amplitudes respectively, then at the moment of $t - \pi/(2\omega_s)$, Eq. (6.1) can be written as

$$\begin{cases} F^{\alpha}\left(t - \dfrac{\pi}{2\omega_1}\right) = F_+\sin(\omega_1 t + \phi_+) - F_-\sin(-\omega_1 t + \phi_-) = F_+^{\beta}(t) - F_-^{\beta}(t) \\ F^{\beta}\left(t - \dfrac{\pi}{2\omega_1}\right) = -F_+\cos(\omega_1 t + \phi_+) + F_-\cos(-\omega_1 t + \phi_-) = -F_+^{\alpha}(t) + F_-^{\alpha}(t) \end{cases} \tag{6.2}$$

Combining Eqs. (6.1) and (6.2), the positive and negative sequence components can be found as

$$\begin{cases} F^{\alpha}_{+}(t) = 0.5\left[F^{\alpha}(t) - F^{\beta}\left(t - \dfrac{\pi}{2\omega_1}\right)\right] \\ F^{\beta}_{+}(t) = 0.5\left[F^{\beta}(t) + F^{\alpha}\left(t - \dfrac{\pi}{2\omega_1}\right)\right] \\ F^{\alpha}_{-}(t) = 0.5\left[F^{\alpha}(t) + F^{\beta}\left(t - \dfrac{\pi}{2\omega_1}\right)\right] \\ F^{\beta}_{-}(t) = 0.5\left[F^{\beta}(t) - F^{\alpha}\left(t - \dfrac{\pi}{2\omega_1}\right)\right] \end{cases} \tag{6.3}$$

Its separation process is as follows (Fig. 6.2).

The separation results of the above two positive and negative sequence separation methods are shown in Fig. 6.3.

In general, the filter/trap-based positive sequence separation method has a smoother dynamic process without large transient shocks, but its dynamic process is relatively slow and the filter/trap design is more complex. The positive and negative sequence separation method based on 1/4 synchronous period delay calculation has a shorter dynamic process, faster response and simple implementation, but its transient process has large overshoot.

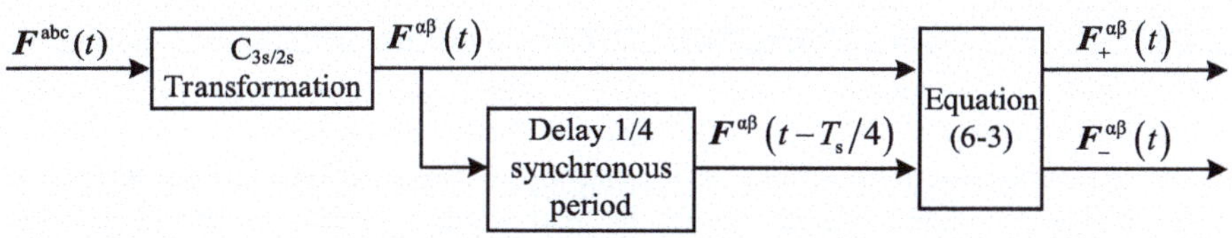

Fig. 6.2 Flow chart of the delay-based separation method for positive and negative sequences

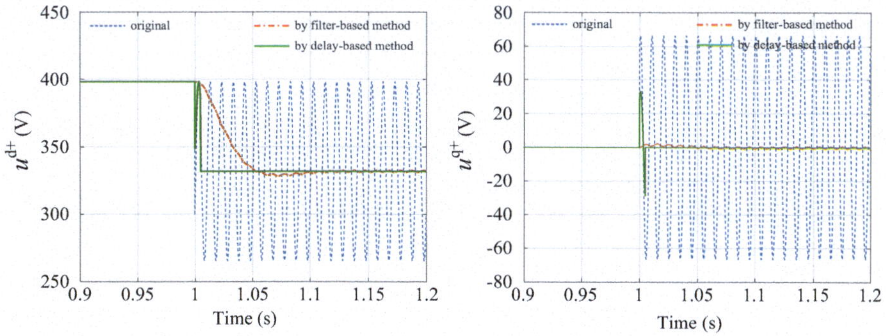

Fig. 6.3 Waveforms of asymmetrical grid voltage in the positive synchronous coordinate frame

6.2.2 Phase Lock Loop Under Asymmetrical Grid Conditions

Under asymmetric grid conditions, various vector-based control algorithms are based on field orientation, most commonly the grid voltage positive sequence component orientation, and therefore the phase of the positive sequence component of the grid voltage needs to be accurately observed.

When the grid voltage is symmetrical, the observation of the grid voltage phase can be carried out using both direct calculation and software phase lock loop. When the grid voltage is asymmetrical, the phase information obtained by the two methods is shown in Fig. 6.4. It can be seen that the phase obtained by the direct calculation method has a pulsation of 100 Hz, which is due to the fact that the rotational speed of the grid voltage vector, synthesised by the positive and negative sequence components of the voltage, fluctuates twice in one cycle when the grid voltage is asymmetrical. Therefore, it is the phase information of the synthesised voltage vector that is obtained by the direct calculation method. However, the phase obtained with the phase lock loop is almost pulsation-free at 100 Hz, which is because when the PLL regulator parameters are small, the cut-off frequency of the PLL closed-loop transfer function is low and acts as a filter for the 100 Hz phase pulsation caused by the negative sequence voltage, while the phase of the positive sequence voltage passes through the PLL without loss, although the filtering here is not completely attenuated, but the error is generally acceptable in engineering cases when the asymmetry is not high.

However, in the case of high grid voltage asymmetry, in order to improve the orientation accuracy of the control method and further improve the control performance of the system, a modified phase lock loop is usually used to detect the phase of the positive sequence component of the grid voltage, i.e. the separated positive sequence component of the grid voltage is used as input to the conventional phase lock loop, where the separation method can be either one of those in Sect. 6.2.1 (Figs. 6.5 and 6.6)

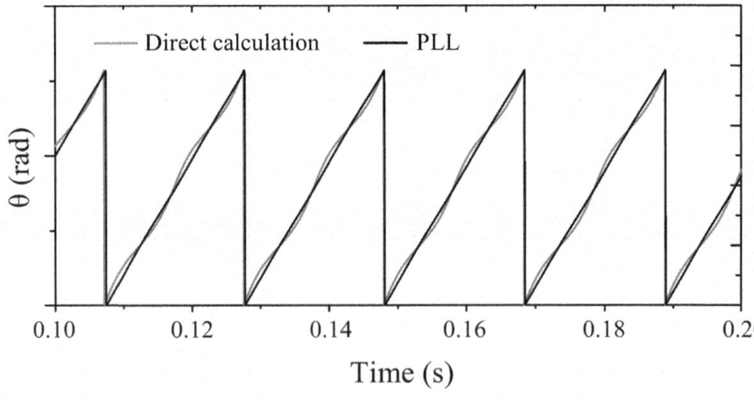

Fig. 6.4 Comparison of two voltage phase observations under asymmetrical grid voltage

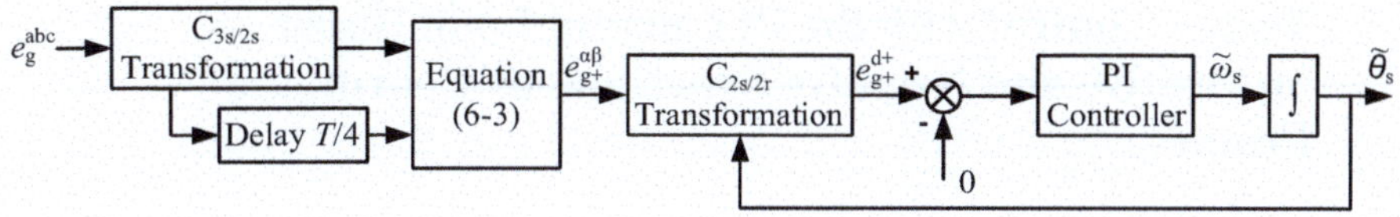

Fig. 6.5 Modified PLL for the positive sequence component of the grid voltage

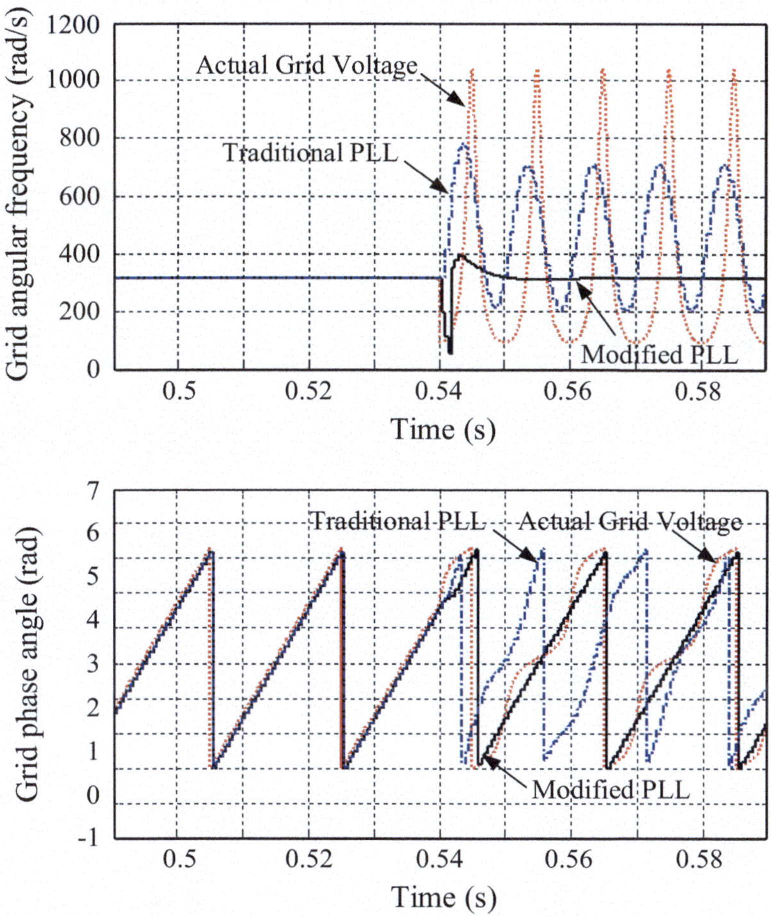

Fig. 6.6 Comparison between the traditional and modified PLLs under asymmetrical grid voltage (A two-phase ground fault occurs at 0.54 s)

6.3 Vector Control with the PI Current Loop Based on Positive and Negative Synchronous Coordinate Frames

6.3.1 Vector Control of the Grid-Side Converter Based on Positive and Negative Synchronous Coordinate Frames

According to the symmetrical component method and Eq. (4.23), an arbitrary electromagnetic vector in a stationary coordinate frame can be expressed as

$$\boldsymbol{F}^{\mathrm{dq}+} = \boldsymbol{F}_{+}^{\mathrm{dq}+} + \boldsymbol{F}_{-}^{\mathrm{dq}+} = \boldsymbol{F}_{+}^{\mathrm{dq}+} + \boldsymbol{F}_{-}^{\mathrm{dq}-} e^{-2\omega_{s}t} \tag{6.4}$$

According to Eqs. (6.4), (2.24) can be rewritten as

$$\begin{cases} u_{\mathrm{g}+}^{\mathrm{d}+} = R_{\mathrm{g}} i_{\mathrm{g}+}^{\mathrm{d}+} + L_{\mathrm{g}} p i_{\mathrm{g}+}^{\mathrm{d}+} - \omega_{1} L_{\mathrm{g}} i_{\mathrm{g}+}^{\mathrm{q}+} + e_{\mathrm{g}+}^{\mathrm{d}+} \\ u_{\mathrm{g}+}^{\mathrm{q}+} = R_{\mathrm{g}} i_{\mathrm{g}+}^{\mathrm{q}+} + L_{\mathrm{g}} p i_{\mathrm{g}+}^{\mathrm{q}+} + \omega_{1} L_{\mathrm{g}} i_{\mathrm{g}+}^{\mathrm{d}+} + e_{\mathrm{g}+}^{\mathrm{q}+} \end{cases} \tag{6.5}$$

$$\begin{cases} u_{\mathrm{g}-}^{\mathrm{d}-} = R_{\mathrm{g}} i_{\mathrm{g}-}^{\mathrm{d}-} + L_{\mathrm{g}} p i_{\mathrm{g}-}^{\mathrm{d}-} + \omega_{1} L_{\mathrm{g}} i_{\mathrm{g}-}^{\mathrm{q}-} + e_{\mathrm{g}-}^{\mathrm{d}-} \\ u_{\mathrm{g}-}^{\mathrm{q}-} = R_{\mathrm{g}} i_{\mathrm{g}-}^{\mathrm{q}-} + L_{\mathrm{g}} p i_{\mathrm{g}-}^{\mathrm{q}-} - \omega_{1} L_{\mathrm{g}} i_{\mathrm{g}-}^{\mathrm{d}-} + e_{\mathrm{g}-}^{\mathrm{q}-} \end{cases} \tag{6.6}$$

The combination of Eqs. (4.39), (4.40), (6.5) and (6.6) is the mathematical model of the grid-side converter under asymmetrical grid conditions.

From Eq. (6.5), the output voltage of the grid-side current-loop controller in the positive synchronous frame can be obtained as

$$\begin{cases} u_{\mathrm{g}+}^{\mathrm{d}+*} = C_{\mathrm{gi}+}^{\mathrm{d}+}\left(i_{\mathrm{g}+}^{\mathrm{d}+*} - i_{\mathrm{g}+}^{\mathrm{d}+}\right) + u_{\mathrm{gc}+}^{\mathrm{d}+} \\ u_{\mathrm{g}+}^{\mathrm{q}+*} = C_{\mathrm{gi}+}^{\mathrm{q}+}\left(i_{\mathrm{g}+}^{\mathrm{q}+*} - i_{\mathrm{g}+}^{\mathrm{q}+}\right) + u_{\mathrm{gc}+}^{\mathrm{q}+} \end{cases} \tag{6.7}$$

where $C_{\mathrm{gi}+}^{\mathrm{d}+}$ and $C_{\mathrm{gi}+}^{\mathrm{q}+}$ are the PI controllers on the d+ and q+ axes for the grid-side current loop, respectively, $i_{\mathrm{g}+}^{\mathrm{d}+*}$ and $i_{\mathrm{g}+}^{\mathrm{q}+*}$ are reference values of the grid-side current on the d+ and q+ axes, respectively, $u_{\mathrm{g}+}^{\mathrm{d}+*}$ and $u_{\mathrm{g}+}^{\mathrm{q}+*}$ are the reference values of the control voltage outputted by the grid-side converter on the d+ and q+ axes, respectively, $u_{\mathrm{gc}+}^{\mathrm{d}+} = -\omega_{1} L_{\mathrm{g}} i_{\mathrm{g}+}^{\mathrm{q}+} + e_{\mathrm{g}+}^{\mathrm{d}+}$ and $u_{\mathrm{gc}+}^{\mathrm{q}+} = \omega_{1} L_{\mathrm{g}} i_{\mathrm{g}+}^{\mathrm{d}+} + e_{\mathrm{g}+}^{\mathrm{q}+}$ are the components of the feedforward compensation term for the grid-side converter output voltage on the d+ and q+ axes, respectively.

Similarly, the output voltage of the grid-side current-loop controller in the negative synchronous coordinate frame can be obtained from Eq. (6.6) as

$$\begin{cases} u_{g-}^{d-*} = C_{gi-}^{d-}\left(i_{g-}^{d-*} - i_{g-}^{d-}\right) + u_{gc-}^{d-} \\ u_{g-}^{q-*} = C_{gi-}^{q-}\left(i_{g-}^{q-*} - i_{g-}^{q-}\right) + u_{gc-}^{q-} \end{cases} \tag{6.8}$$

where C_{gi-}^{d-} and C_{gi-}^{q-} are the PI controllers on the d− and q− axes for the grid-side current loop, respectively, i_{g-}^{d-*} and i_{g-}^{q-*} are reference values of the grid-side current on the d− and q− axes, respectively, u_{g-}^{d-*} and u_{g-}^{q-*} are the reference values of the control voltage outputted by the grid-side converter on the d− and q− axes, respectively, $u_{gc-}^{d-} = \omega_1 L_g i_{g-}^{q-} + e_{g-}^{d-}$ and $u_{gc-}^{q-} = -\omega_1 L_g i_{g-}^{d-} + e_{g-}^{q-}$ are the components of the feedforward compensation term for the grid-side converter output voltage on the d− and q− axes, respectively.

The output of the DC-bus voltage controller for the grid-side converter can still use the result shown in (3.25), but where the grid-side power will be defined by Eq. (4.39). Thus there is

$$\begin{cases} P_g^* = P_{\text{gconst}}^* + P_{2\omega 1}^* \\ Q_g^* = Q_{\text{gconst}}^* + Q_{2\omega 1}^* \end{cases} \tag{6.9}$$

If the negative sequence current injected into the grid by the grid-side converter is 0,[1] i.e. $i_{g-}^{d-*} = i_{g-}^{q-*} = 0$, considering only the average grid-side power, according to Eq. (6.9) it can be found that

$$\begin{cases} i_{g+}^{d+*} = P_g^* \Big/ e_{g+}^{d+} \\ i_{g+}^{q+*} = -Q_g^* \Big/ e_{g+}^{d+} \\ i_{g-}^{d-*} = 0 \\ i_{g-}^{q-*} = 0 \end{cases} \tag{6.10}$$

In summary, the control of the grid-side converter based on positive and negative synchronous rotational coordinate frames is shown in Fig. 6.7.

6.3.2 *Vector Control of the DFIG Based on Positive and Negative Synchronous Coordinate Frames*

According to Eqs. (6.4), (2.19) and (2.20) can be rewritten as

[1] There are various options for the control objectives of the grid-side converter under asymmetrical grid conditions, and here only the control objective of eliminating the negative sequence current on the grid-side is taken as an example, while the selection and implementation of other control objectives of the grid-side converter will be discussed in the subsequent part of this chapter.

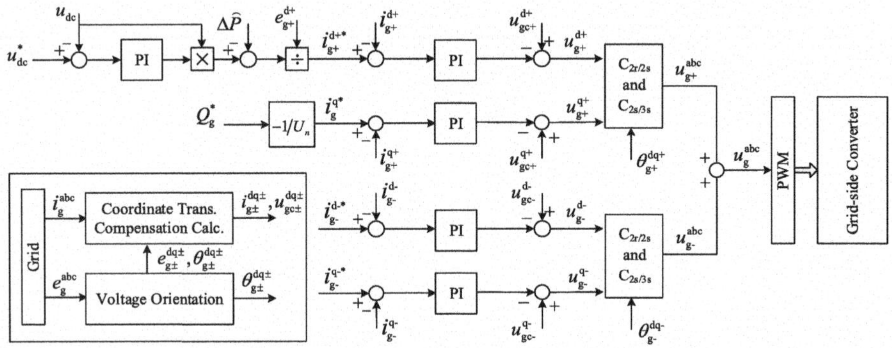

Fig. 6.7 Control diagram of the grid-side converter based on positive and negative synchronous frames

$$\begin{cases} u_{s+}^{d+} = R_s i_{s+}^{d+} + p\psi_{s+}^{d+} - \omega_1 \psi_{s+}^{q+} \\ u_{s+}^{q+} = R_s i_{s+}^{q+} + p\psi_{s+}^{q+} + \omega_1 \psi_{s+}^{d+} \\ u_{r+}^{d+} = R_r i_{r+}^{d+} + p\psi_{r+}^{d+} - \omega_{sl}^{+} \psi_{r+}^{q+} \\ u_{r+}^{q+} = R_r i_{r+}^{q+} + p\psi_{r+}^{q+} + \omega_{sl}^{+} \psi_{r+}^{d+} \end{cases} \tag{6.11}$$

$$\begin{cases} \psi_{s+}^{d+} = L_s i_{s+}^{d+} + L_M i_{r+}^{d+} \\ \psi_{s+}^{q+} = L_s i_{s+}^{q+} + L_M i_{r+}^{q+} \\ \psi_{r+}^{d+} = L_r i_{r+}^{d+} + L_M i_{s+}^{d+} \\ \psi_{r+}^{q+} = L_r i_{r+}^{q+} + L_M i_{s+}^{q+} \end{cases} \tag{6.12}$$

$$\begin{cases} u_{s-}^{d-} = R_s i_{s-}^{d-} + p\psi_{s-}^{d-} + \omega_1 \psi_{s-}^{q-} \\ u_{s-}^{q-} = R_s i_{s-}^{q-} + p\psi_{s-}^{q-} - \omega_1 \psi_{s-}^{d-} \\ u_{r-}^{d-} = R_r i_{r-}^{d-} + p\psi_{r-}^{d-} + \omega_{sl}^{-} \psi_{r-}^{q-} \\ u_{r-}^{q-} = R_r i_{r-}^{q-} + p\psi_{r-}^{q-} - \omega_{sl}^{-} \psi_{r-}^{d-} \end{cases} \tag{6.13}$$

$$\begin{cases} \psi_{s-}^{d-} = L_s i_{s-}^{d-} + L_M i_{r-}^{d-} \\ \psi_{s-}^{q-} = L_s i_{s-}^{q-} + L_M i_{r-}^{q-} \\ \psi_{r-}^{d-} = L_r i_{r-}^{d-} + L_M i_{s-}^{d-} \\ \psi_{r-}^{q-} = L_r i_{r-}^{q-} + L_M i_{s-}^{q-} \end{cases} \tag{6.14}$$

where $\omega_{sl}^{+} = \omega_1 - \omega_r$ and $\omega_{sl}^{-} = \omega_1 + \omega_r$, the combination of Eqs. (6.11) and (6.12) is the mathematical model of the DFIG in the positive sequence synchronous frame, the combination of Eqs. (6.13) and (6.14) is the mathematical model of the DFIG in the negative sequence synchronous frame, the two models and Eqs. (4.34) and (4.33) compose the mathematical model of the DFIG under asymmetrical grid voltage conditions.

Substituting Eqs. (6.12) and (6.14) into Eqs. (6.11) and (6.13) respectively, the voltage equation of the DFIG rotor winding can be written as

$$\begin{cases} u_{r+}^{d+} = R_r i_{r+}^{d+} + \sigma L_r p i_{r+}^{d+} - \omega_{sl}^{+} \psi_{r+}^{q+} + \dfrac{L_M}{L_s} p \psi_{s+}^{d+} \\ u_{r+}^{q+} = R_r i_{r+}^{q+} + \sigma L_r p i_{r+}^{q+} + \omega_{sl}^{+} \psi_{r+}^{d+} + \dfrac{L_M}{L_s} p \psi_{s+}^{q+} \end{cases} \tag{6.15}$$

$$\begin{cases} u_{r-}^{d-} = R_r i_{r-}^{d-} + \sigma L_r p i_{r-}^{d-} + \omega_{sl}^{-} \psi_{r-}^{q-} + \dfrac{L_M}{L_s} p \psi_{s-}^{d-} \\ u_{r-}^{q-} = R_r i_{r-}^{q-} + \sigma L_r p i_{r-}^{q-} - \omega_{sl}^{-} \psi_{r-}^{d-} + \dfrac{L_M}{L_s} p \psi_{s-}^{q-} \end{cases} \tag{6.16}$$

where the positive and negative synchronous rotation coordinate frames are oriented on the basis of the positive and negative sequence grid voltages, respectively. When the grid voltage is stable, the differential term of the stator flux linkage can be considered to be zero and the simplified rotor winding voltage equation is

$$\begin{cases} u_{r+}^{d+} = R_r i_{r+}^{d+} + \sigma L_r p i_{r+}^{d+} + E_{r+}^{d+} \\ u_{r+}^{q+} = R_r i_{r+}^{q+} + \sigma L_r p i_{r+}^{q+} + E_{r+}^{d+} \end{cases} \tag{6.17}$$

$$\begin{cases} u_{r-}^{d-} = R_r i_{r-}^{d-} + \sigma L_r p i_{r-}^{d-} + E_{r-}^{d-} \\ u_{r-}^{q-} = R_r i_{r-}^{q-} + \sigma L_r p i_{r-}^{q-} + E_{r-}^{d-} \end{cases} \tag{6.18}$$

where $E_{r+}^{d+} = -\omega_{sl}^{+} \psi_{r+}^{q+} + L_M p \psi_{s+}^{d+} / L_s$ and $E_{r+}^{q+} = \omega_{sl}^{+} \psi_{r+}^{d+} + L_M p \psi_{s+}^{q+} \Big/ L_s$ are the components of the cross-coupling term for the rotor current on the d+ and q+ axes, respectively, $E_{r-}^{d-} = \omega_{sl}^{-} \psi_{r-}^{q-} + L_M p \psi_{s-}^{d-} / L_s$ and $E_{r-}^{q-} = -\omega_{sl}^{-} \psi_{r-}^{d-} + L_M p \psi_{s-}^{q-} \Big/ L_s$ are the components of the cross-coupling term for the rotor current on the d− and q− axes, respectively. Both above cross-coupling terms can be considered as external voltage disturbances to the rotor winding of the generator.

According to Eqs. (6.17) and (6.18) the rotor-current closed-loop controller output in the positive and negative synchronous coordinate frames can be obtained as

$$\begin{cases} u_{r+}^{d+*} = C_{ri+}^{d+} \left(i_{r+}^{d+*} - i_{r+}^{d+} \right) + u_{rc+}^{d+} \\ u_{r+}^{q+*} = C_{ri+}^{q+} \left(i_{r+}^{q+*} - i_{r+}^{q+} \right) + u_{rc+}^{q+} \end{cases} \tag{6.19}$$

$$\begin{cases} u_{r-}^{d-*} = C_{ri-}^{d-} \left(i_{r-}^{d-*} - i_{r-}^{d-} \right) + u_{rc-}^{d-} \\ u_{r-}^{q-*} = C_{ri-}^{q-} \left(i_{r-}^{q-*} - i_{r-}^{q-} \right) + u_{rc-}^{q-} \end{cases} \tag{6.20}$$

where C_{ri+}^{d+}, C_{ri+}^{q+}, C_{ri-}^{d-} and C_{ri-}^{q-} are the PI controllers for the rotor current loop on the d+ , q+ , d− and q− axes, respectively, i_{r+}^{d+*}, i_{r+}^{q+*}, i_{r-}^{d-*} and i_{r-}^{q-*} are the reference values for the rotor current on the d+, q+, d− and q− axes, respectively, u_{r+}^{d+*}, u_{r+}^{q+*}, u_{r-}^{d-*} and u_{r-}^{q-*} are the reference values for the control voltage on the d+, q+, d− and q− axes, respectively.

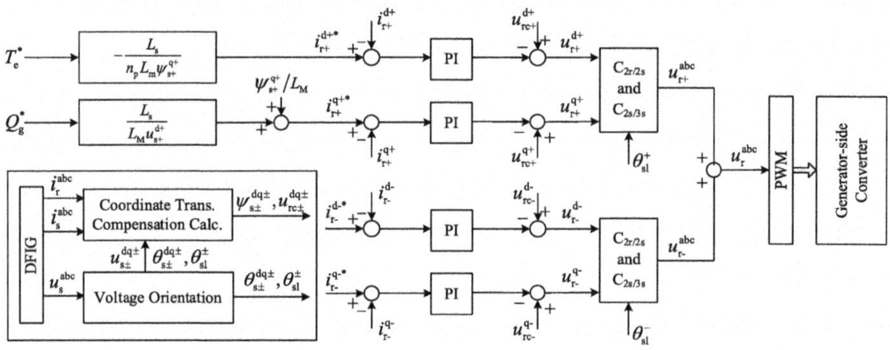

Fig. 6.8 Control diagram of the DFIG based on positive and negative synchronous frames

At this point, the electromagnetic torque and stator side power of the DFIG are defined by Eqs. (4.34) and (4.33), i.e. there are

$$\begin{cases} T_e^* = T_{\text{econst}}^* + T_{e2\omega 1}^* \\ Q_e^* = Q_{\text{econst}}^* + Q_{e2\omega 1}^* \end{cases} \tag{6.21}$$

If the negative sequence current injected into the grid by the generator stator is 0,[2] i.e. $i_{\text{s}-}^{\text{d}-*} = i_{\text{s}-}^{\text{q}-*} = 0$, considering only the average electromagnetic torque of the generator and the average reactive power of the stator, according to Eq. (6.23) it can be found that

$$\begin{cases} i_{\text{s}+}^{\text{d}+*} = T_{\text{e}}^* \big/ n_{\text{p}} \psi_{\text{s}+}^{\text{q}+} \\ i_{\text{s}+}^{\text{q}+*} = -Q_{\text{s}}^* \big/ u_{\text{s}+}^{\text{d}+} \\ i_{\text{s}-}^{\text{d}-*} = 0 \\ i_{\text{s}-}^{\text{q}-*} = 0 \end{cases} \Rightarrow \begin{cases} i_{\text{r}+}^{\text{d}+*} = -T_{\text{e}}^* L_{\text{s}} \big/ n_{\text{p}} L_{\text{M}} \psi_{\text{s}+}^{\text{q}+} \\ i_{\text{r}+}^{\text{q}+*} = \left(\psi_{\text{s}+}^{\text{q}+} + Q_{\text{s}}^* L_{\text{s}} \big/ u_{\text{s}+}^{\text{d}+} \right) \big/ L_{\text{M}} \\ i_{\text{r}-}^{\text{d}-*} = 0 \\ i_{\text{r}-}^{\text{q}-*} = \psi_{\text{s}-}^{\text{q}-} \big/ L_{\text{M}} \end{cases} \tag{6.22}$$

In summary, the control of the DFIG based on positive and negative synchronous rotational coordinate frames is shown in Fig. 6.8.
综上, 基于正反向同步旋转坐标系的双馈风力发电机系统控制框图如图 6.8 所示.

[2] There are various options for the control objectives of the DFIG under asymmetrical grid conditions, and here only the control objective of eliminating the negative sequence current on the stator side is taken as an example, while the selection and implementation of other control objectives of the DFIG will be discussed in the subsequent part of this chapter.

6.4 Vector Control with the PIR/MFPIR Current Loop Based on the Positive Synchronous Coordinate Frame

From Sects. 4.2 and 4.3, it can be seen that under asymmetrical grid conditions, the negative sequence component of the grid voltage will persistently introduce a disturbance of 2 times the grid synchronous frequency in the DFIG-based wind power system in the positive synchronous coordinate frame, and if the transient processes caused by the change from symmetrical to asymmetrical grid voltage are taken into account, then there will also be a disturbance of the grid synchronous frequency caused by the DC component of the generator stator flux linkage in the system at the same time. In other words, if the transient characteristics of asymmetrical grid faults are taken into account, the DFIG-based wind power system will have a self-decaying disturbance of 1x the synchronous frequency and a persistent disturbance of 2x the synchronous frequency. That means that the methods discussed in Sect. 6.3 are not able to deal effectively with transient processes where the grid voltage changes between symmetrical and asymmetrical (they are more suitable for the case where the grid voltage is in an asymmetrical steady state). A typical solution is to introduce the PI controller in the stationary coordinate frame to suppress the disturbance caused by the DC component of the stator flux linkage during the transient process, which will result in 3 coordinate frames (positive synchronous coordinate frame, negative synchronous coordinate frame and stationary coordinate frame) and 12 current loops (6 on the generator side and 6 on the grid side) in the control system at the same time, requiring a large number of positive and negative sequence components to be separated, rotating coordinates to be transformed and controller parameters to be tuned, thus increasing the complexity of the algorithm and making the practical implementation difficult.

Using the mathematical model of the DFIG described by Eqs. (4.25), (4.26), (4.27) and (4.28) and the mathematical model of the grid-side converter described by Eqs. (4.35), (4.36) and (4.40), the control of the DFIG-based wind power system can be carried out directly in the positive synchronous coordinate frame, which can to some extent reduce the amount of positive and negative sequence separation and rotational coordinate transformation and simplify the control algorithm. In order to control both the positive and negative components of the system in the positive synchronous coordinate frame, it is necessary to use the PIR controller described in Sect. 5.2.2 with its resonant frequency at 2x grid synchronous frequency to suppress the effect of the negative sequence components on the system [6, 7]. The MFPIR controller can also be used to control the positive sequence, negative sequence and zero-sequence/DC-bias quantities of the system simultaneously in the positive synchronous coordinate frame by setting the resonance frequency points at 1x grid synchronous frequency and 2x grid synchronous frequency, thus effectively handling transient processes in the asymmetrical voltage dip.

The transfer function and block diagram of the MFPIR controller are shown in Eq. (6.23) and Fig. 6.9 respectively, where $\omega_r|_{r=0,1,\cdots,N-1}$ are N resonance points.

$$G_{\mathrm{MFPIR}}(s) = k_{\mathrm{p}} + \frac{k_{\mathrm{i}}}{s} + \sum_{\mathrm{r}=0}^{N-1} \frac{k_{\mathrm{r}} s}{s^2 + \omega_{\mathrm{r}}^2} \tag{6.23}$$

Referring to Chap. 3, Sect. 3.1, the positive-synchronous-frame-based control diagrams of the PIR/MFPIR-based generator-side converter and the grid-side converter can be obtained as shown in Figs. 6.10 and 6.11.

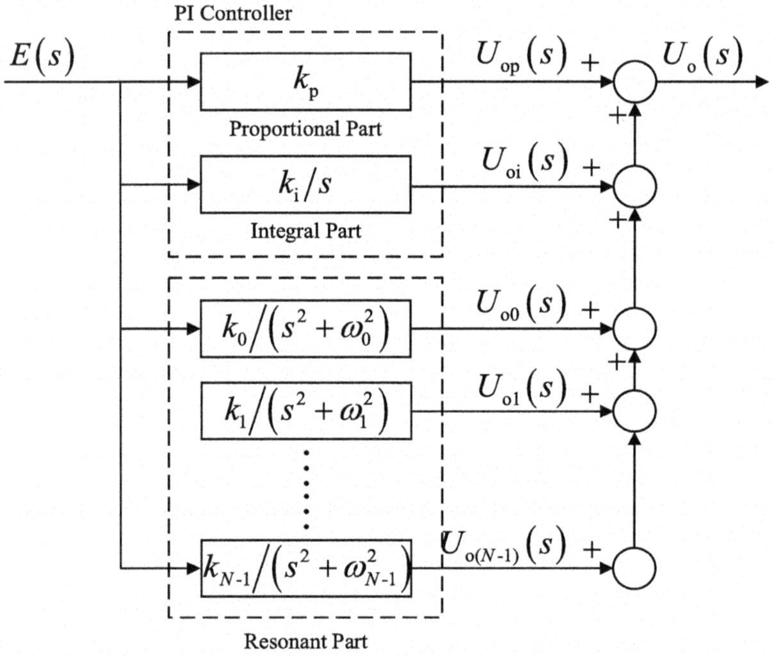

Fig. 6.9 Block diagram of the MFPIR controller

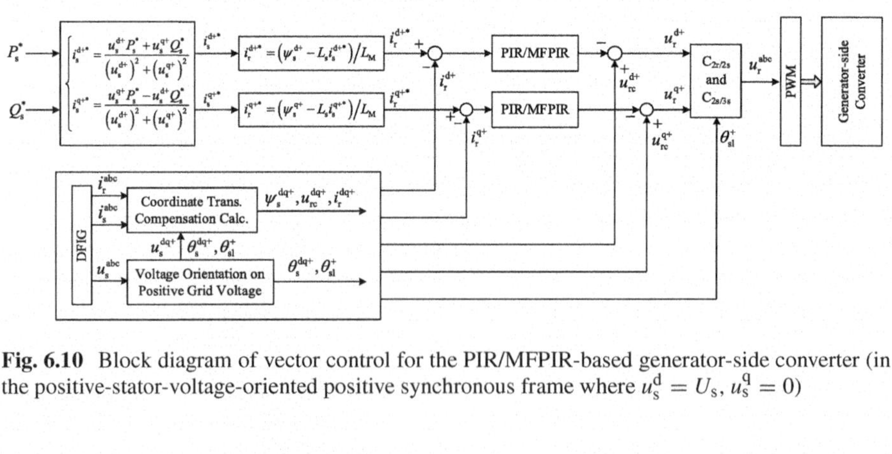

Fig. 6.10 Block diagram of vector control for the PIR/MFPIR-based generator-side converter (in the positive-stator-voltage-oriented positive synchronous frame where $u_{\mathrm{s}}^{\mathrm{d}} = U_{\mathrm{s}}$, $u_{\mathrm{s}}^{\mathrm{q}} = 0$)

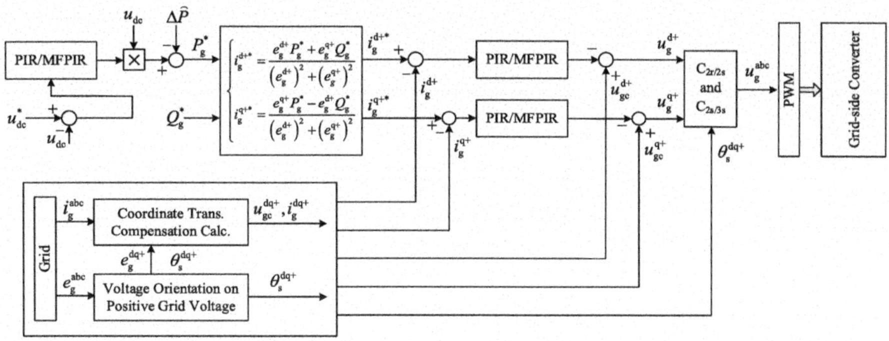

Fig. 6.11 Block diagram of vector control for the PIR/MFPIR-based grid-side converter (in the positive-grid-voltage-oriented positive synchronous frame where $e_g^d = E_g$, $e_g^q = 0$)

6.5 Model Predictive Control Under Asymmetrical Grid Conditions

In this section, the MPC method for asymmetrical grid conditions will be discussed. The method is able to control both the positive and negative sequence components of the system in the positive synchronous coordinate system without decoupling the positive and negative sequences of the wind power system, simplifying the control structure, reducing the number of controllers and controller parameters, facilitating practical implementation and providing a good parameter robustness due to the characteristics of the MPC itself.

6.5.1 Model Predictive Control of the Grid-Side Converter Under Asymmetrical Grid Conditions

In Sect. 3.3.3, the derivation process for the grid-side current loop and the DC-bus voltage of the grid-side converter is based on the complete mathematical models of the grid-side current and the DC-bus voltage. Therefore, it is only necessary to modify the operational coordinate frame from the original grid-voltage oriented synchronous coordinate frame to the synchronous coordinate frame oriented to the positive sequence component of the grid voltage, then the control method derived will still be largely applicable under asymmetrical grid conditions, and to make it more adaptable to asymmetrical grid voltage conditions, the original control method can be improved as follows.

(1) Adding Grid Voltage Prediction to the MPC Strategy for the Grid-side Current Loop

In the derivation of the MPC strategy for the grid-side current loop in Sect. 3.3.1, a two-step prediction model for the grid-side current is developed in order to eliminate

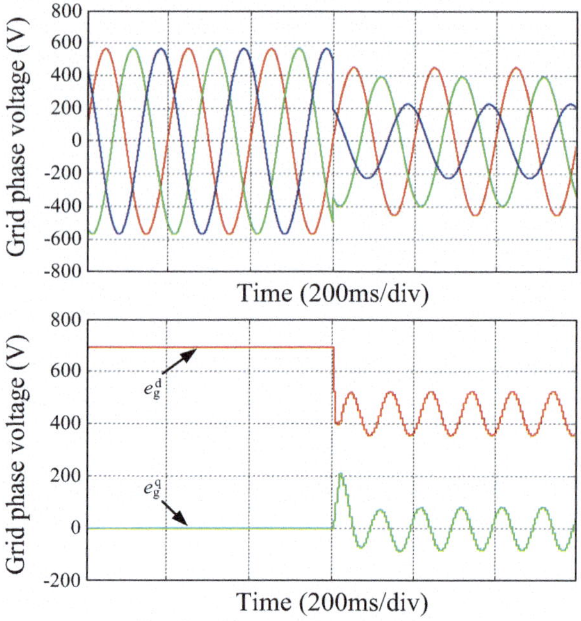

Fig. 6.12 d-axis and q-axis components of the grid voltage under symmetrical and asymmetrical grid conditions

the control delay of the digital controller and improve the system response bandwidth, as shown in Eqs. (3.78) and (3.79). Wherein the prediction of the grid-side current at the moment of $(k+1)T_s$ requires the grid voltage of $\boldsymbol{e}_g^{dq+}(k+1)$ at the moment of $(k+1)T_s$. Under symmetrical grid conditions, it is considered that $\boldsymbol{e}_g^{dq+}(k+1) = \boldsymbol{e}_g^{dq+}(k)$. This assumption will no longer be true when the grid voltage is asymmetrical, and the comparison is shown in Fig. 6.12. Therefore, the prediction of $\boldsymbol{e}_g^{dq+}(k+1)$ needs to be made at the moment of kT_s.

Figure 6.13 shows the prediction method for the grid voltage. Firstly, the positive and negative sequence components of the grid voltage at the current sampling moment ($\boldsymbol{e}_{g\pm}^{\alpha\beta}(k)$) can be obtained according to Eq. (6.3), and the positive and negative sequence components of the grid voltage in the positive synchronous coordinate frame at the current sampling moment ($\boldsymbol{e}_{g+}^{dq+}(k)$ and $\boldsymbol{e}_{g-}^{dq+}(k)$) are obtained by park transformation of $\boldsymbol{e}_{g\pm}^{\alpha\beta}(k)$ according to the angular information obtained from the positive-sequence grid-voltage orientation. For the stable asymmetrical grid voltage, in the positive synchronous coordinate frame, the positive sequence component of the grid voltage is considered to remain constant, while the negative sequence component will rotate negatively at 2 times the synchronous angular frequency of the grid, and will turn through $2\omega_s T_s$ electrical angle to the negative direction by the next sampling moment, so the predicted value of the negative sequence component of the grid voltage at the next sampling moment is obtained as shown in Eq. (6.24). Then the predicted values of the positive and negative sequence components of the grid voltage are superimposed to obtain the predicted value of the grid voltage at the next

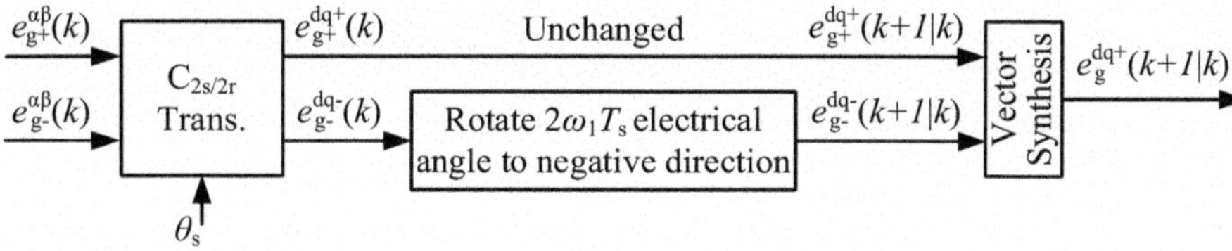

Fig. 6.13 Prediction diagram of the grid voltage under asymmetrical faults

sampling moment as shown in Eq. (6.25). The grid angular frequency ω_s and angle θ_s used in Fig. 6.13 and in Eq. (6.25) are replaced by the $\tilde{\omega}_s$ and $\tilde{\theta}_s$, respectively, obtained from the phase lock loop in Fig. 6.5.

$$\begin{cases} e^{d+}_{g-}(k+1|k) = e^{d+}_{g-}(k)\cos(2\omega_1 T_s) + e^{q+}_{g-}(k)\sin(2\omega_1 T_s) \\ e^{q+}_{g-}(k+1|k) = e^{q+}_{g-}(k)\cos(2\omega_1 T_s) - e^{d+}_{g-}(k)\sin(2\omega_1 T_s) \end{cases} \tag{6.24}$$

$$\begin{cases} e^{d+}_{g}(k+1|k) = e^{d+}_{g+}(k) + e^{d+}_{g-}(k)\cos(2\omega_1 T_s) + e^{q+}_{g-}(k)\sin(2\omega_1 T_s) \\ e^{q+}_{g-}(k+1|k) = e^{q+}_{g+}(k) + e^{q+}_{g-}(k)\cos(2\omega_1 T_s) - e^{d+}_{g-}(k)\sin(2\omega_1 T_s) \end{cases} \tag{6.25}$$

Replace all the grid voltage predicted for the moment of $(k+1)T_s$ in the algorithm introduced in Sect. 3.3.1.2 by the grid voltage prediction $\boldsymbol{e}^{dq+}_{g}(k+1|k)$ in Eq. (6.25).

(2) Using the Full Expression of the Grid-side Power in the MPC Strategy for the DC-bus Voltage Loop

In the MPC strategy for the DC-bus voltage loop in Chap. 3, Sect. 3.3.1.3, the power expression as Eq. (3.73) is used for the calculation of the grid-side active power. This power expression is obtained based on the normal grid voltage condition that the q-axis component of the grid voltage is 0. When the grid voltage is asymmetrical, the condition that the q-axis component of the grid voltage is 0 in the positive synchronous coordinate frame oriented to the positive-sequence grid voltage no longer holds (as in Fig. 6.12), so that Eq. (3.73) is no longer an accurate expression for the grid-side power. The similar problem also arises in the interface function as Eq. (3.91) between the DC-bus voltage loop and the grid-side current loop, for which the full expression of the grid-side power needs to be used in the above process.

The complete relation between grid-side power and grid-side current is shown in Eq. (2.25), which can be reformulated in the positive-sequence grid-voltage-oriented synchronous coordinate system as

$$\begin{cases} P_g = e^{d+}_{g} i^{d+}_{g} + e^{q+}_{g} i^{q+}_{g} \\ Q_g = e^{q+}_{g} i^{d+}_{g} - e^{d+}_{g} i^{q+}_{g} \end{cases} \tag{6.26}$$

The interface function between the DC-bus voltage loop and the grid-side current loop for asymmetrical grid voltage conditions can be obtained from the above equation as

$$\begin{cases} i_{\mathrm{g}}^{\mathrm{d}+*} = \dfrac{e_{\mathrm{g}}^{\mathrm{d}+} P_{\mathrm{g}}^{*} + e_{\mathrm{g}}^{\mathrm{q}+} Q_{\mathrm{g}}^{*}}{e_{\mathrm{g}}^{\mathrm{d}+^{2}} + e_{\mathrm{g}}^{\mathrm{q}+^{2}}} \\ i_{\mathrm{g}}^{\mathrm{q}+*} = \dfrac{e_{\mathrm{g}}^{\mathrm{q}+} P_{\mathrm{g}}^{*} - e_{\mathrm{g}}^{\mathrm{d}+} Q_{\mathrm{g}}^{*}}{e_{\mathrm{g}}^{\mathrm{d}+^{2}} + e_{\mathrm{g}}^{\mathrm{q}+^{2}}} \end{cases} \tag{6.27}$$

Replace all grid-side power expressions in Sect. 3.3.1.3 in the algorithm by Eq. (6.26) and replace the interface function as Eq. (3.91) in Sect. 3.3.1.3 by Eq. (6.27).

By rewriting the algorithm in Sect. 3.3.1 into the positive-sequence grid-voltage-oriented coordinate frame, and with the two improvements mentioned above, the algorithm flow of the MPC for each part of the grid-side converter under asymmetrical grid voltage can be obtained as shown in Fig. 6.14, and the overall control diagram is shown in Fig. 6.15.

6.5.2 *Model Predictive Control of the DFIG Under Asymmetrical Grid Conditions*

In Chap. 3, Sect. 3.3.2, in the derivation of the rotor-current loop and stator-power loop of the generator-side converter, the corresponding MPC algorithm is obtained by simplifying the mathematical model of the DFIG through grid voltage orientation, with ideal grid conditions as a precondition. However, the method in Sect. 3.3.2 neglects the dynamic processes of the generator stator flux linkage and is therefore not well suited to cope with the adverse effects of the natural component of the stator flux linkage during the symmetrical grid voltage dip, making it difficult to achieve the requirements related to uninterrupted operation in the event of grid faults, where improvements are needed. In Sect. 5.2.3, the MPC strategy for the DFIG that takes into account the dynamic processes of the stator flux linkage is derived, which relies on the full mathematical model of the DFIG for the design of the rotor current loop. Therefore, it is only necessary to rewrite the relevant algorithm in Sect. 5.2.3 from the original grid voltage-oriented synchronous coordinate frame to the positive-sequence grid-voltage-oriented synchronous coordinate frame, then the derived control method will still be adaptable under asymmetrical grid conditions, and to make it more adaptable to asymmetrical grid voltage conditions, it can be improved as follows.

(1) Adding Grid Voltage Prediction to the MPC Strategy for the Rotor Current Loop

Similar to the MPC strategy for the grid-side current loop, the MPC strategy for the rotor current loop in Sect. 5.2.3 still needs to use the grid voltage value $\boldsymbol{e}_{\mathrm{g}}^{\mathrm{dq}+}(k+1)$ at the moment of $(k+1)T_{\mathrm{s}}$ to implement the relevant algorithm at the moment of kT_{s}. Therefore, the prediction of the grid voltage needs to be added to the MPC strategy for the rotor current loop, and the prediction process is shown in Fig. 6.11. Therefore,

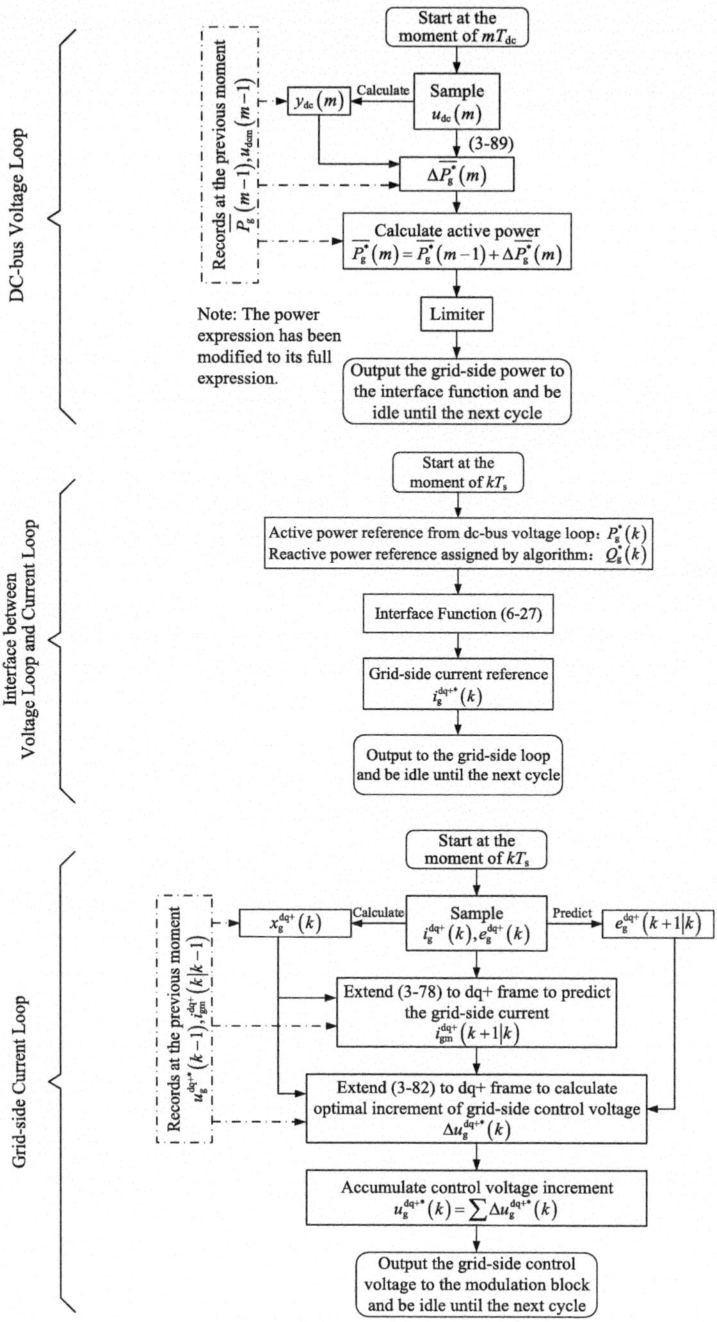

Fig. 6.14 MPC flow of the grid-side converter under asymmetrical grid conditions

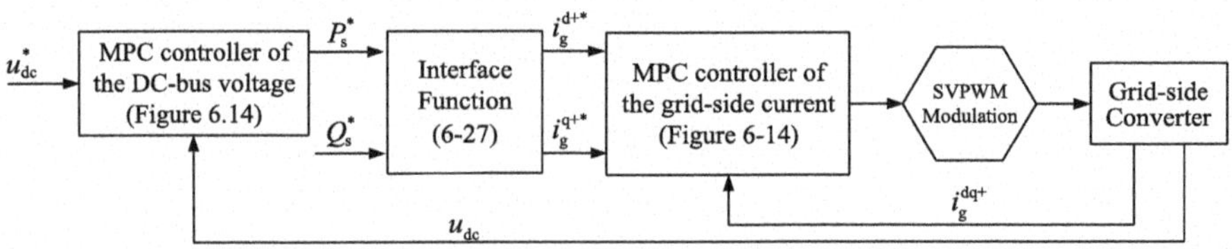

Fig. 6.15 Control diagram of the grid-side converter based on MPC under asymmetrical grid conditions

the grid voltage prediction $\boldsymbol{e}_g^{dq+}(k+1|k)$ obtained by Eq. (6.25) is used to replace all the grid voltage at the moment of kT_s in the algorithm in Sect. 5.2.3.

(2) Using the Full Expression of the Stator Power

In the method described in Sect. 5.2.3, the derivation of the stator power reference to the rotor current reference is simplified by taking symmetrical grid voltages as a precondition, which will no longer be accurate and reasonable under asymmetrical grid voltage conditions, so that the above process needs to be re-derived using the full expression of the stator power.

The full relation between stator power and stator current is shown in Eq. (2.22). In the positive-sequence grid-voltage-oriented synchronous coordinate frame, where the stator voltage is considered to be the grid voltage, the stator power can be rewritten as

$$\begin{cases} P_s = e_g^{d+} + i_s^{d+} + e_g^{q+} + i_s^{q+} \\ Q_s = e_g^{q+} + i_s^{d+} + e_g^{d+} + i_s^{q+} \end{cases} \tag{6.28}$$

The relation between the stator power reference and the stator current reference under asymmetrical grid voltage conditions can be obtained from the above equation as

$$\begin{cases} i_s^{d+*} = \dfrac{e_g^{d+} P_s^* + e_g^{q+} Q_s^*}{e_g^{d+2} + e_g^{q+2}} \\ i_s^{q+*} = \dfrac{e_g^{q+} P_s^* + e_g^{q+} Q_s^*}{e_g^{d+2} + e_g^{q+2}} \end{cases} \tag{6.29}$$

Substituting Eq. (4.30) into the stator voltage equation in Eq. (4.25), where the stator voltage is the grid voltage, and ignoring the effect of stator resistance, it gives

$$\begin{cases} \psi_s^{d+} = \dfrac{1}{\omega_1}\left(2e_{g+}^{q+} - e_g^{q+}\right) \\ \psi_s^{q+} = -\dfrac{1}{\omega_1}\left(2e_{g+}^{d+} - e_g^{d+}\right) \end{cases} \tag{6.30}$$

where the grid voltage $\boldsymbol{e}_{\mathrm{g}}^{\mathrm{dg}+}$ and its positive sequence component $\boldsymbol{e}_{\mathrm{g}}^{\mathrm{dg}+}$ in the positive coordinate frame can be obtained from the phase lock loop and park transformation shown in Fig. 6.5.

Combining Eqs. (4.26), (6.29) and (6.30), the conversion relation between the stator power reference and the rotor current reference under asymmetrical grid voltage conditions can be obtained as

$$\begin{cases} i_{\mathrm{r}}^{\mathrm{d}+*} = \dfrac{1}{L_{\mathrm{M}}}\left[\dfrac{1}{\omega_1}\left(2e_{\mathrm{g}+}^{\mathrm{q}+} - e_{\mathrm{g}}^{\mathrm{q}+}\right) - \dfrac{L_{\mathrm{s}}}{e_{\mathrm{g}}^{\mathrm{d}+^2} + e_{\mathrm{g}}^{\mathrm{q}+^2}}\left(e_{\mathrm{g}}^{\mathrm{d}+}P_{\mathrm{s}}^* + e_{\mathrm{g}}^{\mathrm{q}+}Q_{\mathrm{s}}^*\right)\right] \\ i_{\mathrm{r}}^{\mathrm{q}+*} = \dfrac{1}{L_{\mathrm{M}}}\left[-\dfrac{1}{\omega_1}\left(2e_{\mathrm{g}+}^{\mathrm{d}+} - e_{\mathrm{g}}^{\mathrm{d}+}\right) - \dfrac{L_{\mathrm{s}}}{e_{\mathrm{g}}^{\mathrm{d}+^2} + e_{\mathrm{g}}^{\mathrm{q}+^2}}\left(e_{\mathrm{g}}^{\mathrm{q}+}P_{\mathrm{s}}^* - e_{\mathrm{g}}^{\mathrm{d}+}Q_{\mathrm{s}}^*\right)\right] \end{cases} \tag{6.31}$$

Replace Eq. (5.31) in Sect. 5.2.3 by Eq. (6.31) for the corresponding rotor current reference value obtained from the stator power reference value.

By rewriting the algorithm in Sect. 5.2.3 into the positive-sequence grid-voltage-oriented coordinate frame, and with the two improvements mentioned above, the algorithm flow of the MPC for each part of the generator-side converter under asymmetrical grid voltage can be obtained as shown in Fig. 6.16, and the overall control diagram is shown in Fig. 6.17 (Fig. 6.18).

When a two-phase ground fault occurs on the grid, causing the voltage at the generator terminal of the DFIG-based wind power system to dip to 65% of the rated value, and the voltage of the third phase remains unchanged, the control effect of the above MPC method in a typical 1.5 MW DFIG-based wind power system is shown in Figs. 6.19 and 6.20.

As can be seen from the above results, compared to traditional vector control, the MPC under asymmetrical grid conditions discussed in this section can effectively suppress the effects of the negative sequence components on the system, thereby improving the stator and rotor current waveforms and reducing stator and rotor current pulsation, stator power pulsation, electromagnetic torque pulsation and DC-bus voltage pulsation, thus improving system performance and facilitating uninterrupted operation under grid fault conditions.

Figure 6.17 shows the MPC method with a close loop of rotor current and an open loop of stator power. To further stabilise the power flow injected into the grid under asymmetrical grid conditions by the DFIG-based wind power system, reference can be made to the method in Sect. 3.3.2.2 to introduce a close loop of stator power to form a double-loop MPC structure with an inner loop of rotor current and an outer loop of stator power, the simulation results of which are shown in Figs. 6.21 and 6.22.

As can be seen from the results in Figs. 6.21 and 6.22, when a double-close-loop control structure is used, the stator and rotor currents obtained by both the traditional vector control and the MPC in this section are no longer sinusoidal under asymmetrical grid conditions, and both contain harmonics of two times the grid synchronous frequency in the rotor current, and both have a certain degree of distortion in the

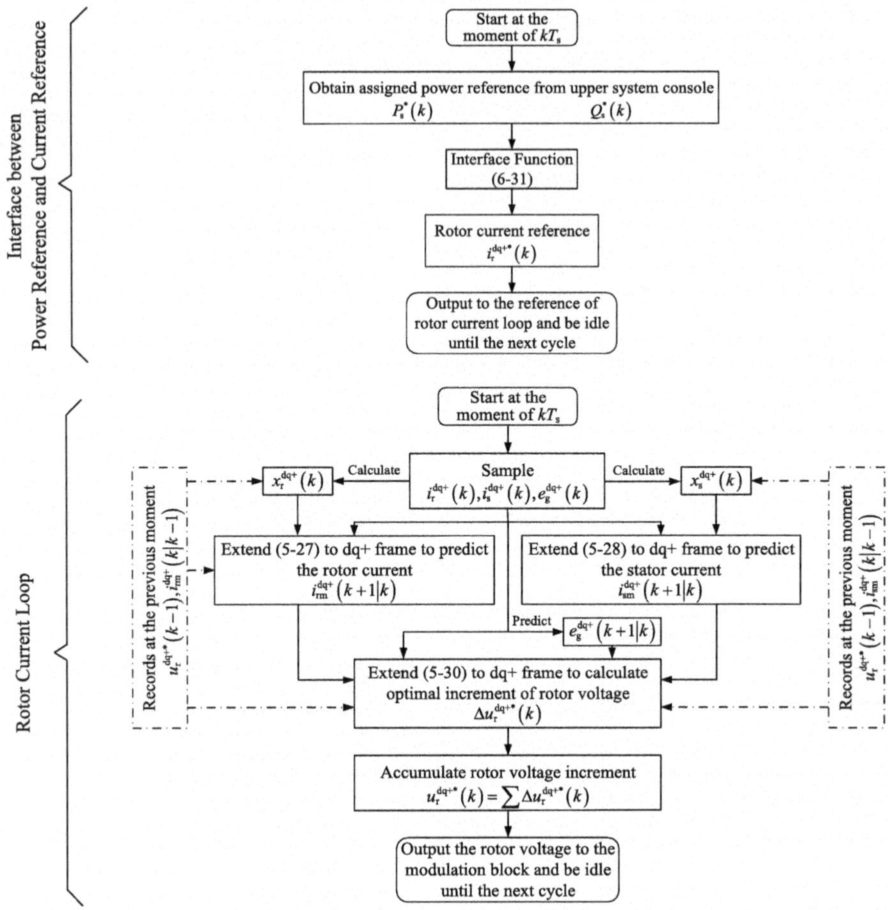

Fig. 6.16 MPC flow of the generator-side converter under asymmetrical grid conditions

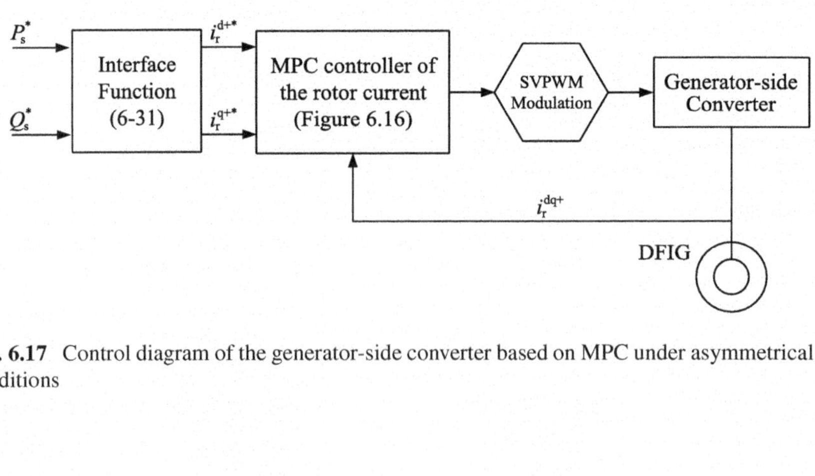

Fig. 6.17 Control diagram of the generator-side converter based on MPC under asymmetrical grid conditions

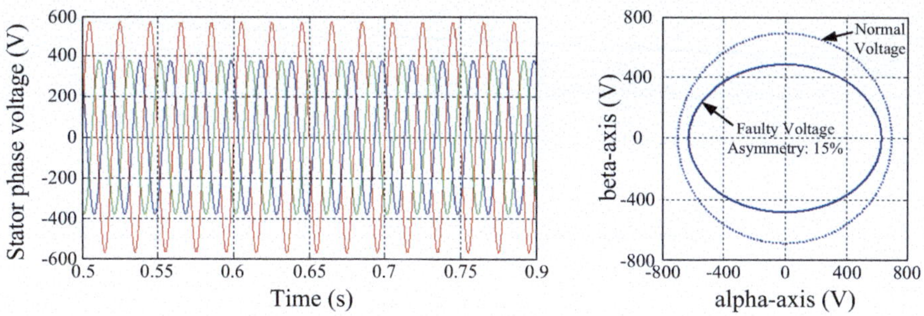

Fig. 6.18 Generator terminal voltage of the DFIG-based wind power system (Left: waveforms in time domain; Right: trajectories in vector space)

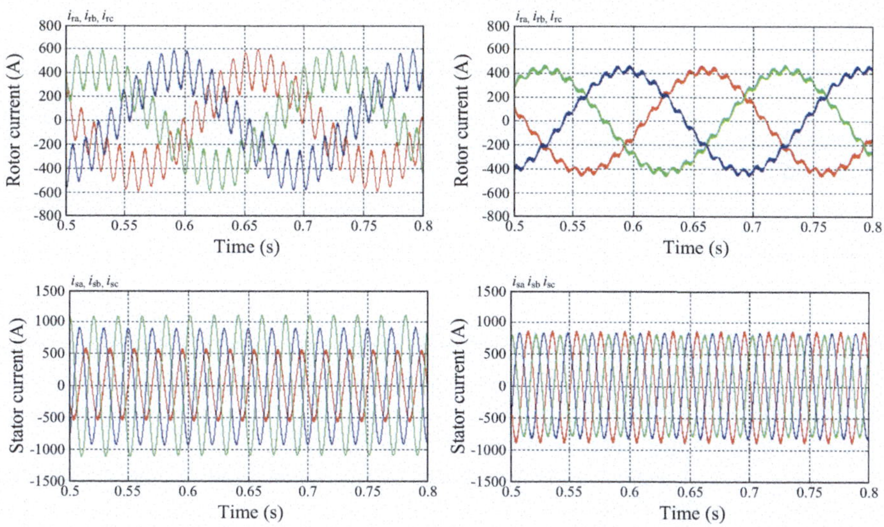

Fig. 6.19 Rotor current under the asymmetrical grid voltage (Left: Traditional VC; Right: Proposed MPC)

stator current, but the causes of the results are essentially different. The traditional vector control loses effective control of the rotor current because of the inability to effectively suppress the negative sequence components of the system, resulting in a large number of harmonics in the rotor current and, through the electromagnetic coupling between the stator and rotor, distorted stator currents on the stator side. The MPC in this section injects specific harmonic components into the rotor current to suppress the stator power pulsation. The magnitude of the injected harmonics is calculated and strictly controlled by the rotor current loop, and under this control the current generated on the stator side ensures that a constant three-phase stator power is injected into the grid. The comparison between the MPC in this section and the traditional vector control also shows that the MPC in this section has lower rotor

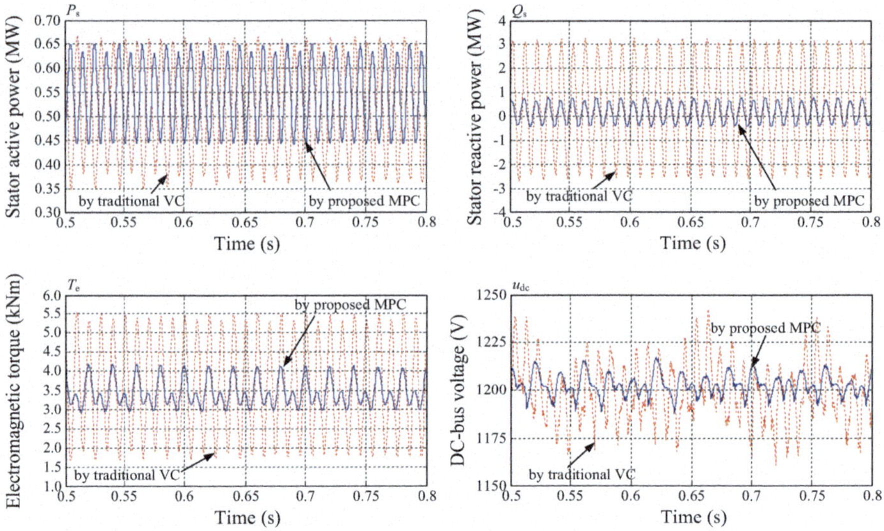

Fig. 6.20 Stator power, electromagnetic torque and DC-bus voltage of the DFIG-based wind power system under the asymmetrical grid voltage

harmonic content, less stator current distortion and lower stator and rotor current peaks than the traditional vector control, which facilitates uninterrupted operation of the system.

Comparing the simulation results of the single-close-loop MPC with those of the double-close-loop MPC, it can be seen that the single-close-loop MPC system has better sinusoidality and symmetry of the stator current, while the double-close-loop MPC system has better stability of the stator power, due to the different close-loop structures of the two systems and the resulting different final control objectives. This result also shows that there is a contradiction in the physical nature between the sinusoidal symmetry of the stator current and the stability of the stator power in the DFIG-based wind power system under asymmetrical grid conditions, the internal mechanism of which and the resulting diversity of control objectives of the DFIG-based wind power system will be discussed in the next section.

6.6 Multi-objective Control Under Asymmetrical Grid Conditions

The working state of the DFIG-based wind power system can be measured from different perspectives using different performance indexes, such as electromagnetic torque, stator power and stator current, etc. Wherein: the smaller the electromagnetic torque pulsation, the smaller the mechanical impact and stress of the system; the smaller the stator power pulsation, the smaller the impact of the system on the

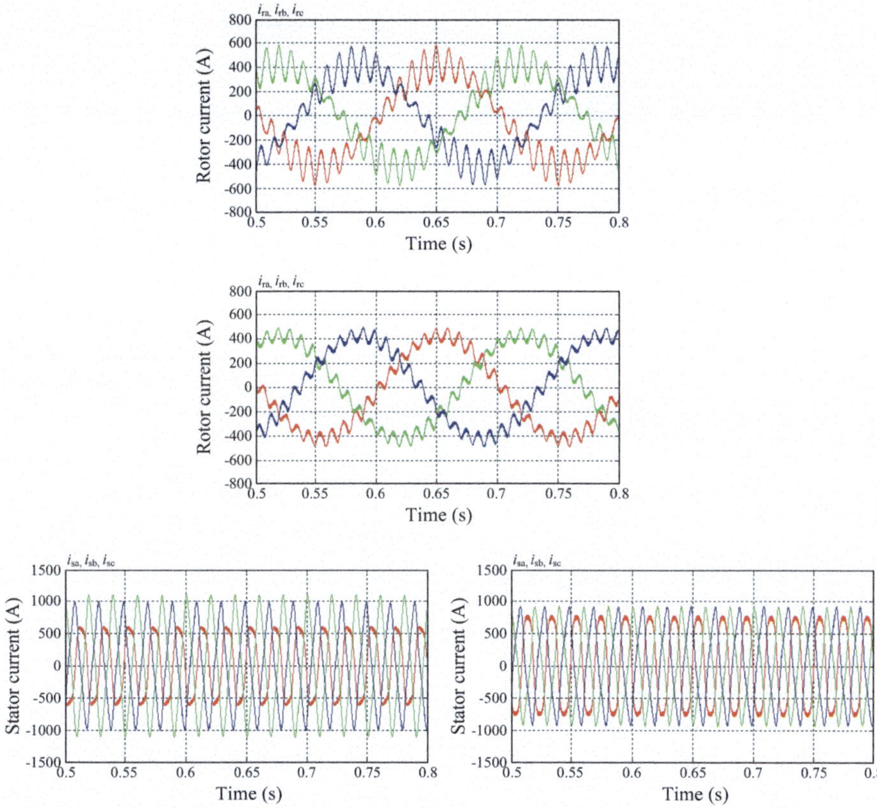

Fig. 6.21 Rotor current under the asymmetrical grid voltage with a double-close-loop control structure (Left: Traditional VC; Right: Proposed MPC)

grid; the lower the harmonic content of the stator current, the higher the quality of the power injected into the grid. Currently, a control strategy usually focuses on only one performance index and uses it as the control target. Under normal grid conditions, the different control objectives of different strategies can be considered as consistent. For example: direct torque control takes the electromagnetic torque as the control objective, while direct power control takes the stator power as the control objective. If the grid voltage is normal and the wind speed is stable, the system whose electromagnetic torque is controlled as stable and pulsation-free will also have stable and pulsation-free stator power, and vice versa. However, when the asymmetrical fault occurs in the grid voltage, the relation between the different performance indexes becomes complex and no longer consistent, or even conflict with each other, which requires the system to make trade-offs among the different performance indexes according to the operating conditions and user requirements [8–11]. This is why it is necessary to analyse the relation between the various control objectives of the

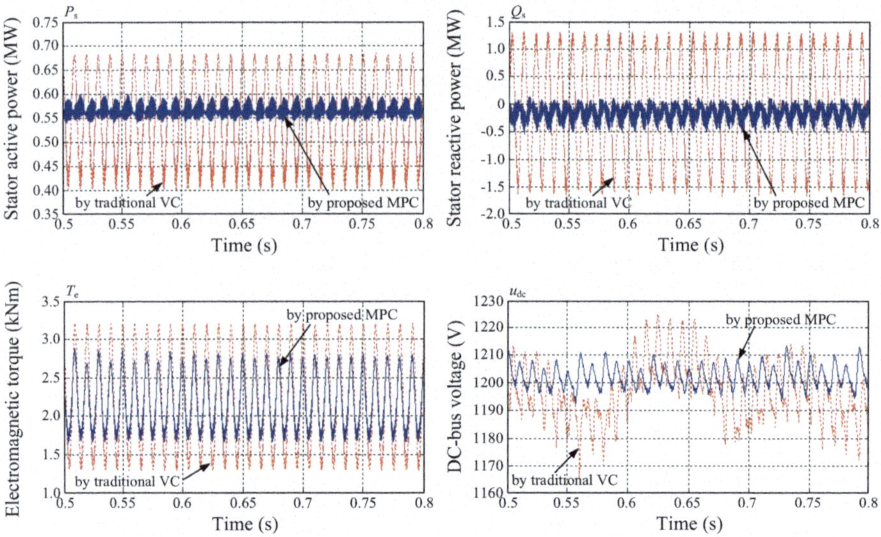

Fig. 6.22 Stator power, electromagnetic torque and DC-bus voltage of the DFIG-based wind power system under the asymmetrical grid voltage with a double-close-loop control structure

DFIG-based system under asymmetrical grid faults and to implement the control of multi-objective in a way that is as simple and intuitive as possible.

6.6.1 Control Objectives for the DFIG-Based Wind Power System Under Asymmetrical Grid Conditions

6.6.1.1 Control Objectives for the Grid-Side Converter

1. Analysis of Control Objectives based on the Calculation Method

Under ideal grid conditions, the control objective of the grid-side converter can be understood as the stable and pulsation-free grid-side active and reactive power and the sinusoidal symmetrical grid-side current (i.e. no negative sequence component). According to Eq. (4.39), the above control objective of the grid-side converter under asymmetrical grid conditions can be expressed as

$$\begin{cases} P_{\text{gconst}} = P_{\text{g}}^{*} \\ P_{\text{g}2\omega_1 \cdot \cos} = 0 \\ P_{\text{g}2\omega_1 \cdot \sin} = 0 \\ Q_{\text{gconst}} = Q_{\text{g}}^{*} \\ Q_{\text{g}2\omega_1 \cdot \cos} = 0 \\ Q_{\text{g}2\omega_1 \cdot \sin} = 0 \\ i_{\text{g}-}^{\text{d}-} = 0 \\ i_{\text{g}-}^{\text{q}-} = 0 \end{cases} \Rightarrow \begin{cases} e_{\text{g}+}^{\text{d}+} i_{\text{g}+}^{\text{d}+} + e_{\text{g}+}^{\text{q}+} i_{\text{g}+}^{\text{q}+} + e_{\text{g}-}^{\text{d}-} i_{\text{g}-}^{\text{d}-} + e_{\text{g}-}^{\text{q}-} i_{\text{g}-}^{\text{q}-} = P_{\text{g}}^{*} & (1) \\ e_{\text{g}+}^{\text{d}+} i_{\text{g}-}^{\text{d}-} + e_{\text{g}+}^{\text{q}+} i_{\text{g}-}^{\text{q}-} + e_{\text{g}-}^{\text{d}-} i_{\text{g}+}^{\text{d}+} + e_{\text{g}-}^{\text{q}-} i_{\text{g}+}^{\text{q}+} = 0 & (2) \\ e_{\text{g}+}^{\text{d}+} i_{\text{g}-}^{\text{q}-} - e_{\text{g}+}^{\text{q}+} i_{\text{g}-}^{\text{d}-} - e_{\text{g}-}^{\text{d}-} i_{\text{g}+}^{\text{q}+} + e_{\text{g}-}^{\text{q}-} i_{\text{g}+}^{\text{d}+} = 0 & (3) \\ e_{\text{g}+}^{\text{q}+} i_{\text{g}+}^{\text{d}+} - e_{\text{g}+}^{\text{d}+} i_{\text{g}+}^{\text{q}+} + e_{\text{g}-}^{\text{q}-} i_{\text{g}-}^{\text{d}-} - e_{\text{g}-}^{\text{d}-} i_{\text{g}-}^{\text{q}-} = Q_{\text{g}}^{*} & (4) \\ e_{\text{g}+}^{\text{q}+} i_{\text{g}-}^{\text{d}-} - e_{\text{g}+}^{\text{d}+} i_{\text{g}-}^{\text{q}-} + e_{\text{g}-}^{\text{q}-} i_{\text{g}+}^{\text{d}+} - e_{\text{g}-}^{\text{d}-} i_{\text{g}+}^{\text{q}+} = 0 & (5) \\ e_{\text{g}+}^{\text{d}+} i_{\text{g}-}^{\text{d}-} + e_{\text{g}+}^{\text{q}+} i_{\text{g}-}^{\text{q}-} - e_{\text{g}-}^{\text{d}-} i_{\text{g}+}^{\text{d}+} - e_{\text{g}-}^{\text{q}-} i_{\text{g}+}^{\text{q}+} = 0 & (6) \\ i_{\text{g}-}^{\text{d}-} = 0 & (7) \\ i_{\text{g}-}^{\text{q}-} = 0 & (8) \end{cases} \quad (6.32)$$

Considering that the grid voltage is not affected by the control algorithm, Eq. (6.32) contains a total of four degrees of freedom (i.e. $i_{\text{g}+}^{\text{d}+}$, $i_{\text{g}+}^{\text{q}+}$, $i_{\text{g}-}^{\text{d}-}$ and $i_{\text{g}-}^{\text{q}-}$), which indicates that at most four of the eight equations in Eq. (6.32) can be satisfied simultaneously. Taking into account the physical meaning of the mathematical model itself, the control objectives for the grid-side converter can be:

Objective 1: the grid-side active power is stable with no 2x synchronous frequency pulsation, i.e. (1)(2)(3)(4) in Eq. (6.32) are satisfied.
Objective 2: the grid-side reactive power is stable with no 2x synchronous frequency pulsation, i.e. (1)(4)(5)(6) in Eq. (6.32) are satisfied.
Objective 3: the grid-side current is sinusoidal and symmetrical with no negative sequence component, i.e. (1)(4)(7)(8) in Eq. (6.32) are satisfied.

2. Analysis of Control Objectives based on the Negative Sequence Power Method

Combining the positive and negative sequence components of the grid current in Eq. (4.39), the expression for the grid-side power can be rewritten as

$$\begin{cases} P_{\text{g}} = P_{\text{g}+} + P_{\text{g}-} \\ Q_{\text{g}} = Q_{\text{g}+} + Q_{\text{g}-} \end{cases} \quad (6.33)$$

where $P_{\text{g}+}, P_{\text{g}-}, Q_{\text{g}+}$ and $Q_{\text{g}-}$ are defined terms introduced in this section for analysis convenience: $P_{\text{g}+}$ and $Q_{\text{g}+}$ are positive sequence grid-side active power and positive sequence grid-side reactive power, respectively, to express the power generated by the positive sequence grid voltage and the grid-side current; $P_{\text{g}-}$ and $Q_{\text{g}-}$ are negative sequence grid-side active power and negative sequence grid-side reactive power, respectively, to express the power generated by the negative sequence grid voltage and the grid-side current. The specific expressions of the above positive and negative sequence grid-side power are as follows:

$$\begin{cases} P_{\text{g}+} = e_{\text{g}+}^{\text{d}+} i_{\text{g}}^{\text{d}+} + e_{\text{g}+}^{\text{q}+} i_{\text{g}}^{\text{q}+} \\ P_{\text{g}-} = \left(e_{\text{g}-}^{\text{d}-} i_{\text{g}}^{\text{d}+} + e_{\text{g}-}^{\text{q}-} i_{\text{g}}^{\text{q}+} \right) \cos(2\omega_1 t) + \left(e_{\text{g}-}^{\text{q}-} i_{\text{g}}^{\text{d}+} - e_{\text{g}-}^{\text{d}-} i_{\text{g}}^{\text{q}+} \right) \sin(2\omega_1 t) \end{cases} \quad (6.33\text{a})$$

$$\begin{cases} Q_{g+} = e_{g+}^{q+} i_g^{d+} - e_{g+}^{d+} i_g^{q+} \\ Q_{g-} = \left(e_{g-}^{q-} i_g^{d+} - e_{g-}^{d-} i_g^{q+}\right) \cos(2\omega_1 t) - \left(e_{g-}^{d-} i_g^{d+} + e_{g-}^{q-} i_g^{q+}\right) \sin(2\omega_1 t) \end{cases} \quad (6.33b)$$

From Eq. (6.33) and the related definition of positive and negative sequence power, it can be seen that, under certain grid conditions, the grid-side active and reactive power and their related mathematical expressions of positive and negative sequence power of the grid-side converter have a total of 2 degrees of freedom, i.e. the d−axis and q-axis components of the grid-side current, and therefore, under asymmetrical grid conditions, the following conclusions can be drawn about the operation of the grid-side converter in the DFIG-based wind power system:

(1) If the positive sequence grid-side power P_{g+} and Q_{g+} are constant, then the grid-side current vector is constant in the positive synchronous coordinate frame and is sinusoidal and symmetrical in the time domain, but a pulsation of 2x the grid synchronous frequency will occur in both the grid-side power P_{g+} and Q_{g+}, causing a certain disturbance to the DC-bus voltage and the total power injected into the grid by the system.
(2) If the grid-side power P_{g+} and Q_{g+} are constant, then the grid-side active and reactive power is stable without pulsation, but the positive sequence grid-side power a and a will appear 2x the grid synchronous frequency pulsation, which indicates that the grid-side current is pulsating at the same frequency in the positive synchronous coordinate frame, manifested in the time domain as distortion of the grid-side current and deterioration of the power quality.
(3) The positive sequence grid-side power (P_g, Q_g) and the grid-side power (P_{g+}, Q_{g+}) cannot be controlled to a non-zero constant value at the same time, i.e. the grid-side converter cannot operate in a certain operating state that allows it to meet both performance requirements of the sinusoidal and symmetrical grid-side current and the stable and pulsation-free grid-side power simultaneously.

From the above three conclusions, it can be seen that the grid-side converter under asymmetrical grid conditions has a variety of operating states with different performance indexes, and accordingly, the control objectives of the control strategy also show diversity, which can be summarised as follows:

Objective I: sinusoidal and symmetrical grid-side current.
Objective II: stable and pulsation-free grid-side active and reactive power.
Objective III: a compromised objective between Objective I and Objective II.

6.6.1.2 Control Objectives for the Generator-Side Converter

1. Analysis of Control Objectives based on the Calculation Method

Under ideal grid conditions, the control objective of the generator-side converter can be understood as the stable and pulsation-free stator active and reactive power and electromagnetic torque of the DFIG and the sinusoidal and symmetrical stator current (i.e. no negative sequence component). According to Eqs. (4.33) and (4.34), the above control objective of the generator-side converter under asymmetrical grid conditions can be expressed as

$$\begin{cases} P_{\text{sconst}} = P_{\text{s}}^{*} \\ P_{\text{s}2\omega_1 \cdot \cos} = 0 \\ P_{\text{s}2\omega_1 \cdot \sin} = 0 \\ Q_{\text{sconst}} = Q_{\text{s}}^{*} \\ Q_{\text{s}2\omega_1 \cdot \cos} = 0 \\ Q_{\text{s}2\omega_1 \cdot \sin} = 0 \\ T_{\text{econst}} = T_{\text{e}}^{*} \\ T_{\text{e}2\omega_1 \cdot \cos} = 0 \\ T_{\text{e}2\omega_1 \cdot \sin} = 0 \\ i_{\text{s}-}^{\text{d}-} = 0 \\ i_{\text{s}-}^{\text{d}-} = 0 \end{cases} \Rightarrow \begin{cases} u_{\text{s}+}^{\text{d}+} i_{\text{s}+}^{\text{d}+} + u_{\text{s}+}^{\text{q}+} i_{\text{s}+}^{\text{q}+} + u_{\text{s}-}^{\text{d}-} i_{\text{s}-}^{\text{d}-} + u_{\text{s}-}^{\text{q}-} i_{\text{s}-}^{\text{q}-} = P_{\text{s}}^{*} & (1) \\ u_{\text{s}+}^{\text{d}+} i_{\text{s}-}^{\text{d}-} + u_{\text{s}+}^{\text{q}+} i_{\text{s}-}^{\text{q}-} + u_{\text{s}-}^{\text{d}-} i_{\text{s}+}^{\text{d}+} + u_{\text{s}-}^{\text{q}-} i_{\text{s}+}^{\text{q}+} = 0 & (2) \\ u_{\text{s}+}^{\text{d}+} i_{\text{s}-}^{\text{q}-} - u_{\text{s}+}^{\text{q}+} i_{\text{s}-}^{\text{d}-} - u_{\text{s}-}^{\text{d}-} i_{\text{s}+}^{\text{q}+} + u_{\text{s}-}^{\text{q}-} i_{\text{s}+}^{\text{d}+} = 0 & (3) \\ u_{\text{s}+}^{\text{q}+} i_{\text{s}+}^{\text{d}+} - u_{\text{s}+}^{\text{d}+} i_{\text{s}+}^{\text{q}+} + u_{\text{s}-}^{\text{q}-} i_{\text{s}-}^{\text{d}-} - u_{\text{s}-}^{\text{d}-} i_{\text{s}-}^{\text{q}-} = Q_{\text{s}}^{*} & (4) \\ u_{\text{s}+}^{\text{q}+} i_{\text{s}-}^{\text{d}-} - u_{\text{s}+}^{\text{d}+} i_{\text{s}-}^{\text{q}-} + u_{\text{s}-}^{\text{q}-} i_{\text{s}+}^{\text{d}+} - u_{\text{s}-}^{\text{d}-} i_{\text{s}+}^{\text{q}+} = 0 & (5) \\ u_{\text{s}+}^{\text{d}+} i_{\text{s}-}^{\text{d}-} + u_{\text{s}+}^{\text{q}+} i_{\text{s}-}^{\text{q}-} - u_{\text{s}-}^{\text{d}-} i_{\text{s}+}^{\text{d}+} - u_{\text{s}-}^{\text{q}-} i_{\text{s}+}^{\text{q}+} = 0 & (6) \\ u_{\text{s}-}^{\text{d}-} i_{\text{s}-}^{\text{d}-} + u_{\text{s}-}^{\text{q}-} i_{\text{s}-}^{\text{q}-} - u_{\text{s}+}^{\text{d}+} i_{\text{s}+}^{\text{d}+} - u_{\text{s}+}^{\text{q}+} i_{\text{s}+}^{\text{q}+} = T_{\text{e}}^{*} n_{\text{p}} / \omega_{\text{s}} & (7) \\ u_{\text{s}-}^{\text{d}-} i_{\text{s}+}^{\text{d}+} + u_{\text{s}-}^{\text{q}-} i_{\text{s}+}^{\text{q}+} - u_{\text{s}+}^{\text{d}+} i_{\text{s}-}^{\text{d}-} - u_{\text{s}+}^{\text{q}+} i_{\text{s}-}^{\text{q}-} = 0 & (8) \\ u_{\text{s}+}^{\text{q}+} i_{\text{s}-}^{\text{d}-} + u_{\text{s}-}^{\text{q}-} i_{\text{s}+}^{\text{d}+} - u_{\text{s}+}^{\text{d}+} i_{\text{s}-}^{\text{q}-} - u_{\text{s}-}^{\text{d}-} i_{\text{s}+}^{\text{q}+} = 0 & (9) \\ i_{\text{s}-}^{\text{d}-} = 0 & (10) \\ i_{\text{s}-}^{\text{q}-} = 0 & (11) \end{cases} \tag{6.34}$$

Under grid-connected conditions, the DFIG stator voltage is considered to be the grid voltage and is not affected by the control algorithm. Accordingly, Eq. (6.34) has four degrees of freedom (i.e. $i_{\text{s}+}^{\text{d}+}$, $i_{\text{s}+}^{\text{q}+}$, $i_{\text{s}-}^{\text{d}-}$ and $i_{\text{s}-}^{\text{q}-}$), which indicates that at most four of the eleven equations in Eq. (6.34) can be satisfied simultaneously. Taking into account the physical meaning of the mathematical model itself, the control objectives for the generator-side converter can be:

Objective 1: the stator active power is stable with no 2x synchronous frequency pulsation, i.e. (1)(2)(3)(4) in Eq. (6.34) are satisfied.
Objective 2: the stator reactive power is stable with no 2x synchronous frequency pulsation, i.e. (1)(4)(5)(6) in Eq. (6.34) are satisfied.
Objective 3: the electromagnetic torque is stable with no 2x synchronous frequency pulsation, i.e. (4)(7)(8)(9) in Eq. (6.34) are satisfied.
Objective 4: the stator current is sinusoidal and symmetrical with no negative sequence component, i.e. (1)(4)(10)(11) in Eq. (6.34) are satisfied.

2. Analysis of Control Objectives based on the Negative Sequence Power Method

Combining the positive and negative sequence components of the stator current in Eq. (4.33) and considering the stator voltage to be the grid voltage, the expression for the stator power can be rewritten as

$$\begin{cases} P_{\text{s}} = P_{\text{s}+} + P_{\text{s}-} \\ Q_{\text{s}} = Q_{\text{s}+} + Q_{\text{s}-} \end{cases} \tag{6.35}$$

where P_{s+}, P_{s-}, Q_{s+}, and Q_{s-} are defined terms introduced in this section for analysis convenience: P_{s+} and Q_{s+} are positive sequence stator active power and positive sequence stator reactive power, respectively, to express the power generated by the positive sequence grid voltage and the stator current; P_{s-} and Q_{s+} are negative sequence stator active power and negative sequence stator reactive power, respectively, to express the power generated by the negative sequence grid voltage and the stator current. The specific expressions of the above positive and negative sequence stator power are as follows:

$$\begin{cases} P_{s+} = e_{g+}^{d+} i_s^{d+} + e_{g+}^{q+} i_s^{q+} \\ P_{s-} = \left(e_{g-}^{d-} i_s^{d+} + e_{g-}^{q-} i_s^{q+}\right) \cos(2\omega_1 t) + \left(e_{g-}^{q-} i_s^{d+} - e_{g-}^{d-} i_s^{q+}\right) \sin(2\omega_1 t) \end{cases} \tag{6.35a}$$

$$\begin{cases} Q_{s+} = e_{g+}^{q+} i_s^{d+} - e_{g+}^{d+} i_s^{q+} \\ Q_{s-} = \left(e_{g-}^{q-} i_s^{d+} - e_{g-}^{d-} i_s^{q+}\right) \cos(2\omega_1 t) - \left(e_{g-}^{d-} i_s^{d+} + e_{g-}^{q-} i_s^{q+}\right) \sin(2\omega_1 t) \end{cases} \tag{6.35b}$$

The air-gap power transferred from the rotor to the air gap in the DFIG can be expressed as

$$P_{Te} = T_e \omega_s / n_p \tag{6.36}$$

Equation (6.36) shows that for a DFIG with a fixed number of pole pairs, A is the form of the power expression relating only to the electromagnetic torque. Substituting Eq. (4.34) into Eq. (6.36), it gives

$$P_{Te} = P_{s+} - P_{s-} \tag{6.37}$$

From Eqs. (6.36) and (6.37) and the related definition of positive and negative sequence power, it can be seen that, under certain grid conditions, the stator active and reactive power, air-gap power and their related mathematical expressions of positive and negative sequence power have a total of 2 degrees of freedom, i.e. the d-axis and q-axis components of the stator current, and therefore, under asymmetrical grid conditions, the following conclusions can be drawn about the operation of the DFIG:

(1) If the air-gap power P_{Te} and stator reactive power Q_s are constant values, the electromagnetic torque and stator reactive power are stable without pulsation, but at this time the stator active power P_s and positive sequence stator power P_{s+}, Q_{s+} will appear 2x grid synchronous frequency pulsation, while the stator active power pulsation causes an active power disturbance to the grid, and the positive sequence stator power pulsation indicates that the stator current appears a same-frequency pulsation in the positive synchronous coordinate frame, which is expressed in the time domain as stator current distortion and power quality deterioration.
(2) If the positive sequence stator power P_{s+} and Q_{s+} are constant values, then the stator current vector is constant in the positive synchronous coordinate frame

and shows the sinusoidal and symmetrical stator current in the time domain, but at this time a pulsation of 2x grid synchronous frequency will occur in both the stator power P_s, Q_s and air-gap power P_{Te}, where the stator power pulsation causes a power disturbance to the grid, and the air-gap power pulsation indicates a same-frequency pulsation of the electromagnetic torque, causing a mechanical shock to the power generation system.

(3) If the stator power P_s and $Q_s$are constant values, the stator active power and reactive power are stable without pulsation, but at this time the positive sequence stator power P_{s+}, Q_{s+} and air gap power P_{Te} will appear 2x grid synchronous frequency pulsation, while the positive sequence stator power pulsation indicates that the stator current appears a same-frequency pulsation in the positive synchronous coordinate frame, which is expressed in the time domain as stator current distortion and power quality deterioration, and the air-gap power pulsation indicates a same-frequency pulsation in the electromagnetic torque, causing a mechanical shock to the power generation system.

(4) The air-gap power P_{Te}, positive sequence stator power (P_{s+}, Q_{s+}) and stator power (P_s, Q_s) cannot be controlled to a non-zero constant value at the same time, i.e. the DFIG cannot operate in a certain operating state that allows it to meet the three performance requirements of the stable and pulsation-free electromagnetic torque, the sinusoidal and symmetrical stator current and the stable and pulsation-free stator power simultaneously.

From the above four conclusions, it can be seen that the DFIG under asymmetrical grid conditions has a variety of operating states with different performance indexes, and accordingly, the control objectives of the control strategy also show diversity, which can be summarised as follows:

Objective I: stable and pulsation-free electromagnetic torque and stator reactive power.
Objective II: sinusoidal and symmetrical grid-side current.
Objective III: stable and pulsation-free stator active and reactive power.
Objective IV: A compromised objective among Objective I, II and III.

In summary, the control objective of the DFIG-based wind power system under asymmetrical grid conditions will no longer be single. Different control objectives will allow the system to operate in different states and thus show different performance characteristics to adapt to different users or operating conditions, allowing the system to show more flexibility and adaptability.

6.6.2 Multi-objective Control Based on Negative Sequence Power Injection

As can be seen from the analysis in the previous section, when analysing the control objectives of the DFIG-based wind power system under asymmetrical grid conditions, the results obtained by the calculation method and the negative-sequence power method are not identical, and the control objectives obtained by the calculation method are overall weaker than the control objectives obtained by the negative sequence method, e.g. the calculation method considers that the stator active and reactive power of the DFIG cannot be stabilised at the same time, while the negative sequence method considers that they can be stabilised at the same time. The calculation method considers the various control objectives of the DFIG-based wind power system to be discrete and without intermediate states, whereas the negative sequence method considers the control objectives to be continuous and to have any intermediate states. These differences result from a hidden precondition in the calculation method that the controller does not have a good AC reference response (e.g. a typical PI controller), whereas the negative sequence method does not have this precondition.

Based on the calculation method to realise multi-objective control, it is only necessary to select the equation set from Eqs. (6.32) and (6.34) that can satisfy the current control objective to solve, and assign the obtained reference value of each control quantity to the corresponding control algorithm, such as using the calculated reference value of positive and negative sequence current as the reference value of positive and negative sequence current of the vector control in the positive and negative synchronous coordinate frames to realise each different control objective. This method is simple, easy to implement, and can be achieved by PI controllers, but the quantity and quality of control objectives that can be achieved by this method are more limited, and its calculation process is related to the system parameters, and the control accuracy is more affected by the system parameters. Multi-objective control based on the negative sequence method can achieve a greater number and higher quality of control objectives, cope with more complex operating conditions, rely on close loop control and the control results are less likely to be affected by system parameters. Although it requires the controller to have some ability to respond to AC quantities, this is easily achieved with the help of the MPC strategy for asymmetrical grid conditions discussed earlier in this chapter. The next part of this section will focus on the multi-objective control strategy based on the negative sequence power injection method.[3]

[3] Multi-objective control based on the negative sequence power injection method can be implemented with a variety of controllers that have some nonlinear response capability, and the MPC method is discussed here as an example, while in practice it can be implemented by any algorithm with sufficient negative sequence power regulation capability.

6.6.2.1 Multi-objective Control for the Grid-Side Converter

According to Eq. (6.33), the expression of the grid-side power reference can be redefined as

$$\begin{cases} P_{\mathrm{g}}^{*} = P_{\mathrm{g}+}^{*} + \lambda P_{\mathrm{g}-}^{*} \\ Q_{\mathrm{g}}^{*} = Q_{\mathrm{g}+}^{*} + \lambda Q_{\mathrm{g}-}^{*} \end{cases} \tag{6.38}$$

where $P_{\mathrm{g}+}^{*}$ and $P_{\mathrm{g}-}^{*}$ are the grid-side positive and negative sequence active power references, respectively, $Q_{\mathrm{g}+}^{*}$ and $Q_{\mathrm{g}-}^{*}$ are the grid-side positive and negative sequence reactive power references, respectively, and λ is the weight factor for the grid-side negative sequence power, used to regulate the grid-side negative sequence power content, with its value range as [0, 1].

Substituting the expressions of grid-side positive and negative sequence power in Eq. (6.33) into Eq. (6.38), the grid-side current reference can be obtained as

$$\begin{cases} i_{\mathrm{g}}^{\mathrm{d}+*} = \dfrac{F_1 P_{\mathrm{g}}^{*} + F_2 Q_{\mathrm{g}}^{*}}{F_1 F_4 + F_2 F_3} \\ i_{\mathrm{g}}^{\mathrm{q}+*} = \dfrac{F_3 P_{\mathrm{g}}^{*} - F_1 Q_{\mathrm{g}}^{*}}{F_1 F_4 + F_2 F_3} \end{cases} \tag{6.39}$$

where the specific expressions for F_1, F_2, F_3 and F_4 are as follows.

$$\begin{cases} F_1 = e_{\mathrm{g}+}^{\mathrm{d}+} + \lambda e_{\mathrm{g}-}^{\mathrm{d}-} \cos\theta_{\pm} + \lambda e_{\mathrm{g}-}^{\mathrm{q}-} \sin\theta_{\pm} \\ F_2 = e_{\mathrm{g}+}^{\mathrm{q}+} + \lambda e_{\mathrm{g}-}^{\mathrm{q}-} \cos\theta_{\pm} - \lambda e_{\mathrm{g}-}^{\mathrm{d}-} \sin\theta_{\pm} \\ F_3 = e_{\mathrm{g}+}^{\mathrm{q}+} + \lambda e_{\mathrm{g}-}^{\mathrm{q}-} \cos\theta_{\pm} - \lambda e_{\mathrm{g}-}^{\mathrm{d}-} \sin\theta_{\pm} \\ F_4 = e_{\mathrm{g}+}^{\mathrm{d}+} + \lambda e_{\mathrm{g}-}^{\mathrm{d}-} \cos\theta_{\pm} + \lambda e_{\mathrm{g}-}^{\mathrm{q}-} \sin\theta_{\pm} \end{cases} \tag{6.39a}$$

By replacing the interface part between the power reference and the current reference in Fig. 6.15 with Eq. (6.39), the control diagram of the MPC strategy for the grid-side converter with multi-objective control capability can be obtained as shown in Fig. 6.23.

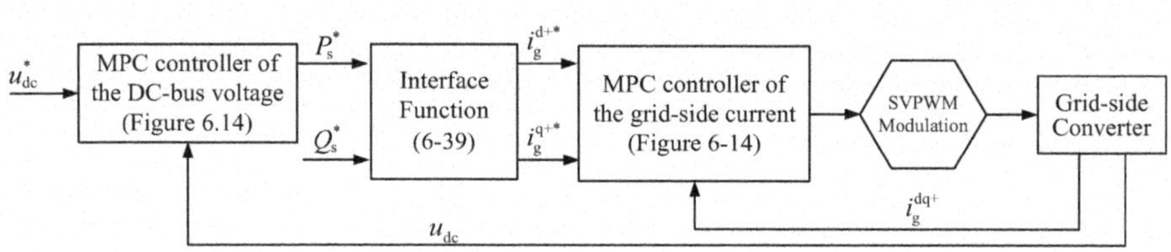

Fig. 6.23 Control diagram of the grid-side converter based on MPC with multi-objective control capability

In Fig. 6.23, the grid-side MPC controller ensures that the grid-side current follows its reference given in Eq. (6.39), while the definition of the grid-side power reference given in Eq. (6.38) shows that the control strategy in Fig. 6.23 is capable of achieving the following control characteristics when the weight factor λ is taken to different values:

(1) When $\lambda = 0$, the power reference values are $P_{\mathrm{g}}^* = P_{\mathrm{g}+}^*$ and $Q_{\mathrm{g}}^* = Q_{\mathrm{g}+}^*$. The strategy in Fig. 6.23 will ensure that the grid-side positive sequence active and reactive power of the grid-side converter are their reference values by controlling the grid-side current to follow its reference value, i.e. $P_{\mathrm{g}+}^*$ and $Q_{\mathrm{g}+}^*$ remain constant and the grid-side converter achieves its Objective I.
(2) When $\lambda = 1$, the power reference values are $P_{\mathrm{g}}^* = P_{\mathrm{g}+}^* + P_{\mathrm{g}-}^*$ and $Q_{\mathrm{g}}^* = Q_{\mathrm{g}+}^* + Q_{\mathrm{g}-}^*$. The strategy in Fig. 6.23 will ensure that the grid-side active and reactive power of the grid-side converter are their reference values by controlling the grid-side current to follow its reference value, i.e. P_{g}^* and Q_{g}^* remain constant and the grid-side converter achieves its Objective II.
(3) When λ is any other value in the range [0, 1], the negative sequence power component of the grid-side power reference will vary between the two scenarios (1) and (2) above, and the strategy in Fig. 6.23, by controlling the grid-side current to follow its reference value, achieves Objective III of the grid-side converter.

6.6.2.2 Multi-objective Control for the Generator-Side Converter

According to Eq. (6.35), the expression of the stator power reference can be redefined as

$$\begin{cases} P_{\mathrm{s}}^* = P_{\mathrm{s}+}^* + \mu P_{\mathrm{g}-}^* \\ Q_{\mathrm{s}}^* = Q_{\mathrm{s}+}^* + |\mu| Q_{\mathrm{s}-}^* \end{cases} \tag{6.40}$$

where $P_{\mathrm{s}+}^*$ and $P_{\mathrm{s}-}^*$ are the stator positive and negative sequence active power references, respectively, $Q_{\mathrm{s}+}^*$ and $Q_{\mathrm{s}-}^*$ are the stator positive and negative sequence reactive power references, respectively, and μ is the weight factor for the stator negative sequence power, used to regulate the stator negative sequence power content, with its value range as $[-1, +1]$.

Substituting the expressions of stator positive and negative sequence power in Eq. (6.35) into Eq. (6.40), the stator current reference can be obtained as

$$\begin{cases} i_{\mathrm{s}}^{\mathrm{d}+*} = \dfrac{G_1 P_{\mathrm{s}}^* + G_2 Q_{\mathrm{s}}^*}{G_1 G_4 + G_2 G_3} \\ i_{\mathrm{s}}^{\mathrm{q}+*} = \dfrac{G_3 P_{\mathrm{s}}^* - G_1 Q_{\mathrm{s}}^*}{G_1 G_4 + G_2 G_3} \end{cases} \tag{6.41}$$

where the specific expressions for G_1, G_2, G_3 and G_4 are as follows.

$$\begin{cases} G_1 = e_{g+}^{d+} + \mu e_{g-}^{d-} \cos\theta_{\pm} + \mu e_{g-}^{q-} \sin\theta_{\pm} \\ G_2 = e_{g+}^{q+} + \mu e_{g-}^{q-} \cos\theta_{\pm} - \mu e_{g-}^{d-} \sin\theta_{\pm} \\ G_3 = e_{g+}^{q+} + |\mu| e_{g-}^{q-} \cos\theta_{\pm} - |\mu| e_{g-}^{d-} \sin\theta_{\pm} \\ G_4 = e_{g+}^{d+} + |\mu| e_{g-}^{d-} \cos\theta_{\pm} + |\mu| e_{g-}^{q-} \sin\theta_{\pm} \end{cases} \tag{6.41a}$$

By combining Eqs. (4.26), (6.29) and (6.41), the conversion relation between the stator power reference and the rotor current reference can be obtained as

$$\begin{cases} i_{r}^{d+*} = \dfrac{1}{L_M}\left[\dfrac{1}{\omega_1}\left(2e_{g+}^{q+} - e_{g}^{q+}\right) - L_s \dfrac{G_1 P_s^* + G_2 Q_s^*}{G_1 G_4 + G_2 G_3}\right] \\ i_{r}^{q+*} = \dfrac{1}{L_M}\left[-\dfrac{1}{\omega_1}\left(2e_{g+}^{d+} - e_{g}^{d+}\right) - L_s \dfrac{G_3 P_s^* - G_1 Q_s^*}{G_1 G_4 + G_2 G_3}\right] \end{cases} \tag{6.42}$$

By replacing the interface part between the power reference and the current reference in Fig. 6.17 with Eq. (6.42), the control diagram of the MPC strategy for the generator-side converter with multi-objective control capability can be obtained as shown in Fig. 6.24.

In Fig. 6.24, the generator-side MPC controller ensures that the rotor current follows its reference given in Eq. (6.42), while the definition of the stator power reference given in Eq. (6.40) shows that the control strategy in Fig. 6.24 is capable of achieving the following control characteristics when the weight factor μ is taken to different values:

(1) When $\mu = -1$, the power reference values are $P_s^* = P_{s+}^* - P_{s-}^*$ and $Q_s^* = Q_{s+}^* + Q_{s-}^*$. The strategy in Fig. 6.24 will ensure that the air-gap power and stator reactive power of the DFIG are their reference values by controlling the rotor current to follow its reference value, i.e. P_{Te} and Q_s remain constant and the generator-side converter achieves its Objective I.

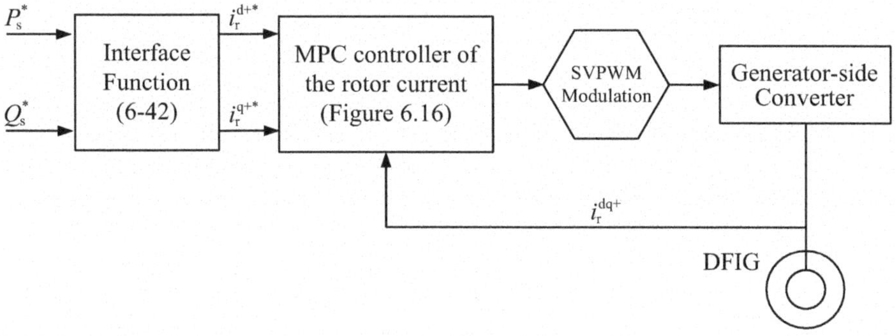

Fig. 6.24 Control diagram of the generator-side converter based on MPC with multi-objective control capability

(2) When $\mu = 0$, the power reference values are $P_s^* = P_{s+}^*$ and $Q_s^* = Q_{s+}^*$. The strategy in Fig. 6.24 will ensure that the stator positive sequence active and reactive power of the DFIG are their reference values by controlling the rotor current to follow its reference value, i.e. P_{s+} and Q_{s+} remain constant and the generator-side converter achieves its Objective II.
(3) When $\mu = 1$, the power reference values are $P_s^* = P_{s+}^* + P_{s-}^*$ and $Q_s^* = Q_{s+}^* + Q_{s-}^*$. The strategy in Fig. 6.24 will ensure that the stator active and reactive power of the DFIG are their reference values by controlling the rotor current to follow its reference value, i.e. P_s and Q_s remain constant and the generator-side converter achieves its Objective III.
(4) When μ is any other value in the range $[-1, +1]$, the negative sequence power component of the stator power reference will vary among the three scenarios (1), (2) and (3) above, and the strategy in Fig. 6.24, by controlling the rotor current to follow its reference value, achieves Objective IV of the generator-side converter.

In summary, the negative sequence power injection method can control the content of negative sequence power in the grid-side power of the grid-side converter and the stator power of the DFIG respectively by independent adjustment of two simple parameters, thus realising continuous, smooth and online switching between multiple control objectives of the grid-side converter and the generator-side converter, with a simple control method and a clear physical meaning (Fig. 6.25).

6.6.2.3 Coordinated Control of the Grid-Side Converter and the Generator-Side Converter

In Sects. 6.6.2.1 and 6.6.2.2, a multi-objective control strategy based on negative sequence power injection for the grid-side converter and the generator-side converter under asymmetrical grid conditions is discussed for the independent regulation of the operational performance of the grid-side converter and the generator-side converter, respectively. As a whole, the coordination between the control objectives of the grid-side converter and the generator-side converter in order to optimise the performance of the entire DFIG-basd wind power system to meet the specific needs of complex operating conditions is a complex and practical issue.

The power flow in the DFIG-based wind power system is shown in Fig. 6.26, where the total active/reactive power injected into the grid by the system is the sum of its stator active/reactive power and the grid-side active/reactive power, respectively, and the active power injected into the DC bus by the system is the difference between the rotor active power and the grid-side active power, i.e. there is

$$\begin{cases} P_{\text{total}} = P_s + P_g \\ Q_{\text{total}} = Q_s + Q_g \\ P_{\text{dc}} = P_r - P_g \end{cases} \tag{6.43}$$

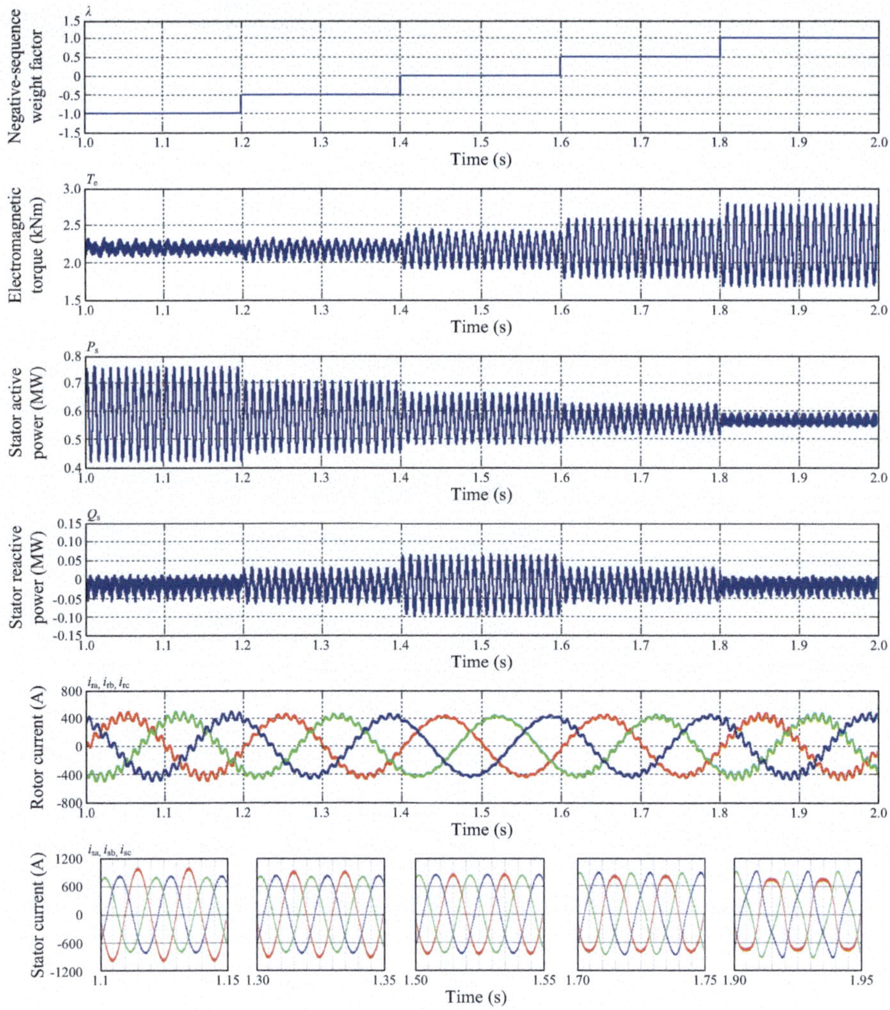

Fig. 6.25 Simulation results of online switching for multiple control objectives in the generator-side converter

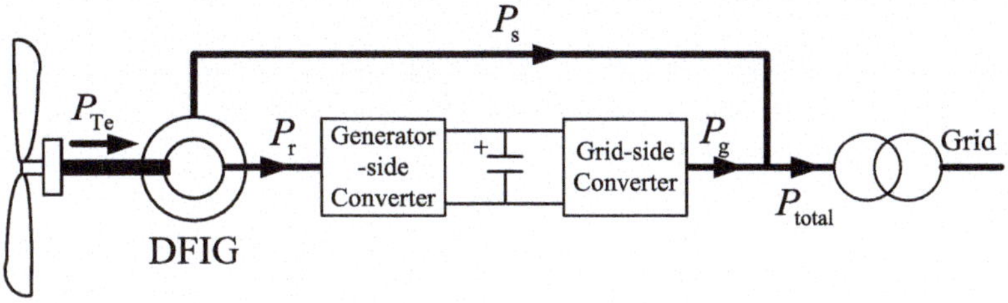

Fig. 6.26 Power flow in the DFIG-based wind power system

Under asymmetrical grid conditions, the following power expressions can be obtained from Eqs. (4.33), (4.34), (4.39) and (6.43).

$$\begin{aligned} P_{total} &= P_s + P_g \\ &= \left(P_{sconst} + P_{gconst}\right) \\ &\quad +\left(P_{s2\omega1\cdot\cos} + P_{g2\omega1\cdot\cos}\right)\cos(2\omega_1 t) \\ &\quad +\left(P_{s2\omega1\cdot\sin} + P_{g2\omega1\cdot\sin}\right)\sin(2\omega_1 t) \end{aligned} \tag{6.44}$$

$$\begin{aligned} Q_{\text{total}} &= Q_{\text{s}} + Q_{\text{g}} \\ &= \left(Q_{\text{sconst}} + Q_{\text{gconst}}\right) \\ &\quad +\left(Q_{\text{s}2\omega1\cdot\cos} + Q_{\text{g}2\omega1\cdot\cos}\right)\cos(2\omega_1 t) \\ &\quad +\left(Q_{\text{s}2\omega1\cdot\sin} + Q_{\text{g}2\omega1\cdot\sin}\right)\sin(2\omega_1 t) \end{aligned} \tag{6.45}$$

$$\begin{aligned} P_{\text{dc}} &= P_{\text{r}} - P_{\text{g}} = (P_{\text{Te}} - P_{\text{s}}) - P_{\text{g}} \\ &= \left(T_{\text{econst}}\omega_{\text{r}} - P_{\text{sconst}} - P_{\text{gconst}}\right) \\ &\quad +\left(T_{\text{e}2\omega1\cdot\cos}\omega_{\text{r}} - P_{\text{s}2\omega1\cdot\cos} - P_{\text{g}2\omega1\cdot\cos}\right)\cos(2\omega_1 t) \\ &\quad +\left(T_{\text{e}2\omega1\cdot\sin}\omega_{\text{r}} - P_{\text{s}2\omega1\cdot\sin} - P_{\text{g}2\omega1\cdot\sin}\right)\sin(2\omega_1 t) \end{aligned} \tag{6.46}$$

To improve the grid stability, it is generally required that the power injected into the grid from the DFIG-based wind power system is stable, i.e. in Eqs. (6.44) and (6.45), there are

$$\begin{cases} P_{\text{s}2\omega1\cdot\cos} + P_{\text{g}2\omega1\cdot\cos} = 0 \\ P_{\text{s}2\omega1\cdot\sin} + P_{\text{g}2\omega1\cdot\sin} = 0 \end{cases} \tag{6.47}$$

$$\begin{cases} Q_{\text{s}2\omega1\cdot\cos} + Q_{\text{g}2\omega1\cdot\cos} = 0 \\ Q_{\text{s}2\omega1\cdot\sin} + Q_{\text{g}2\omega1\cdot\sin} = 0 \end{cases} \tag{6.48}$$

Substituting Eqs. (6.47) and (6.48) into Eq. (6.46), it gives

$$\begin{aligned} P_{\text{dc}} &= \left(T_{\text{econst}}\omega_{\text{r}} - P_{\text{sconst}} - P_{\text{gconst}}\right) \\ &\quad +T_{\text{e}2\omega1\cdot\cos}\omega_{\text{r}}\cos(2\omega_1 t) + T_{\text{e}2\omega1\cdot\sin}\omega_{\text{r}}\sin(2\omega_1 t) \end{aligned} \tag{6.49}$$

Equation (6.49) shows that if the 2x grid frequency pulsation component of the electromagnetic torque can be eliminated at this point, the active power injected into the DC bus by the system can be stabilised, thus eliminating the possible 2x grid frequency pulsation on the DC-bus voltage.

In summary, from two perspectives: grid stability and converter control performance, the DFIG-based wind power system should work with: stable electromagnetic torque without 2x frequency pulsation to reduce mechanical stress and vibration fatigue of the wind power system; smooth injection of total active and reactive power into the grid without 2x frequency pulsation to improve grid stability; stable DC-bus voltage without 2x frequency pulsation to improve capacitor lifetime and fault-free operation.[4] In combination with Sects. 6.6.2.1 and 6.6.2.2, the respective control objectives under coordinated control of the grid-side converter and the generator-side converter are:

(1) The generator-side converter is used to achieve a stable and pulsation-free electromagnetic torque and stator reactive power, i.e. its Objective I, in which case the stator active power will appear a 2x grid frequency pulsation.
(2) The grid-side converter is used to eliminate the stator active power pulsation in order to achieve a stable and pulsation-free injection of total power into the grid from the system, and its Objective III.

References

1. National Grid, Grid Code Connection Conditions [EB/OL]. (2008-01-01) [2013-03-26]. http://www.nationalgrid.com
2. Y. Zhou, P. Bauer, J.A. Ferreira et al., Operation of grid-connected DFIG under unbalanced grid voltage condition. IEEE Trans. Energy Convers. **24**(1), 240–246 (2009)
3. C. Ng, L. Ran, J. Bumby, Unbalanced-grid-fault ride-through control for a wind turbine inverter. IEEE Trans. Ind. Appl. **44**(3), 845–856 (2008)
4. T. Brekken, N. Mohan, A novel doubly-fed induction wind generator control scheme for reactive power control and torque pulsation compensation under unbalanced grid voltage conditions, in *Power Electronics Specialist Conference*, vol. 2 (2003), pp. 760–764
5. T.K.A. Brekken, N. Mohan, Control of a doubly fed induction wind generator under unbalanced grid voltage conditions. IEEE Trans. Energy Convers. **22**(1), 129–135 (2007)
6. J. Kearney, M.F. Colon, E. Coyle, The application of multi frequency resonant controllers in a DFIG to improve performance by reducing unwanted power and torque pulsations and reducing current harmonics, in *45th International Universities Power Engineering Conference* (2010), pp. 1–6
7. M. Mohseni, M. Mesbah, S. Islam et al., A novel current regulator for DFIG wind turbines with enhanced performance under unbalanced supply voltage condition, in *Power and Energy Society General Meeting* (2010), pp. 1–8
8. D. Santos-Martin, J.L. Rodriguez-Amenedo, S. Arnaltes, Providing ride-through capability to a doubly fed induction generator under unbalanced voltage dips. IEEE Trans. Power Electron. **24**(7), 1747–1756 (2009)
9. H. Jiabing, H. Yikang, W. Hongsheng, Coordinated control of grid- and rotor-side converters of doubly-fed induction generator under unbalanced network voltage conditions. Proc. CSEE **9**, 97–104 (2010)
10. Z. Lu, Research on the operation for doubly fed induction generator system to ride-through grid faults. Ph.D. thesis (Beijing Jiaotong University, Beijing, China, 2012)
11. M. Hongwei, Research on model predictive control and grid fault ride-through strategies of doubly fed induction generator wind turbine system. Ph.D. thesis (Tsinghua University, Beijing, China, 2013)

[4] 由本节前面的论述可知，直流母线电压稳定无 2 倍频脉动是电磁转矩和注入电网总有功功率稳定的自然结果.

Chapter 7
The PMSG-Based Wind Power System and Its Weak-Grid Operation Control

The typical structure of the PMSG-based wind power system is shown in Fig. 7.1. In this case, the generator-side converter is connected to the generator stator and realises generator control by regulating the stator current. The grid-side converter is used to control the DC-bus voltage and feeds the full power into the grid to achieve decoupled active and reactive power control. Between the grid-side converter and the grid there is usually a filter structure to suppress the harmonic currents caused by the converter.

The PMSG-based wind power system uses a full-scale power converter to isolate the PMSG from the grid to a certain extent, reducing the impact of grid faults on generator operation, and its uninterrupted operation under faulty grid conditions is relatively easy to achieve. However, under weak grid conditions, the voltage at the wind power system access point will fluctuate due to the cable impedance, when the wind power system outputs different power, and therefore cannot give a stable voltage and frequency reference. At the same time, when wind power accounts for a large proportion of the grid capacity, the wind power system needs to help stabilise the voltage and frequency of the grid in the same way as a conventional synchronous generator, i.e. the wind power system is required to be able to deliver the active and reactive power required by the grid. This operation mode is also applicable in the case of islanded operation.

This chapter first introduces the dynamic mathematical model of the PMSG-based wind power system on various coordinate frames and its vector control method under ideal grid conditions, and then investigates the characteristics of the PMSG-based wind power system and its operating mode under weak grid conditions. This requires that the active power captured by the wind turbine must be balanced with the power required by the grid. On the generator side, two methods of regulating the power captured by the wind turbine are discussed, one by regulating the generator speed and the other by regulating the pitch angle of the blades. On the grid side, the converter needs to determine the amount of active and reactive power output according to the state of the grid. This chapter presents a method for the grid-side converter to regulate

H. Ma et al., *The Control Principle of Wind Power Generation System*,
https://doi.org/10.1007/978-981-99-6083-5_7

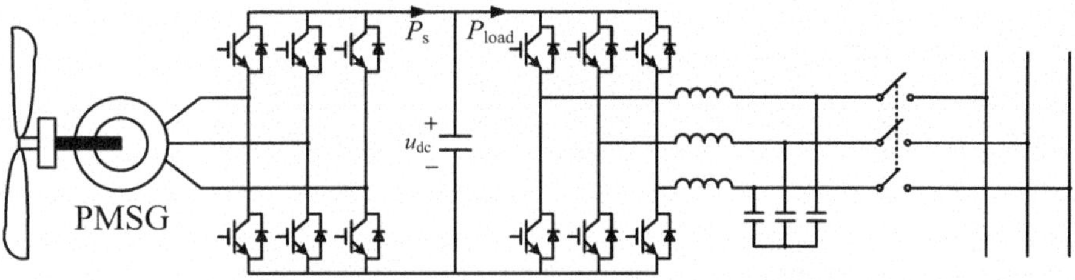

Fig. 7.1 Typical topology of the PMSG-based wind power system

the grid frequency and amplitude using the droop characteristic curve, and to achieve equal power sharing among multiple converters.

7.1 Mathematical Model and Control Principle of the PMSG-Based Wind Power System

As can be seen from Fig. 7.1, the PMSG-based wind power system has the same grid-side converter structure as the DFIG-based wind power system (including the grid-side converter and filter), for which the mathematical model and control method can be referred to the relevant content discussed earlier in this book. This section focuses on the mathematical model of the PMSG and the vector control method of the PMSG-based wind power system under ideal grid conditions.

7.1.1 Mathematical Model of the PMSG

In order to simplify the modelling process, the following assumptions are usually made for the mathematical model of the PSMG [1]:

(1) Neglecting the space harmonics, the three-phase windings are considered symmetrical and the excitation and armature reaction magnetic fields are sinusoidally distributed in space.
(2) Neglecting magnetic circuit saturation, the self- and mutual inductances of each winding are considered to be constant.
(3) The core losses, hysteresis losses and eddy current losses are neglected.
(4) The effect of frequency changes and temperature changes on the winding resistance is not considered.
(5) No damping winding is present in the rotor.
(6) The electrical conductivity of the generator permanent rotor is zero.

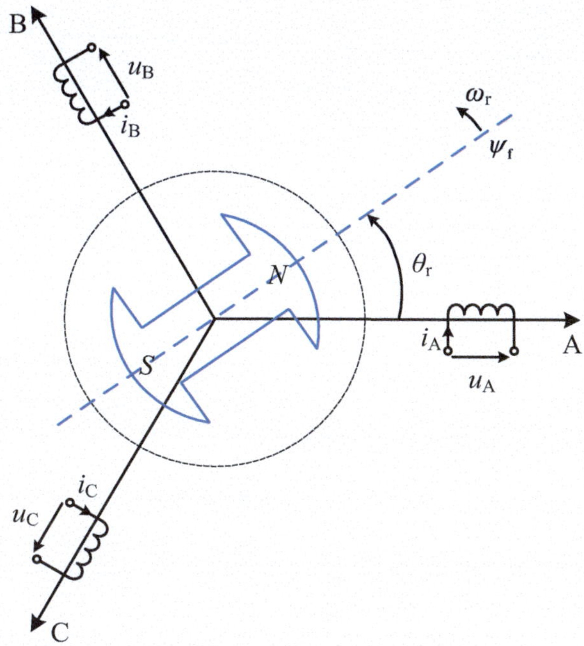

Fig. 7.2 Winding model of the PMSG in 3-phase stationary frame (ABC frame)

At this point, the winding model of the PMSG in the three-phase stationary coordinate system (i.e. ABC coordinate system) is shown in Fig. 7.2.

The mathematical model of the PMSG shown in Fig. 7.2 can be composed of the following voltage equation, flux linkage equation, torque equation and motion equation.

(1) **Voltage Equation**

As can be seen in Fig. 2.5, the voltage balance equation for the three-phase stator winding is

$$\begin{bmatrix} u_A \\ u_B \\ u_C \end{bmatrix} = \begin{bmatrix} R_s & 0 & 0 \\ 0 & R_s & 0 \\ 0 & 0 & R_s \end{bmatrix} \begin{bmatrix} i_A \\ i_B \\ i_C \end{bmatrix} + p \begin{bmatrix} \psi_A \\ \psi_B \\ \psi_C \end{bmatrix} \tag{7.1}$$

where p is the differential operator, equivalent to d/dt; u_A, u_B, u_C are the instantaneous stator phase voltages; i_A, i_B, i_C are the instantaneous stator phase currents; ψ_A, ψ_B, ψ_C are the instantaneous stator phase flux linkages; R_s is the stator phase resistance.

(2) **Flux Linkage Equation**

As shown in Fig. 7.3, the flux linkage in the stator A-phase winding consists of the self-induced flux linkage ψ_{AA} (including the A-phase mutual induced flux linkage

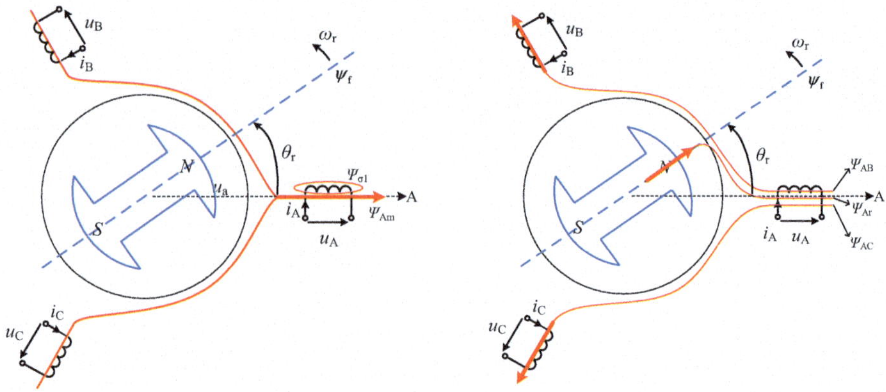

(a) Self-induced Flux Linkages (b) Mutual-induced Flux Linkages

Fig. 7.3 Space distribution of flux linkages in stator A-phase winding of the PMSG[1]

ψ_{Am} and the leakage flux linkage $\psi_{\sigma 1}$) produced by the A-phase current and the mutual induced flux linkage (including ψ_{AB}, ψ_{AC}, ψ_{Af}) produced by the B-phase, C-phase currents and the rotor permanent magnet, i.e. there are

$$\psi_{\text{AA}} = \psi_{\text{Am}} + \psi_{\sigma 1} \tag{7.2}$$

$$\psi_{\text{A}} = \psi_{\text{AA}} + \psi_{\text{AB}} + \psi_{\text{AC}} + \psi_{\text{Af}} \tag{7.3}$$

Similarly, the flux linkage equation of the PMSG can be obtained as following:

$$\begin{aligned} \begin{bmatrix} \psi_{\text{A}} \\ \psi_{\text{B}} \\ \psi_{\text{C}} \end{bmatrix} &= \begin{bmatrix} \psi_{\text{AA}} + \psi_{\text{AB}} + \psi_{\text{AC}} + \psi_{\text{Af}} \\ \psi_{\text{BA}} + \psi_{\text{BB}} + \psi_{\text{BC}} + \psi_{\text{Bf}} \\ \psi_{\text{CA}} + \psi_{\text{CB}} + \psi_{\text{CC}} + \psi_{\text{Cf}} \end{bmatrix} \\ &= \begin{bmatrix} L_{\text{AA}} & L_{\text{AB}} & L_{\text{AC}} \\ L_{\text{BA}} & L_{\text{BB}} & L_{\text{BC}} \\ L_{\text{CA}} & L_{\text{CB}} & L_{\text{CC}} \end{bmatrix} \begin{bmatrix} i_{\text{A}} \\ i_{\text{B}} \\ i_{\text{C}} \end{bmatrix} + \begin{bmatrix} \cos\theta_{\text{r}} \\ \cos(\theta_{\text{r}} + 120°) \\ \cos(\theta_{\text{r}} - 120°) \end{bmatrix} \psi_{\text{f}} \end{aligned} \tag{7.4}$$

where L_{AA}, L_{BB}, L_{CC} are the self-inductance of each respective winding, the remaining inductance parameters are the mutual inductance between each stator winding, and i_{A}, i_{B}, i_{C} are the phase currents of each stator winding respectively.

[1] This figure is only a schematic illustration that shows the coupling of flux linkages rather than the actual space distribution of flux linkages.

(3) **Torque Equation and Motion Equation**

Similar to the torque and motion equations in the DFIG in Sect. 2.1, the torque and motion equations in the PMSG can be obtained as

$$T_e = n_p \psi_f \left[i_A \sin\theta_r + i_B \sin(\theta_r + 120°) + i_C \sin(\theta_r - 120°) \right] \tag{7.5}$$

The motion equation of the DFIG can thus be found as

$$T_e = T_L + \frac{J}{n_p} \frac{d\omega_r}{dt} \tag{7.6}$$

where T_L is the load torque, ω_r is the rotor electrical angular velocity and J is the load inertia.

In summary, Eqs. (7.1), (7.4), (7.5) and (7.6) are the mathematical model of the PMSG in the ABC coordinate frame (ABC-frame).

Similarly, by applying the coordinate transformation represented by Eqs. (2.15a) and (2.15c) to the mathematical model of the PMSG in ABC-frame, the mathematical models of the PSMG in αβ-frame and dq-frame can be obtained as

$$\begin{cases} u_s^\alpha = R_s i_s^\alpha + p\psi_s^\alpha \\ u_s^\beta = R_s i_s^\beta + p\psi_s^\beta \end{cases} \tag{7.7}$$

$$\begin{cases} \psi_s^\alpha = L_s^\alpha i_s^\alpha + \psi_f \cos\theta_r \\ \psi_s^\beta = L_s^\beta i_s^\beta + \psi_f \sin\theta_r \end{cases} \tag{7.8}$$

$$T_e = n_p \psi_f \left(i_s^\beta \cos\theta_r - i_s^\alpha \sin\theta_r \right) \tag{7.9}$$

$$\begin{cases} u_s^d = R_s i_s^d + p\psi_s^d - \omega_r \psi_s^q \\ u_s^q = R_s i_s^q + p\psi_s^q + \omega_r \psi_s^d \end{cases} \tag{7.10}$$

$$\begin{cases} \psi_s^d = L_s^d i_s^d + \psi_f \\ \psi_s^q = L_s^q i_s^q \end{cases} \tag{7.11}$$

$$T_e = n_p \left[\psi_f i_s^q + \left(L_s^d - L_s^q \right) i_s^d i_s^q \right] \tag{7.12}$$

where L_s^α, L_s^β, L_s^d and L_s^q are the α-axis, β-axis, d-axis and q-axis components of the stator inductance, respectively. In a non-salient pole generator: $L_s^\alpha = L_s^\beta$, $L_s^d = L_s^q$; in a salient pole generator: $L_s^\alpha \neq L_s^\beta$, $L_s^d \neq L_s^q$. Equations (7.7), (7.8), (7.9) and (7.6) are the mathematical model of the PMSG in the αβ-frame and Eqs. (7.10), (7.11), (7.12) and (7.6) are the mathematical model of the PMSG in the dq-frame.

7.1.2 Vector Control of the PMSG-Based Wind Power System

Vector control of the PMSG is usually obtained based on the rotor-flux-oriented synchronous coordinate frame. Substituting the flux linkage Eq. (7.11) into the stator voltage Eq. (7.10), it gives

$$\begin{cases} u_s^d = R_s i_s^d + L_s^d p i_s^d - L_s^q \omega_r i_s^q \\ u_s^q = R_s i_s^q + L_s^q p i_s^q + L_s^d \omega_r i_s^d + \omega_r \psi_f \end{cases} \tag{7.13}$$

The output of the stator current-loop controller can be obtained according to Eq. (7.13) as

$$\begin{cases} u_s^{d*} = C_{si}^d \left(i_s^{d*} - i_s^d \right) + u_{sc}^d \\ u_s^{q*} = C_{si}^q \left(i_s^{q*} - i_s^q \right) + u_{sc}^q \end{cases} \tag{7.14}$$

where C_{si}^d and C_{si}^q are the d-axis and q-axis PI controllers of the stator current loop respectively, i_s^{d*} and i_s^{q*} are the d-axis and q-axis stator current reference values respectively, u_s^{d*} and u_s^{q*} are the d-axis and q-axis control voltage reference values of the generator-side converter respectively, $u_{sc}^d = -L_s^q \omega_r i_s^q$ and $u_{sc}^q = L_s^d \omega_r i_s^d + \omega_r \psi_f$ are the d-axis and q-axis feedforward compensation terms, respectively.

According to Eqs. (7.11), (7.12) and (7.14), the control diagram of the rotor-flux-oriented vector control for the PMSG-based wind power system is shown in Fig. 7.4. In this diagram, the outer control loop is the speed loop and the inner loop is the torque (current) loop. The generator is mainly controlled by the q-axis current component, and the d-axis current component can be used to regulate the excitation current, thus achieving unit power factor operation and reducing losses. Both the speed and current loops are regulated by PI controllers. The speed of the generator varies with wind speed according to the MPPT requirements, where the MPPT algorithm and the derivation of the vector control of the grid-side converter can be found in the relevant chapters earlier in this book.

7.2 Operational Characteristics of the PMSG-Based Wind Power System Under Weak Grid Conditions

7.2.1 Comparison Between Strong and Weak Grid Working Modes

The basic concepts of strong and weak grids were introduced in Sect. 1.3.2 of the Introduction. A strong grid is the grid with infinite capacity, where the grid can be seen as an ideal voltage source and its voltage frequency and amplitude do not vary

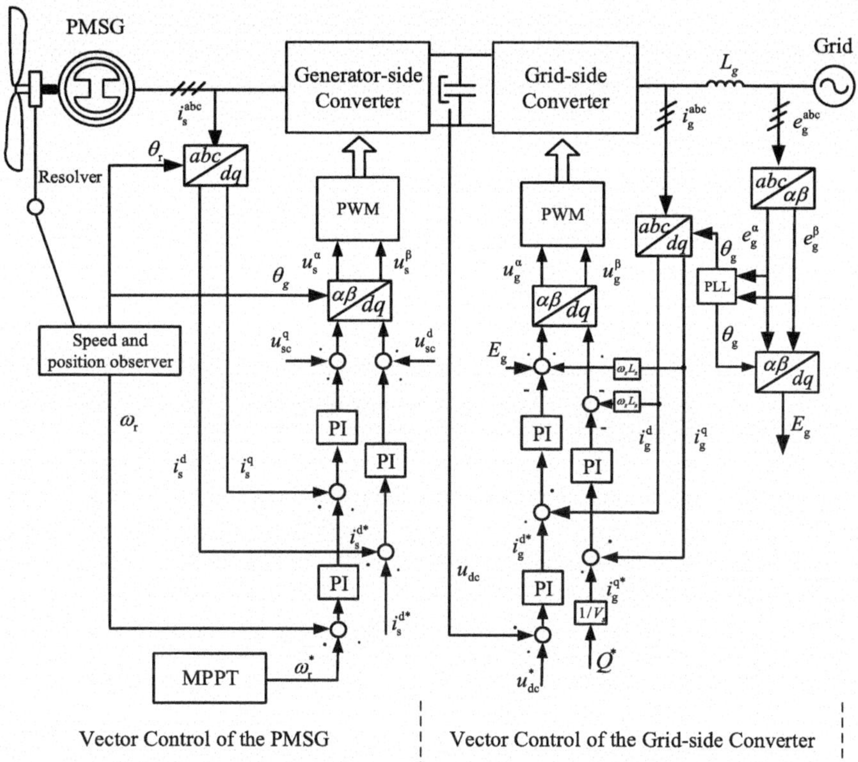

Fig. 7.4 Control diagram of the rotor-flux-oriented vector control for the PSMG-based wind power system

with the load power. Currently, both the DFIG and PMSG are used as a current source to feed the grid with the maximum power that can be captured by the wind turbine, and the grid considers the wind power system as an ordinary load. This operation mode is feasible in the case of a strong grid where wind power does not account for a large proportion of the grid capacity, but in the case of a weak grid where wind power accounts for a relatively large proportion, there are problems:

Firstly, due to the influence of the transmission line impedance, the grid is no longer an ideal voltage source for the wind power system, and the voltage at the wind power system access point will fluctuate with the power as the wind power system outputs different power. In this case the grid cannot give a stable voltage reference, i.e. the grid angle θ_s and amplitude E_s in Fig. 7.4 will fluctuate and the grid-side converter cannot continue to feed power into the grid as a current source according to the operating mode in Fig. 7.4. The grid-side converter therefore needs to operate as a voltage source with its own assigned electrical angle (from virtual phase lock loop) and amplitude reference to enable regulation of the output voltage amplitude and frequency.

Secondly, when wind power accounts for a large proportion of the grid capacity, the wind power system needs to inject the required active and reactive power into the grid in order to regulate the grid voltage frequency and amplitude. For active power, the wind turbine provides as much active power as the grid (load) requires, and no longer provides the maximum power that the wind turbine can capture (in MPPT mode) as it does under strong grid conditions. That is, the wind turbine cannot operate in the MPPT mode shown in Fig. 7.4 to capture the maximum wind power but should instead determine the amount of power captured according to the needs of the grid (load). For example, if the grid (load) requires a power of 600 kW, then the wind turbine should operate at Point B1 or B2 in Fig. 7.5 (not at Point A), balanced with the load power. If the power captured by the wind turbine is not balanced with the load, this will cause the DC-bus voltage of the converter in Fig. 7.1 to increase or decrease, thus risking the normal operation of the converter. The amount of active power required by how much the current grid frequency deviates from its rated value. A simple method is, according to the current grid frequency and droop curve, to calculate the amount of active power that the converter needs to output. The setting of the droop curve is described in detail in Sect. 7.5. The above analysis is also applicable to the isolated operation of the wind power system, where the wind power system is the only supply source and the power captured and outputted by the wind turbine must be balanced with the load power. In summary, the active power needs to be transferred from the generator side to the load side, so the power captured by the wind turbine needs to be balanced with the power required by the grid (load). In addition, the grid-side converter can achieve regulation of the grid voltage amplitude by injecting reactive power into the grid. For the full-scale power converter in Fig. 7.1, the reactive power is mainly circulated within the grid-side converter, so the wind turbine's ability to output reactive power to the grid depends mainly on the current capacity of the grid-side converter, and the control is relatively simple, so this section focuses on the regulation of active power. Table 7.1 gives a comparison between the control methods of the system controller, the generator-side converter and the grid-side converter in the strong and weak grid modes, respectively.

7.2.2 *Active Regulation of Active Power Under Weak Grid Conditions*

As can be seen from the above analysis, under weak grid conditions, the grid-side converter injects active power into the grid according to the grid's demand for active power, which is provided by the wind power captured by the wind turbine and transferred from the generator side to the grid side, so the generator side needs to regulate the capture of wind power to balance with the active power injected into the grid by the grid-side converter. There are two main methods of regulating the power captured by the wind turbine: firstly, by regulating the generator speed to change the captured power, as shown in Fig. 7.5, where the power captured by the wind turbine

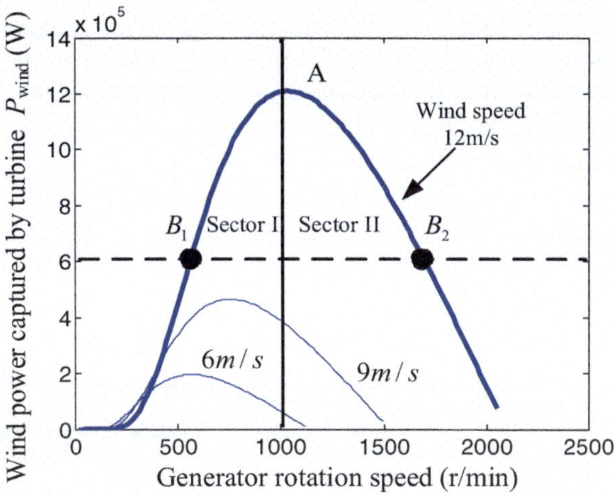

Fig. 7.5 Power characteristic curve of the wind turbine [2]

Table 7.1 Comparison between strong and weak grid modes

	Strong grid mode	Weak grid mode
System controller	MPPT Give the generator speed reference according to wind speed and MPPT algorithm	Determine the amount of wind power captured by the wind turbine and the corresponding working point according to the active power required by the grid (load)
Generator-side converter	Control the generator speed using the vector control with speed-loop structure	Control the generator speed in the same way as the strong grid mode
Grid-side converter	Feed the maximum power captured by the turbine into the grid as a current source, where the grid gives the voltage and frequency references	Operate as a voltage source and determine the amount of active and reactive injected into the grid according to the current grid voltage frequency and amplitude and droop curve of the system

varies with the generator speed. Secondly, by regulating the pitch angle of the blades, where the larger the pitch angle, the lower the power captured by the wind turbine, and the smaller the pitch angle, the higher the power captured by the wind turbine. The balance of active power in the system is illustrated in Fig. 7.6.

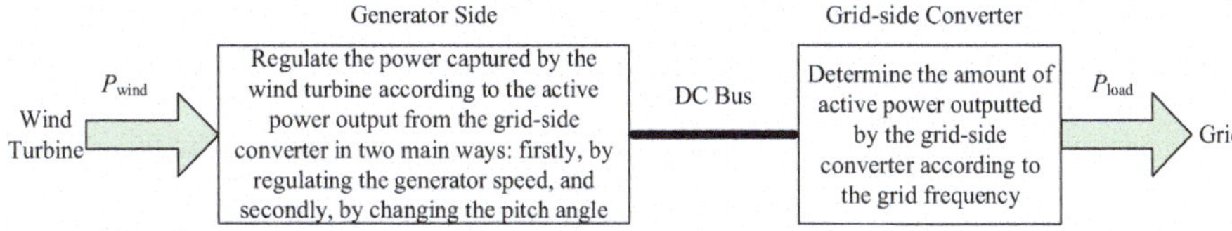

Fig. 7.6 Balance of active power in the PMSG-based wind power system

The regulation of the generator speed can change the value of C_p and thus the amount of power captured by the wind turbine. The kinetic energy stored and released by the wind turbine during acceleration and deceleration, respectively, can also be used to regulate the output power. Under strong grid conditions, the wind turbine also finds the maximum power point by regulating the rotation speed, so the generator speed control in a strong grid can be well adapted to a weak grid. The regulation of the pitch angle can be operated in conjunction with the regulation of the generator speed to achieve effective control of the power captured by the wind turbine.

7.3 Generator-Speed-Regulation-Based Method for Control of Wind-Turbine-Capturing Power

This section will discuss how to regulate the generator speed to control the power captured by the wind turbine under weak grid conditions, thus achieving a balance between the power captured by the wind turbine and the power required by the grid [3].

7.3.1 Effect of Generator Speed on Wind-Turbine-Capturing Power

As can be seen in Fig. 7.5, the power captured by the wind turbine can be controlled by regulating the generator speed and each wind turbine power curve consists of two sectors: Sector I and Sector II. In Sector I the power curve has a positive slope and in Sector II the curve has a negative slope. For a given load power, e.g. 600 kW, the wind turbine has two possible operating points, B1 and B2 in Fig. 7.5. The actual operating point of the generator can be determined according to the generator output power variation ΔP_s and the speed direction variation $\Delta\omega_\mathrm{r}$. In Sector I, the output power increases with increasing speed; in Sector II, the output power decreases with increasing speed. The exact sector in which the generator operates can be obtained from Eq. (7.15):

$$\begin{cases} \Delta P_\mathrm{s} \cdot \Delta\omega_\mathrm{r} > 0 & \text{Sector I} \\ \Delta P_\mathrm{s} \cdot \Delta\omega_\mathrm{r} < 0 & \text{Sector II} \end{cases} \tag{7.15}$$

The generator output power P_s can be expressed as

$$P_\mathrm{s} = u_\mathrm{s}^\mathrm{d} i_\mathrm{s}^\mathrm{d} + u_\mathrm{s}^\mathrm{q} i_\mathrm{s}^\mathrm{q} \tag{7.16}$$

where the definitions of the variables are the same as in Sect. 7.1.

For a given load power, the generator can be operated in either Sector I or Sector II for different considerations. When the generator is operating in Sector I, it has a low speed, which corresponds to low mechanical losses and stresses. When the generator is operating in Sector II, it has a high-level mechanical energy that can be released or stored to balance fluctuations in load power. Both sectors should be considered when designing the control system. In addition, the operating point may change from one sector to the other during a rapid change in load or wind speed [2]. When wind speed is high, the wind turbine is expected to operate in Sector I in order to keep the generator speed within the rated operating range.

As mentioned earlier, the power captured by the wind turbine needs to be balanced with the power injected into the grid by the grid-side converter (the power required by the grid). According to Fig. 7.1, the relation between the power a captured by the wind turbine and the power an injected into the grid by the grid-side converter can be determined by observing the DC-bus voltage of the converter. If the power injected into the grid is greater than the power captured by the wind turbine, then the DC-bus voltage is lower than the its reference and the generator speed needs to be regulated to increase the captured wind power. Conversely, if the power injected into the grid is less than the power captured by the wind turbine, the DC-bus voltage will be higher than its reference and the generator speed needs to be regulated to reduce the captured wind power. Therefore, the power balance between the generator side and the grid side is transformed into the DC-bus voltage control issue. The more stable the DC-bus voltage is, the better the balance between the power outputted by the generator and the power injected into the grid by the grid-side converter. Figure 7.7 gives a schematic diagram of DC-bus voltage control. The DC-bus voltage of the converter is detected to obtain the required generator output power and then the required wind turbine speed is determined, according to the power characteristic curve of the wind turbine, to obtain the appropriate operating point. The generator speed control is achieved by means of vector control as described in the previous section. Therefore, the control objective is the DC-bus voltage and the control variable is the generator speed. In the following section, the mathematical model of the DC bus will be established to find the relation between the DC-bus voltage and the generator speed.

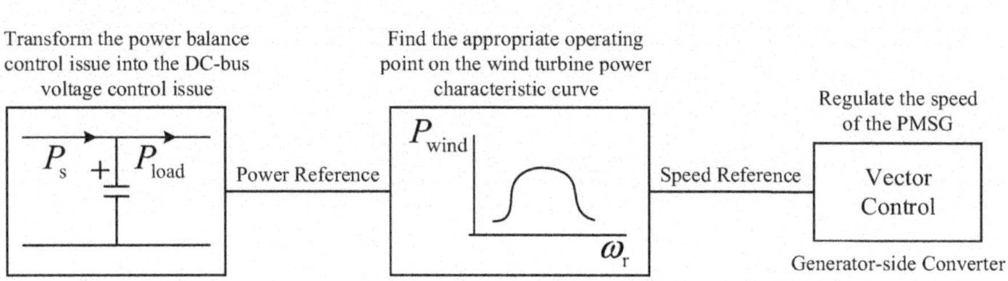

Fig. 7.7 Schematic diagram of DC-bus voltage control by regulating generator speed

7.3.2 Mathematical Model of the DC Bus

In order to establish a proper DC-bus voltage control system, it is necessary to find the relation between generator speed and DC-bus voltage as well as the relevant system parameters.

If the converter losses are ignored, the converter DC-bus voltage equation can be expressed as Eq. (7.17).

$$P_s - P_{load} = u_{dc} i_{dc} = u_{dc} C \frac{du_{dc}}{dt} \tag{7.17}$$

where u_{dc} and i_{dc} are the converter DC-bus voltage and current in Fig. 7.1, respectively, C is the DC-bus capacitance and P_s is the generator output power. In the steady state, P_s and the wind-turbine-capturing power P_{wind} are equal (neglecting losses); during acceleration or deceleration of the generator, P_s and P_{wind} are not equal, and the difference between them is the variation in mechanical power due to the acceleration or deceleration of the generator, as shown in Eq. (7.18), J is the rotation inertia of the system, and the electrical and mechanical losses are neglected here.

$$P_{wind} - P_s = \omega_r J \frac{d\omega_r}{dt} \tag{7.18}$$

It is important to note here that the generator speed and the wind turbine speed have a definite relation and are equal in the case of a direct-drive wind turbine. For systems with a gearbox, the gear ratio determines the ratio of the wind turbine speed to the generator speed. In this chapter, the generator speed ω_r represents both the wind turbine speed and the generator speed and the gear ratio is considered to have been taken into account.

The wind turbine model can be described by Eq. (2.36) and, according to Eqs. (2.36), (7.17) and (7.18), it can be seen that the generator output power P_s can be controlled by regulating the generator speed, thus controlling the DC-bus voltage. To obtain the relation between the DC-bus voltage u_{dc} and the generator speed ω_r, C_p is considered to be a function of ω_r,[2] and Eqs. (2.36), (7.17) and (7.18) can be combined to obtain Eq. (7.19).

$$\frac{1}{2} C_p(\omega_r) \rho S v^3 - P_{load} - u_{dc} C \frac{du_{dc}}{dt} = J \omega_r \frac{d\omega_r}{dt} \tag{7.19}$$

Equation (7.19) presents the relation between the DC-bus voltage and the generator speed, in which there are nonlinear terms: $C_p(\omega_r)$ and two second-order terms. If the load (grid) power P_{load} is considered as a disturbance, Eq. (7.19) can be linearised at the given rated operating points (DC bus voltage u_{dc0}, generator speed ω_{r0} and

[2] As can be seen in Sect. 2.2.3, C_p is a function of the tip-speed ratio/speed and the pitch angle and is considered here to be a function of speed by assuming that the pitch angle is constant.

wind speed v_0) to give the transfer function between the DC-bus voltage $u_{dc}(s)$ and generator speed $\omega_r(s)$, which can be expressed in the frequency domain in the form shown in Eq. (7.20).

$$\frac{U_{dc}(s)}{\omega_r(s)} = \frac{k - J\omega_{r0}s}{u_{dc0}Cs} = \frac{J\omega_{r0}}{u_{dc0}C}\frac{(-s + \frac{k}{J\omega_{r0}})}{s} \tag{7.20}$$

where k is the slope rate of the wind turbine power characteristic curve near the operating point in Fig. 7.5. From Fig. 7.5 it can be seen that if the wind turbine is operating in Sector I, $k > 0$ and the system will have a right half plane zero (RHZ), which is generally low in frequency and depends on the values of k and J. As can be seen from the subsequent analysis, this RHZ increases the control complexity and limits the system's performance. On the other hand, in Sector II, $k < 0$ and the transfer function will have a left half plane zero (LHZ), which makes the system relatively simple to control.

Given the RHZ in Sector I and its control complexity, the focus here is on the design of the controller in Sector I. The design of the controller in Sector II can follow the design steps in Sector I. The control performance in both sectors is verified and the differences in control and mechanical power flow in the two sectors are also discussed subsequently.

As a result of the above analysis, the control structure, as shown in Fig. 7.8, can be used to achieve control of the DC-bus voltage under weak grid conditions. The outer loop is the DC-bus voltage loop, where the required generator speed ω_r^* is produced by the voltage loop controller and regulates the generator speed via the inner speed loop.

In Fig. 7.8, the speed control of the PMGS is similar to the control method used in a strong grid. By changing the generator speed ω_r, the wind-turbine-capturing power P_{wind} will change according to the $P_{wind} - \omega_r$ curve in Fig. 7.5. At the same time, the generator output power P_s will also change with P_{wind} according to the law of Eq. (7.18), thus compensating for the load power variation and thus maintaining a constant DC-bus voltage.

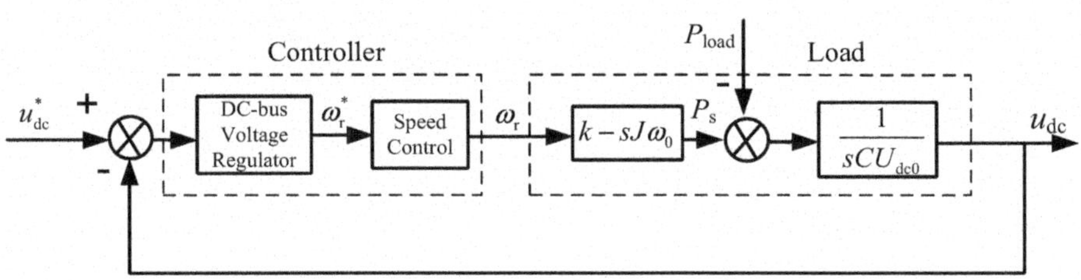

Fig. 7.8 DC-bus voltage loop

7.3.3 Design of the DC-Bus Voltage Controller

The design of the DC-bus voltage controller should take into account the system performance requirements. There are two main performance indexes:

Firstly, the range of DC-bus voltage variation. During an increase or decrease in load (grid) power, the dc-bus voltage dip or overshoot should be between the minimum voltage U_{dcmin} and the maximum voltage U_{dcmax}, where U_{dcmin} is determined by the grid-side AC voltage amplitude and U_{dcmax} must not exceed the withstand voltage of the power electronic devices and the dc-bus capacitors. Equation (7.21) gives the limits of U_{dcmin} and U_{dcmax}, where M_{max} is the maximum modulation ratio, U_{line} is the effective value of the grid-side AC line voltage and $U_{cap,device}$ is the maximum voltage that the power electronic devices and DC-bus capacitors can withstand. It should be noted here that there are other constraints to be considered in the choice of U_{dcmin} and U_{dcmax}, such as inrush currents, and only the constraints in Eq. (7.21) are considered here.

$$\begin{aligned} U_{dcmin} \cdot M_{max} &> \sqrt{2} U_{line} \\ U_{dcmax} &< U_{cap,device} \end{aligned} \tag{7.21}$$

Secondly, the recovery time of the DC-bus voltage, i.e. the time it takes for the DC-bus voltage to return to its reference value when the load changes. Generally speaking, it is desirable to have a fast system response. In this section, the recovery time is also used as a performance index, but no specific limit is given, as the system can still operate normally within the voltage tolerance. In the subsequent controller design, both of the above performance indexes are used to evaluate the controller performance.

In Sector I, in order to design a more efficient voltage regulator, the physical relation among the DC-bus voltage, the load power, the generator and the converter needs to be considered. It is also necessary to find the physical factors that limit the control performance of the system when the load varies or the wind speed varies, which are analysed below.

The case of a voltage dip due to an increase in load power is firstly considered. Assuming that the load power P_{load} has an increase of ΔP_{load} at time t_1, as shown in Fig. 7.9, and according to the wind turbine power characteristics in Sector I in Fig. 7.5, the generator should increase its speed to increase the captured wind power to meet the load power requirement. During the acceleration process, part of the captured wind power P_{wind} will be converted into mechanical power P_{mec} and the rest will be converted into generator output power P_s, which is fed into the DC bus through the generator-side converter and thus transferred to the load side. The mechanical power P_{mec} corresponds to the storage of mechanical energy, the amount of which depends on the rotation inertia of the system, which is generally big for large-capacity wind turbines.

The power flow of the system during acceleration can be illustrated in Fig. 7.10. In terms of the power relation, when the wind turbine is operating in Sector I, the

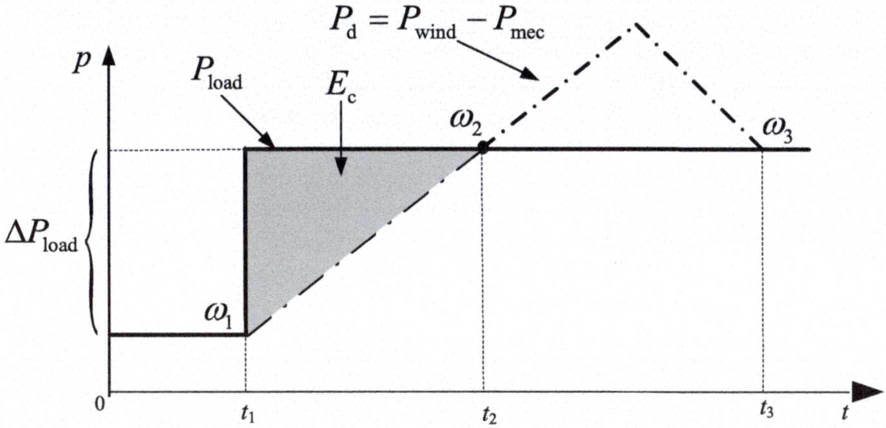

Fig. 7.9 Analysis on DC-bus voltage dip (assuming constant $d\omega_r/dt$)

generator should not accelerate too fast, otherwise the large $d\omega_r/dt$ will result in most of the wind power P_{wind} being converted into mechanical power P_{mec}, which reduces the power P_s that can be fed to the DC bus, although the wind power P_{wind} increases with speed in Sector I.

Therefore, the generator speed control in Sector I requires a trade-off between increasing the captured wind power and increasing the stored mechanical power. There should be an optimal acceleration rate $d\omega_r/dt$ that maximises the generator output power P_s. The energy shaded in Fig. 7.9, which is caused by the difference between the load power P_{load} and the generator output power P_s during the sudden

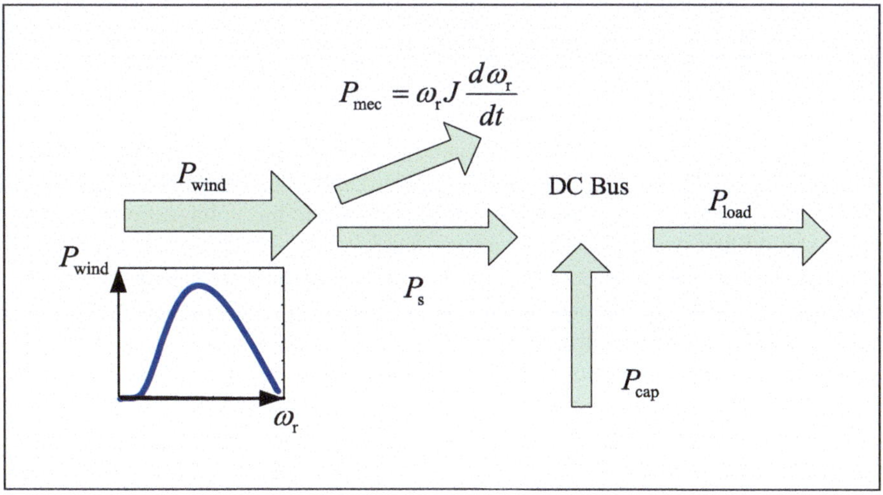

Fig. 7.10 Power flow during the generator acceleration

loading process and is also resposible for the DC-bus voltage dip, needs to be supplemented by the energy stored in DC-bus capacitors. For a given value of the DC bus capacitance, the smaller the shaded area, the smaller the DC-bus voltage dip. To reduce the shaded area, the generator output power needs to have the maximum slope dP_s/dt. For simplicity of analysis, assuming the wind turbine power variation with speed $dP_{wind}/d\omega_r$ and the acceleration $d\omega_r/dt$ to be constants, $dP_{wind}/d\omega_r = k$ and $d\omega_r/dt = k_1$ are defined. The change rates of the wind power captured by the wind turbine and the mechanical power can be expressed in the form of Eqs. (7.22a) and (7.22b), respectively. The difference between the above two change rates is the change rate of the generator output power dP_s/dt as shown in Eq. (7.23).

$$\frac{dP_{wind}}{dt} = \frac{dP_{wind}}{d\omega_r} \cdot \frac{d\omega_r}{dt} = k \cdot \frac{d\omega_r}{dt} = k \cdot k_1 \tag{7.22a}$$

$$\frac{dP_{mec}}{dt} = \frac{d(\omega_r J \frac{d\omega_r}{dt})}{dt} = k_1^2 J \tag{7.22b}$$

$$\frac{dP_s}{dt} = \frac{d(P_{wind} - P_{mec})}{dt} = k \cdot k_1 - k_1^2 J \tag{7.23}$$

As can be seen from Eq. (7.23), the change rate of the generator output power dP_s/dt depends on the slope of the wind turbine power characteristic curve k and the rotation inertia of the wind turbine J. When the load increases, a larger slope k corresponds to a faster change of the wind turbine power, which reduces the DC-bus voltage dip and recovery time. On the other hand, a larger inertia J will result in more mechanical (kinetic) energy being stored during acceleration, which means that less energy is fed into the DC bus and the load side.

As shown in Eq. (7.23), for given parameters k and J, by choosing the appropriate acceleration $d\omega_r/dt(k_1)$, the maximum generator output power change rate dP_s/dt can be achieved, thus reducing the dc-bus voltage dip. The relation between the generator output power change rate dP_s/dt and the generator acceleration $d\omega_r/dt$ is shown in Fig. 7.11.

It can be seen from Fig. 7.11 that the generator output power change rate dP_s/dt has a maximum value $dP_{s(max)}/dt$, which also corresponds to an optimal acceleration $d\omega_r/dt_{(optimal)}$. Since the maximum dP_s/dt reduces the dc-bus voltage dip, the control system should find the optimal acceleration $d\omega_r/dt_{(optimal)}$. It should be noted that a very important aspect for any controller design is to find the physical constraints that limit the optimal control performance of the system. Through the above analysis, the constraints that limit the control of the DC-bus voltage are k and J.

A higher wind speed implies a larger slope k and a larger $dP_{s(max)}/dt$, as shown in Fig. 7.11a. A smaller rotation inertia J corresponds to less mechanical energy stored during acceleration and therefore a larger $dP_{s(max)}/dt$, as shown in Fig. 7.11b. These analyses are in agreement with the system transfer function obtained in Eq. (7.20). In Sector I, the system performance is mainly limited by the (RHZ), which is located at the very frequency determined by k and J. In short, a low wind speed and a

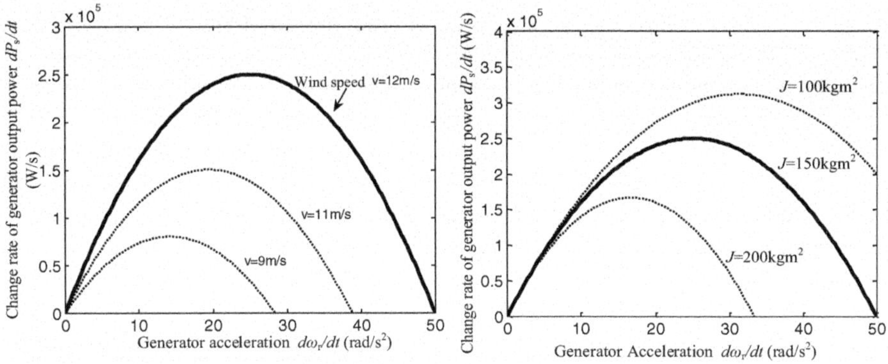

Fig. 7.11 Relation between the generator output power change rate and the generator acceleration (Left: with different wind speeds; Right: with different rotation inertias)

high rotation inertia will slow down the system response. A similar analysis is also applicable to the case of decreasing load and increasing DC-bus voltage in Sector I, where there should be an optimal negative acceleration $d\omega_r/dt$, corresponding to a minimum dP_s/dt.

It is important to note that for the wind turbine operating in Sector II, the increase in load power P_{load} triggers the wind turbine to slow down and thus increases the captured wind power P_{wind}. At the same time, the deceleration of the system reduces the amount of mechanical energy stored, which is converted into electrical energy and discharged to the DC bus and load side, thus helping to cope with the increase in load power and maintain the DC-bus voltage. Therefore, in Sector II, the control bandwidth of the system is not subject to similar constraints as in Sector I. The same analysis can be applied to the case of decreasing load and increasing DC-bus voltage in Sector II.

Based on the analysis above, the DC-bus voltage controller in Fig. 7.8 needs to be able to dynamically find the maximum rate of power change dP_s/dt. The PI regulator, with proportional and integral parameters obtained from the small-signal model near the rated operating point, can achieve the desired performance only near the rated operating point, but cannot guarantee that the optimal dP_s/dt is always reached during a large sudden load change, which will affect and limit the control performance, resulting in a larger fluctuation in the DC-bus voltage and a longer recovery time. For the better dynamic performance during a larger load variation, a compound adaptive PI regulator can be obtained according to the energy flow relation illustrated in Figs. 7.9 and 7.10.

Figure 7.12 shows the flow chart of the compound adaptive DC-bus-voltage regulator, which is described as follows: firstly, the sector in which the wind turbine speed is located is determined according to the method in Eq. (7.15), and if it is in sector I, the control method is divided into two parts according to the magnitude of the error between the actual DC-bus voltage and the DC-bus voltage reference (marked as DC-bus voltage deviation here). When the DC-bus voltage deviation is smaller than the tolerance Δ, fixed values K_p and K_i, obtained from the small-signal model near the rated point to ensure the steady-state control accuracy, are used for the proportional and integral parameters of the PI regulator, respectively. For the operating point in Fig. 7.5 (in Sector I where power $P = 600$ kW, speed $\omega_0 = 62.8$ rad/s, $k = 13000$ Nm and wind speed $v_o = 12$ m/s), the parameters of the PI regulator are chosen as $K_p = 0.3$ and $K_i = 0.05$. When the DC-bus voltage deviation is greater than Δ, the parameters of the PI regulator are adjusted to improve the dynamic performance of the system. Specifically, at each control cycle the proportional parameter K_p has a small increment ΔK_p to vary the acceleration of the generator $d\omega_r/dt$. Here the control cycle of the DC-bus voltage loop is much longer than that of the speed and torque loops. As K_p increases, the proportional part of the DC-bus voltage regulator is adjusted accordingly and depending on the polarity of the DC-bus voltage deviation, the reference to the speed loop will increase or decrease. For the positive DC-bus voltage deviation (voltage dips), the output of the DC-bus voltage regulator will increase the reference value of the generator speed, thus increasing the generator speed faster, i.e. with a larger $d\omega_r/dt$. For the negative DC-bus voltage deviation (voltage overshoot), the speed reference will be reduced, resulting in a faster reduction of the generator speed, i.e. with a more negative $d\omega_r/dt$. In the above two cases, the greater power change rate $|dP_s/dt|$ is made, as expected in the previous discussion. By continuing to increase the value of K_p until $|dP_s/dt|$ is no longer increased, the maximum power change rate and the optimal $d\omega_r/dt$ are obtained. $|dP_s/dt|$ can be calculated by measuring the output voltage and current of the generator-side converter and the power expression is shown in Eq. (7.16). If $|dP_s/dt|$ is decreased according to the calculated result, then K_p is reduced and thus $|d\omega_r/dt|$, so that the optimal operating point in Fig. 7.11 is found. When the DC-bus deviation is less than its tolerance Δ, K_p returns to its initial value, which is determined by the small-signal model.

For the operation of the wind turbine in Sector II, according to the previous analysis, the mechanical energy stored in the wind turbine will help to compensate for the load power variation and, in terms of the transfer function, there is no RHZ limiting the response of the system. The desired performance can therefore be achieved by a simple PI regulator (with fixed parameters). The bandwidth of the DC-bus voltage control loop in Sector II can be higher than that in Sector I, limited only by the bandwidth of the inner loop and the capacity of the system.

No matter whether the wind turbine is operating in Sector I or Sector II, the system has a certain control bandwidth. When the load power changes within the control bandwidth, the power balance can be achieved by regulating the generator speed. When the load power changes dramatically, the energy storage part in the DC bus of the converter is needed to compensate for the load power variation, which will be discussed below (Table 7.2).

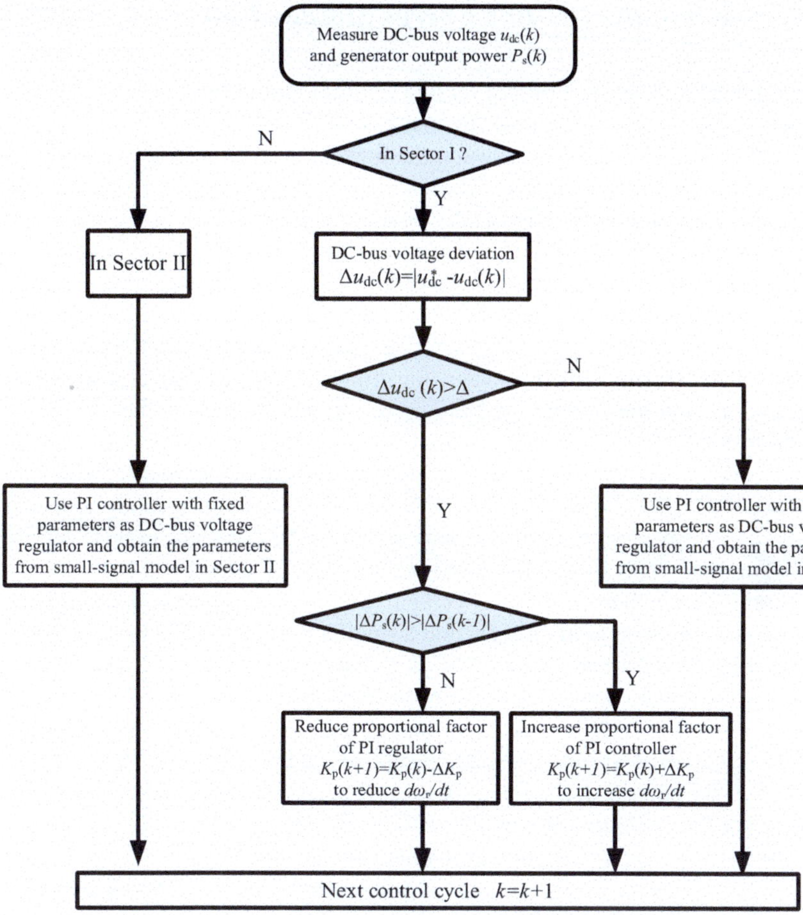

Fig. 7.12 Flow chart of the compound adaptive regulator for DC-bus voltage

Table 7.2 Simulation results of a 1.5WM wind power system with a 30% step increase in load power

Wind speed	8 m/s		10 m/s		12 m/s	
Performance	Voltage dip (V)	Recovery time (s)	Voltage dip (V)	Recovery time (s)	Voltage dip (V)	Recovery time (s)
Adaptive regulator	117	20	110	16	105	12
PI regulator	135	29	126	24	120	20

7.3.4 Selection of the DC-Bus Energy Storage Unit

The DC bus energy storage unit in Fig. 7.1 has an important influence on the control performance of the system, serving as an energy buffer to provide or absorb active power when there is an imbalance between the power captured by the wind turbine and that required by the grid (load). The DC-bus voltage controller designed in Sect. 7.3.3 has a certain control bandwidth, beyond which the load power variation will cause an imbalance between the power captured by the wind turbine and the load power, which needs to be compensated for by the energy storage unit on the DC bus. As mentioned in the previous section, the system response in Sector I is governed by the slope of the wind turbine power characteristic curve k and the rotation inertia J, both of which also determine the frequency location of RHZ. A large-capacity wind turbine system usually has a large rotation inertia, so that the control bandwidth of the system is limited. For slow changes in load power (within the control bandwidth), the generator speed and power can be regulated according to the method given in Sect. 7.3.3 and the DC-bus voltage can be well controlled; for sudden changes in load power, only the energy storage of the DC-bus, which is the capacitor in this section, can be relied upon to cope with transient power imbalances. Clearly, in Sector II the system responds faster than it does in Sector I, with a wider control bandwidth, a greater ability to cope with load changes and a smaller energy storage unit required.

Therefore, the DC-bus capacitance will determine the variation of the DC-bus voltage during a sudden change in load power. Here the wind turbine operating in Sector I with a sudden load increase is still analysed as an example. According to Fig. 7.9, when the load power P_{load} is suddenly added a at moment of t_1, the generator output power P_{s} will increase at a rate of $k_2 = dP_{\text{s}}/dt$, as shown in Eq. (7.24). The time required for the generator output power to reach balance with the load power is a. The DC-bus voltage dip is caused by the imbalance between the generator output power and the load power, as shown in the shaded area in Fig. 7.9. The shaded area E_{c} can be obtained from Eq. (7.24), and this energy needs to be supplemented by the DC-bus capacitor, which can be expressed by Eq. (7.27) in combination with the formula of the capacitor storage energy)

$$E_{\text{c}} = \frac{1}{2}\Delta P_{\text{load}} T = \frac{1}{2}\frac{\Delta P_{\text{load}}^2}{k_2} \tag{7.24}$$

$$\frac{1}{2}Cu_{\text{dc}}^2(t_1) - \frac{1}{2}Cu_{\text{dc}}^2(t_2) = \frac{1}{2}\frac{\Delta P_{\text{load}}^2}{k_2} \tag{7.25}$$

According to the allowed DC-bus voltage dip $u_{\text{dc}}(t_1) - u_{\text{dc}}(t_2)$ and the load power variation P_{load}, the required DC-bus capacitance can be determined by Eq. (7.26).

$$C = \frac{\Delta P_{\text{load}}^2}{k_2(u_{\text{dc}}^2(t_1) - u_{\text{dc}}^2(t_2))} \tag{7.26}$$

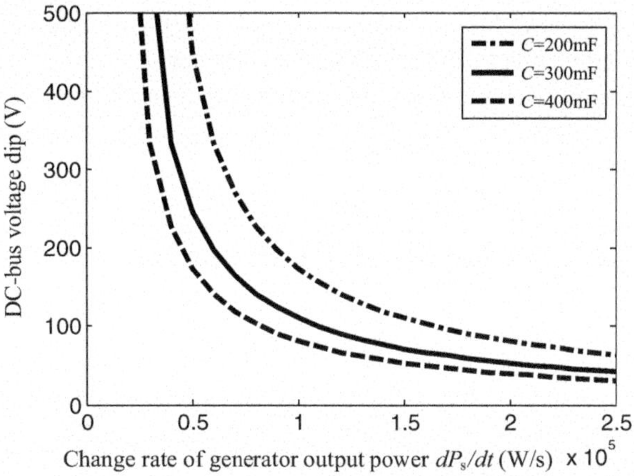

Fig. 7.13 Relation between DC-bus voltage dip and change rate of generator output power

Figure 7.13 shows the relation between the DC-bus voltage dip and the change rate of generator output power with different DC-bus capacitance in a wind power system. It can be seen that a sufficiently large DC-bus capacitance and energy storage is required to cope with transient changes in the load; the larger the DC-bus capacitance, the smaller the DC-bus voltage dip; the faster the generator output power change, the smaller the DC-bus voltage dip. These conclusions also apply when the load power is reduced and the DC-bus voltage rises. If a capacitor with $C = 300$ mF is used, for the case of $dP_s/dt = 150\,\text{kW/s}$, the DC-bus voltage dip is around 100 V according to Fig. 7.13 (simulation results are given in Fig. 7.15).

It should be noted here that the DC-bus capacitance are in the range of a few hundred mF, which is relatively large compared to other converter applications, such as motor drives for the same power class. By calculation, for electrolytic capacitors, the volume of a capacitor bank for 300 mF and 900 V is approximately 0.18 m^3, which is acceptable for a 1.5 MW converter. At the same time, if the voltage values of the generator and the DC bus are increased, the required DC-bus capacitance can be reduced, but not the required energy storage capacity and therefore the volume of the capacitor bank. Of course, other energy storage devices, such as the battery and the supercapacitor, can be better used as energy storage units. If the wind turbine is operated in Sector II, the required energy storage is also significantly reduced, and here the worst-case scenario is considered, i.e. the situation in Sector I. From an energy point of view, in Sector II, the acceleration and deceleration of the wind turbine corresponds to the storage and release of mechanical energy respectively, which is the process desired by the load power, i.e. the mechanical energy stored and released during acceleration and deceleration, respectively, can help to cope with the load power change. In Sector I, the process is reversed, when the load requires more power, the wind turbine accelerates at this point, resulting in some of the energy being stored in mechanical energy. Therefore, in Sector I, the energy storage unit on the DC bus buffers the mechanical energy and the required DC-bus energy storage is larger. Therefore, a sufficient number of DC-bus energy storage units are required to ensure reliable operation of the wind turbine in Sector I.

7.3.5 Other Working Conditions

(1) Effect of Wind Speed Change

Most of the analysis above focuses on the case where the wind speed is constant. When the wind speed varies, the controller needs to regulate the generator speed to maintain the DC-bus voltage. For a given load power, there are two possible scenarios when the wind speed changes: firstly, the operating point is in the same sector before and after the wind speed changes. Secondly, the operating point changes from one sector to another, which can occur when the wind speed changes rapidly.

The first case is shown in Fig. 7.14a, where the wind turbine starts at Point A, in Sector I, with a load power of 600 kW and a wind speed of 12 m/s. Then the wind speed decreases from 12 to 10 m/s, so that the operating point of the wind turbine changes to B. Subsequently, under the regulation of the controller, the operating point of the wind turbine moves to Point C, in order to balance the load power, and all these operating points are located in Sector I.

The second case is shown in Fig. 7.14b, where the wind speed increases from 10 to 12 m/s, causing the operating point to change from Point A to Point B (from Sector II to Sector I), and the controller will make the generator operate at Point C in Sector I in order to balance the load power.

In both cases above, the DC-bus voltage can be effectively controlled.

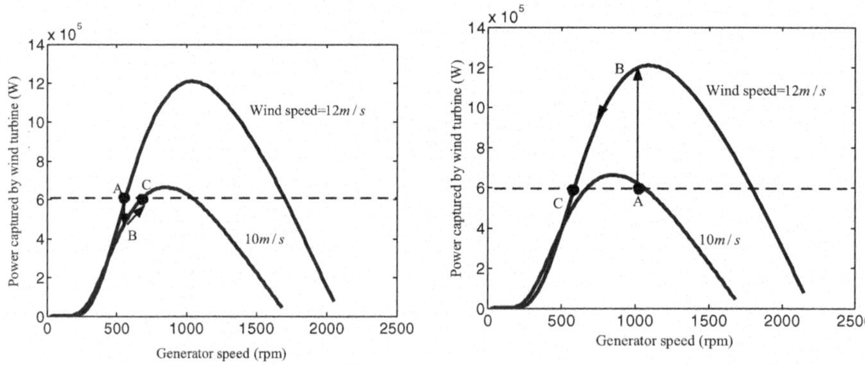

(a) Wind speed is decreasing Operating point is in Sector I

(b) Wind speed is increasing Operating point shifts from Sector II to Sector I

Fig. 7.14 Change of operating point with variable wind speed

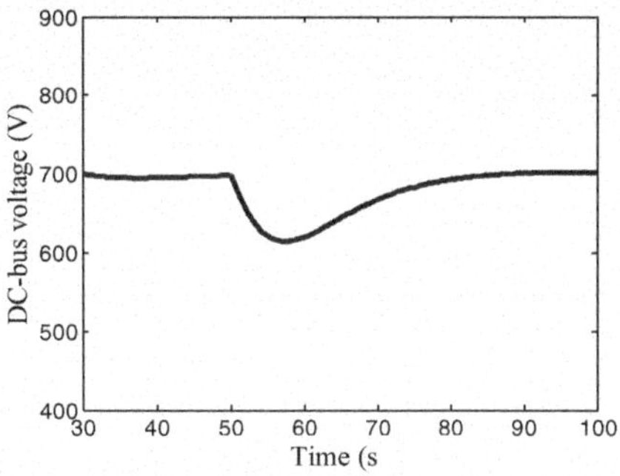

(a) 1st case

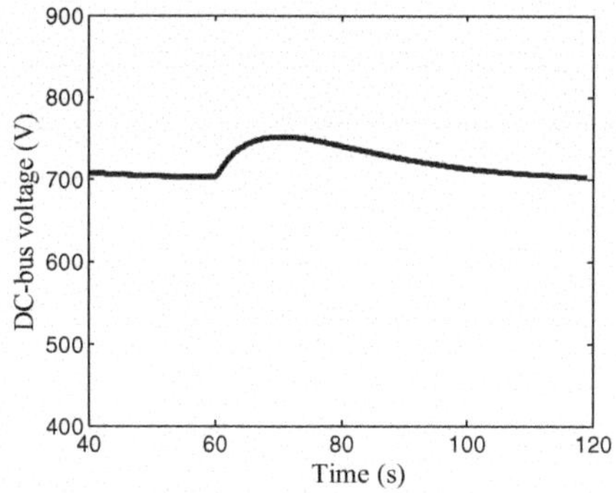

(b) 2nd case

Fig. 7.15 Simulation results of DC-bus voltage with a wind speed change

(2) Switching between Strong Grid Mode and Weak Grid Mode

As mentioned previously, the wind turbine operates at different operating points in the strong and weak grid modes, and the phase lock loop of the grid angle can be used to determine whether the grid is a strong or weak.

The DC-bus voltage controller for the weak grid presented in this chapter also enables MPPT control in the case of a strong grid. Under strong grid conditions, the DC-bus voltage is controlled by the grid-side converter. On the generator side, it is only necessary to set the DC-bus voltage to a value higher than the DC-bus voltage controlled by the grid-side converter, without changing the controller structure, and the controller will automatically regulate the generator speed to find the maximum power point and capture as much wind power as possible to transition to the MPPT state.

On the other hand, under the weak grid conditions, the reference of the DC-bus voltage returns to its normal value, and the power of the generator is then regulated to transition the generator from the MPPT state to the weak grid state where the generator power equals to the required load power (reducing the excess power). Figure 7.16 depicts a schematic of the transition from the weak grid state to the MPPT state and the corresponding simulation results. It is important to note here that the anti-windup characteristics of the voltage and speed regulators as well as their initial values need to be set appropriately to cope with the switching between these two modes.

(3) Range Requirements for System Parameters

The normal operation of the system requires the relevant parameters to meet certain requirements.

When the wind power system is operating in a weak grid, a sudden reduction in load power or an increase in wind speed can lead to an increase in the DC-bus voltage. The control methods described in this chapter combined with some DC-bus energy

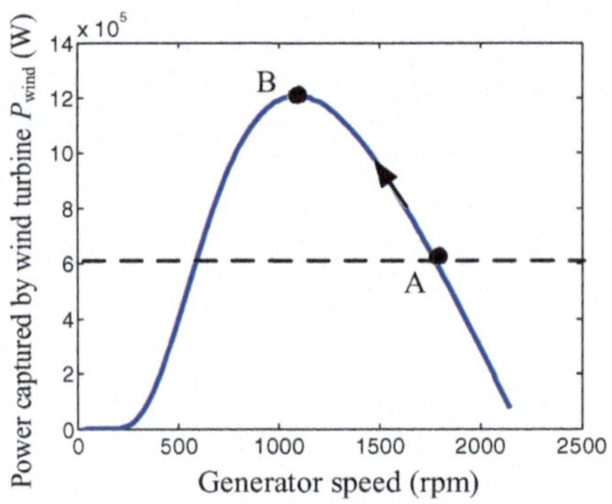

(a) change of operating point

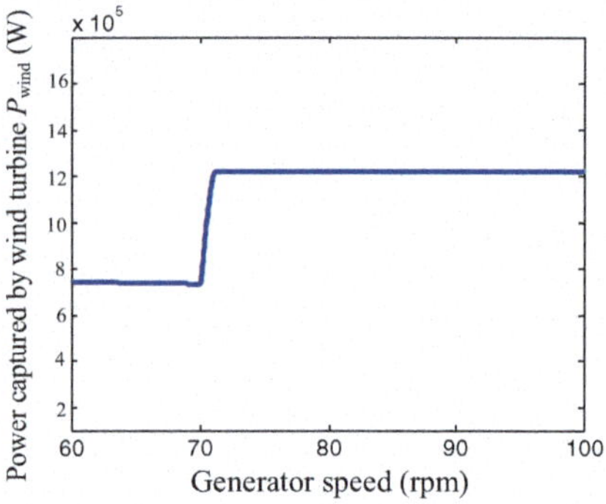

(b) change of power captured by wind turbine P_{wind}

Fig. 7.16 Switching from weak grid mode to strong grid mode (MPPT)

storage units will reduce the voltage increase, but certain protection measures, such as the crowbar circuit connected in parallel to the DC bus, are still needed to prevent overvoltage situations. These protect against extreme situations and thus reduce the number of energy storage units on the DC bus.

Considering the efficiency and compatibility of the system, the speed range of the generator in this chapter is the same for both strong and weak grid cases, but there are two issues to be noted: firstly, for a higher wind speed, the generator speed cannot be higher than the rated speed to regulate the captured wind energy due to the speed limit of the generator and mechanical stresses, implying that the wind turbine power characteristic curve in Fig. 7.5 does not have a complete Sector II. In this case, it is desirable to operate the wind turbine in Sector I to obtain the complete adjustable power range, together with the pitch angle regulation. Secondly, in practical system operation, if the wind speed is high enough, meaning that the wind power captured by the wind turbine can be balanced with the load power, the power can be regulated by the method in this paper. Conversely, if the wind speed is too low, the maximum captured wind power may not be sufficient to meet the load power requirements. In this case, the generator will output the possible maximum wind power and the DC-bus voltage will dip, of which the magnitude depends on the energy storage unit on the DC bus. In this case, the DC-bus voltage will be well supported if there are batteries or other energy storage units on the DC-bus.

7.4 Pitch-Angle-Regulation-Based Method for Control of Wind-Turbine-Capturing Power

The analysis in the previous section considers the case where the pitch angle of the wind turbine blade is fixed at 0° and only the generator speed is variable to simplify the analysis. In large wind turbines, the pitch angle can generally be controlled to regulate the wind energy captured by the wind turbine. It can therefore be used in conjunction with speed regulation to cope with the load power change under weak power conditions. Currently, the pitch angle can change as fast as 10–20 degrees per second and the pitch control can help stabilise the DC-bus voltage [4].

According to Eq. (2.36), the wind energy utilisation coefficient C_{p} is a function of two variables, the generator speed and the pitch angle, and the following analysis focuses on the effect of the pitch angle β on wind energy capture. A typical wind energy utilisation coefficient equation is used here for the study [5], as shown in Eq. (7.27).

$$\begin{cases} C_p(\lambda,\beta) = 0.22(\frac{116}{\lambda_i} - 0.4\beta - 5)e^{(-\frac{12.5}{\lambda_i})} \\ \lambda_i = (\frac{1}{\lambda+0.08\beta} - \frac{0.035}{\beta^3+1})^{-1} \\ \lambda = \frac{2\pi Rn}{v} = \frac{\omega_{\mathrm{r}} R}{v} \end{cases} \tag{7.27}$$

The variables are defined in the same way as in Eq. (2.36), R is the radius of the wind turbine, n is the speed of the wind turbine, ω_{r} is the angular frequency of the wind turbine and λ is the tip-speed ratio, defined as the ratio of the blade-tip linear velocity to the wind speed. The wind turbine power characteristic curve obtained from the above formula is shown in Fig. 7.17. It can be seen that for the same tip-speed ratio, the larger the pitch angle, the smaller the wind energy utilisation coefficient and thus the smaller the captured wind power.

The pitch actuator of a wind turbine can be modeled in first and second order. The models will vary according to the specific wind turbine type, depending on the design of the wind turbine blade and the dynamic performance of the pitch mechanism (motor drive). Here a second-order model is used and the change rate of the pitch angle is limited. The maximum change rate of the pitch angle β is usually between 5 and 10 degrees per second during normal operation and can reach 10–20 degrees per second in case of emergency. The model of the pitch angle actuator can be represented as a second-order system as shown in Eq. (7.28).

$$\beta = \frac{c}{as^2 + bs + c}\beta^* \tag{7.28}$$

In Eq. (7.28), $a = 1, b = 5, c = 28$, where the pitch angle is regulated to control the power captured by the wind turbine to balance with the load power and thus control the DC-bus voltage. So the outer control loop is the DC-bus voltage loop and the inner loop is the pitch angle loop.

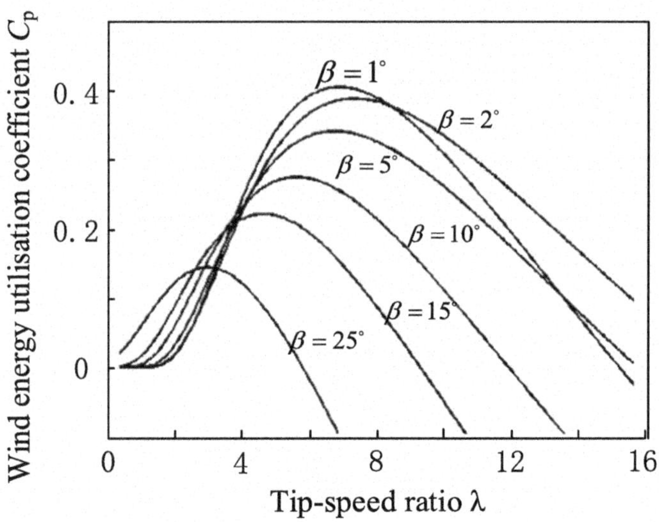

Fig. 7.17 Relation among wind energy utilisation coefficient, pitch angle and tip-speed ratio

Figure 7.18 illustrates the control diagram of the pitch angle. When the wind turbine is in normal operation, the selection switch in the diagram is located at the top and the controller regulates the pitch angle according to the rotation speed. When the wind turbine is started, the pitch angle gradually decreases as the generator speed rises until the blades are fully expanded with $\beta = 0$. When the generator speed is higher than the rated speed, the wind turbine starts to pitch down and the pitch angle gradually increases, reducing the power captured by the wind turbine and thus protecting the wind turbine and the tower system [6].

When the wind turbine is operated under weak grid conditions, the selection switch in Fig. 7.18 is located below and the DC-bus voltage loop generates the pitch angle reference via the PI regulator. When the DC-bus voltage is higher than its reference value, the pitch angle increases and the power captured by the wind turbine decreases;

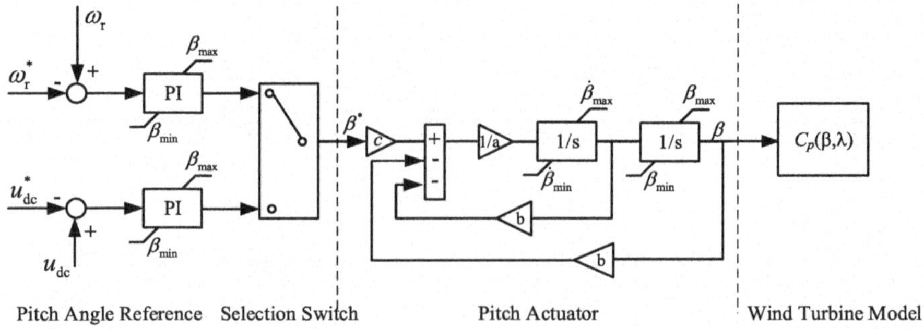

Fig. 7.18 Control diagram of the pitch angle

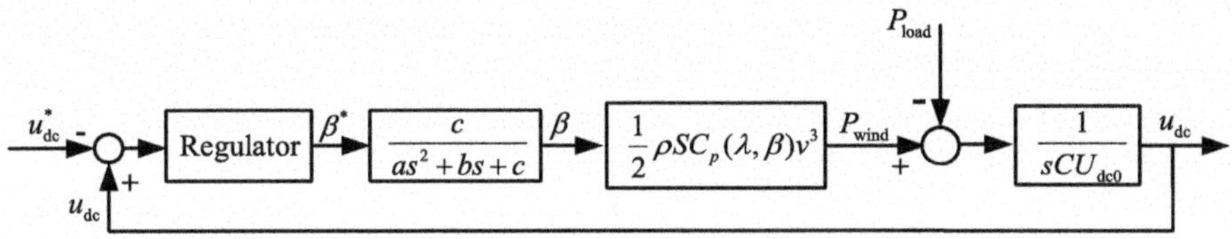

Fig. 7.19 Control diagram of the DC-bus voltage

conversely, if the DC-bus voltage is lower than its reference value, the pitch angle decreases and the power captured by the wind turbine increases. And the final control of the pitch angle is achieved by means of the pitch actuator. According to the wind turbine power characteristic curve, the power of the wind turbine is regulated to balance with the load power. The output of the PI regulator limits the pitch angle variation range and the maximum change rate of the pitch angle $\dot{\beta}_{max}$ is also taken into account in the pitch actuator [7] (Figs. 7.19 and 7.20).

In a practical system, pitch angle regulation and generator speed regulation should work together to cope with the load power change. The speed regulation method make full use of the mechanical energy stored in the wind turbine to cope with the load change while varying the wind energy utilisation coefficient C_p, where there is a good dynamic response, especially in Sector II, and therefore the rapid change in load power can be coped with well, but attention needs to be paid to the speed range of the generator. Considering the complexity of the pitch actuator and the associated mechanical stresses, it is desired to reduce the pitch angle regulation. Given that the pitch angle can be regulated at a rate of 10–20 degrees per second, which is also relatively fast, in the event of large fluctuations in load (as can be observed from the DC-bus voltage), the pitch angle and generator speed can be regulated together to cope with the load change.

7.5 Mathematical Model and Control of the Grid-Side Converter Under Weak Grid Conditions

7.5.1 Control Method for Single Grid-Side Converter

The above sections describe the control methods for the generator-side converter as well as the pitch angle, with the objective of obtaining a stable DC-bus voltage and achieving a balance of active power in the system. The grid-side converter, on the other hand, needs to help stabilise the voltage and frequency of the grid and, in the case of a weak grid, to operate as a voltage source. For the operation of a single wind turbine, the regulation of the output voltage and frequency by the power electronic converter needs to be considered. For parallel operation of several wind turbines, the distribution of active and reactive power between the multiple converters also needs to be considered.

Figure 7.21 shows the structure of the grid-side converter, where the converter output power is delivered to the local load and the grid through a filter.

According to the voltage and current marked in Fig. 7.20, the mathematical model of the grid-side converter can be established in the synchronous rotation frame. u_{ga}, u_{gb}, u_{gc} are the A-phase, B-phase and C-phase converter output voltage respectively, i_{ao}, i_{bo}, i_{co} are the A-phase, B-phase and C-phase converter output currents respectively, and u_{ao}, u_{bo}, u_{co} are the A-phase, B-phase and C-phase output voltage through the filter. The mathematical model of the voltage loop can be established according to the relation between the voltage and current of the filter capacitor, as shown in Eq. (7.29).

$$\frac{d}{dt}\begin{bmatrix} u_{\mathrm{o}}^{\mathrm{d}} \\ u_{\mathrm{o}}^{\mathrm{q}} \end{bmatrix} = \begin{bmatrix} 0 & \omega_1 \\ -\omega_1 & 0 \end{bmatrix} \cdot \begin{bmatrix} u_{\mathrm{o}}^{\mathrm{d}} \\ u_{\mathrm{o}}^{\mathrm{q}} \end{bmatrix} + \frac{1}{3C_{\mathrm{f}}}\begin{bmatrix} i_{\mathrm{o}}^{\mathrm{d}} \\ i_{\mathrm{o}}^{\mathrm{q}} \end{bmatrix} - \frac{1}{3C_{\mathrm{f}}}\begin{bmatrix} i_{\mathrm{g}}^{\mathrm{d}} \\ i_{\mathrm{g}}^{\mathrm{q}} \end{bmatrix} \tag{7.29}$$

According to the relation between the voltage and current on the filter inductor and coordinate transformation as Eq. (2.15), the mathematical model of the current loop can be obtained as Eq. (7.30).

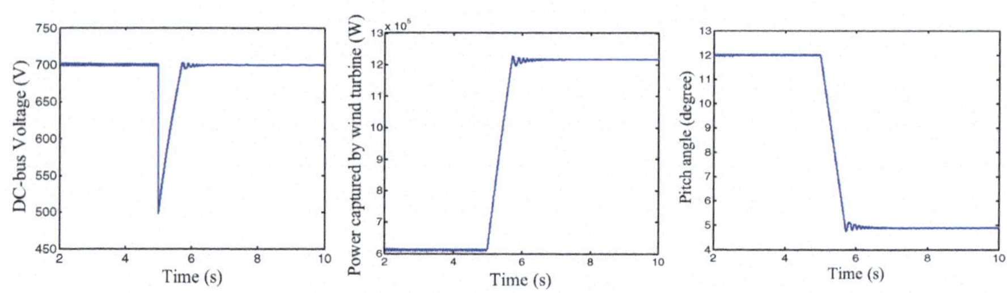

Fig. 7.20 Simulation results of the DC-bus voltage controlled by pitch angle regulation

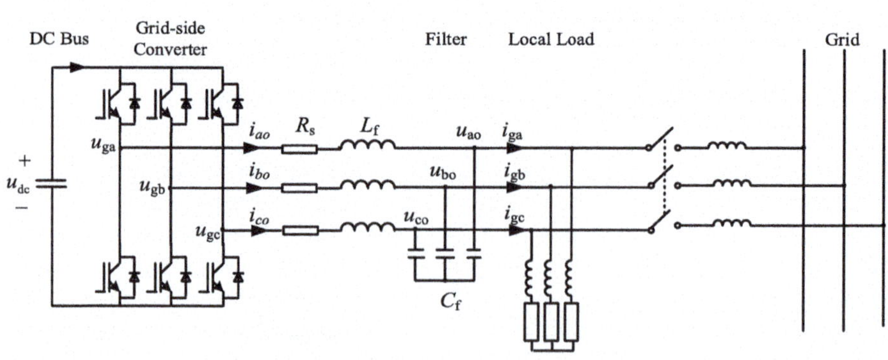

Fig. 7.21 Structure of the grid-side converter

$$\frac{d}{dt}\begin{bmatrix} i_{\mathrm{o}}^{\mathrm{d}} \\ i_{\mathrm{o}}^{\mathrm{q}} \end{bmatrix} = \begin{bmatrix} -\frac{R_{\mathrm{s}}}{L_{\mathrm{f}}} & \omega_1 \\ -\omega_1 & -\frac{R_{\mathrm{s}}}{L_{\mathrm{f}}} \end{bmatrix} \cdot \begin{bmatrix} i_{\mathrm{o}}^{\mathrm{d}} \\ i_{\mathrm{o}}^{\mathrm{q}} \end{bmatrix} - \frac{1}{L_f}\begin{bmatrix} u_{\mathrm{o}}^{\mathrm{d}} \\ u_{\mathrm{o}}^{\mathrm{q}} \end{bmatrix} + \frac{1}{L_f}\begin{bmatrix} u_{\mathrm{g}}^{\mathrm{d}} \\ u_{\mathrm{g}}^{\mathrm{q}} \end{bmatrix} \tag{7.30}$$

If the grid-voltage-oriented method is used, i.e. the grid voltage vector is oriented on the d-axis in the synchronous rotation frame, $u_{\mathrm{o}}^{\mathrm{d}} = U_{\max}$ and $u_{\mathrm{o}}^{\mathrm{q}} = 0$ can be found, where $U_{\max}$ is the output voltage amplitude. Substituting Eq. (7.30) into the current-loop model, it gives

$$\begin{cases} L_{\mathrm{f}}\frac{di_{\mathrm{o}}^{\mathrm{d}}}{dt} = -R_{\mathrm{s}}i_{\mathrm{o}}^{\mathrm{d}} + \omega_{\mathrm{s}}L_{\mathrm{f}}i_{\mathrm{o}}^{\mathrm{q}} - U_{\max} + u_{\mathrm{g}}^{\mathrm{d}} \\ L_{\mathrm{f}}\frac{di_{\mathrm{o}}^{\mathrm{q}}}{dt} = -R_{\mathrm{s}}i_{\mathrm{o}}^{\mathrm{q}} - \omega_{\mathrm{s}}L_{\mathrm{f}}i_{\mathrm{o}}^{\mathrm{d}} + u_{\mathrm{g}}^{\mathrm{q}} \end{cases} \tag{7.31}$$

By taking out the cross-coupling term, decoupling the d-axis and q-axis and using the PI regulator, a close loop of current can be achieved. The outer voltage loop also use the PI controller, where the load current can be processed as a disturbance or added to the control system as a feedback term. In summary, a double-close-loop control system with an outer voltage loop and an inner current loop can be obtained, as shown in Fig. 7.22. As can be seen, in the voltage loop, $i_{\mathrm{g}}^{\mathrm{dq}}$ is superimposed on the output of the voltage loop as the compensation term for the load current. If the load current is not detected in the practical system, it can be considered as a disturbance to the system, which is not compensated for and is only eliminated by the PI regulator. Here, the transformation from the dq-frame to the ABC-frame requires an angle. Under weak grid conditions, the grid cannot give a stable voltage angle reference, so the converter needs to have its own PLL (virtual PLL) to generate the angle of the converter output voltage.

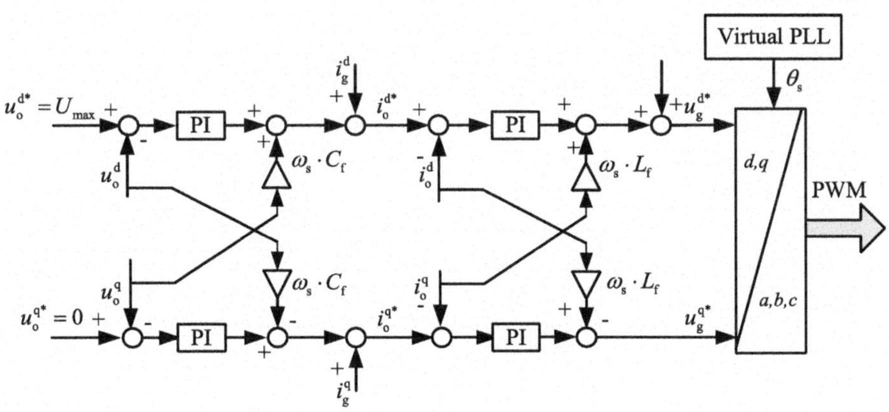

Fig. 7.22 Control diagram of the grid-side converter under weak grid conditions

7.5.2 Control Method for Parallel Grid-Side Converters

When there are multiple voltage source converters operating in parallel (corresponding to multiple wind turbines) in a system, an appropriate method is required to achieve equal power sharing of the parallel converters, which means the load power is distributed proportionally to the capacity of each converter. In a conventional multi-synchronous-generator system, power equalisation can be achieved due to the droop relation between active power and output frequency and the droop relation between reactive power and output voltage amplitude in the synchronous generators themselves and their control systems. How the droop characteristics of a conventional synchronous generator can be simulated in a power electronic converter will be discussed below.

Taking two converters connected in parallel as an example, the simplified schematic diagram is shown in Fig. 7.23. The sum of output and line impedances of Converter 1 is $Z_1 = r_1 + jX_1 = R_{Z1}\angle\theta_1$, and the sum of output and line impedances of Converter 2 is $Z_2 = r_2 + jX_2 = R_{Z2}\angle\theta_2$. r_1 and r_2 are the line equivalent resistances, X_1 and X_2 are the equivalent inductances, U_{10} and U_{20} are the no-load output voltages of Converter 1 and Converter 2 respectively, U is the parallel AC bus voltage, σ_1 and σ_2 are the phase angle differences between the output voltage of the respective converter module and the common bus voltage.

The output current of the Converter n (n = 1, 2) is

$$\boldsymbol{I}_{\mathrm{n}} = \frac{U_{\mathrm{n}0}\angle\delta_{\mathrm{n}} - U\angle 0}{R_{\mathrm{Zn}}\angle\theta_{\mathrm{n}}} \tag{7.32}$$

The output power is

$$S_{\mathrm{n}} = U_{\mathrm{n}0}\angle\delta_{\mathrm{n}} \cdot (\boldsymbol{I}_{\mathrm{n}})^* = P_{\mathrm{n}} + jQ_{\mathrm{n}} \tag{7.33}$$

where P_{n} is the output active power of Converter n, Q_{n} is the output reactive power of Converter n. Combining Eqs. (7.32) and (7.33) gives the expression for active and reactive power, as shown in Eq. (7.34).

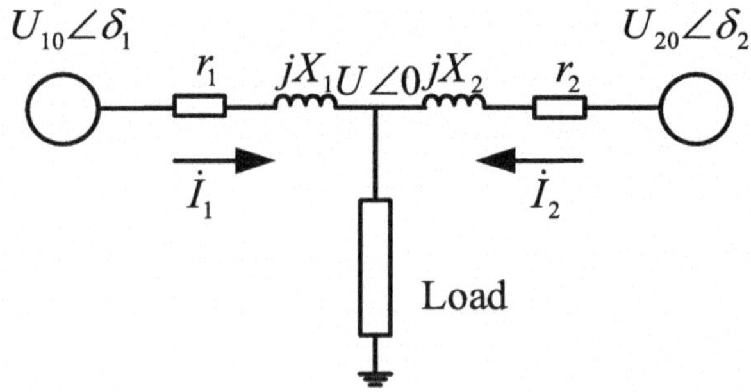

Fig. 7.23 Simplified schematic diagram of two converters connected in parallel

$$
\begin{aligned}
P_{\mathrm{n}} &= \frac{1}{R_{\mathrm{Zn}}}[(U_{\mathrm{n0}}U\cos\delta_{\mathrm{n}} - U^2)\cos\theta_{\mathrm{n}} + U_{\mathrm{n0}}U\sin\delta_n\sin\theta_n] \\
Q_{\mathrm{n}} &= \frac{1}{R_{\mathrm{Zn}}}[(U_{\mathrm{n0}}U\cos\delta_{\mathrm{n}} - U^2)\sin\theta_{\mathrm{n}} - U_{\mathrm{n0}}U\sin\delta_n\cos\theta_n]
\end{aligned} \tag{7.34}
$$

From Eq. (7.34), when the sum of the output and line impedances is purely inductive, i.e. $\theta_{\mathrm{n}} = 90°$, there are

$$
\begin{aligned}
P_{\mathrm{n}} &= \frac{U_{\mathrm{n0}}U\sin\delta_{\mathrm{n}}}{X_{\mathrm{n}}} \\
Q_n &= \frac{(U_{\mathrm{n0}}U\cos\delta_{\mathrm{n}} - U^2)}{X_{\mathrm{n}}}
\end{aligned} \tag{7.35}
$$

According to Eq. (7.35), when δ_{n} is small, the magnitude of the active power output from the Converter n is mainly determined by the power angle δ_{n}, which can be adjusted by the output frequency of the Converter n. The reactive power output by the Converter n depends mainly on the amplitude of the output voltage U_{n0}, so the reactive power can be adjusted by the amplitude of the output voltage of the Converter [8]. Based on the above two characteristics, the conventional method for synchronous generators based on a droop characteristic can be used to achieve equalisation of the active and reactive power. The expression for the droop characteristic is shown in Eq. (7.36):

$$
\begin{cases}
\omega_{\mathrm{n}} = \omega_1^* - m_{\mathrm{p}\omega}P_{\mathrm{n}} \\
U_{\mathrm{n0}} = U_0^* - n_{\mathrm{qE}}Q_{\mathrm{n}}
\end{cases} \tag{7.36}
$$

where U_0^* and ω_1^* are the output voltage amplitude and angular frequency of the converter with no load respectively, $m_{\mathrm{p}\omega}$ and n_{qE} are the slope of the droop curves. The slope of the droop curve should be set differently depending on the converter capacity. The larger the converter capacity, the smaller the slope, meaning a large output power variation for small frequency and voltage amplitude fluctuations; conversely, the smaller the converter capacity, the larger the slope. Therefore, in order to achieve the power equalisation between different-capacity converters, the coefficient (slope) of the droop curve can be chosen as

$$
\begin{aligned}
k_1 \cdot S_1 = k_2 \cdot S_2 = k_3 \cdot S_3 ..... = k_{\mathrm{n}} \cdot S_{\mathrm{n}} \\
h_1 \cdot S_1 = h_2 \cdot S_2 = h_3 \cdot S_3 ..... = h_{\mathrm{n}} \cdot S_{\mathrm{n}}
\end{aligned} \tag{7.37}
$$

where k is the slope of the active power droop curve, h is the slope of the reactive power droop curve, and S is the capacity of the converter.

Figure 7.24 illustrates the droop curves of two converters with different capacities. Wherein, Fig. 7.24a represents the active power and the corresponding output frequency of the converters. The output frequency of a converter is regulated according to the active power output in order to achieve the distribution of active power between different converters. In Fig. 7.24a, ω_1 corresponds to the rated frequency 50 Hz, P_1 and P_2 are the rated active power of respective converter, where P_2 is greater than P_1. It can be seen that the higher the rated power, the smaller the slope of the droop curve, and ω_s^* represents the output frequency of converters with no load. Figure 7.24b gives the relation between the reactive power and the voltage amplitude.

A complete grid-side converter control system is shown in Fig. 7.25. The outermost loop is the power control loop, where the converter calculates the output active and reactive power by detecting the voltage and current at its own output. Then, according to the droop curve, the required output voltage frequency and amplitude are obtained to realise the power distribution. The inner loop is a double-close-loop control system for the voltage converter, where the output voltage of the converter is controlled to track the voltage reference according to the output voltage frequency and amplitude references given by the power loop.

In addition, an important difference between the power electronic converter and the conventional synchronous generator is that the power electronic converter has a faster dynamic response. Considering the impact of the power electronic converter on the stability of the power system in a wind power system, the dynamic characteristics of the power electronic converter are expected to be the same as those of the

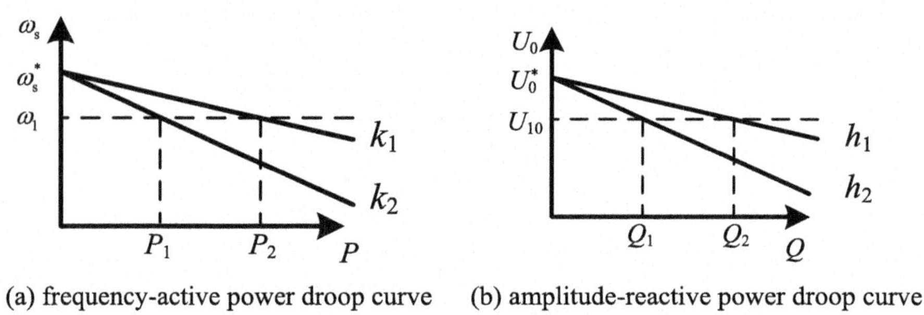

Fig. 7.24 Droop characteristic curve

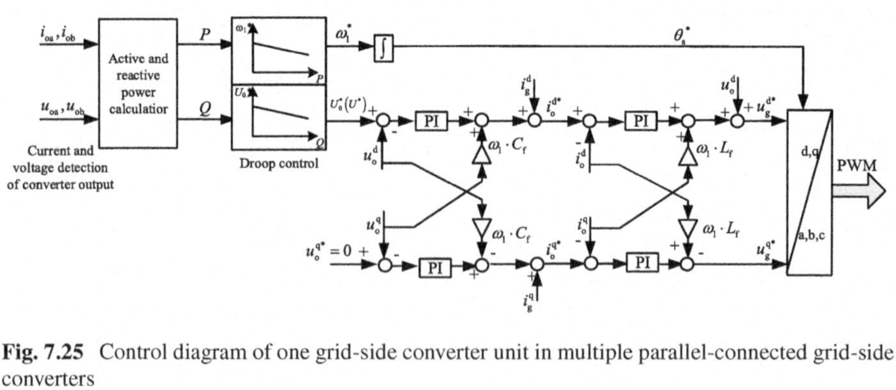

Fig. 7.25 Control diagram of one grid-side converter unit in multiple parallel-connected grid-side converters

conventional synchronous generator. A synchronous generator model can be added to the controller to slow down the dynamics of the power electronic converter to match the characteristics and operation modes of other generators on the grid.

References

1. L. Yongdong, *Digital Control Systems for AC Motors*, 3rd edn. (China Machine Press, Beijing, China, 2004)
2. L. Ran, J.R. Bumby, P.J. Tavner, Use of turbine inertia for power smoothing of wind turbines with a DFIG, in *International Conference on Harmonics and Quality of Power* (2004), pp. 106–111
3. Y. Xibo, Research on high power permanent magnet wind generator system and high voltage converter control. Ph.D. Thesis, Tsinghua University, Beijing, China, 2010
4. J.F. Conroy, R. Watson, Low-voltage ride-through of a full converter wind turbine with permanent magnet generator. IET Renew. Power Gener. **1**(3), 182–189 (2007)
5. Z. Yu, Design and analysis of direct drive wind turbine systems. Ph.D. Thesis, Tsinghua University, Beijing, China, 2006
6. Y. Xingjia, X. Zuoxia, L. Yingming et al., Coordinate control strategy research of variable speed adjustment pitch wind generation set. Acta Energiae Solaris Sinica **30**(5), 639–644 (2009)
7. G. Hua, Y. Geng, Output power level control of variable-speed variable-pitch wind generators. Proc. CSEE **28**(25), 130–137 (2008)
8. L. Xinchun, Research on control for parallel operation of UPS with no control interconnection. Ph.D. Thesis, Huazhong University of Science and Technology, Wuhan, China, 2003

Chapter 8
Medium/High-Voltage PMSG-Based Wind Power System Technologies

With the gradual increase in the stand-alone capacity of wind turbines, 3–5 MW wind turbines are becoming the dominant models for onshore wind turbines, while the capacity of offshore wind turbines has already exceeded 7 MW and is expected to reach 10 MW by 2023.[1] At current voltage and current levels in power electronics, low-voltage converters with capacities of 500 kVA and above are typically constructed by parallel connected devices or modules to meet the current requirements. The cost of transmission cables for the transmission of the power generated by the wind turbine from the top of the tower to the bottom of the tower often depends on the current level. With a voltage level of 690 V, the current for a 2 MW system will reach around 1700 A, which usually requires multiple sets of transmission cables to be connected in parallel to transmit the power and also results in line losses and voltage drops. Therefore, the use of medium/high-voltage converters is the trend for large-capacity wind turbines.

This chapter firstly discusses an ideal converter structure for large-capacity medium/high-voltage wind power systems—the diode clamped three-level converter—and investigates some of the key issues in its application, including the modulation method with midpoint voltage, switching losses and common-mode voltage control capability, and the analysis and study of typical faults; then a modular cascaded converter based on the multi-phase PMSG is discussed for the medium/high-voltage conversion of large-capacity wind turbines, especially for the topology with no boost transformer, and its mathematical model and control method are discussed in detail.

---

[1] According to the latest reports by the time the book became available, the first 10 WM offshore permanent magnet wind power generator with fully independent intellectual property rights in China was successfully developed at Dongfang Electric Group in August 2019, with a full-scale power converter supplied by CEEG Group Co., Ltd., and completed on-load experiments.

H. Ma et al., *The Control Principle of Wind Power Generation System*,
https://doi.org/10.1007/978-981-99-6083-5_8

8.1 A Medium/High-Voltage Wind Power System Based on the Three-Level Converter

8.1.1 Topology and Control of the Three-Level Wind Power System

The wind power system using a diode-clamped three-level converter is shown in Fig. 8.1. The PMSG in this figure is rated at 3.5 kV and can be either a low-speed direct-drive generator or a high-speed generator with a gearbox. The three-level converter in Fig. 8.1 has a back-to-back topology with four switching tubes per bridge arm where the power electronic devices are IGBTs, and of course other power electronic devices such as IGCTs can be used. The midpoint of the DC bus is earthed through the impedance, the generator-side converter has an output filter, and the grid-side converter usually uses the leakage reactance of the transformer as a filter inductor.

Throughout the discussions in this chapter, the grid is considered to be a strong grid (ideal voltage source) with the generator and converter acting as current sources to feed power into the grid. In this case, the grid-side converter maintains a constant DC-bus voltage and delivers active and reactive power to the grid, while the generator-side converter achieves MPPT by controlling the generator speed, which is the same as the control method in Chap. 7. Obviously, if the grid is weak, the generator and converter system can still be controlled using the methods in Chap. 7. As the full-scale power converter isolates the grid from the generator, the performance of the full-scale power converter directly affects the performance of the interface between the generator and the grid.

Here, the multi-level PWM modulation algorithm is an important part of the study of the multi-level converter in Fig. 8.1. SVPWM (Space Vector Pulse Width Modulation) requires the selection of a suitable voltage vector from a large number of switching states, as well as the determination of the action time of the relevant vectors and the selection of redundant vectors, making the operation more complex. CPWM (Carrier Pulse Width Modulation) is simpler, as the switching signal is obtained by comparing the modulating waveform with the carrier waveform, where the key

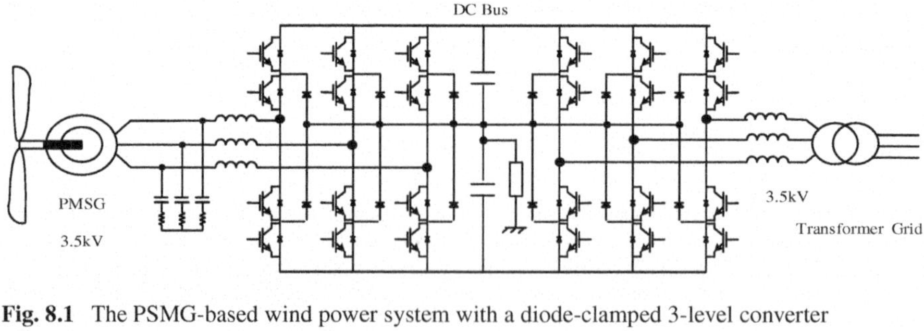

Fig. 8.1 The PSMG-based wind power system with a diode-clamped 3-level converter

issue is how to find the right modulating waveform, which will be studied in detail in this section. At the same time, SVPWM and CPWM are equivalent with each other, where the bridge between them is the zero-sequence component [1]. The basic requirements for a multilevel PWM algorithm are twofold: the first is the requirement to meet the load control objective, to achieve the control of the load (grid) current by generating the required output voltage; the second is the requirement for the performance of the converter itself, such as midpoint voltage balancing, common mode voltage suppression, improved voltage utilisation efficiency, which is mainly achieved through the adjustment of the zero sequence component [2].

8.1.2 Optimisation of Control Objectives for the Three-Level Converter

8.1.2.1 CPMW and the Zero-Sequence Component

1. Zero-sequence Component Injection

The basic modulation method of CPWM is to compare the reference waveform with the carrier waveform and then obtain the trigger signal for the corresponding switch of the converter, thus completing the modulation of the converter. The general expression for the CPWM reference waveform is shown in Eq. (8.1).

$$u_i(t) = u_i^*(t) + c_i(t) \tag{8.1}$$

where u_i^* is the fundamental-frequency component and c_i is the zero-sequence component. u_i^* is a three-phase symmetrical sinusoidal signal, which can be a voltage reference in open-loop control or the output of a current loop in a closed-loop system. c_i is the quantity injected into the three-phase reference signal simultaneously and equally, so it will not be present in the line voltage and does not affect the desired output line voltage, but it can be used as a control variable to achieve some of the control objectives of the converter itself. The study in this section is on the three-level converter, but considering the generality of the algorithm and the application of more multi-level count converters in wind power generation, the analysis in the multi-level converter is also given in the section.

For an N-level diode-clamped converter, if the negative terminal of the DC bus is selected as the zero voltage reference and the $1/(N-1)$ of the DC-bus voltage is selected as the base value, the output of the phase voltage can be normalised between 0 and $(N-1)$, as shown in Eq. (8.2).

$$0 \le u_i(t) = u_i^*(t) + c_i(t) \le N-1 \tag{8.2}$$

Accordingly, the range of the zero-sequence component that can be superimposed on each phase is

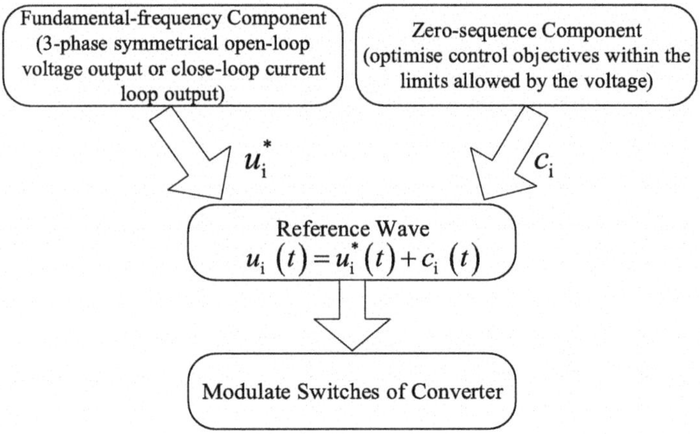

Fig. 8.2 Schematic of reference wave synthesis

$$-u^*_{\min}(t) \le c_i(t) \le N - 1 - u^*_{\max}(t) \tag{8.3}$$

where $u^*_{\min}(t) = \min(u^*_a(t), u^*_b(t), u^*_c(t))$ and $u^*_{\max}(t) = \max(u^*_a(t), u^*_b(t), u^*_c(t))$ are the minimum and maximum values of the fundamental-frequency components respectively. Equation (8.3) indicates that the superimposed zero-sequence component must not cause the final synthesis voltage $u_i(t)$ to exceed the range of Eq. (8.2). The zero-sequence component needs to be selected within the range allowed by Eq. (8.3) and then superimposed on the reference signal to achieve different control objectives. A diagram of zero-sequence component injection is shown in Fig. 8.2. The fundamental-frequency component and the zero-sequence component are synthesised to obtain the reference waveform, which then modulates the corresponding converter switch.

In order to make the following analysis clearer, a three-level converter is used as an example. Converters with more levels can be analysed in a similar way. For a three-level converter, $N = 3$, combining Eqs. (8.2) and (8.3) gives Eqs. (8.4) and (8.5).

$$0 \le u_i = u_i^*(t) + c_i(t) \le 2 \tag{8.4}$$

$$-u^*_{\min}(t) \le c_i(t) \le 2 - u^*_{\max}(t) \tag{8.5}$$

2. Modulation of Converter Switches

This last step in Fig. 8.2 is made very easy using the CPWM to modulate the converter switches. The advantage of using $1/(N - 1)$ of the -DC bus voltage as the base value for the voltage reference $u_i(t)$ is that it simplifies the modulation of the converter considerably. In a three-level converter, if half of the DC-bus voltage is chosen as the base value, then the output phase voltage reference value with respect to the negative DC-bus terminal can be normalised between 0 and 2.

The integer part of the voltage reference represents the voltage layer of the output voltage and the fractional part represents the duty cycle, thus simplifying the calculation of the duty cycle. Taking the three-level structure in Fig. 8.3 as an example and considering the modulation of the A-phase switches, the actions of switches S_{a1} and S_{a3}, S_{a2} and S_{a4} are complementary, i.e. S_{a1} turns on, then S_{a3} turns off and vice versa. If the reference value of the A-phase voltage is between 0 and 1, then the voltage layer $l_a = 0$ (i.e. the integer part is 0) means that switch S_{a1} is always off, S_{a3} is always on and S_{a2} and S_{a4} are modulated according to the value of the duty cycle (fractional part). Similarly, the voltage layer $l_a = 1$ corresponds to the case where S_{a2} is always on, S_{a4} is always off, and S_{a1} and S_{a3} are then modulated according to the magnitude of the duty cycle. If the A-phase voltage has a normalised value of 0.8, then $l_a = 0$, the switch S_{a1} is always off, S_{a3} is always on, S_{a2} is turned on at $T_1 = 0.1$ and off at $T_4 = 0.9$, as shown in Fig. 8.4, with S_{a4} and S_{a2} complementary. The modulation process for B and C phases is the same as for A phase.

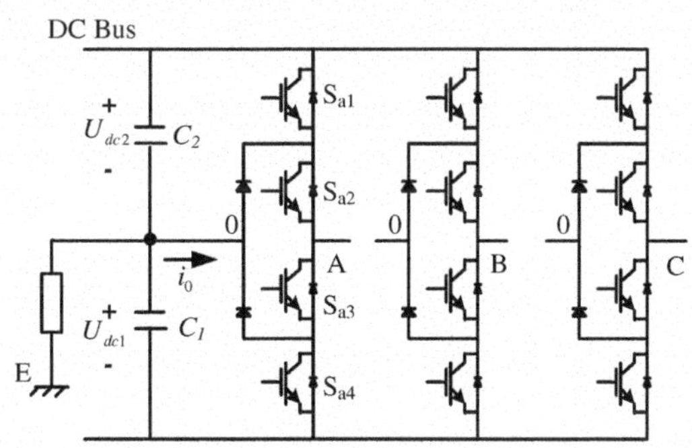

Fig. 8.3 Basic topology of a 3-level converter

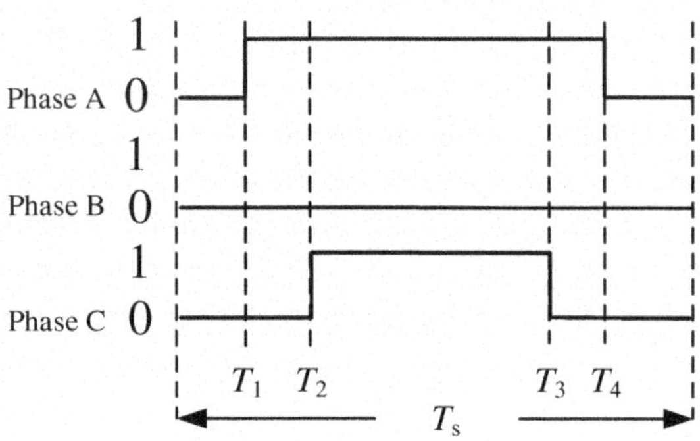

Fig. 8.4 Converter output voltage of each phase in one PWM cycle

8.1.2.2 Control Objective Optimisation and Zero-Sequence Component Injection

1. Control of the Midpoint Voltage of the DC Bus

The control of the midpoint voltage of the DC bus is a very important issue in the diode-clamped converter. In this section, the corresponding control objective and control method will be derived by analysing the relation between the midpoint voltage and the zero-sequence component of the DC bus on the basis of the CPWM modulation.

In order to reduce the fluctuations in the DC-bus midpoint voltage of the converter in Fig. 8.3, the control objective can be chosen in the form shown in Eq. (8.6):

$$\min U = \left| U_{dc1} - \frac{U_{dc1} + U_{dc2}}{2} \right| \tag{8.6}$$

where U is the control objective, U_{dc1} and U_{dc2} are the voltages across the upper and lower capacitors of the DC bus in Fig. 8.3. Equation (8.6) indicates that it is desired that the voltage on the lower capacitor be as close as possible to half the DC-bus voltage, making the voltages on the upper and lower capacitors equal so as to ensure that the midpoint voltage is balanced. The variation of the midpoint voltage ΔU_{dc1} during each switching cycle can be expressed as

$$\Delta U_{dc1} = \frac{i_0 \times T_s}{C} \tag{8.7}$$

where i_0 is the DC-bus midpoint current in Fig. 8.3, T_{s} is the switching period and C is the value of the half DC-bus capacitance. Each bridge arm of the converter has three states, i.e. the output is connected to the DC-bus positive point, the DC-bus negative point and the DC-bus midpoint. Only in the last state, the output current affects the balance of the midpoint voltage, because the current flows out or in from the midpoint, while in the other two states the current flows out or in from the DC-bus positive or negative points. When any phase output is connected to the DC-bus midpoint, the midpoint current flows in or out depending on the direction of the load current, thus affecting the midpoint voltage of the dc bus. Assuming that the midpoint current i_0 is constant over a switching cycle, the average midpoint current over a switching cycle is the product of each phase current $i_{\mathrm{i}}(\mathrm{i} = \mathrm{a, b, c})$ and the corresponding midpoint voltage action time [3]. The practical phase current can be measured by the current sensor and the action time of the midpoint voltage can be easily determined from the modulation algorithm given above. If the integer part of the reference voltage $u_{\mathrm{i}}(\mathrm{i} = \mathrm{a, b, c})$ is 0 and the voltage layer $l_{\mathrm{i}} = 0$, then the action time of the midpoint voltage is the fractional part of the reference voltage $u_{\mathrm{i}}(\mathrm{i} = \mathrm{a, b, c})$, i.e. $\mathrm{S_{a2}}$ and $\mathrm{S_{a3}}$ are on. On the other hand, if the voltage layer $l_{\mathrm{i}} = 1$, then the midpoint voltage has an action time of 1 minus the fractional part of u_{i}. In this way, the average midpoint current over one switching cycle can be expressed as

$$\overline{i_0} = \sum_{\text{i=a, b, c}} i_i \times ((1 - \text{Int}(u_i)) \times \text{Frac}(u_i) + \text{Int}(u_i) \times (1 - \text{Frac}(u_i))) \qquad (8.8)$$

where $\text{Int}(u_i)$ represents the integer part of u_i in the normalised system and $\text{Frac}(u_i)$ represents the fractional part of u_i.

The control objective is to keep the voltage on capacitor C_1 as close as possible to half the DC-bus voltage in the next cycle, as shown in Eq. (8.6). The voltage fluctuation on capacitor C_1 in the next cycle can be predicted by Eq. (8.7), which shows that the midpoint current will determine the fluctuation of capacitor voltage. It is clear from Eq. (8.8) that the average midpoint current is influenced by the voltage reference u_i which, according to Eq. (8.2), can be changed by regulating the zero-sequence component. The control objective can therefore be rewritten as Eq. (8.9), where n is the discretised time and the control variable is the zero-sequence component, which needs to be regulated within the allowed range.

$$\begin{aligned} &\min U = \left| U_{dc1}(n-1) - \tfrac{U_{dc1}(n-1)+U_{dc2}(n-1)}{2} + \Delta U_{dc1}(n) \right| \\ &\text{Constraints:} \quad -u^*_{\min}(t) \le c_i(t) \le 2 - u^*_{\max}(t) \end{aligned} \qquad (8.9)$$

With the above analysis, the control method for the DC-bus midpoint voltage can be described as follows: (1) In each control cycle, the phase currents of the three phases and the voltages on the upper and lower capacitors of the DC-bus are measured. The fundamental component of the reference voltage u^*_i is given by the current inner loop; (2) the range of the zero-sequence component that can be superimposed is calculated according to Eq. (8.5); (3) within this range, the optimal zero-sequence component is found such that the control objective as in Eq. (8.9) is minimised.

A simple way of implementing Step (3) above is as follows. Several values of the zero-sequence component are picked evenly within the range allowed for the zero-sequence component. By superimposing c_i of each zero-sequence component onto the fundamental component u^*_i, the voltage reference value u_i can be obtained, and according to Eqs. (8.7) and (8.8), the voltage variation at the midpoint of the DC bus for the next switching cycle can be predicted, and the minimum value of Eq. (8.9) and the corresponding zero-sequence component can also be obtained. The zero-sequence component that minimises the objective function in Eq. (8.9) is finally selected as the zero-sequence component for the synthesis of the reference voltage. For an N-level converter, the control objective in Eq. (8.6) becomes that the voltage of each capacitor is as close as possible to the $1/(N-1)$ DC bus voltage. The sum of the difference between the $N-1$ capacitor voltages and the $1/(N-1)$ DC bus voltage can then be selected as the objective function, and the control objective is to minimise this function by selecting the proper zero-sequence component.

The search algorithm above only finds the optimal value among the zero-sequence components selected for testing. The control performance of the midpoint voltage is influenced by the number of test points, and exactly how many points are selected

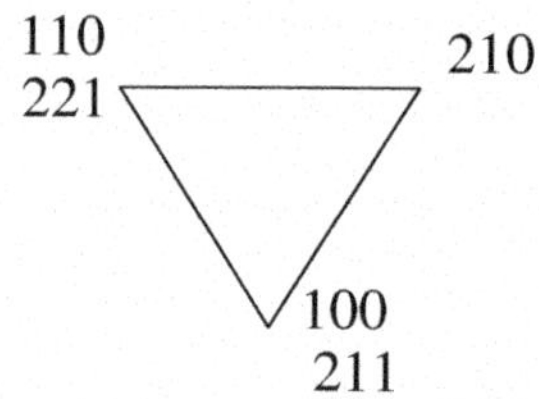

Fig. 8.5 A vector triangle in vector space of a 3-level converter

depends on the DSP's computing speed and the control cycle. This control method can be seen as a form of predictive control.

2. Implementation of the Five-segment Method

Within one PWM switching cycle, the switches in all three or only two phases of the bridge arms act. The mode in which the switches in all three phases act is often referred to as seven-segment type, while the mode in which the switches in two phases act is referred to as five-segment type. The seven-segment PWM offers better harmonic performance, while the five-segment PWM reduces switching losses due to the reduction of switching action in one phase.

In the SVPWM, the reference vector is usually synthesised by selecting the three vectors on the vertices of the vector triangle where the reference vector is located. Taking a vector triangle in the vector space of a three-level converter as an example, as shown in Fig. 8.5, the three vectors on the vertices may act in the sequence of 100-110-210, 110-210-211, 210-211-221. where 0, 1 and 2 represent the corresponding phase outputs being clamped to the DC-bus negative, mid- and positive points respectively.

By selecting the appropriate zero-sequence component, the CPWM also implements a five-segment PWM algorithm. In this case, the zero-sequence component is superimposed on the phase with the maximum duty cycle each time, thus covering all possible sequences of voltage vector actions. The magnitude of the zero-sequence component injected can be selected as 1 minus the maximum value of the duty cycles, so that the phase with the maximum duty cycle does not act, thus implementing the five-segment algorithm.

The phase with the maximum duty cycle can be determined by comparing the fractional part of the reference voltage ui in each phase. Figure 8.6a–c, corresponding to 100-110-210, 110-210-211 and 210-211-221 respectively, give the possible sequence of switching states for the voltage reference in Fig. 8.5 during a switching cycle. Figure 8.6d gives a schematic representation of the different sequences obtained by adding a proper zero-sequence component. In Fig. 8.6a, Phase B has the maximum duty cycle. If the zero-sequence component corresponding to the shaded part in Fig. 8.6d is added to the three-phase voltage simultaneously, then the switching state sequence changes from (a) 100-110-210 to (b) 110-210-211, which equivalently implements the selection of the vector sequence in SVPWM (selection of different redundant vectors), so the selection process of the redundant vectors is said to be the process of superimposing the zero-sequence components with different magnitudes. In the next step, in order to balance the midpoint voltage of the DC bus, it is necessary to search for the optimal sequence that can balance the midpoint voltage.

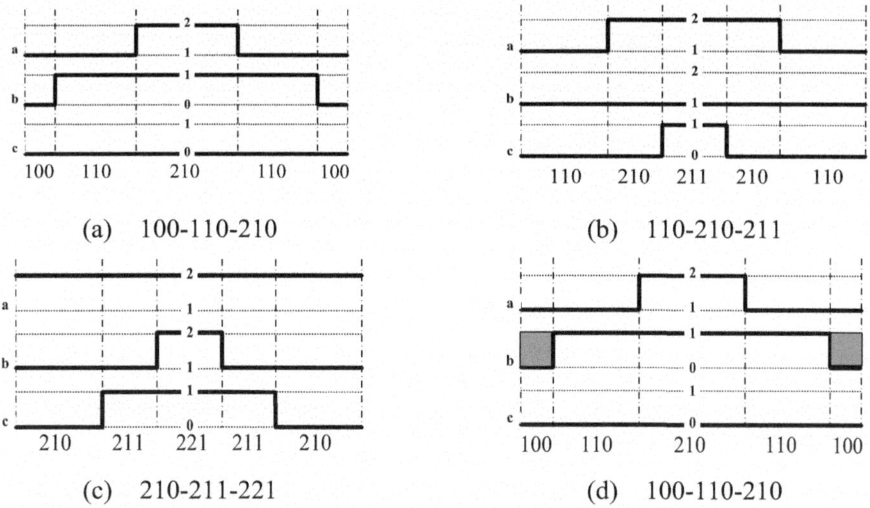

Fig. 8.6 Possible switching sequences of converter phase voltage output in one PWM cycle

All possible voltage sequences can be tested by adding or subtracting appropriate zero-sequence components, from which the voltage sequence that minimises the Eq. (8.9) is selected. This is the optimal voltage vector sequence that balances the midpoint voltage whilst achieving five-segment algorithm. The value of the zero-sequence component corresponding to this sequence is added to the fundamental frequency component to form the final modulated signal. It is important to note that the added zero-sequence component still needs to satisfy the range in Eq. (8.5). The five-segment method based on zero-sequence component injection can also be applied to $N\,(N > 3)$ level converters, although there will be more feasible switching sequences for N-level converters (Fig. 8.7).

3. Common-mode Voltage Suppression

An important research issue for power electronics converters using PWM modulation is the suppression and cancellation of the common-mode voltage, the problems caused by which are becoming more and more prominent as the switching frequency increases and the voltage level increases. In a wind power system, the common-mode voltage can cause damage to the generator shaft, insulation breakdown and EMI interference problems [5, 6]. There are both hardware and software methods for suppressing the common-mode voltage, and this section focuses on the software approach, which investigates how to suppress the common-mode voltage by superimposing the zero-sequence component. In contrast to the conventional approach, this section uses a reasonable selection of zero-sequence components to suppress the common-mode voltage to within 1/6 of the DC-bus voltage without attenuating the maximum amplitude of the output voltage.

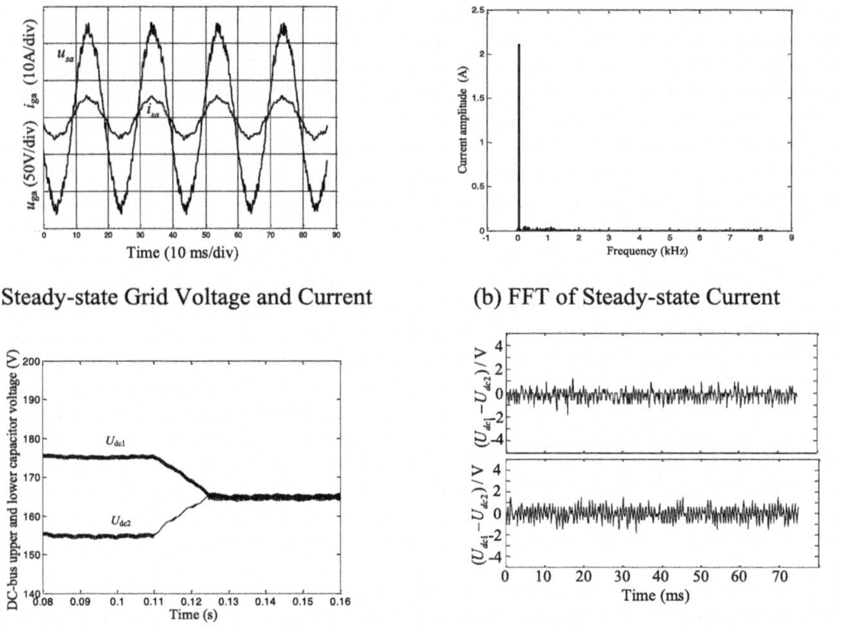

(a) Steady-state Grid Voltage and Current
(b) FFT of Steady-state Current
(c) Balance Process of Capacitor Voltages
(d) Voltage Difference between Upper and Lower Capacitor Voltages

Fig. 8.7 Experimental results of vector control based on the modulation method proposed in this section

The common-mode voltage is defined as the voltage between the midpoint O of the DC bus and the midpoint of the generator or transformer, as shown in Eq. (8.10).

$$U_{cm} = (U_{AO} + U_{BO} + U_{CO})/3 \tag{8.10}$$

where U_{cm} is the common-mode voltage, U_{AO}, U_{BO} and U_{CO} are the output voltages of each phase with respect to the midpoint O of the dc bus, as shown in Fig. 8.3. As mentioned earlier, each phase has three different switching states (0, 1, 2), corresponding to the output voltage of each phase clamped to the DC-bus positive point ($+U_{\mathrm{dc}}/2$), midpoint (0) and negative point ($+U_{\mathrm{dc}}/2$) respectively. The relationship between the common-mode voltage and the sum of the three-phase voltage layers (the integer part of the reference voltage u_{i} is respective the voltage layer) is derived below to find the appropriate zero-sequence component to achieve the suppression of the common-mode voltage.

Since the normalised value of the output voltage for each phase ranges from 0 to 2, the sum of the three-phase output-voltage layers is between 0 and 4, i.e. $0 \leq (l_{\mathrm{a}} + l_{\mathrm{b}} + l_{\mathrm{c}}) \leq 4$, where l_{a}, l_{b} and l_{c} are the voltage layers for Phase A, Phase B and Phase C respectively. Here the analysis is carried out using a five-segment PWM as an example, as shown in Fig. 8.4, where two phases will change the output voltage

Table 8.1 Relation between sum of three-phase voltage layers and the common-mode voltage

Sum of 3-phase voltage layers	Possible common-mode volage
$l_a + l_b + l_c = 4$	$+U_{dc}/2$, $+U_{dc}/3$, $+U_{dc}/6$
$l_a + l_b + l_c = 3$	$+U_{dc}/3$, $+U_{dc}/6$, 0
$l_a + l_b + l_c = 2$	$+U_{dc}/6$, 0, $-U_{dc}/6$
$l_a + l_b + l_c = 1$	0, $-U_{dc}/6$, $-U_{dc}/3$
$l_a + l_b + l_c = 0$	$-U_{dc}/6$, $-U_{dc}/3$, $-U_{dc}/2$

levels during a switching cycle. The relation between the common-mode voltage and the sum of the three-phase output-voltage layers (the starting voltage level of each phase in Fig. 8.4 is the voltage layer of each phase) can be obtained by calculation. In Fig. 8.4, the sum of the three-phase voltage layers is $0 + 0 + 0 = 0$. Between 0 and T_1, the common-mode voltage is

$$\left(\left(-\frac{U_{dc}}{2}\right)+\left(-\frac{U_{dc}}{2}\right)+\left(-\frac{U_{dc}}{2}\right)\right)/3=-\frac{U_{dc}}{2} \tag{8.11}$$

Between T_1 and T_2, the voltage level of Phase A jumps from 0 to 1 (from the DC-bus negative point jumps to the DC-bus midpoint), then the common-mode voltage becomes

$$\left(0+\left(-\frac{U_{dc}}{2}\right)+\left(-\frac{U_{dc}}{2}\right)\right)/3=-\frac{U_{dc}}{3} \tag{8.12}$$

Between T_2 and T_3, the common-mode voltage becomes $-U_{dc}/6$. For the different sums of voltage layers, the values of the common-mode voltage in one switching cycle can all be calculated as above and summarised in Table 8.1, where it can be seen that the amplitude of the common-mode voltage in one switching cycle is always less than or equal to $U_{dc}/6$ only when $l_a + l_b + l_c = 2$.

In the proposed CPWM algorithm based on zero-sequence component injection, a constraint of $l_a + l_b + l_c = 2$ can be added to the selection of the zero-sequence components, and only those zero-sequence components that satisfy this condition can be added to the fundamental frequency component to form the final reference voltage, thus ensuring that the amplitude of the common-mode voltage does not exceed $U_{dc}/6$. Through such a process, the 19 vectors desired in the equivalent SVPWM to produce a common-mode voltage less than a are selected from all 27 vectors, as shown in Fig. 8.8, where the vectors corresponding to the shaded common-mode voltages are not utilized [4]. It should be noted that the control objective here is to suppress the common-mode voltage and does not guarantee a balanced midpoint voltage. For the back-to-back converter, when the common-mode voltage suppression algorithm is applied to the generator-side converter, the grid-side converter can use the midpoint balance algorithm to achieve control of the midpoint voltage. Figure 8.9 illustrates the experimental results about the common-mode voltage.

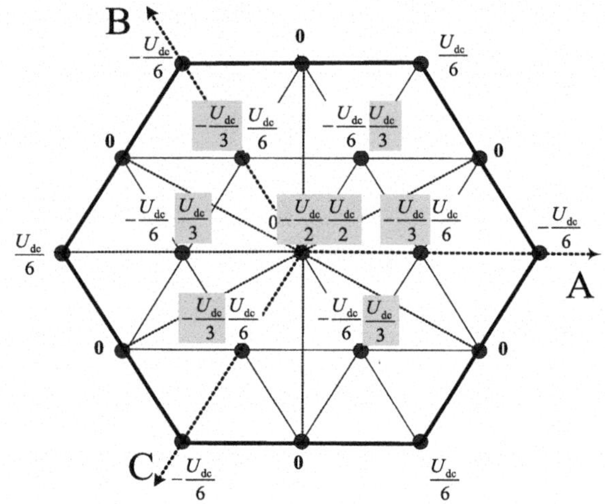

Fig. 8.8 Distribution of voltage vectors and common-mode voltages in vector space

8.1.3 *Fault Detection and Protection of the Three-Level Converter*

Because of the harsh environment in which wind turbine systems are located and the high maintenance costs, they are expected to have more sophisticated fault detection and protection functions. This is particularly important for the diode-clamped three-level converter for wind power generation, as this topology is more complex and has a higher number of power electronic devices compared to the two-level converter, increasing the number of fault points. In the following, the single-phase earth fault and the short-circuit fault the diode-clamped three-level converter are investigated, and detection and protection methods are given.

8.1.3.1 Detection of the Earth Fault

1. Description and Detection Method of the Earth Fault

In the large-capacity three-level wind power system, the grid-side converter is usually connected to the grid via a step-up transformer and the generator is not earthed at its midpoint. The only earth point in the system is the DC-bus midpoint O grounded through impedance, as shown in Fig. 8.10. This section discusses the detection of a single-phase earth fault in Fig. 8.10.

Firstly, the voltage change across the midpoint earth impedance voltage U_{OE} is analysed before and after the occurrence of an earth fault. During normal system operation, the common-mode voltage generated by the PWM algorithm will drop between the midpoint O of the DC bus and the midpoint of the generator (or transformer). On the generator side, for example, the generator case is earthed and there

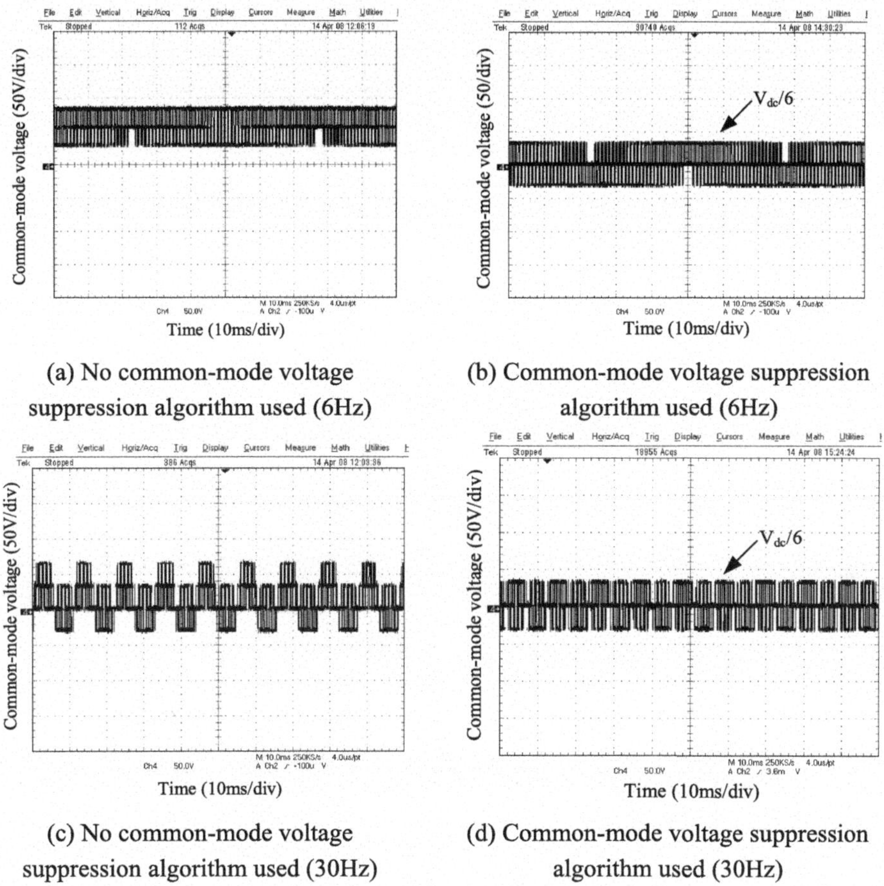

Fig. 8.9 Experimental results about the common-mode voltage

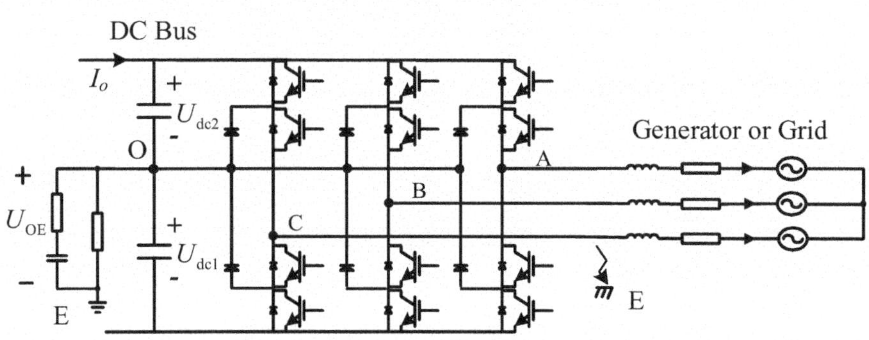

Fig. 8.10 Schematic of a single-phase earth fault in the diode-clamped 3-level converter

is stray capacitance between the midpoint of the generator and its case, where the DC-bus midpoint voltage to ground, U_{OE}, is the voltage split of the system common-mode voltage between the midpoint ground impedance and the stray capacitance. As the value of the stray capacitance is generally small and its equivalent impedance is large, U_{OE} is a very small part of the system common-mode voltage and can be approximated as zero. When an earth fault occurs, the value of U_{OE} changes due to the change of the earth circuit. For example, when an earth fault occurs on phase B of the generator side, the output of Phase B will be directly connected to earth and the U_{OE} becomes $-U_{BO}$ (i.e. the output voltage of Phase B relative to the midpoint of the DC bus is taken backwards) and the value of U_{OE} will be very different from the normal operation. Therefore, the occurrence of an earth fault can be determined by detecting the voltage U_{OE} across the midpoint impedance.

Figure 8.11a gives the simulation results before and after an earth fault occurs in Phase B at 70 ms. The simulation system has a capacity of 1 kW, a DC-bus voltage of 320 V, a midpoint earth impedance of 30 kΩ with a resistor and a resistive-capacitive branch connected in parallel, as shown in Fig. 8.11, where the resistance and capacitance in the parallel branch are 20 Ω and 10 nF respectively. Before the earth fault occurs, only a small portion of the common-mode voltage is applied to the earth impedance. When the earth fault occurs, the B-phase PWM waveform with an amplitude of 160 V appears on the earth impedance. To facilitate fault detection, Fig. 8.12b gives the waveform of $-U_{OE\_f}$ after inverting the U_{OE} and passing it through a low-pass filter. Figure 8.12c gives the modulated waveform of the Phase B (with respect to the midpoint of the DC bus) and it can be seen that the modulated waveform of Phase B is consistent with $-U_{OE\_f}$ when the earth fault occurs.

The earth fault detection procedure described in this section is shown in Fig. 8.12 and is described as follows: (1) The voltage U_{OE} on the earth impedance is low-pass

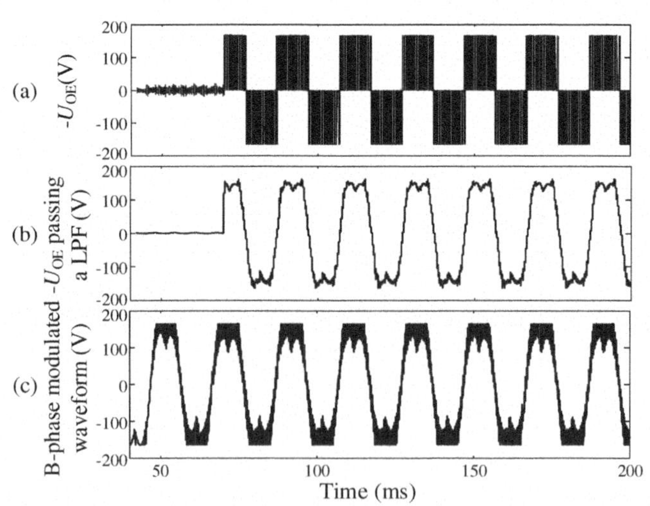

Fig. 8.11 Simulation waveforms when an earth fault occurs in phase B

Fig. 8.12 Flow chart for earth fault detection

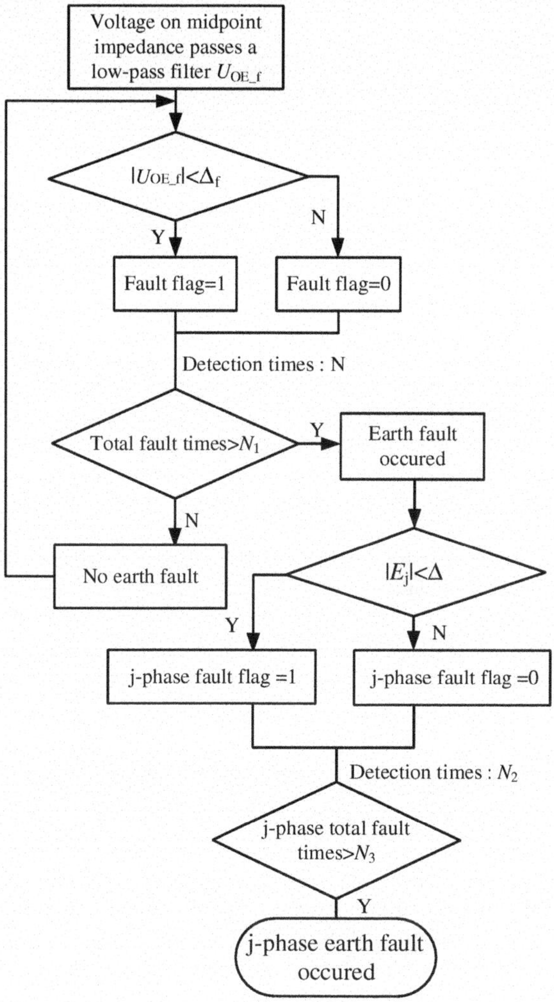

filtered to obtain U_{OE_f}, for which the absolute value is taken, and then compared with a pre-set threshold value Δ_f (the magnitude of the threshold value depends on the value of the stray impedance and the earth impedance; usually 20% of the DC-bus voltage can be chosen) to determine whether an earth fault has occurred. (2) If an earth fault has occurred, the U_{OE_f} is then compared with the modulated waveform of each phase. Assuming that the modulated waveform of each phase is U_{jo}^* (j = A, B, C) with respect to the midpoint of the DC bus, the error between $-U_{OE\_f}$ and U_{jo}^* is $E_j = U_{jo}^* - (-U_{OE\_f})$, if the absolute value of E_j is less than the set threshold Δ_f, then an earth fault has occurred in the corresponding phase. In engineering practice, multiple sampling is usually carried out and the occurrence of a fault is confirmed only if more than a certain number of faults have been detected continuously. The values of N,

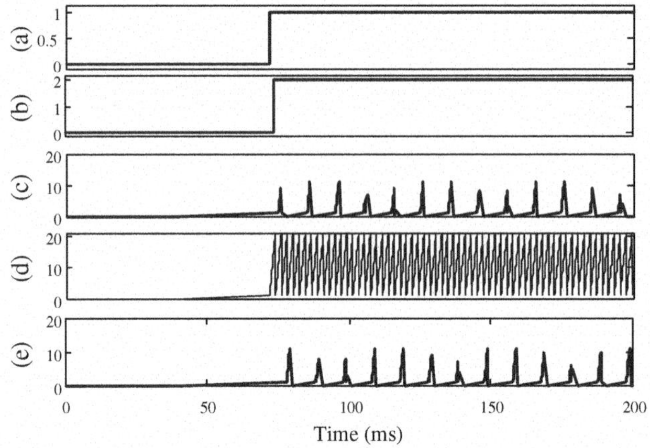

Fig. 8.13 Simulation results for earth fault detection

N_1, N_2, N_3 and Δ_f in Fig. 8.12 can be adjusted depending on the sensitivity and noise disturbance.

Figure 8.13a gives the simulation results for earth fault detection, it can be seen that the earth fault occurring at 70 ms can be successfully detected (0 for no fault occurred and 1 for a fault occurred), where $N = 30$ and $N_1 = 25$. Figure 8.13b shows that an earth fault occurred in Phase B (1 for Phase A, 2 for Phase B and 3 for Phase C). Figure 8.13c–e give the counts of the fault flags for each phase in 20 samples with the threshold $N_3 = 15$. Only Phase B always has the count of fault flag greater than 15, indicating that an earth fault has occurred in Phase B.

In addition, if an earth fault occurs in the DC Bus (positive or negative point), the DC-bus midpoint voltage U_{OE} becomes half the DC-bus voltage. By checking the value of U_{OE}, the DC-bus earth fault can also be detected.

It is important to note here that in Fig. 8.13, although only Phase B has an earth fault, part of the fault signals are also detected in Phases A and C, as shown in Fig. 8.13c, e. This is mainly because in a three-phase system, the modulated waveforms of two phases may be equal at some point in time, as shown in the circles in Fig. 8.14; therefore, even though a fault has occurred in Phase B, at some point in time, the fault may also be detected in Phases A and C (illustrated as the parts in the circles). This problem can be solved by sampling multiple times: if a fault is detected in Phase B all the time, it means that a fault has indeed occurred in Phase B. The number of faults detected in the non-faulty phase is mainly related to the assigned error tolerance Δ. The smaller the Δ, the fewer faults will be detected in the non-faulty phase. Δ should also be assigned taking into account the noise interference in the field.

2. Design of the Midpoint Earth Impedance

The earth circuit of the midpoint-clamped three-level converter is shown in Fig. 8.15, where R_{1N}, R_{2N} and C_{2N} constitute the midpoint earth impedance and C_M represents

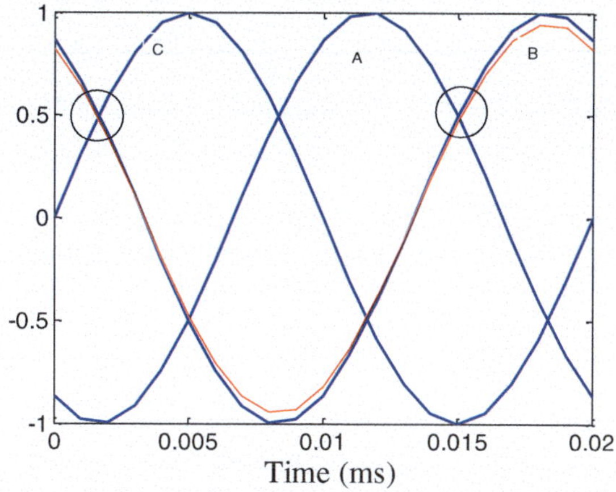

Fig. 8.14 Schematic of fault detection

the stray capacitance between the generator midpoint and earth. The design of the midpoint earth impedance should follow the following principles:

(1) The midpoint earth impedance should be large enough to limit the fault current in the event of a single-phase earth fault.
(2) The midpoint earth impedance should be less than the stray impedance between the rest of the converter and earth for the purpose of earthing.
(3) When there is no earth fault, the common-mode voltage shared on the midpoint earth impedance should be as small as possible in order to detect the occurrence of an earth fault.

In the design, resistor R_{1N} is used as the DC and low-frequency impedance, C_{2N} is used as the high-frequency impedance and R_{2N} is used to limit the capacitor current.

When an earth fault occurs, the current flowing through the midpoint earth impedance to the midpoint O of the DC bus can be expressed by Eq. (8.13)

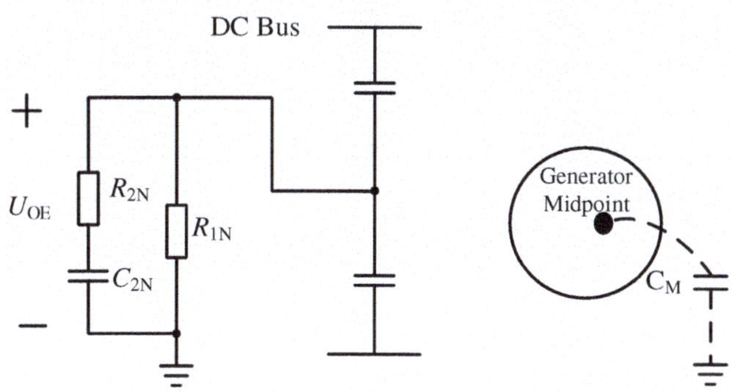

Fig. 8.15 Earth impedance on generator side in wind power system based on the midpoint-clamped 3-level converter

$$I_{\text{mid}} = \frac{U_{\text{OE}}}{Z} \tag{8.13}$$

where I_{mid} is the DC-bus midpoint current and Z is the midpoint earth impedance. When a single-phase earth fault occurs, U_{OE} is the voltage of the faulted phase with respect to the midpoint of the DC bus; when a DC bus earth fault occurs, U_{OE} is half of the DC-bus voltage. If the capacitor $C_{2\text{N}}$ is assumed to filter out the high frequency component of the voltage on the midpoint earth impedance, then the low-frequency current value at fault is determined by the resistor $R_{1\text{N}}$, as shown in Eq. (8.14).

$$I_{\text{mid}} = \frac{U_{\text{OE}}}{R_{1\text{N}}} \tag{8.14}$$

For the single-phase earth fault and the DC bus earth fault, the maximum value of U_{OE} is half the value of the DC-bus voltage. Depending on the capacity of the converter, the maximum permitted midpoint current I_{mid} can be determined. Therefore, the minimum value of $R_{1\text{N}}$ can be given by Eq. (8.15)

$$R_{1N} = \frac{1}{2}\frac{U_{\text{DC}}}{I_{mid}} \tag{8.15}$$

In normal operation, the voltage U_{OE} across the midpoint earth impedance will be the voltage splitting of the common-mode voltage between the stray capacitor C_{M} and the midpoint earth impedance. For the midpoint-clamped converter, the common-mode voltage depends on the PWM algorithm and the possible values are $\pm U_{\text{dc}}/2$, $\pm U_{\text{dc}}/3$, $\pm U_{\text{dc}}/6$,0. For the common-mode voltage at frequencies above the switching frequency, its distribution depends on the values of $C_{2\text{N}}$ and C_{M}, where $C_{2\text{N}}$ should be sufficiently large (much larger than C_{M}) to allow the detection algorithm to work properly.

The above analysis focuses on the generator side as an example. When a back-to-back converter configuration is used, the grid side should also be considered, in particular the stray impedance between the step-up transformer and earth. The above analysis of generator-side earth faults can also be applied to the case of grid-side earth faults.

8.1.3.2 Device Short-Circuit Fault and Capacitor Overvoltage Protection

Short-circuit and open-circuit faults in power electronics are common forms of faults in converters. In the case of the midpoint-clamped three-level converter, the short-circuit fault in the power electronic device of inner bridge arm (IGBT or IGCT) or in the clamping diode will lead to a doubled voltage on the DC-bus capacitor, which may damage the DC-bus capacitor and the power electronic device. For example, in Fig. 8.16, if the inner-bridge-arm device on Phase A of the grid-side converter is short-circuited, then the power supply to Phase A will be connected directly to the midpoint of the DC bus via the clamping diode. It is considered here that when the device short-circuit fault occurs, the protection of the system will be triggered, all normal power electronic devices will be switched off and the load will be removed. After the fault has occurred, the current flows in the path shown in the thick line in Fig. 8.16. The line voltage Uba will be added directly to the diodes $D_{b1} - D_{b2} - D_{a5}$, while U_{ca} will be added directly to $D_{c1} - D_{c2} - D_{a5}$. After rectification, the peak line voltages U_{ba} and U_{ca} will appear on the DC bus capacitor C_1. During normal operation, the magnitude of the line voltage (e.g. U_{ba} or U_{ca}) will be shared between capacitors C_1 and C_2. In faulty operation, the voltage amplitude of capacitor C_1 will become twice as high as in normal operation. A similar situation will occur when the short-circuit fault occurs in the clamping diode.

In general, the doubling of the DC-bus voltage can occur both on the grid side and on the generator side in Fig. 8.1. In a high-capacity wind turbine system, the grid side is usually equipped with contactors, and when the device short-circuit fault occurs, disconnection of the contactors usually takes the time of several grid cycles (in several tens of milliseconds), during which the grid voltage is still present and can lead to the doubling of the DC-bus voltage. For the generator side, the voltage at the generator terminal will be maintained due to the rotation of the PMSG, which will also lead to the doubling of the DC-bus voltage. Here the grid side will mainly be discussed as an example.

The capacitor overvoltage phenomenon in Fig. 8.17 is highly likely to damage capacitors and power electronics. There are two main methods to protect the capacitor from overvoltage: (1) the method by clamping the capacitor voltage and (2) the method by disconnecting the charging circuit. The method by disconnecting the charging circuit is to connect an additional controllable switch in series with the charging circuit (between the midpoint of the DC bus and the clamping diode). When the system is operating normally, the additional switch is closed and when the fault

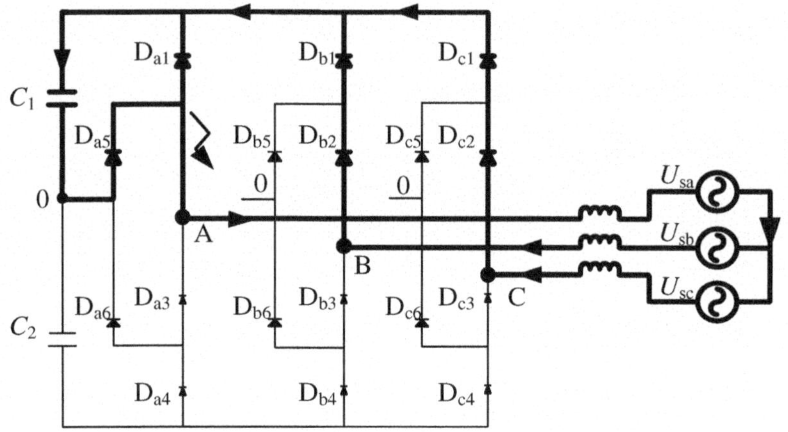

Fig. 8.16 Current path after short circuit in the device of inner bridge arm

occurs, the switch is opened, disconnecting the capacitor from the charging circuit, as shown by switch S_1 in Fig. 8.18. However, this controllable additional switch increases the losses during normal operation of the system. Therefore, this section will focus on the voltage-clamped method by controlling the resistor connected in parallel to the DC bus capacitor. A circuit diagram for this method is given in Fig. 8.19a.

During normal system operation, switches S_1 and S_2 are open and the protection circuit does not function. When an overvoltage is detected on the DC-bus capacitor, the switch closes and the corresponding resistor is connected in parallel to both terminals of the capacitor, thus discharging the capacitor and clamping the voltage. The energy released across the resistor depends on the parameters of the system, such as the input voltage, input inductance and clamped voltage. Figure 8.19b gives a schematic diagram of the equivalent charging and discharging circuit for the capacitor. The relation between the capacitor voltage U_C and the input line voltage U_l can be expressed in Eq. (8.16).

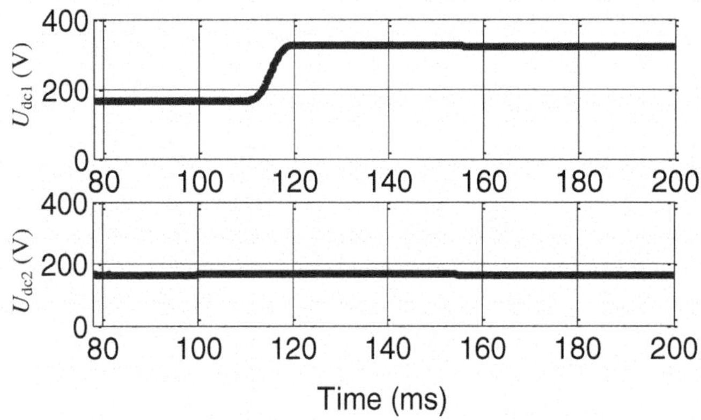

Fig. 8.17 DC-bus voltage before and after short circuit in the device of inner bridge arm

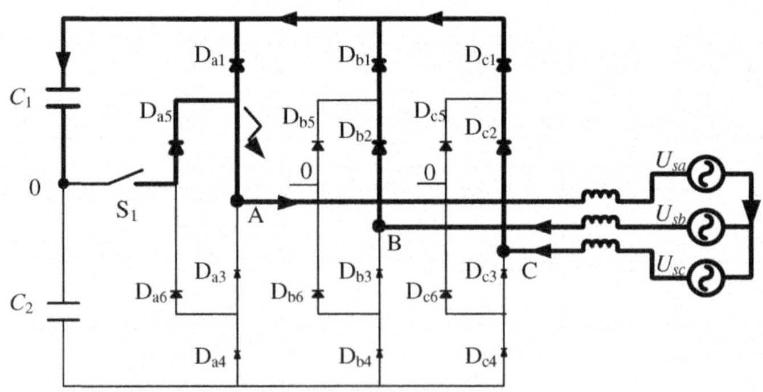

Fig. 8.18 The method by disconnecting the charging circuit

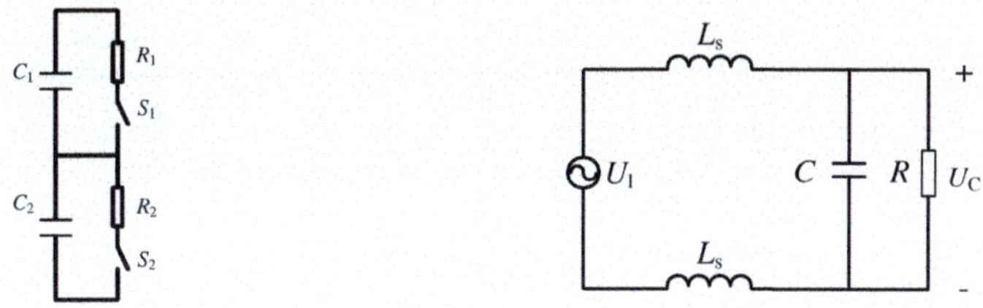

(a) The capacitor-voltage-clamped circuit

(b) Simplified equivalent charging and discharging circuit of the capacitor

Fig. 8.19 The method by clamping the capacitor voltage

$$\frac{U_C}{U_{ll\_p}} \approx \frac{1}{1 - 2\omega_1^2 L_s C + j\frac{2\omega_1 L_s}{R}} \tag{8.16}$$

where L_s is the input inductor and R is the clamping resistor connected in parallel to both terminals of the capacitor. Assuming that the capacitor voltage is clamped at 250 V, the relation between the capacitor power to be released and the value of the input inductance is shown as the dashed line in Fig. 8.20, which also gives the case of clamping at 220 V as a comparison. As can be seen from the figure, a higher clamped voltage and a larger input inductance will reduce the amount of power to be released by the clamping resistor.

The maximum clamped voltage depends on the withstand voltage of the capacitors and power electronics. Depending on the clamped voltage U_{clamp} and the released power P_{clamp}, the resistance of the clamping resistor can be determined by Eq. (8.17).

$$R = \frac{U_{\text{clamp}}}{P_{\text{clamp}}} \tag{8.17}$$

Fig. 8.20 Relation between capacitor power to be released and input inductance

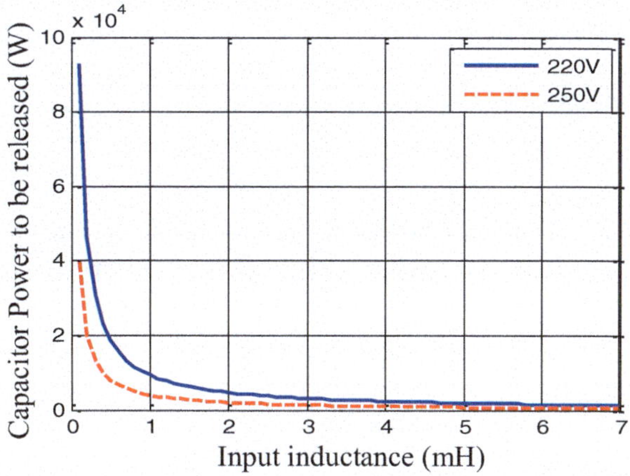

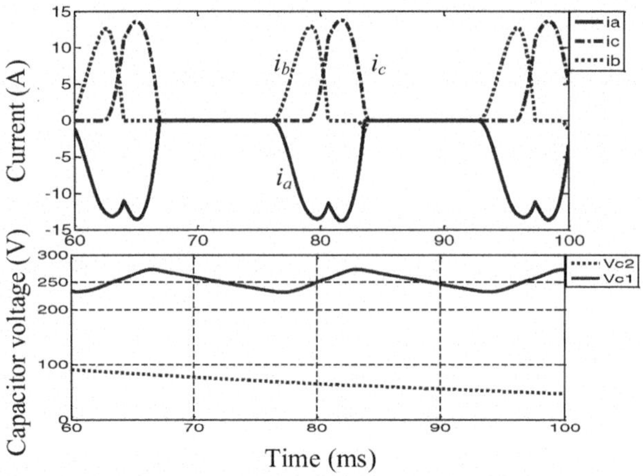

Fig. 8.21 Simulation results of the resistor-clamped method

In simulation, the clamped voltage is chosen to be 250 V and the input inductor is 5 mH. According to Fig. 8.20, the clamping resistor power is around 800 W and the required resistance is 78 Ω. In the actual system, according to the charging loop in Fig. 8.19, the overvoltage phenomenon is generated only half of the time per cycle, resulting in a pulsation of the capacitor voltage. The simulation results using the resistor-clamped method are given in Fig. 8.21, where the voltage of capacitor C_1 is clamped at around 250 V after the fault occurs.

To further improve the performance of the system, a varistor can be used, whose resistance can vary with voltage, the higher the voltage value, the lower the resistance value, exactly in line with the requirement of the capacitor power to be released. Figure 8.22 gives the simulation results using the NTE varistor 1V115. It can be seen that both the maximum DC-bus voltage and the maximum input current are reduced compared to Fig. 8.21. When selecting a clamping resistor or varistor, it is also necessary to consider the operating time of the grid-side contactor, as this determines the length of the fault-state time and thus the energy required to be consumed by the clamping resistor.

8.2 The Transformerless Medium/High-Voltage PMSG-Based Wind Power System

The previous section discussed a midpoint-clamped medium/high-voltage converter topology, but considering the voltage withstand level of the current devices, the topology is not suitable for the applications with power output at the voltage level

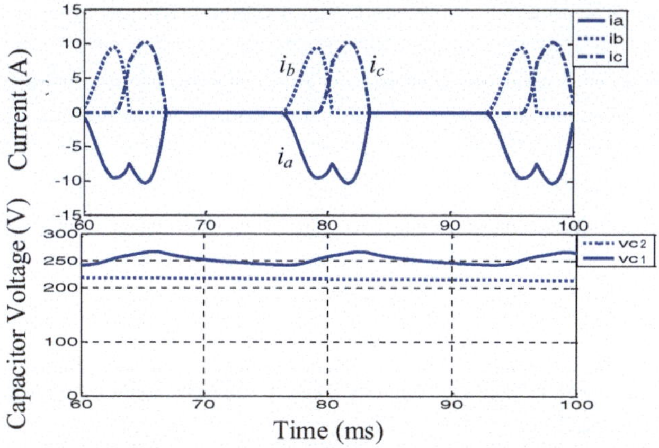

Fig. 8.22 Simulation results of the varistor-clamped method

of 6 kV and above. For the medium/high-voltage conversion problems of large-capacity wind turbines, especially topologies that can eliminate the need for step-up transformers, a modular cascaded converter based on a multi-phase PMSG is discussed in this section.

8.2.1 Structure of the Transformerless Medium/High-Voltage PMSG-Based Wind Power System

8.2.1.1 Introduction of the Overall System Topology

The converter topology discussed in this section is shown in Fig. 8.23. Individual windings of the generator, insulated from each other, provide isolated supplies to each low-voltage module of the cascaded converter, with each module having rectifier, boost and inverter functions. Multiple low-voltage modules are connected in series to form one phase of the medium/high-voltage converter, and the three single-phase medium/high-voltage converters are connected in a star pattern to form the three-phase converter. As can be seen from Fig. 8.23, the topology studied in this section is very similar to the cascaded H-bridge converter used in the motor drive field, except that in the motor drive field a transformer with multiple secondary windings (zigzag transformer) is required to provide multiple isolated power supplies to each low-voltage module, whereas in this section a multiple-winding PMSG is used instead of the phase-shifting transformer to provide the multiple power supplies. At the same time, the direct connection of the generator and converter to the medium/high-voltage grid raises the requirements for the generator insulation, which should match the voltage level of the grid. In addition, for offshore wind power applications, where the system capacity is generally large and maintenance costs are high, it is desirable that the converter be able to operate redundantly in the event of a fault, so that the multi-phase PMSG and modular structure in Fig. 8.23 are ideally suited.

Figure 8.23b gives the topology of each low-voltage module, consisting of the generator-side rectifier, the DC bus and the grid-side H-bridge inverter. The generator-side rectifier is responsible for power factor correction (PFC) of the winding current

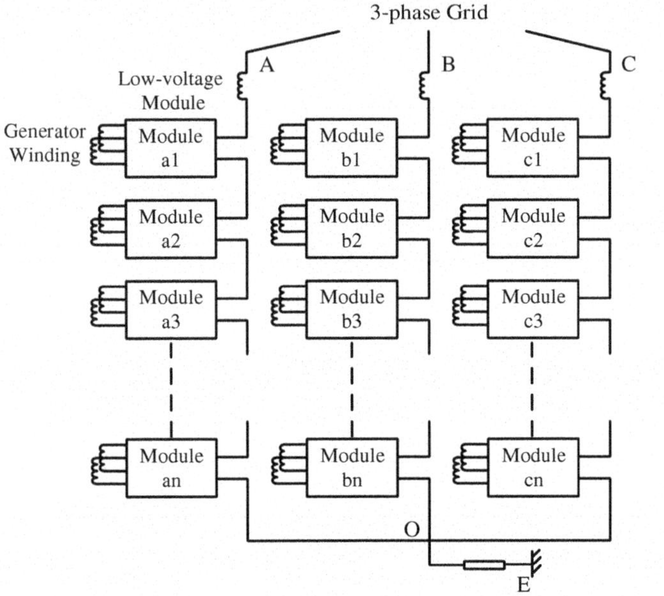

(a) Topology of the Transformerless Cascaded Converter

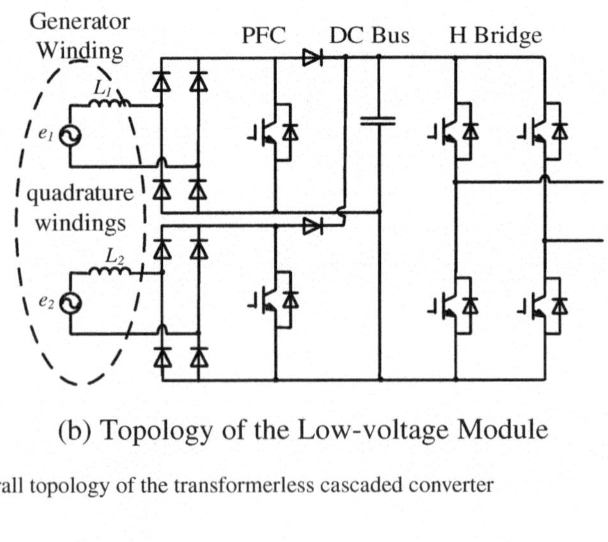

(b) Topology of the Low-voltage Module

Fig. 8.23 Overall topology of the transformerless cascaded converter for unit power factor operation and has a certain boost function, thus extending the speed range of the generator. The inverter employs vector control to regulate the active and reactive power injected into the grid, and phase-shifted carrier PWM technology to modulate the inverter switches. This topology and control method will be investigated in detail below.

8.2.1.2 Topology of the Generator-Side Rectifier in a Low-Voltage Module

Variable-speed wind turbines generally have three operating zones: when both generator speed and power have not reached rated values, the system maintains a constant tip-speed ratio and runs in the MPPT state as wind speed changes; when generator speed reaches rated speed value and power has not yet reached rated power value, the system maintains constant speed (rated speed) operation and generator power continues to grow as wind speed increases; when both generator speed and power have reached rated values, the system maintains a constant speed and constant power state.

The AC/DC/AC low-voltage module of the novel wind power generation system discussed in this section is supplied by multiple isolated windings of a PMSG. When the system is operating in MPPT, the generator speed varies in proportion to the wind speed, resulting in the magnitude of the inductivce motive force of each winding of the PMSG also varying in proportion to the wind speed, thus the voltage at the generator terminal is also in constant change. At the same time, the generator speed in a direct-drive wind power system is generally low, resulting in a very low voltage at the generator terminal, so that the diode rectifier circuit alone cannot obtain a sufficiently high DC-bus voltage. Therefore, a suitable topology and control strategy is needed to stabilise the DC-bus voltage and achieve a voltage boost. In addition, the rectifier circuit should also have the function of improving the current waveform, increasing the power factor and reducing generator losses. In order to reduce the number of individual generator windings required, a single-phase rectifier circuit topology can be considered here [7].

The main topologies of a single-phase rectifier are full-bridge, half-bridge, single-tube PFC, and bridgeless PFC circuits, which have their own characteristics [8]. Taking into account the characteristics of wind power applications: power flows in one direction (from the generator side to the grid side) and the generator operates at unit power factor to reduce losses; also, in order to reduce the number of switching devices, the single-switch PFC structure is finally chosen in this section.

The single-switch PFC structure is more commonly used in traditional switching power supply circuits, usually in the form of an "uncontrolled rectifier + boost chopper", as shown in Fig. 8.24a [9]. In wind power generation, although this circuit can achieve the purpose of boost and unit power factor operation, a DC storage inductor is required on the DC side, which is used to store and release magnetic field energy when the boost chopper is in operation. As a result of the unipolar pulsating current flowing in the inductor, its bias magnetisation is very serious and it is very easy

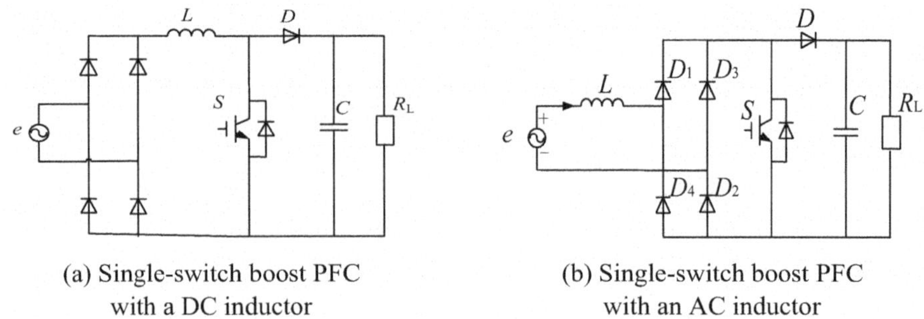

Fig. 8.24 Single-switch boost PFC circuits

to be saturated, so a large air-gap or even a air-core inductor has to be used, making the inductor large and costly. This is particularly problematic for wind turbines with higher power, where the inductor current becomes correspondingly larger.

Here the armature inductor of the generator is used as the AC-side energy storage inductor instead of an additional DC-side inductor, thus saving the size and cost of the system, as shown in Fig. 8.24b. This topology is also capable of boost and unit power factor operation, and the operating principle of this circuit is analysed below.

8.2.1.3 Mathematical Model and Parameter Selection of the Generator-Side Rectifier in a Low-Voltage Module

1. Mathematical Model of the Grid-side Rectifier

The structure of a single-phase, single-switch PFC circuit using the generator armature inductor is shown in Fig. 8.25a. In order to use the armature inductor to replace the DC inductor in a conventional boost chopper, it is first necessary to analyse the energy storage process of the AC inductor. In the structure shown in Fig. 8.25a, *e* and *L* represent the inductive motive force and armature inductance of the individual windings of the PMSG respectively, D_1–D_4 consist of the single-phase full-wave uncontrolled rectifier bridge, S is the chopper switch, *C* is the filter capacitor and R_L represents the equivalent DC load (in this case the actual load is the H-bridge unit). When current $i \geq 0$ is applied, D_1 and D_2 are on and the circuit topology in operation is shown in Fig. 8.25b; when current $i < 0$ is applied, D_3 and D_4 are on and the circuit topology in operation is shown in Fig. 8.25c. Thus, the circuit topology can be equated with Fig. 8.25d for the entire period of the inductive motive force e. It has the same function as a conventional PFC circuit using a DC energy storage inductor, but the energy storage inductor *L* flows AC current, so it is possible to unify the energy storage inductor in the rectifier with the armature inductor of the generator, resulting in a reduced size and weight and lower cost of the rectifier.

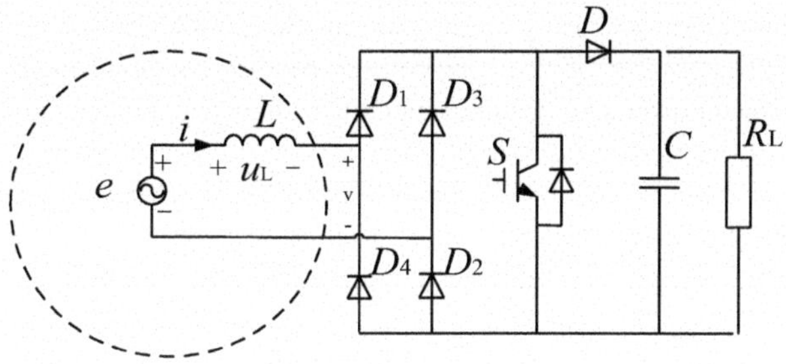

(a) Circuit Structure of the Grid-side Rectifier

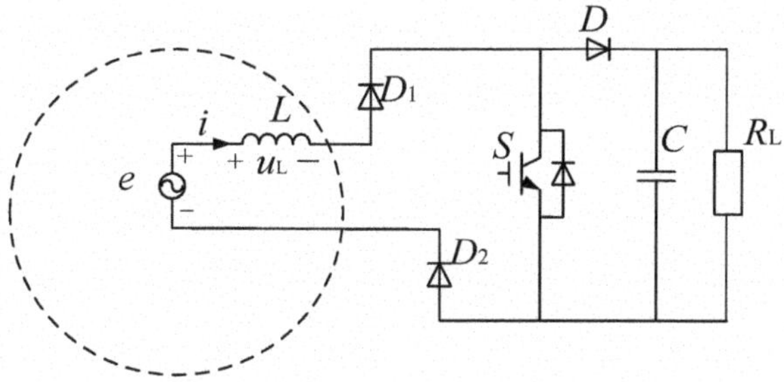

(b) Equivalent Circuit with a Positive Current

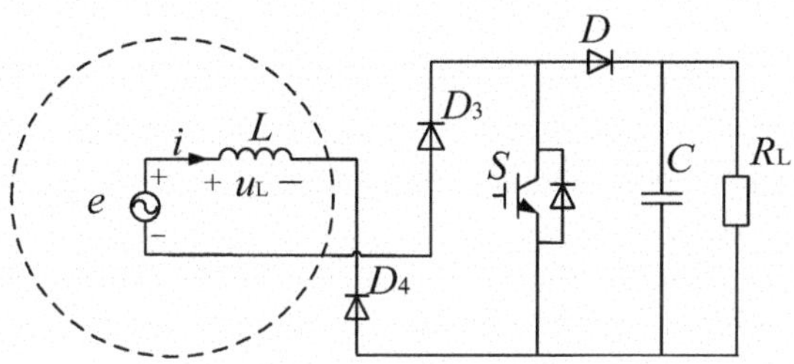

(c) Equivalent Circuit with a Negative Current

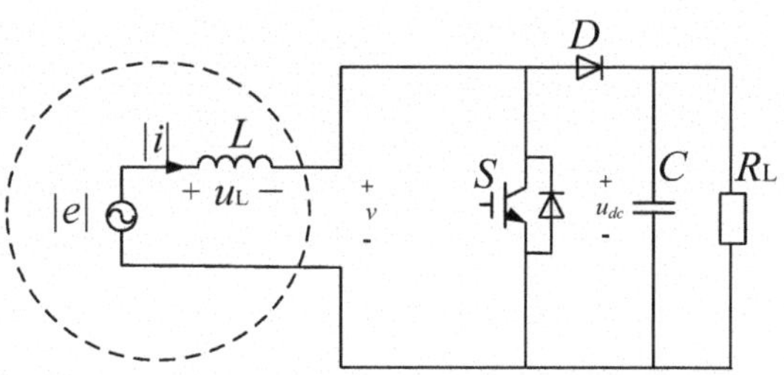

(d) Equivalent Circuit of the Grid-side Rectifier

Fig. 8.25 PFC circuit using the generator armature inductor

Based on the above operating conditions of the circuit, the following mathematical model of the rectifier of the low-voltage module is developed and the corresponding control method and parameters of the energy storage elements (inductors and capacitors) of the circuit are derived.

According to Fig. 8.25d, let T_c be the switching period, f_c be the switching frequency and k be the duty cycle.

When $0 \le t \le kT_c$, S is on, and the voltage applied to the inductor L is

$$u_L = L\frac{\mathrm{d}|i|}{\mathrm{d}t} = |e| \tag{8.18}$$

The current rising ripple is

$$\Delta|i| = \frac{|e|}{L}kT_c \tag{8.19}$$

When $kT_c < t \le T_c$, S is off, and the voltage applied to the inductor L is

$$u_L = L\frac{\mathrm{d}|i|}{\mathrm{d}t} = |e| - u_{dc} \tag{8.20}$$

The current falling ripple is

$$\Delta|i| = \frac{u_{dc} - |e|}{L}(1-k)T_c \tag{8.21}$$

The expression of duty cycle k in Eq. (8.21) contains both fundamental and harmonic components, and is a rather complex expression related to factors such as inductive motive force, inductance, winding current, DC-bus voltage and switching frequency when the circuit is operating. For ease of calculation, the expression for duty cycle k is simplified by approximating the current rising ripple as equal to the current falling ripple in a chopping cycle, and there is

$$\frac{|e|}{L}kT_c = \frac{u_{dc} - |e|}{L}(1-k)T_c \tag{8.22}$$

Therefore, the duty cycle of chopping PWM can be written approximately as

$$d_{c1} = k = \frac{u_{dc} - |e|}{u_{dc}} \tag{8.23}$$

2. Selection of Induction Parameter

In the rectifier circuit shown in Fig. 8.25, the energy storage inductor L and the filter capacitor C are two very important components. In order to fulfil the function of the circuit, their parameter values must meet a certain range, and this section begins with the analysis of the inductor parameters, which mainly determine the current ripple and the current over-zero distortion in the single-switch PFC circuit itself.

The minimum value of the inductance is determined by the current ripple: the higher the armature inductance of the generator, the lower the current ripple. Combining Eqs. (8.19) and (8.23), the current ripple can be approximated as

$$\Delta i = \frac{e}{L} kT_{\mathrm{c}} = \frac{e}{L} \frac{u_{\mathrm{dc}} - e}{u_{\mathrm{dc}}} T_{\mathrm{c}} \tag{8.24}$$

When and only when $e = u_{\mathrm{dc}} - e$, i.e. $k = 0.5$, the current ripple achieves its maximum value as

$$\Delta i_{\max} = \frac{u_{\mathrm{dc}}}{4L} T_{\mathrm{c}} = \frac{u_{\mathrm{dc}}}{4Lf_{\mathrm{c}}} \tag{8.25}$$

If it is required that $\Delta i_{\max}$ does not exceed δ times the peak current I_m, the energy storage induction satisfies

$$L \geq \frac{u_{\mathrm{dc}}}{4\delta I_{\mathrm{m}} f_{\mathrm{c}}} \tag{8.26}$$

Therefore, Eq. (8.26) is the minimum inductance value derived from the current ripple limit.

The maximum value of the energy storage inductance parameter is determined by the current over-zero distortion angle: the larger the inductance value, the larger the current over-zero distortion angle. The topology of the single-switch PFC circuit determines that there is a current over-zero distortion problem, as can be seen in Fig. 8.25, where the direction (polarity) of the rectifier output voltage is determined by the direction of the current. If the current direction is positive, then the rectifier can only output a positive voltage. This positive voltage and the inductive electromotive force of the generator acting on the armature inductor determine the waveform of the current. If the current direction is negative, then the rectifier can only output a negative voltage. If the generator armature current and electromotive force are kept in phase (unit power factor), then the vector diagram on the rectifier side is shown in Fig. 8.26, where E is the electromotive force of the generator and U is the output voltage of the rectifier. Figure 8.27 gives a sketch of the over-zero distortion. Ideally, the relation between the rectifier output voltage and the winding current is as shown in Fig. 8.27, but the direction of the output voltage of the single-switch PFC circuit is governed by the direction of the current. For example, in Fig. 8.27, when the over-zero current changes from positive to negative, the output voltage u of the rectifier is still required to be positive, but as the current becomes negative, the circuit topology changes to the situation in Fig. 8.25c, where the voltage u can only be negative, and so causes an over-zero distortion in the current waveform, and the same distortion occurs at the moment when the over-zero current changes from negative to positive.

Figure 8.28 shows the simulation waveform for the current over-zero distortion, where the angle of the over-zero distortion is defined as α. The relation between

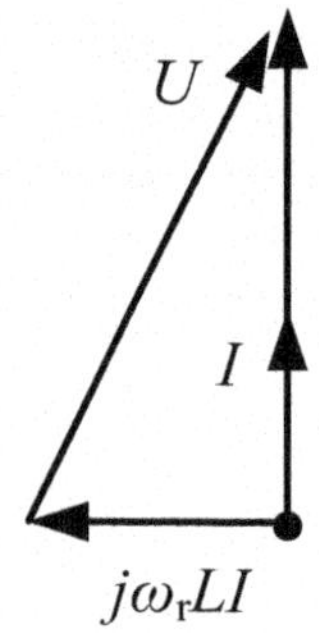

Fig. 8.26 Vector diagram of single-switch PFC circuit

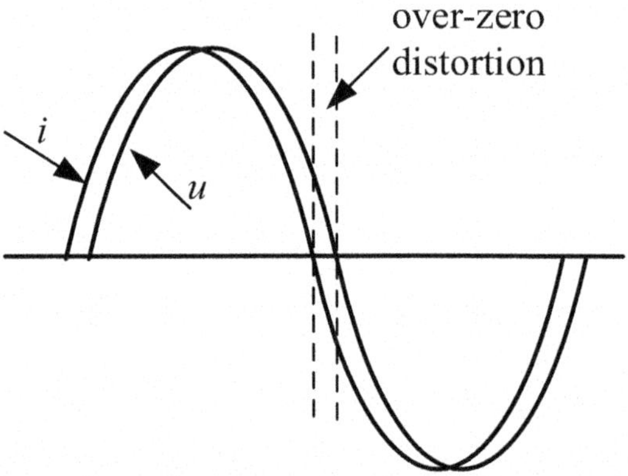

Fig. 8.27 Current over-zero distortion in PFC circuit

the over-zero distortion and the armature inductance of the generator is shown in Eq. (8.27).

$$\alpha = 2arctg\left(\frac{\omega_r L I_m}{E_m}\right) \tag{8.27}$$

where ω_r is the stator angular frequency of the generator, E_m and I_m are the amplitude of the electromotive force and current in the generator winding respectively, and L is the armature inductance. Equation (8.27) shows that, with the same values of ω_r, E_m and I_m, the current over-zero distortion angle will increase with the increase of inductance L, which in turn will cause the power factor to drop, so according to the system requirements for the over-zero distortion angle limit, the maximum value of inductance L can be determined as shown in Eq. (8.28).

$$L \leq \frac{E_m}{\omega_r I_m} tg \frac{\alpha_{cr}}{2} \tag{8.28}$$

In selecting the generator inductance according to the principles given in Eqs. (8.26) and (8.28), the range of inductance can first be determined from the

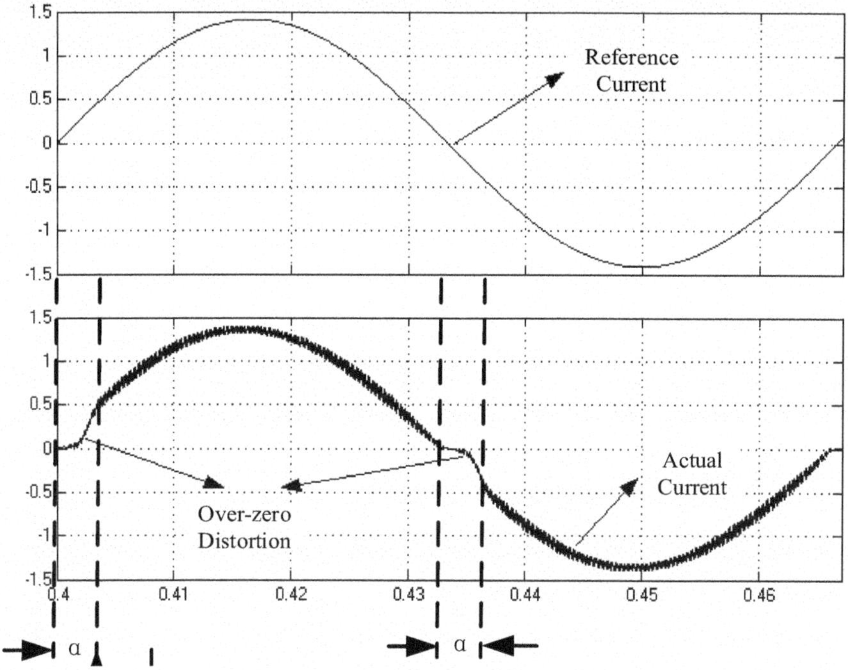

Fig. 8.28 Simulation for current over-zero distortion in PFC circuit

electromotive force, current and angular frequency of the winding in the rated state. In actual wind turbine operation with variable speed, below the rated wind speed, the system is operated in MPPT and the power absorbed by the wind turbine is proportional to the third power of the rotation speed. Since a varies proportionally to the speed ag, a varies proportionally to the square of the speed. From Eq. (8.27), in the rated state, if the selected inductance can make the over-zero distortion angle meet the requirements, when the generator speed decreases from the rated speed, the distortion angle will drop at a faster rate and will still meet the requirements; from Eq. (8.26), the interruption occurs when the current drops to δI_m.

In addition to the above two factors, the requirements of the generator short-circuit current on the armature inductance should also be taken into account when carrying out the generator inductance design.

3. Selection of Capacitance Parameter

The main function of the DC-bus capacitor is to filter out the voltage ripple on the DC bus. From the rectifier side, when the generator winding current is in phase with the electromotive force, the power fed to the DC bus has an AC component in addition to the DC component because it is a single-phase rectifier circuit. The frequency of the AC component is two times the stator frequency of the generator and requires the DC-bus capacitor to suppress the DC-bus voltage fluctuations caused by it. Let the fundamental instantaneous values of the electromotive force and armature current be as follows.

$$e = E_m \sin(\omega_r t), i = I_m \sin(\omega_r t) \quad (8.29)$$

The power input from the generator side to the DC bus can be expressed as

$$\begin{aligned} p = ei &= E_m \sin(\omega_r t) \cdot I_m \sin(\omega_r t) \\ &= \frac{E_m I_m}{2} - \frac{E_m I_m}{2} \cos(2\omega_r t) \end{aligned} \quad (8.30)$$

The first term of the expression for power in Eq. (8.30) is the direct flow power and the second term is the pulsating power with two times the frequency of the generator stator. Assuming that the DC-bus voltage U_{dc} is constant (its ripple is controlled within a certain range), then the expression for the current flowing to the DC-bus capacitor and load side is

$$i_o = \frac{p}{U_{dc}} = \frac{E_m I_m}{2U_{dc}} - \frac{E_m I_m}{2U_{dc}} \cos(2\omega_r t) \quad (8.31)$$

The second term in Eq. (8.31) is the pulsating current and this current will cause the capacitor voltage to fluctuate with a fluctuating value of

$$\begin{aligned} \Delta u &= -\frac{1}{C} \int_0^t \frac{E_m I_m}{2U_{dc}} \cos(2\omega_r \varphi) d\varphi \\ &= -\frac{P_{avg}}{2\omega_r C U_{dc}} \sin(2\omega_r t) \end{aligned} \quad (8.32)$$

In Eq. (8.32), $P_{avg} = E_m I_m / 2$ is the average power. It can be seen that the frequency of the DC-bus voltage fluctuation is two times the generator stator frequency, proportional to the average power P_{avg} and inversely proportional to the angular frequency, the filter capacitance and the DC-bus voltage. If the amplitude of the DC-bus voltage ripple is required to be no more than U_{dc} times the average voltage σ, then the filter capacitance should satisfy:

$$C \geq \frac{P_{avg}}{2\omega_r \sigma U_{dc}^2} \quad (8.33)$$

It is important to note here that due to the variable speed operation of the generator, which usually has a low stator frequency ω_r, according to Eq. (8.33), a larger capacitor is required to filter out the voltage fluctuations on the DC bus, which inevitably increases the size and cost of the system, and the lifetime of the capacitor will also affect the reliability of the converter. A method of suppressing power pulsations by parallelly connecting DC sides of two rectifiers connected to the windings with an electrical phase difference of 90° is given below.

4. Suppression of Rectifier-side Power Pulsation

Both the rectifier and inverter sides of the low-voltage module are single-phase circuits, so their power is pulsating at two times the fundamental frequency. Compared to the grid frequency (50 or 60 Hz), the frequency of the generator stator is generally lower, so the DC-bus filter capacitance required is also larger. The power pulsations on the generator side are eliminated below by winding in parallel.

The output power of the single-phase winding and its rectifier is shown in Eq. (8.30). If two windings with a 90° phase difference are rectified and then connected in parallel, the expression for the power fed to the DC bus at this point is shown in Eq. (8.34).

$$\begin{aligned} p_{\text{dc-link}}(t) &= E_{\text{m}} \sin(\omega_{\text{r}} t) I_m \sin(\omega_{\text{r}} t) + E_{\text{m}} \sin\left(\omega_{\text{r}} t + \frac{\pi}{2}\right) I_m \sin\left(\omega_{\text{r}} t + \frac{\pi}{2}\right) \\ &= E_{\text{m}} I_m \end{aligned} \tag{8.34}$$

Equation (8.34) shows that, by connecting the windings and their rectifiers in parallel with each other with a phase difference of 90°, the pulsation in Eq. (8.30) of two times the stator frequency can be eliminated, leaving only the DC component, which achieves a constant power fed from the generator to the DC bus, and therefore can greatly reduce the filter capacitance on the DC bus. Equation (8.34) can be understood in this way: the output pulsating power of the two quadrature windings, which differ in phase by 90° from each other, after rectification, differs by 180° from each other, and at this point is connected in parallel to eliminate the pulsating component of the power. The topology of the rectifier after parallel connection of the quadrature windings is shown in Fig. 8.24b.

8.2.1.4 Modeling of the H-Bridge Inverter in a Low-Voltage Module

The part of the low-voltage module that is connected to the grid is the H-bridge inverter, which will be modelled and analysed below. In the H-bridge inverter structure shown in Fig. 8.29, let the switching function of the switch S_1 be s_1, which is defined as follows:

$$s_1 = \begin{cases} 1, & S_1 \text{ or its reverse-paralleled diode is on} \\ 0, & S_3 \text{ or its reverse-paralleled diode is on} \end{cases} \tag{8.35}$$

The switching functions s_2, s_3 and s_4 of each of the other switches are similarly defined, and the switching functions of the two switches on the same bridge arm are complementary.

The input-output relation of the H-bridge described by the switching function is

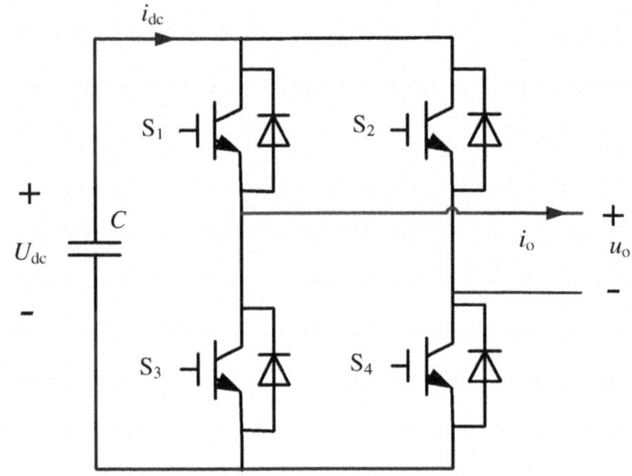

Fig. 8.29 Structure of H-bridge inverter

$$\begin{cases} u_o = (s_1 - s_2)U_{dc} \\ i_{dc} = (s_1 - s_2)i_o \end{cases} \tag{8.36}$$

Define the switching function as

$$s = s_1 - s_2 \tag{8.37}$$

Then Eq. (8.36) can be rewritten as

$$\begin{cases} u_o = sU_{dc} \\ i_{dc} = si_o \end{cases} \tag{8.38}$$

Define the average switching function over a switching cycle as follows.

$$d = \overline{S(t)} = \frac{1}{T}\int_{t-T}^{t} S(\varphi)\mathrm{d}\varphi \tag{8.39}$$

Since the switching frequency of the switches in each arm of the H-bridge is much greater than the frequencies of the output-side fundamental voltage and current and the DC-bus voltage fluctuations, it can be assumed that the DC-bus voltage and output current remain constant during a switching cycle, and then Eq. (8.38) can be simplified to an average model as shown in Eq. (8.40):

$$\begin{cases} \overline{u_o} = \frac{1}{T}\int_{t-T}^{T} s(\varphi)U_{dc}\mathrm{d}\varphi = dU_{dc} \\ \overline{i_{dc}} = \frac{1}{T}\int_{t-T}^{T} s(\varphi)i_o\mathrm{d}\varphi = di_o \end{cases} \tag{8.40}$$

The fundamental wave is the dominant component in the switching function represented by Eq. (8.37). If only the fundamental wave in the switching function is considered, the duty cycle in Eq. (8.39) can be expressed directly as

$$d = K_1 \cos(\omega_s t + \theta_1) \tag{8.41}$$

where ω_1 is the output (grid) angular frequency of the H-bridge inverter, let the fundamental expression for the output current in Fig. 8.29 be

$$i_o = I_{om} \sin(\omega_1 t + \varphi_1) \tag{8.42}$$

where I_{om} and φ_1 denote the amplitude and phase angle of the output current fundamental wave, respectively. Substituting Eqs. (8.41) and (8.42) into Eq. (8.40), it gives

$$\begin{aligned} \overline{i_{dc}} &= K_1 \sin(\omega_1 t + \theta_1) I_{om} \sin(\omega_1 t + \varphi_1) \\ &= \frac{K_1 I_{om}}{2} \cos(\theta_1 - \varphi_1) - \frac{K_1 I_{om}}{2} \cos(2\omega_1 t + \theta_1 + \varphi_1) \end{aligned} \tag{8.43}$$

In Eq. (8.43), the first term of the DC-bus current is the direct flow component; the second term is the pulsating component with two times the frequency of the grid. Since the H-bridge inverters are cascaded and connected to the grid, a voltage pulsation of 100 Hz is generated on the dc bus for a 50 Hz frequency grid. This 2x grid-frequency power pulsation in the single-phase circuit is generated by the same mechanism as the single-phase power pulsation on the generator side, except that here the fundamental frequency is the grid frequency, which is usually higher than the stator frequency of the generator and can be filtered out using the relatively small capacitor.

The DC-bus voltage can be calculated according to Eq. (8.43) as

$$\begin{aligned} u_{dc} &= \frac{1}{C} \int_0^t \frac{K_1 I_{om}}{2} (\cos(\theta_1 - \varphi_1) - \cos(2\omega_1 t + \theta_1 + \varphi_1))\, dt \\ &= U_{dc} - \frac{K_1 I_{om}}{4\omega_1 C} \sin(2\omega_1 t + \theta_1 + \varphi_1) \end{aligned} \tag{8.44}$$

If the amplitude of the DC-bus voltage fluctuations due to the H-bridge inverter circuit is required to be no more than ρ times the average DC-bus voltage U_{dc}, then the value of the DC-bus capacitance needs to satisfy:

$$C \geq \frac{K_1 I_{om}}{4\omega_1 \rho U_{dc}} \tag{8.45}$$

Since the rectifier side suppresses the power pulsation by means of parallel connection of the quadrature windings, the DC-bus capacitor can be selected only for the

100 Hz pulsation on the grid side shown in Eq. (8.45). In addition, the selection of capacitors needs to take into account the noise of the switching frequency.

8.2.1.5 Mathematical Model of the Multi-phase PMSG

The topology discussed in this section uses a multi-pole, multi-phase PMSG. The multi-phase generator makes uninterrupted operation under fault conditions possible, so that when one phase of the generator or one module of the converter fails, the rest of the system can operate normally. At the same time, the multi-phase generator also has the advantage of lower torque harmonics. Based on the derivation in this section "Suppression of Rectifier-side Power Pulsation", it is desirable that the generator has multiple pairs of quadrature windings to eliminate the power pulsation from the generator side. At the same time, the armature inductance of the generator has a significant impact on the performance of the system, so the generator needs to be modelled and analysed.

For a multi-phase PMSG with m-phase balanced windings, let the space harmonic number of its magnetic motive force be v, with the following relation:

$$v = km + 1 \tag{8.46}$$

where k is a positive or negative integer including zero, and m is the actual number of phases of the generator. When the number of phases of the generator is m, the windings can be connected in two ways: One is that the windings are connected with an electrical angle of $360/m$ between neighbouring phases, i.e. the windings are evenly distributed in space, for a 6-phase generator, the electrical angle between the neighbouring windings is $360/6 = 60°$ and the number of the magnetic motive force harmonics is $v = km+1 = 6\ m+1$. The other one is that the windings are connected with an electrical angle of $180/m$, i.e. the windings are not evenly distributed in space, for a 6-phase generator, the electrical angle between the neighbouring windings is $180/6 = 30°$ and the number of the magnetic motive force harmonics is $v = 2\ km + 1 = 12\ m + 1$, which doubles the number of harmonics and is equivalent to the harmonic performance of a generator with 12 evenly distributed phases [10]. The winding connection with a $180/m$ electrical angle between neighbouring phases is essentially equivalent to doubling the number of phases, with a reduction in the number of harmonics and good generator performance.

For the multi-phase PMSG studied in this section for wind power generation, it is desired to have multiple pairs of windings differing by 90 electrical degrees from each other, which creates certain requirements on the number of phases of the generator.

Taking a 6-phase PMSG as an example, quadrature windings can occur whenever there is a difference of 30° between the neighbouring windings, as shown in Fig. 9.30. For the $180/m$ connection, any generator with an integer multiple of 6 phases can have quadrature windings, e.g. 12-phase, 18-phase, 24-phase and 30-phase generators.

For the 360/*m* connection, the windings are evenly distributed in space, and in order to form quadrature windings, the number of generator phases is required to be a multiple of 4. Then, according to the requirements of the H-bridge cascade side (3 phases), the number of generator phases should be a multiple of 6, so the number of generator phases should be a multiple of 12, the lowest common multiple of 4 and 6.

In order to analyse the variation of the generator electromotive force with speed and the interrelationship of the flux linkages between the windings, the following mathematical model of the multi-phase PMSG in the stationary coordinate frame is given:

Here the m-phase PMSG is considered to be an ideal generator, i.e. it satisfies the following assumptions [10]:

(1) The effects of saturation, eddy currents, magnetic hysteresis and skin effects are neglected.
(2) The inductive electromotive force and the air-gap magnetic field are distributed sinusoidally, and the harmonics of the magnetic field are not taken into account.
(3) The permanent magnet has a constant magnetic motive force, i.e. the equivalent excitation current is constant.
(4) Three-phase stator windings are distributed in space in a symmetrical star style.
(5) There is no damping winding on the rotor.
(6) Effects of frequency and temperature on generator parameters are not taken into account.

Under the above premises, the equations for the stator-side flux linkage and the voltage of the PMSG in the stationary coordinate frame can be expressed as

$$\psi_{\mathrm{m}} = L_{\mathrm{ss}} i_{\mathrm{m}} + \psi_{\mathrm{r}} \cdot F_{\mathrm{m}}(\theta) \tag{8.47}$$

$$u_{\mathrm{m}} = R_{\mathrm{m}} \cdot i_{\mathrm{m}} + p\psi_{\mathrm{m}} \tag{8.48}$$

where

$$\psi_{\mathrm{m}} = \begin{bmatrix} \psi_{\mathrm{A}} \\ \psi_{\mathrm{B}} \\ \psi_{\mathrm{C}} \\ \vdots \\ \psi_{\mathrm{m}} \end{bmatrix}, \ i_{\mathrm{m}} = \begin{bmatrix} i_{\mathrm{A}} \\ i_{\mathrm{B}} \\ i_{\mathrm{C}} \\ \vdots \\ i_{\mathrm{m}} \end{bmatrix}, \ u_{\mathrm{m}} = \begin{bmatrix} u_{\mathrm{A}} \\ u_{\mathrm{B}} \\ u_{\mathrm{C}} \\ \vdots \\ u_{\mathrm{m}} \end{bmatrix}, \ F_{\mathrm{m}}(\theta) = \begin{bmatrix} \sin\theta \\ \sin(\theta + \frac{1}{m} \cdot 2\pi) \\ \sin(\theta + \frac{2}{m} \cdot 2\pi) \\ \vdots \\ \sin(\theta + \frac{m-1}{m} \cdot 2\pi) \end{bmatrix}$$

$$R_{\rm m} = \begin{bmatrix} R & & & & \\ & R & & & \\ & & R & & \\ & & & \ddots & \\ & & & & R \end{bmatrix}_{m\times m}$$

$$L_{\rm ss} = L_{\rm ms}\begin{bmatrix} 1 & \cos(\frac{1}{m}2\pi) & \cos(\frac{2}{m}2\pi) & \cdots & \cos(\frac{m-1}{m}2\pi) \\ \cos(\frac{1}{m}2\pi) & 1 & \cos(\frac{1}{m}2\pi) & \cdots & \cos(\frac{m-2}{m}2\pi) \\ \cos(\frac{2}{m}2\pi) & \cos(\frac{1}{m}2\pi) & 1 & \cdots & \cos(\frac{m-3}{m}2\pi) \\ \vdots & \vdots & \vdots & \ddots & \vdots \\ \cos(\frac{m-1}{m}2\pi) & \cos(\frac{m-2}{m}2\pi) & \cos(\frac{m-3}{m}2\pi) & \cdots & 1 \end{bmatrix} + L_{\rm ls}\begin{bmatrix} 1 & & & & \\ & 1 & & & \\ & & 1 & & \\ & & & \ddots & \\ & & & & 1 \end{bmatrix}$$

where $\psi_{\rm m}$ indicates the stator flux linkage of the m-phase PMSG, $i_{\rm m}$ is the stator current, $\psi_{\rm r}$ is the rotor permanent magnet flux linkage, $L_{\rm ss}$ is the inductance matrix, including self and mutual inductances, $L_{\rm ms}$ and $L_{\rm ls}$ are the mutual and leakage inductances of the stator winding of any phase, um indicates the generator terminal voltage, $R_{\rm m}$ is the stator resistance, and $F_{\rm m}(\theta)$ indicates the distribution of the stator winding in space, which can be distributed according to 360/m connection or 180/m connection. For different generator winding distributions in space, it is sufficient to determine the expression for $F_{\rm m}(\theta)$ according to the actual winding space location.

Equation (8.47) shows that the flux linkage of each phase of the stator winding consists of the self-induced flux linkage of that phase of the stator winding (including the leakage flux linkage), the mutual induced flux linkage from the other phases of the stator winding, and the mutual induced flux linkage from the rotor winding to the stator winding. The space position of the mutual induced flux linkage between the stator windings is relatively fixed, while the rotor-to-stator mutual induced flux linkage is related to the rotor position.

Equation (8.48) represents the stator voltage equation for a PMSG, where the electromotive force is expressed as $p\psi_m$, which includes the electromotive force induced in the stator winding by the rotor permanent magnet.

$$E = \psi_{\rm r} p F_{\rm m}(\theta) \tag{8.49}$$

Taking the first phase in the expression $F_{\rm m}(\theta)$ as an example, i.e. Phase A in the stator winding, the induced electromotive force generated by the rotor permanent magnet in this phase is

$$E = \psi_{\mathrm{r}} p \sin\theta = \psi_{\mathrm{r}} \frac{d \sin\theta}{dt} = \psi_{\mathrm{r}} \frac{d \sin\omega_1 t}{dt} = \psi_{\mathrm{r}} \omega_1 \cos\omega_1 t \tag{8.50}$$

It can be seen that the amplitude of the fundamental electromotive force E generated by the permanent magnet is proportional to the generator speed ω_{r}. In wind power generation, the electromotive force E is always changing because the generator is in variable speed operation, and in order to obtain a stable DC-bus voltage, a boost circuit is required to boost the electromotive force and extend the speed operation range of the generator. The mutual inductance matrix in the generator model can be used to analyse the effect of a faulty low-voltage module on other phases.

For the control of the multi-phase PMSG, a generalised coordinate transformation (generalised park transformation) based on the characteristics of the spatial distribution of the windings can be considered to convert the generator into the two-phase rotating coordinate frame (dq-frame), where the torque and excitation current can be decoupled and controlled by two PI regulators. The adoption of this approach presupposes that the parameter profile of each winding (module) is identical. However, the topology studied in this section has the following two characteristics: firstly, the modular structure requires that the control of each module is relatively independent, so that when a fault occurs in one module, it does not affect the work of the other modules as much as possible; secondly, during the system dynamics, the DC-bus voltages of different modules may be different, and if centralised control based on the generalised park transformation is used, the DC-bus voltage of a specific module cannot be controlled individually. Therefore, this section uses a control method based on the module in Fig. 8.24b, i.e. a set of controllers for each pair of quadrature windings with an outer loop for the DC-bus voltage and an inner loop for the generator-winding current, respectively, and the control method for the rectifier in a low-voltage module is given below.

8.2.2 *Control of the Transformerless Medium/High-Voltage PMSG-Based Wind Power System*

8.2.2.1 Control of the Generator-Side Rectifier in a Low-Voltage Module

This section will discuss the rectifier control method in Fig. 8.24b, the equivalent circuit diagram for the PFC circuit is shown in Fig. 8.31.

Where E is the electromotive force of the generator and U is the output voltage of the rectifier, these two voltage sources act on the armature inductor of the generator to determine the magnitude and direction of the armature current of the generator. The control strategy employed for this circuit is the average current control method. As the energy storage inductor is transferred from the DC side to the generator interior, some adjustments need to be made to the traditional average current control method for the DC-side inductor, and the adjusted control diagram is shown in Fig. 8.32. The control system has an outer DC-bus voltage loop and an inner current loop. The outer voltage loop outputs the current amplitude reference of the current loop through a PI regulator, which is then multiplied by the phase information of the electromotive force of the generator to obtain the reference value of the instantaneous current to achieve unit power factor operation of the generator. The current loop also uses a PI regulator to obtain the control signal for the main switch device. It should be noted here that the phase information of the generator electromovie force is not directly measurable in the topology proposed in this section. The rotor position of the generator can be obtained from the signal of the encoder mounted on the rotor of the PMSG, which is then reconstructed according to the spatial distribution of the generator stator winding to obtain the phase information of the winding electromovie force. Since the two windings are in quadrature, the phase of the second winding can be obtained by shifting the phase of the first winding by 90° (Fig. 8.33).

8.2.2.2 Control of Grid-Side Cascaded H-Bridge Inverter

The grid-side cascaded H-bridge inverter is responsible for transferring the power captured by the wind turbine to the grid. Before modelling the system and designing the control method, the control strategy of the entire wind turbine system is first analysed in order to determine how the grid-side inverter will work.

The whole wind power converter system works as follows: the grid-side inverter gets the active power reference value according to the current wind speed and the MPPT requirements, and the reactive power reference value according to the grid requirements for reactive power, and feeds the active and reactive power to the grid through vector control. At the same time, the rectifier on the generator side in Fig. 8.32 regulates the output power of the generator by detecting the DC-bus voltage of the corresponding low-voltage module, tracking the grid-side power reference and achieving the MPPT. At system start-up, when the wind speed reaches the cut-in wind speed, the generator starts to rotate and the generator-side rectifier is put into operation to control the DC-bus voltage to the reference value. The grid-side contactor is then closed. As the equivalent DC-bus voltage after the cascaded H-bridge is higher than the grid voltage, closing the grid-side contactor does not cause current shocks. The cascaded H-bridge inverter is then started to feed power into the grid. Based on the above functional requirements, the following mathematical model and control method for the grid-side cascaded H-bridge inverter is developed.

1. Mathematical Model of the Cascaded H-bridge Inverter

A mathematical model of an individual H-bridge inverter has been developed in Sect. 8.2.1.4, on the basis of which a cascaded H-bridge inverter can be modelled with a unified control method. Here it is assumed that the grid is a three-phase symmetrical system with a linear grid-side filter inductor and that the effects of on-state voltage drop of switches, switching losses and distributed parameters are neglected.

Equation (8.40) describes an average model of the output voltage u_0 and output current i_o of a H-bridge inverter cell. For an N-cell cascaded system, the output voltage can be expressed as the sum of the output voltages of the N cells:

$$\begin{cases} \overline{u}_{\text{opj}} = d_{\text{pj}} \cdot u_{\text{dc\_pj}} \\ \overline{u}_{\text{op}} = \sum\limits_{j=1}^{N} d_{\text{pj}} \cdot u_{\text{dc\_pj}} \end{cases} \tag{8.51}$$

Where the subscript p represents the Phase p (p = a, b, c), u_{op} is the total output voltage, d_{pj} and $u_{\text{dc\_pj}}$ are the duty cycle and dc-bus voltage of the jth H-bridge inverter respectively. The dc-side output current of the H-bridge inverter is

$$\overline{i}_{\text{dc\_pj}} = d_{\text{pj}} \cdot i_{\text{opj}} \tag{8.52}$$

where $i_{\text{dc\_pj}}$ and i_{opj} are the average DC-bus current and output current of the jth H-bridge inverter. To further obtain the simplified model, the following assumptions are made.

Firstly, as the PFC rectifiers are independent to each other and their function is to stabilise the corresponding DC-bus voltage, the output of a PFC rectifier can be equated to a DC-voltage supply and all DC-bus voltages are considered to be equal in value, i.e.

$$U_{\text{dc}} = u_{\text{dc\_pj}} = u_{\text{dc\_aj}} = u_{\text{dc\_bj}} = u_{\text{dc\_cj}}, \text{j} = 1, 2 \ldots N \tag{8.53}$$

Secondly, for the cascaded H-bridge inverter, this section uses the phase-shifted carrier PWM technique to modulate it so that if the inverter switching frequency is much greater than the grid frequency, the duty cycles of the switching functions of the H-bridge inverters are approximately equal:

$$d_{\text{p}} = d_{\text{p1}} = d_{\text{p2}} = \cdots = d_{\text{pN}}, \text{p} = \text{a, b, c} \tag{8.54}$$

Based on the above two analyses and according to Eq. (8.49), the output voltage per phase of the system can be obtained as

$$u_{\text{op}} = Nd_{\text{p}}U_{\text{dc}}, \quad \text{p} = \text{a, b, c} \tag{8.55}$$

The simplified equivalent circuit of the grid-side cascaded H-bridge inverter is shown in Fig. 8.34.

According to Fig. 8.34, the voltage equation can be written as

$$\begin{cases} u_{\mathrm{oa}} = d_{\mathrm{a}} \cdot NU_{\mathrm{dc}} = R_{\mathrm{s}} i_{\mathrm{sa}} + L_{\mathrm{s}} \dfrac{\mathrm{d}i_{\mathrm{sa}}}{\mathrm{d}t} + u_{\mathrm{sa}} - u_{\mathrm{ng}} \\ u_{\mathrm{ob}} = d_{\mathrm{b}} \cdot NU_{\mathrm{dc}} = R_{\mathrm{s}} i_{\mathrm{sb}} + L_{\mathrm{s}} \dfrac{\mathrm{d}i_{\mathrm{sb}}}{\mathrm{d}t} + u_{\mathrm{sb}} - u_{\mathrm{ng}} \\ u_{\mathrm{oc}} = d_{\mathrm{c}} \cdot NU_{\mathrm{dc}} = R_{\mathrm{s}} i_{\mathrm{sc}} + L_{\mathrm{s}} \dfrac{\mathrm{d}i_{\mathrm{sc}}}{\mathrm{d}t} + u_{\mathrm{sc}} - u_{\mathrm{ng}} \end{cases} \tag{8.56}$$

The instantaneous active and reactive power output from the inverter are

$$\begin{cases} P = \sum\limits_{\mathrm{k=a,b,c}} u_{\mathrm{sk}} i_{\mathrm{sk}} \\ Q = \frac{\sqrt{3}}{3}[(u_{\mathrm{sa}} - u_{\mathrm{sb}}) i_{\mathrm{sc}} + (u_{\mathrm{sb}} - u_{\mathrm{sc}}) i_{\mathrm{sa}} + (u_{\mathrm{sc}} - u_{\mathrm{sa}}) i_{\mathrm{sb}}] \end{cases} \tag{8.57}$$

The mathematical model, which has been developed above, based on the three-phase stationary coordinate frame is clear and intuitive, but the physical quantity to be controlled (current) is an AC quantity. To facilitate controller design, the mathematical model can be converted to the synchronous rotational coordinate frame by means of the synchronous rotational coordinate transformation, which is given below for the grid-side cascaded H-bridge inverter in two-phase synchronous rotational coordinate frame (dq-frame).

The mathematical model of the cascaded H-bridge inverter in two-phase synchronous rotating coordinate frame is shown in Eq. (8.56), where u_{sd}, u_{sq} are the d-axis and q-axis components of the grid voltage respectively, u_{od} and u_{oq} are the d-axis and q-axis components of the inverter output voltage respectively, and i_{sd} and i_{sq} are the d-axis and q-axis components of the grid-side current.

$$L_s \frac{d}{dt}\begin{bmatrix} i_{sd} \\ i_{sq} \end{bmatrix} = \begin{bmatrix} u_{od} \\ u_{oq} \end{bmatrix} - \begin{bmatrix} R_s & -\omega_1 L_s \\ \omega_1 L_s & R_s \end{bmatrix}\begin{bmatrix} i_{sd} \\ i_{sq} \end{bmatrix} - \begin{bmatrix} u_{sd} \\ u_{sq} \end{bmatrix} \tag{8.58}$$

The relation between the output voltage of the inverter and the average switching function is

$$\begin{cases} u_{\mathrm{od}} = d_{\mathrm{d}} \cdot NU_{\mathrm{dc}} \\ u_{\mathrm{oq}} = d_{\mathrm{q}} \cdot NU_{\mathrm{dc}} \end{cases} \tag{8.59}$$

When the d-axis of the synchronous rotating coordinate frame is oriented to A-phase grid voltage, the complete mathematical model of the inverter in the synchronous rotating coordinate frame is

$$\begin{cases} u_{od} = L_s \frac{\mathrm{d}i_{sd}}{\mathrm{dt}} + R_s i_{sd} - \omega_1 L_s i_{sq} + U_s \\ u_{oq} = L_s \frac{\mathrm{d}i_{sq}}{\mathrm{dt}} + R_s i_{sq} + \omega_1 L_s i_{sd} \\ P = i_{\mathrm{sd}} U_s \\ Q = i_{\mathrm{sq}} U_s \end{cases} \tag{8.60}$$

According to Eq. (8.58), when the grid voltage is stable, controlling the d-axis current can control the active power output of the inverter, and controlling the q-axis current can control the reactive power output of the inverter. The controller calculates the d-axis and q-axis components of the required current according to the active and reactive power references and the current grid voltage amplitude. The current loop uses PI regulators to achieve control of the active and reactive currents and deliver power to the grid.

The overall control diagram of the cascaded H-bridge inverter is given in Fig. 8.35. The active power reference P^* is calculated according to the wind speed and MPPT requirements. The rectifier side automatically adjusts the output power of the generator by stabilising the DC-bus voltage; the reactive power is used to achieve the support of the grid voltage. The power reference corresponds to current reference of the current loop and the current controller output is the control voltage reference u_{abc}^*. The cascaded H-bridge inverter is modulated by means of a PWM algorithm, which is described below (Fig. 8.36).

2. Modulation Method of the Grid-side Cascaded H-bridge Inverter

In Fig. 8.35, once the reference voltage u_{abc}^* has been obtained, the cascaded H-bridge inverter needs to be modulated. There are two main modulation methods: fundamental frequency modulation and PWM modulation. The fundamental frequency modulation is mainly used for low switching frequency applications, where the switches only operate twice in a grid power cycle, but this method requires a rotation algorithm to solve the problem of uneven power distribution between the various cascaded modules. The PWM modulation can achieve equal power sharing between different cascaded modules and has a better harmonic performance, here the PWM modulation method is used. The phase-shifted carrier PWM method is a widely used modulation method for cascaded H-bridge type systems, where the duty cycle of each H-bridge module is equal, only staggered in time by a certain angle. For the modulation of the first stage H-bridge module, the modulation signal for the left arm of the H-bridge in Fig. 8.29 is first obtained according to the voltage reference given by the current inner loop, and the modulation signal for the right arm of the H-bridge is 180° different from that of the left arm, so that unipolar modulation can be achieved and harmonic performance can be optimised. The modulation waveforms of the other stages are the same as those of the first stage, but with an angular difference of π/N between the carriers, N being the number of stages of the cascaded module. This enables the multi-level output to be achieved and the corresponding switching frequency harmonics to be raised above $N\omega_c$, ω_c being the switching frequency of the H-bridge inverter device. The simulation results of modulating a 3-stage cascaded H-bridge inverter using the phase-shifted carrier PMW method are given in Fig. 8.37. Figure 8.37a shows the carrier and the modulation signals of the left and right bridge arms of the H-bridge module with a modulation wave frequency of 50 Hz and a carrier frequency of 1.125 kHz, which shows that the modulation signals differ by 180° to each other. Figure 8.37b shows the output voltages of the H-bridge module (in a per unit value system), with each H-bridge module output being 3-level and unipolar modulated. Figure 8.37c shows the carrier signals of the three H-bridge modules with a phase difference of $\pi/3$ between them. Figure 8.37d shows the output voltage of the cascaded three H-bridge modules with an 7-level output voltage of 7, verifying the validity of the phase-shifted carrier PWM. In the specific implementation of the phase-shifted carrier PWM, multiple timers can be used to implement multiple carriers with mutual phase difference π/N. However, considering the synchronisation of the timers and the system resources of the DSP or CPLD, the method of phase-delayed shifting for the pulses is usually used.

8.2.3 *Performance Optimisation of the Transformerless Medium/High-Voltage PMSG-Based Wind Power System*

Sections 8.2.1 and 8.2.2 have investigated the basic operating principles and methods of the multi-phase PMSG and its converter system, where the system can operate properly and transmit the captured wind energy to the grid. Taking into account the actual system operation and performance requirements, this section will further investigate and improve the following aspects. Firstly, improve the generator-side rectifier performance by using the controller in the synchronous rotating coordinate frame to improve the control accuracy of the generator current. Secondly, improve the control bandwidth of the DC-bus voltage loop. Third, discuss how to eliminate DC-bus voltage fluctuations caused by grid-side power pulsation and the effect of the voltage pulsation on the generator side. Fourthly, research into the effect of harmonic currents on generator torque and give the method for connecting the generator winding to the rectifier module. Fifth, consider the uninterrupted operation capability of the system after a module failure and the interaction between the windings. Sixthly, use a module structure with two rectifiers in series and a three-level inverter in order to increase the voltage level of each module and thus reduce the number of stages of modules required and the number of individual windings of the generator.

8.2.3.1 Improvement of the Rectifier Control Performance

In Sects. 8.2.1 and 8.2.2 the rectifier module is made up of two quadrature windings, which is connected in parallel to the DC bus after passing through the PFC boost circuit. Each winding is controlled separately and a PI regulator in the stationary coordinate frame is used as the controller for the current loop, so that two PI regulators are required for each rectifier module. According to the spectral characteristics of the PI regulator, it can ensure a steady-error-free state of the DC component of the controlled object. In this section, since the speed of the wind turbine (stator frequency) varies with the wind speed, the frequencies of the stator electromotive force and the current are variable and the PI regulator cannot give a steady-error-free state of the relevant physical quantity (current) at the stator frequency, so weakening the control performance of the current loop. One solution is to use the park transformation, which is widely used in three-phase motor control, to transform the AC quantity into the direct flow under the synchronous rotating coordinate frame and then control it. The use of PI regulator under the synchronous rotating coordinate frame can give a steady-error-free state of the controlled variable. A pair of quadrature windings of the generator is equivalent to the windings located in a two-phase stationary coordinate frame, which can be transformed into a synchronous rotating coordinate frame for control by using the information of the rotor position.

If the d-axis of the synchronous rotating frame is oriented to the rotor flux linkage, the mathematical model of the quadrature windings and rectifier in Fig. 8.23b in the synchronous rotating frames can be expressed as

$$\begin{bmatrix} V_\mathrm{d} \\ V_\mathrm{q} \end{bmatrix} = \begin{bmatrix} R & 0 \\ 0 & R \end{bmatrix} \begin{bmatrix} i_\mathrm{d} \\ i_\mathrm{q} \end{bmatrix} + Lp \begin{bmatrix} i_\mathrm{d} \\ i_\mathrm{q} \end{bmatrix} + \begin{bmatrix} 0 & -\omega_\mathrm{r} L \\ \omega_\mathrm{r} L & 0 \end{bmatrix} \begin{bmatrix} i_\mathrm{d} \\ i_\mathrm{q} \end{bmatrix} + \begin{bmatrix} 0 \\ \omega_\mathrm{r} \psi_\mathrm{r} \end{bmatrix} \tag{8.61}$$

where U_d and U_q are the d-axis and q-axis components of the generator terminal voltage respectively, i_d and i_q are the d-axis and q-axis components of the generator respectively, ω_r is the stator angular frequency, L is the stator inductance, R is the stator resistance, and ψ_r is the rotor flux linkage. Here, the d-axis and q-axis currents correspond to the generator output active and reactive power, respectively.

The transformation relation between the two-phase stationary and two-phase rotating coordinate frames is

$$\begin{bmatrix} U_\mathrm{d} \\ U_\mathrm{q} \end{bmatrix} = \begin{bmatrix} \cos\theta_\mathrm{r} & -\sin\theta_\mathrm{r} \\ \sin\theta_\mathrm{r} & \cos\theta_\mathrm{r} \end{bmatrix} \begin{bmatrix} u_1 \\ u_2 \end{bmatrix} \tag{8.62}$$

where, θ_r is the rotor position angle and u_1 and u_2 indicate the generator terminal voltage in the 2-phase stationary coordinate frame. Based on the above mathematical models of the generator and rectifier, the control method of the PFC rectifier module in the two-phase rotating coordinate frame can be obtained. The relation between the output voltage on the AC side of the PFC circuit (generator terminal voltage) and the DC-bus voltage in Fig. 8.25a can be expressed as

$$u_\mathrm{n} = (1 - d_\mathrm{n}) u_\mathrm{dc} sign(i_\mathrm{n}) \quad n = 1, 2 \tag{8.63}$$

where u_n indicates the output voltage of the rectifier, d_n indicates the duty cycle of the main PFC switch, u_dc indicates the DC-bus voltage of the module and sign(in) indicates the direction of the winding current, with the direction of the current flowing from the generator to the rectifier being positive. Equation (8.61) shows that the diodes in Fig. 8.20a conduct differently when the current direction is different, so the output voltage of the rectifier is also different.

Figure 8.38 gives a diagram of the control based on the synchronous rotating coordinate frame, including the DC-bus voltage loop and the current loop. The output of the voltage loop is the d-axis current reference, and the q-axis current reference is assigned as zero for unit power factor operation. PI regulators are used for both the voltage and current loops and the rotor position is still obtained from the encoder. The cross-coupling terms to be compensated for in the diagram can be expressed as

$$\begin{aligned} u_{\text{d\_com}} &= -\omega_r L i_q \\ u_{\text{d\_com}} &= \omega_r L i_d + \omega_r \psi_r \end{aligned} \tag{8.64}$$

It is important to note here that a prerequisite assumption for doing a synchronous rotation transformation is that the physical quantity under study is sinusoidal, so that it is a DC quantity when transformed to the corresponding synchronous rotation coordinate frame. For the generator winding used in this section, as the midpoint of the winding is opened and connected out, then there may be a certain amount of 3rd and 5th harmonic components in the winding electromotive force, the magnitude of which depends on the winding method of the stator winding. Therefore, in the synchronous rotating coordinate frame, in addition to the DC component transformed from the fundamental component, there will also be the 2nd harmonic component (transformed from the 3rd harmonic in the stationary coordinate frame) and the 4th harmonic component (transformed from the 5th harmonic in the stationary coordinate frame). Due to the small content of the 5th harmonic, only the influence of the fundamental and 3rd harmonic components can be considered (Fig. 8.39).

In order to control the fundamental and 3rd harmonic components simultaneously, a rotational coordinate transformation based on the fundamental frequency and 3rd harmonic frequency can be used respectively, and then the fundamental and 3rd harmonic components can be controlled under these two rotational coordinate frames respectively. However, certain filters need to be designed to filter out the 2nd harmonic component transformed from the 3rd harmonic component in the fundamental frequency synchronous rotating coordinate frame, and to filter out the component transformed from the fundamental component in the 3x frequency synchronous rotating coordinate frame, which will increase the complexity of the control system. In addition, in this section, the reference current of the generator winding is fundamental sinusoidal, and the torque density of the generator can be further improved if the reference current contains a certain amount of the 3rd harmonic. At the same time, when using the rotating coordinate transformation, attention should also be paid to the effect of current over-zero distortion on the current waveform after the rotation coordinate transformation.

In Sect. 8.3.2, Fig. 8.38 uses a control method based on the stationary coordinate frame, which allows control of the fundamental and 3rd harmonic at the same time due to the high bandwidth of the PI regulator near the low frequencies. If the current reference is the fundamental sinusoidal signal, then the output voltage of the rectifier contains the 3rd harmonic component, which counteracts the 3rd harmonic component of the generator electromotive force. The PI regulator control based on the stationary coordinate frame does not achieve a steady-error-free current control when the generator is running at variable speed, but it eliminates the complex operation of synchronous rotation coordinate transformation for multiple frequencies. Therefore, in practice, if there is no 3rd harmonic in the electromotive force of the generator, then the rotating-coordinate-frame-based control method is more suitable; otherwise, the stationary-coordinate-frame-based control is easier to implement, although with some loss of control accuracy.

In this section, the current reference is given as a sinusoidal signal, if the electromotive force contains the 3rd harmonic, the analysis on the power generated by the interaction between the fundamental sinusoidal current and the 3rd harmonic in the electromotive force is made as follows, where let the generator winding current be

$$i = I_{\mathrm{m}} \sin(\omega_{\mathrm{r}} t) \tag{8.65}$$

The 3rd harmonic in electromotive force can be written as

$$e_3 = E_3 \sin(3\omega_{\mathrm{r}} t) \tag{8.66}$$

Considering the quadrature windings in parallel, the expression for the output power is

$$\begin{aligned} p_3 &= E_3 \sin(3\omega_{\mathrm{r}} t) \cdot I_{\mathrm{m}} \sin(\omega_{\mathrm{r}} t) + E_3 \sin(3\omega_{\mathrm{r}} t + \frac{3}{2}\pi) \cdot I_{\mathrm{m}} \sin(\omega_{\mathrm{r}} t + \frac{\pi}{2}) \\ &= -\frac{E_3 I_{\mathrm{m}}}{2}[\cos(4\omega_{\mathrm{r}} t) - \cos(2\omega_{\mathrm{r}} t)] - \frac{E_3 I_{\mathrm{m}}}{2}[\cos(4\omega_{\mathrm{r}} t) + \cos(2\omega_{\mathrm{r}} t)] \\ &= -E_3 I_{\mathrm{m}} \cos(4\omega_{\mathrm{r}} t) \end{aligned} \tag{8.67}$$

It can be seen that if the electromotive force contains the 3rd harmonic, the power output to the DC bus after the parallel connection of the quadrature windings contains the 4th harmonic, and the magnitude of the harmonic depends on the 3rd harmonic electromotive force and the magnitude of the winding current, as shown in Eq. (8.65).

8.2.3.2 Improvement of the Bandwidth of the DC-Bus Voltage Loop

From Sects. 8.2.1 and 8.2.2, it can be seen that there is a power pulsation of 2 times the fundamental frequency in the single-phase power. Although the rectifier side (generator side) achieves constant power by connecting the quadrature winds in parallel to eliminate the pulsation of 2 times the stator frequency, the grid-side single-phase inverter (H-bridge) still introduces a pulsation of 100 Hz (for 50 Hz grid frequency), causing fluctuations in the DC-bus voltage, where the grid-side power is shown in Eq. (8.66).

$$\begin{aligned} p = u \cdot i &= U_{\mathrm{om}} \cos(\omega_1 t) \cdot I_{\mathrm{om}} \cos(\omega_1 t - \varphi) \\ &= \frac{1}{2} U_{\mathrm{om}} I_{\mathrm{om}} (\cos\varphi + \cos(2\omega_1 t - \varphi)) \end{aligned} \tag{8.68}$$

where U_{om} and I_{om} are the peak values of the output voltage and current of the H-bridge module respectively and φ is the power factor. The 2x grid frequency pulsation on the DC bus will affect the performance of the rectifier module. The rectifier has an outer voltage loop and an inner current loop, and fluctuations in the dc bus voltage will be reflected in the feedback signal of the voltage loop, while the output of the outer voltage loop is the reference for the inner current loop, then there will be a pulsation of the same frequency in the reference value for the inner current loop. This pulsation is particularly noticeable in the current loop when the DC-bus voltage loop has a high bandwidth (to track the grid-side power feed).

The simulation results are given in Fig. 8.40. As shown in Fig. 8.40a, the 100 Hz pulsation of the DC-bus voltage results in the voltage loop output containing 100 Hz harmonic in addition to the DC component. According to the operating principle of the PFC circuit, the output of the voltage loop will be multiplied by the electromotive force to obtain the current loop reference. Since the voltage loop output contains a 100 Hz pulsation, the current loop reference is no longer a sinusoidal signal, but a current reference waveform as shown in Fig. 8.40b. Figure 8.40c shows the actual generator winding current waveform after control by the current loop, which is no longer sinusoidal, but contains harmonics introduced by the voltage loop as well as the current reference. The harmonic currents may have an effect on the generator torque and the losses in the converter and generator.

The harmonic components in the winding current are derived and analysed below. The current loop reference is obtained by multiplying the output of the voltage loop with the phase information of the electromotive force, as shown in Fig. 8.41. Where $\sin(\omega_r t + \theta_r)$ represents the phase information of the current determined by the electromotive force and the pulsating component in the DC-bus voltage can be expressed as $\sin(2\pi \cdot 100 \cdot t + \varphi)$ with a frequency of 100 Hz.

According to Fig. 8.41, the current-loop reference i^* can be expressed as Eq. (8.67).

$$\begin{aligned} i^* =& [K_p \cdot \sin(2\pi \cdot 100 \cdot t + \varphi) + K_i \int_0^t \sin(2\pi \cdot 100 \cdot t + \varphi) dt] \cdot \sin(\omega_r t + \theta_r) \\ =& [K_p \cdot \sin(2\pi \cdot 100 \cdot t + \varphi) + \frac{K_i}{2\pi \cdot 100} \cos(2\pi \cdot 100 \cdot t + \varphi)] \cdot \sin(\omega_r t + \theta_r) \\ =& K_p(-\frac{1}{2})[\cos(2\pi \cdot 100 \cdot t + \varphi + \omega_r t + \theta_r) - \cos(2\pi \cdot 100 \cdot t + \varphi - \omega_r t - \theta_r)] \\ &+ \frac{K_i}{2\pi \cdot 100}(\frac{1}{2})[\sin(2\pi \cdot 100 \cdot t + \varphi + \omega_r t + \theta_r) - \sin(2\pi \cdot 100 \cdot t + \varphi - \omega_r t - \theta_r)] \end{aligned} \tag{8.69}$$

It can be seen from Eq. (8.67) that the DC-bus voltage pulsation causes the winding current reference to contain a frequency component of $2\pi \cdot 100 \pm \omega_r$.

To increase the control bandwidth of the DC-bus voltage loop and to reduce the effect of 100 Hz pulsation on the DC-bus voltage on the winding current and maintain the sinusoidal current waveform, a trap can be added to the feedback path of the voltage loop to filter out the 100 Hz pulsation component of the DC-bus voltage.

The centre frequency of the trap is chosen to be 100 Hz and the typical transfer function for a second order trap is

$$G(s) = A_0 \frac{s^2 + \omega_n^2}{s^2 + \omega_n s/Q + \omega_n^2} \tag{8.70}$$

where A_0 is the gain of the filter, ω_n is the characteristic angular frequency, i.e. the frequency to be filtered out, and Q is the quality factor, representing the frequency selection characteristics of the trap. For the trap used in this section, $A_0 = 1$, and if the frequency to be filtered is 100 Hz, then $\omega_n = 2\pi \cdot 100 = 628$ rad/s. When Q is taken 1 and 10, the frequency domain characteristics of the trap are shown in Fig. 8.42, and the higher the quality factor, the better the frequency selection characteristics. Since the frequency of the grid is relatively fixed, the trap frequency of the trap is fixed and the design is relatively simple.

With the addition of the trap, the output of the DC-bus voltage loop has only the DC component, as shown in Fig. 8.43a, as the 100 Hz pulsation on the feedback path of the DC-bus voltage loop is filtered out. This allows the bandwidth of the DC-bust voltage loop to be further increased without affecting the current waveform, at which point the current loop reference signal is shown in Fig. 8.43b. The corresponding generator winding current waveform is given in Fig. 8.43c, where the generator current is essentially sinusoidal. It should be noted here that the actual DC-bus voltage still pulsates at 100 Hz, only not introduced into the DC-bus voltage loop.

8.2.3.3 Effect of DC-Bus Voltage Fluctuations on Generator Torque

The previous section analysed how the voltage loop bandwidth of the rectifier can be increased without affecting the winding current waveform by incorporating a trap to filter out the pulsating components on the DC-bus feedback path. The accuracy of the processor chip (e.g. DSP) has a direct impact on the performance of the trap when it comes to digital implementation. For the generator side, another important question is whether the harmonic currents in Fig. 8.40c will have an effect on the torque of the generator and whether torque pulsation will be generated if there is no trap. Further, if no torque pulsation is generated, a certain amount of harmonic currents can be considered to be injected on the generator side to counteract the power pulsation from the grid side, which will be analysed below.

For the effect of harmonic currents on generator torque, the main analysis is made from the magnetic field generated by harmonic currents, and let the three-phase voltage and current output from the grid-side cascaded inverter be

$$\text{Phase A} \quad \begin{aligned} u_{o,A} &= U_{om}\cos(\omega_1 t) \\ i_{o,A} &= I_{om}\cos(\omega_1 t - \varphi) \end{aligned} \tag{8.71a}$$

$$\text{Phase B} \quad \begin{aligned} u_{o,B} &= U_{om}\cos(\omega_1 t - 2\pi/3) \\ i_{o,B} &= I_{om}\cos(\omega_1 t - \varphi - 2\pi/3) \end{aligned} \tag{8.71b}$$

$$\text{Phase C}\quad \begin{array}{l} u_{\text{o,C}} = U_{\text{om}}\cos(\omega_1 t - 4\pi/3) \\ i_{\text{o,C}} = I_{\text{om}}\cos(\omega_1 t - \varphi - 4\pi/3) \end{array} \tag{8.71c}$$

Using Phase A as an example, the output power of each H-bridge module can be expressed in the form shown in Eq. (8.66), rewritten here as follows

$$\begin{aligned} p = u \cdot i &= U_{\text{om}}\cos(\omega_1 t) \cdot I_{\text{om}}\cos(\omega_1 t - \varphi) \\ &= \frac{1}{2}U_{\text{om}}I_{\text{om}}(\cos\varphi + \cos(2\omega_1 t - \varphi)) \end{aligned} \tag{8.72}$$

According to the mathematical model of the rectifier in Eq. (8.59) in the synchronous rotating coordinate frame, the active and reactive power output from the pair of quadrature windings of the generator can be expressed as

$$\begin{aligned} P &= \frac{3}{2}Ei_d \\ Q &= \frac{3}{2}Ei_q \end{aligned} \tag{8.73}$$

where E is the magnitude of the electromotive force. From Eq. (8.71), it can be seen that the d-axis current of the winding corresponds to the output power of the H-bridge inverter, and by combining Eqs. (8.70) and (8.71), it gives

$$i_{\text{d}} = \frac{U_{\text{om}}I_{\text{om}}}{3E}[\cos\varphi + \cos(2\omega_1 t - \varphi)] \tag{8.74}$$

From Eq. (8.72) it can be seen that if the power of the generator and the load power are balanced, then the d-axis current of the quadrature windings is shown in Eq. (8.72). Since the d-axis current contains a harmonic component of $2\omega_1$, which means that the generator current contains a harmonic current of $2\omega_1 \pm \omega_r$ in the stationary coordinate frame. The spatial magnetic field generated by the harmonic current with a frequency of $2\omega_1$ in Eq. (8.72) is analysed below.

Let the number of individual windings of the generator be M. Every two windings differing by 90° from each other in electrical phase make a pair, making a total of $M/2$ pairs of quadrature windings. Since the grid side is three-phase, the number of stages of the cascaded H-bridge inverter is $M/2/3$. For phase A, $M/6$ pairs of spatially evenly distributed quadrature windings are connected to the grid-side A-phase modules, with each pair of quadrature windings differing from each other by an electrical angle of $2\pi/(M/6)$. In a synchronous rotating coordinate frame, let the angle between the d-axis and the α-axis of one of the pairs of quadrature windings be θ_r. Based on their spatial positions and the harmonic component ($2\omega_1$) in the current shown in Eq. (8.72), then the spatial harmonic magnetic field f_1 generated by the harmonic current is

$$\begin{aligned} f_1 &= i_{d2} \cdot \cos(\theta_r) \\ i_{d2} &= \tfrac{U_{om} I_{om}}{3E} \cos(2\omega_1 t - \varphi) \end{aligned} \tag{8.75}$$

If the M/6 pairs of windings in Phase A are considered together, the magnetic field f synthesized by the harmonic currents can be expressed as

$$f = \sum_{n=0}^{(M/6)-1} i_{d2} \cdot \cos(\theta_r + n \cdot \frac{2\pi}{M/6}) \tag{8.76}$$

Since these $M/6$ pairs of windings are all connected to modules on the same phase of the grid side, i.e. Phase A, the expressions for i_{d2} are the same, and Eq. (8.74) can be rewritten as

$$f = i_{d2} \sum_{n=0}^{(M/6)-1} \cos\left(\theta_r + n \cdot \frac{2\pi}{M/6}\right) \tag{8.77}$$

The term in Eq. (8.75) satisfies

$$\sum_{n=0}^{(M/6)-1} \cos\left(\theta_r + n \cdot \frac{2\pi}{M/6}\right) = 0 \tag{8.78}$$

According to Eq. (8.76), the expression (8.75) for the synthesised magnetic field of the harmonic current is equal to 0, i.e. $f = 0$. This indicates that the spatially harmonic magnetic field caused by the harmonic current is 0 and does not cause torque pulsations in the generator. Therefore, when connecting the generator winding to the low-voltage module, the spatially evenly distributed generator windings should be connected to the same phase on the grid side so that the effects of the current harmonic caused by the 100 Hz pulsation of the DC-bus voltage on the generator torque can be counteracted. Figure 8.44 gives a diagram of the connection of the generator windings and low-voltage modules, where A, B and C represent Phase A, B and C of the grid, and the spatially evenly distributed windings and their low-voltage modules should be connected to the same phase on the grid side.

8.2.3.4 Suppression of DC-Bus Voltage Fluctuations

As noted in the previous section, the generator winding harmonic current used to counteract the 100 Hz power pulsation on the grid side does not affect the torque of the generator, so it is possible to consider balancing the pulsating power generated directly from the generator side with the power on the grid side. In this way, there would be no 100 Hz fluctuations in the DC-bus voltage and the required DC-bus filter capacitance could thus be reduced. Of course, the harmonic current in the generator windings resulting from this solution may lead to increased losses in the generator and converter. The power required to be generated on the generator side in each low-voltage module and the corresponding d-axis currents are shown in Eqs. (8.70) and (8.72). In Eq. (8.72), both the dc component and the 2nd harmonic component are included. In order to achieve control of the 2nd harmonic component, a resonant regulator can be introduced in the rotating coordinate frame to control the 2nd harmonic current without steady errors.

The control diagram using PIR regulators is shown in Fig. 8.45, compared to Fig. 8.44, where the PI regulators are replaced by the PIR regulators. Both the DC-bus voltage and current loops use the PIR regulators. The 100 Hz pulsation on the DC bus will be reflected in the reference of the current loop by the resonant regulator of the voltage loop, and controlled by the current loop regulator. As the grid side frequency is relatively fixed, the resonant frequency of the resonant regulators can be set at 100 Hz. This method does not require any grid-side information to be known, but only the grid frequency, which is relatively simple.

The simulation results with the PIR regulator are given in Fig. 8.46, where the DC-bus voltage is given at 224 V. Figure 8.46a gives the DC-bus voltage without the PIR regulator, where a 100 Hz pulsation is visible. Figure 8.46b gives the DC-bus voltage before and after the addition of the PIR. At 0.3 s, the DC-bus voltage fluctuation is seen to be well suppressed by the addition of the resonant regulator. Figure 8.46c gives the d-axis current waveform. After adding the PIR regulator, the d-axis current shows a pulsation of 100 Hz, which achieves the compensation of the power on the H-bridge inverter side and verifies the effectiveness of the PIR regulator. Figure 8.46d gives the winding current. After adding the resonant regulator, the current waveform is distorted, although the dc bus voltage fluctuation is reduced. According to the previous derivation, the current distortion does not cause the generator torque to pulsate, but it may increase the losses in the generator and converter.

8.2.3.5 Uninterrupted Operation Capability Under Fault Conditions

A significant advantage of using the multi-phase PMSG and modular construction is the uninterrupted operation capability of the system in the event of a fault. When one of the generator phases or low-voltage converter modules fails, the rest of the system is able to operate normally, in the case of wind power, i.e. to maintain power delivery to the grid (in some cases requiring reduced power operation) and then, when conditions are available, to carry out system servicing and maintenance.

For the converter fault study in Fig. 8.24 can be divided into two parts, firstly the fault study on the grid side (cascaded H-bridge inverter) and secondly the fault study on the generator side (rectifier side).

For grid-side converter faults, reference can be made to the fault protection and analysis methods used for high-voltage cascaded inverters used in the motor drive field. Usually in each H-bridge inverter module there is a switch (e.g. thyristor) connected in parallel with the H-bridge output, which can be used to short-circuit the faulty unit in the event of a fault in one of the H-bridge modules, thus maintaining the normal operation of the rest of the modules [11]. As the cascaded inverter connected to the grid, the output voltage of the inverter still needs to meet the requirements of the grid voltage level. After a fault has occurred, the faulty module is removed and the normal module needs to change the corresponding modulation ratio, or the DC-bus voltage value, to meet the grid voltage requirements. When a fault occurs in one of the low-voltage modules of a phase, in order to maintain the balance of the three-phase system, the mid-point offset technique can be used to adjust the amplitude and phase of the three-phase output voltage to achieve a balanced output of the three-phase voltage [12].

The fault protection of the generator-side converter is somewhat different from the cascaded converter in the motor drive field, where the input side of the converter mostly uses a phase-shifting transformer and an uncontrollable diode rectifier, whereas in this section is the generator and the PFC rectifier boost circuit, with the addition of controllable devices. In the following, the operating state of the system and the method of uninterrupted operation in the event of an open-circuit fault in the main PFC switch are considered.

When an open-circuit fault occurs in the main switch of the PFC circuit, the operation of the rectifier side is changed from PFC rectification to diode uncontrolled rectification. Figure 8.47a gives the change in the DC-bus voltage of the converter module before and after the failure of the PFC main switch. It can be seen that the DC-bus voltage drops from the assigned value of 224 V for PFC rectification to about 130 V (peak of electromotive force) for diode uncontrolled rectification. Figure 8.47b gives the generator winding current waveform where the generator current changes from sinusoidal to diode rectified current. When entering the diode rectification method, the current waveform is rich in harmonics, which can lead to torque pulsations in the generator as well as harmonic losses. However, due to the use of a multi-phase generator, the more phases the generator has, the less a single winding fault will affect the whole generator.

An open-circuit fault in the rectifier-side switch can have an impact on the grid-side inverter operation. As the DC-bus voltage of the faulty module is reduced, if the three-phase symmetrical PWM generation method is still used, this will cause a reduction in the voltage output capability of the faulty phase, which will affect the grid current. Figure 8.48a gives the grid current after a fault has occurred and shows that the current amplitude of the faulty phase is reduced compared to that of the non-faulty phase, which can lead to an imbalance in the three-phase output power of the inverter. The output voltage can be compensated by detecting the DC-bus voltage and dynamically adjusting the modulation ratio of the output voltage of the faulty phase. The compensated current waveform is shown in Fig. 8.48b, which shows that the amplitude of the current in each phase is almost the same.

A further issue to consider on the generator side is the influence of the faulty phase on the other non-faulty phases. When the rectifier of the faulty phase goes into diode rectification, the harmonic components of the generator current will affect the non-faulty phases through the mutual inductance between the generator windings. The mutual inductance matrix of the multi-phase PMSG is shown in the mathematical model in Sect. 8.2.1.5. Depending on the spatial distribution of the windings (different mutual inductance parameters) and the distortion degree of the current waveform in the faulty phase, the effect of the fault on the other non-faulty phases will vary.

The effect of the fault on the other non-faulty phases can be studied by simulation. Figure 8.49 gives the simulation waveforms of the winding current of the 6-phase generator in Fig. 8.30 before and after the fault, where the winding current is at an electrical angle of 30° from the faulty phase. Before the fault, the winding current waveform is shown in Fig. 8.49a and after the fault, its current is shown in Fig. 8.49b. It can be seen that the current waveform of the non-faulty phase is sinusoidal with no significant change after the occurrence of the fault. This is due to the PI regulator used in the current loop of the PFC circuit. When the fault phase winding current has an effect on the induced electromotive force of the non-faulty phase through mutual inductance, the PI regulator can dynamically adjust the rectifier output voltage so that the winding current tracks the reference value of the sinusoidal current and maintains the generator in uninterrupted operation.

In addition, as the converter system is directly connected to the grid, the generator and converter system are required to have certain coping capabilities when the grid fault occurs, such as low-voltage ride-through capability and operation capability under asymmetrical grid conditions. For the cascaded converter system in this section, the control method in the event of a grid fault is similar to that of the two-level full-scale power converter in Fig. 7.1. In the case of a three-phase symmetrical voltage dip, for example, the output current of the grid-side cascade inverter will increase

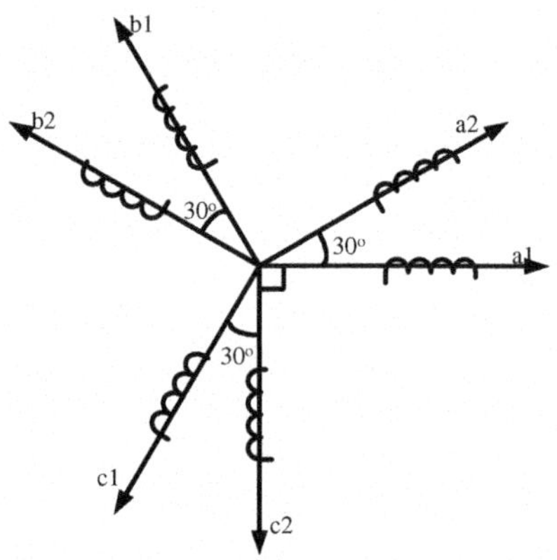

Fig. 8.30 6-phase PMSG with 180/*m* electrical angle between neighbouring phases

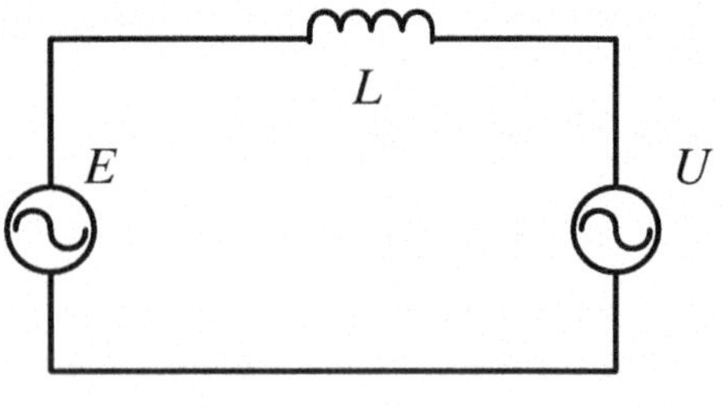

Fig. 8.31 Equivalent circuit of the PFC

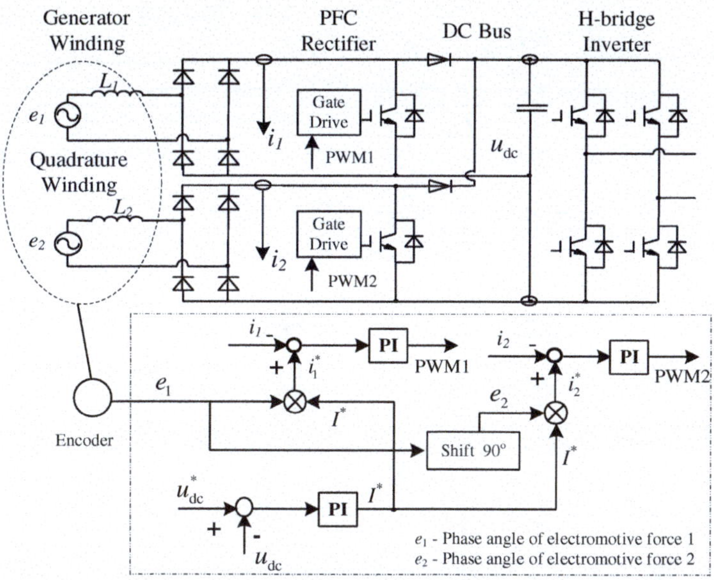

Fig. 8.32 Control diagram of generator-side rectifier in a low-voltage module

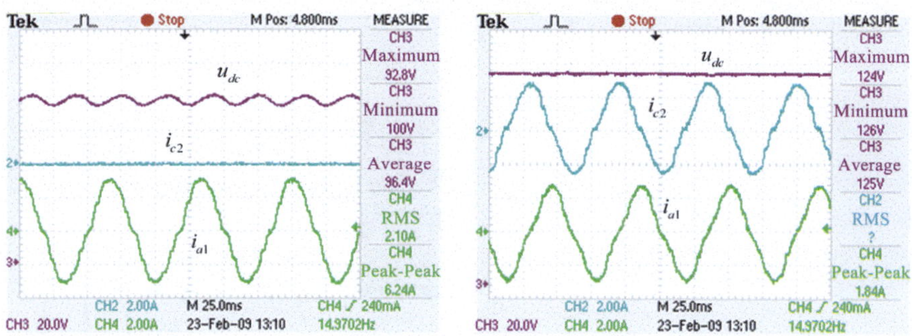

(a) Single-winding PFC Rectifier (b) Quadrature-winding PFC Rectifier

Fig. 8.33 Experimental results of single-winding and quadrature-winding rectifiers

the current loop output to its limit value due to the grid voltage dip. At this point the power captured by the wind turbine is greater than the power fed into the grid by the grid-side inverter and the DC-bus voltage of the low-voltage module will increase. The rise in the DC-bus voltage can be suppressed by controlling the reduction of the current in the PFC rectifier circuit, thereby reducing the power fed to the DC bus. This will result in an increase in wind turbine speed, but the speed range is still controllable due to the short voltage dip time and the high rotational inertia of

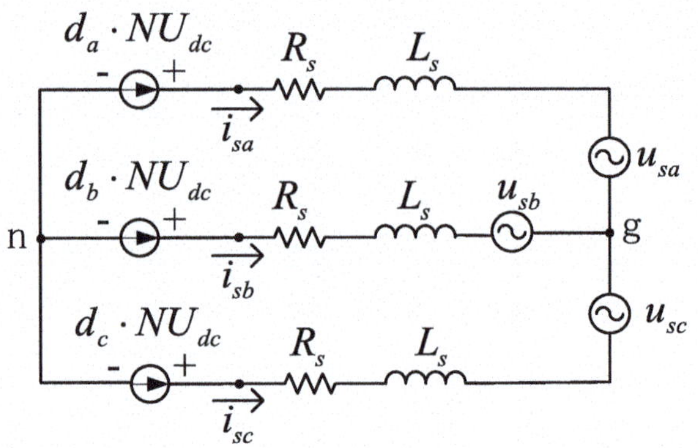

Fig. 8.34 The simplified equivalent circuit of the cascaded H-bridge inverter

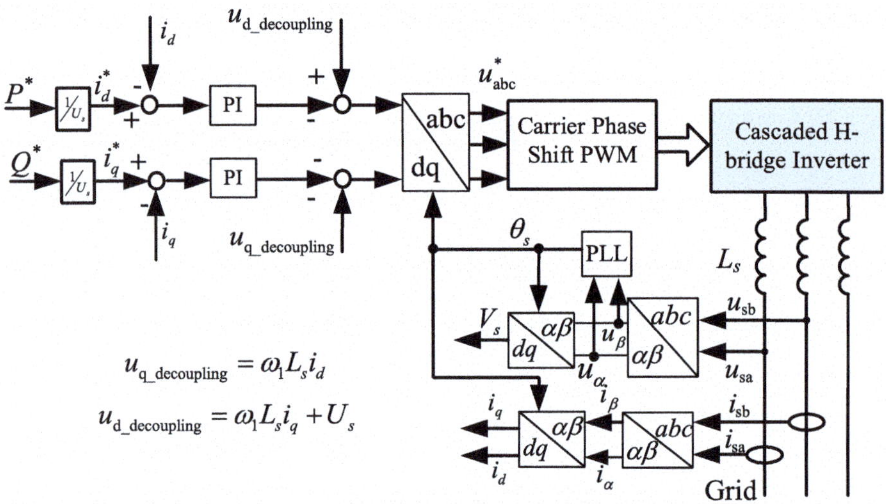

Fig. 8.35 Overall control diagram of the grid-side cascaded H-bridge inverter

the wind turbine. At the same time, the excess power can be released by a resistor connected in parallel to the DC bus to maintain a constant DC-bus voltage.

8.2.4 Module Structure with Series Rectifier and Three-Level Inverter Bridge

In the converter topology studied in this section, multiple sets of low-voltage modules are required to be connected in series in order to achieve a medium/high-voltage

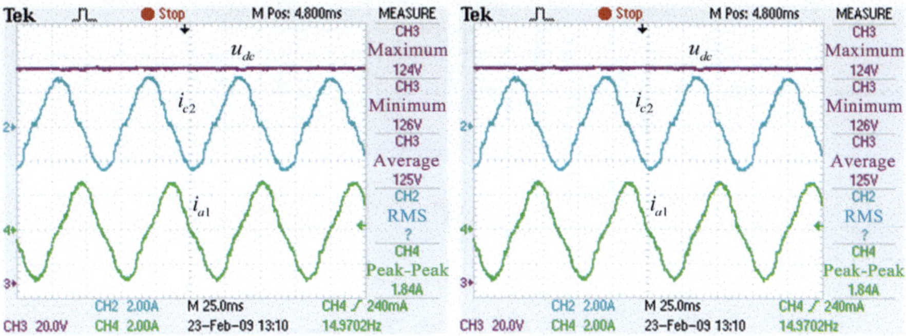

Fig. 8.36 Experimental results of the grid-connected single-phase cascaded H-bridge inverter

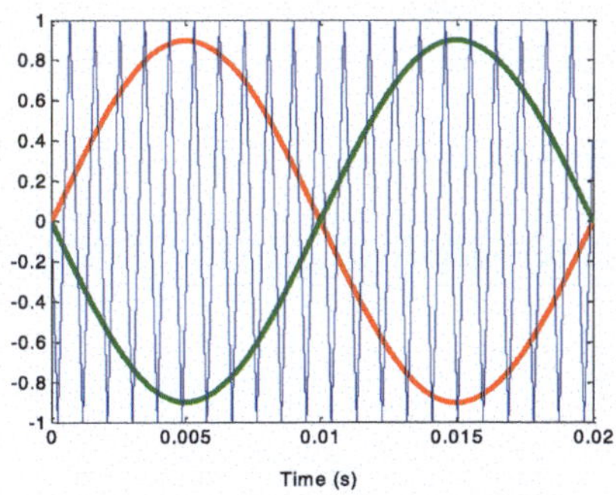

(a) Modulation signals for H-bridge arms

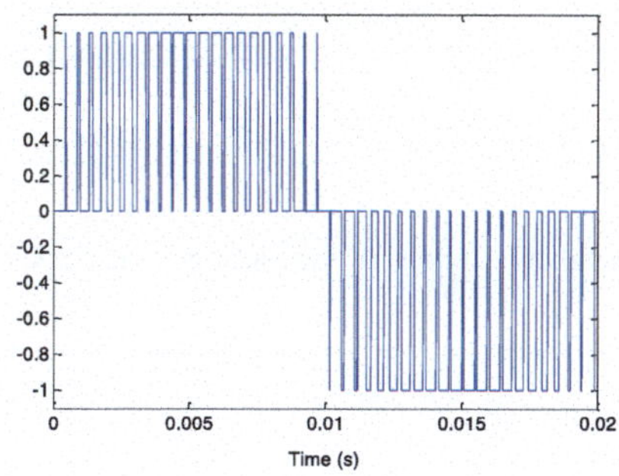

(b) Output of a H-bridge Module

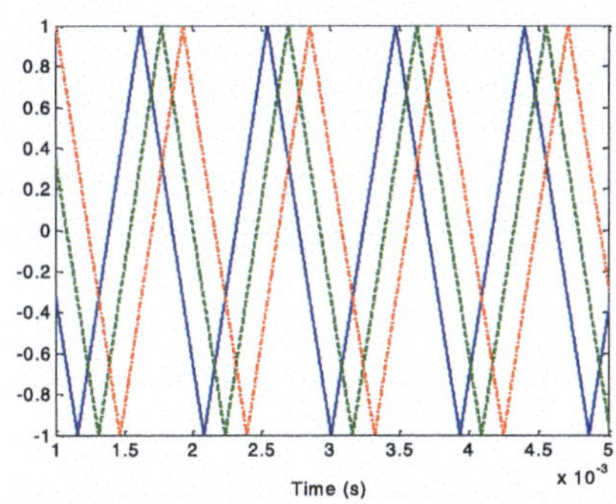

(c) Carriers for different modules

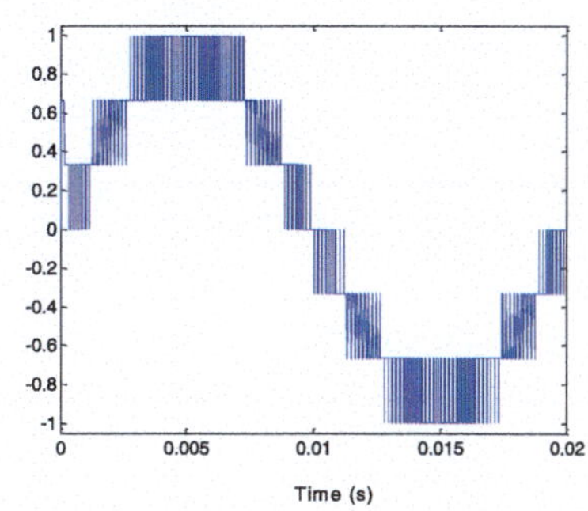

(d) Output voltage of a 3-module cascaded H-bridge inverter

Fig. 8.37 Simulation results of the phase-shifted carrier modulation for a cascaded H-bridge inverter

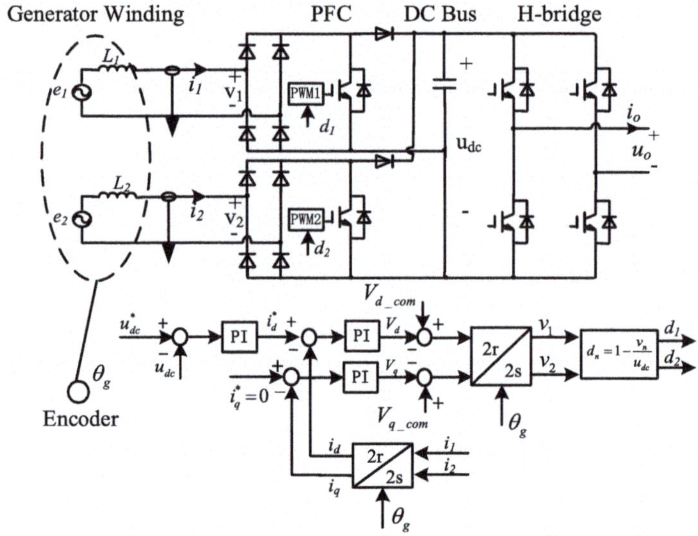

Fig. 8.38 Control of the rectifier module based on the synchronous rotating frame

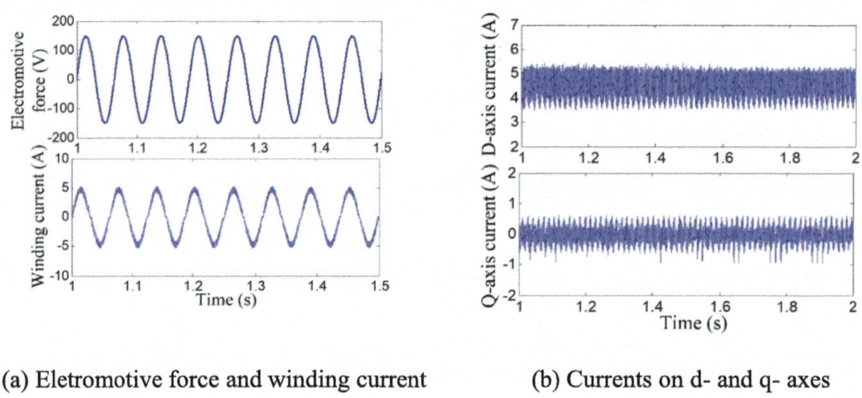

Fig. 8.39 Simulation results of the rectifier module in the synchronous rotating frame

output, which requires a larger number of generator windings to provide power supplies to the low-voltage modules, but the number of individual generator windings is limited. In order to reduce the number of individual generator windings required, the voltage rating of each low-voltage module can be increased, thus reducing the number of stages of cascaded modules. In Sect. 8.2.2.1, the use of two quadrature windings connected in parallel can counteract the power pulsation at 2 times the stator frequency, likewise two such windings connected in series can counteract the power pulsation, the expression for which is still shown in Eq. (8.34). The structure in Fig. 8.50 can therefore be adopted by connecting two rectifier modules in series so

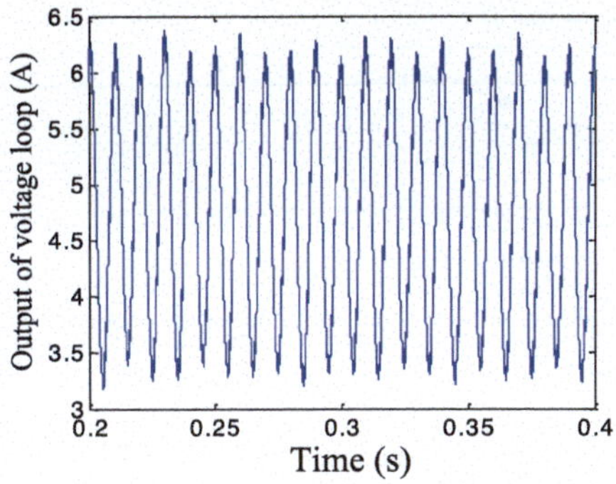

(a) Output of the DC-bus voltage loop

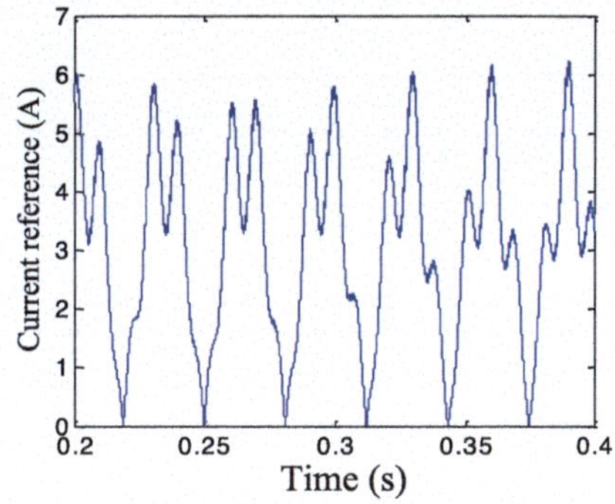

(b) Reference of the current loop

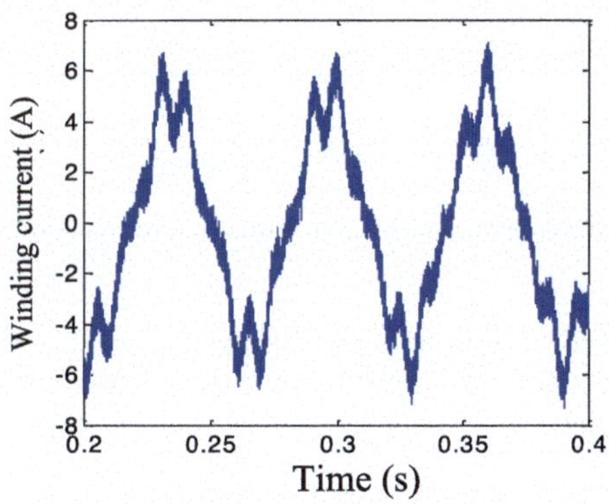

(c) Actual winding current

Fig. 8.40 Effect of DC-bus voltage pulsation on generator winding current

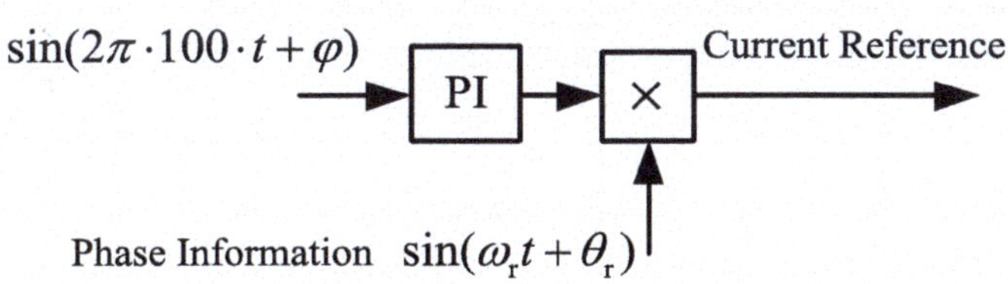

Fig. 8.41 Schematic of the current reference for the current loop

that the voltage rating of each low-voltage module is twice as high as in Fig. 8.24b using a parallel connection. Taking into account the voltage limitations of the power electronics, a diode-clamped three-level (NPC) bridge arm structure can be used on the inverter side accordingly. The voltage stress on each switch of this inverter module is half that of each switch of the H-bridge module, and the output voltage of each module is 5 levels (the output of the H-bridge inverter module is 3 levels). This structure increases the number of output levels while increasing the voltage level of

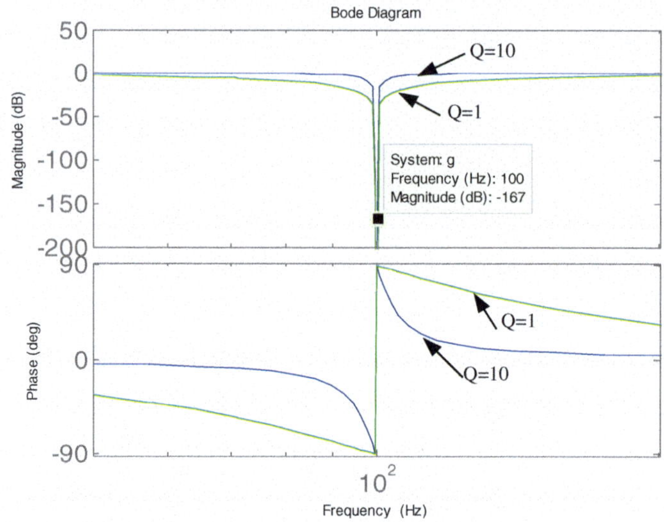

Fig. 8.42 Frequency characteristics of the trap with a centre frequency of 100 Hz

the low-voltage module. At the same withstand voltage level of the devices and grid voltage, this structure allows for half the number of generator windings required than with a parallel structure.

It should be noted that although the total power on the rectifier side of low-voltage modules in this structure is constant, there is still a power pulsation of two times the generator stator frequency at the midpoint O of the DC bus in Fig. 8.50, which needs to be filtered by a larger capacitor. On the other hand, the problem of controlling the midpoint voltage of the NPC inverter is avoided as the midpoint voltage of the DC bus is clamped by two sets of boost rectifier circuits.

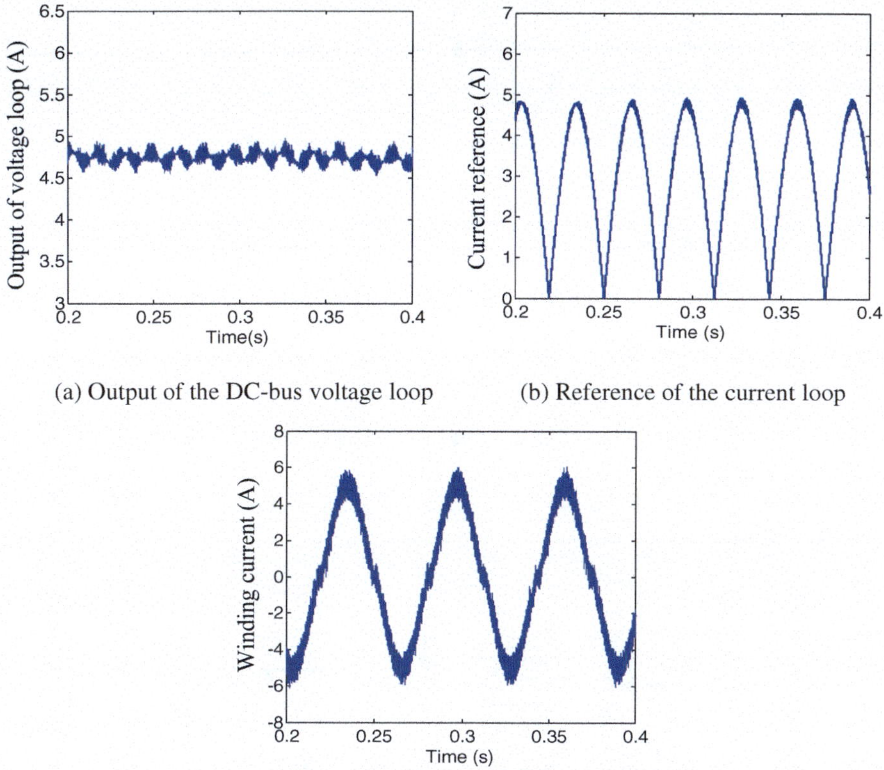

(a) Output of the DC-bus voltage loop (b) Reference of the current loop

(c) Actual winding current

Fig. 8.43 Effect of DC-bus voltage pulsation on generator winding current with a trap-based control structure

To avoid the influence of low-frequency pulsations at the midpoint O of the DC bus on the output voltage of the grid-side inverter, the structure in Fig. 8.51 can also be used. In this structure, the inverter side has the separate DC-bus capacitor and the power transferred to the DC bus from the rectifier side is constant, but this structure needs to solve the problem of balancing the midpoint voltage of the NPC inverter. In terms of fundamental power, the midpoint voltage of this structure is stable over an operating frequency cycle, but considering the differences in the parameters of the devices and energy storage elements and the dynamic processes, methods such as the selection of redundant switching states are required to achieve midpoint voltage balancing [13].

The control strategy for the rectifier side of the module structure in Fig. 8.50 is essentially the same as that for the module in Fig. 8.24b and will not be repeated here. For the control of the inverter side, vector control is still used to regulate the active and reactive power, except that here the control object is the cascaded NPC inverter.

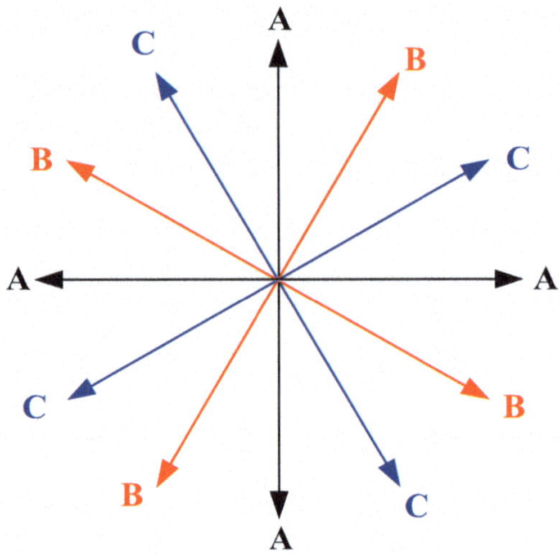

Fig. 8.44 Distribution of the generator windings and its connection to the grid side

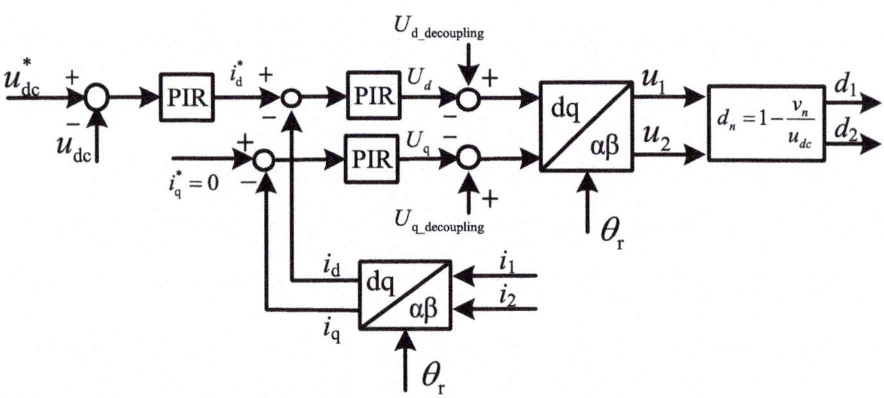

Fig. 8.45 Control diagram of rectifier module by PIR regulators

The modulation method for the cascaded NPC inverter will be discussed below. Since the midpoint of the DC bus in Fig. 8.50 is clamped by the front rectifier bridge, modulation of the entire inverter side can still be achieved simply by using the phase-shifted carrier PWM. In this section, the phase-shifted carrier method from the

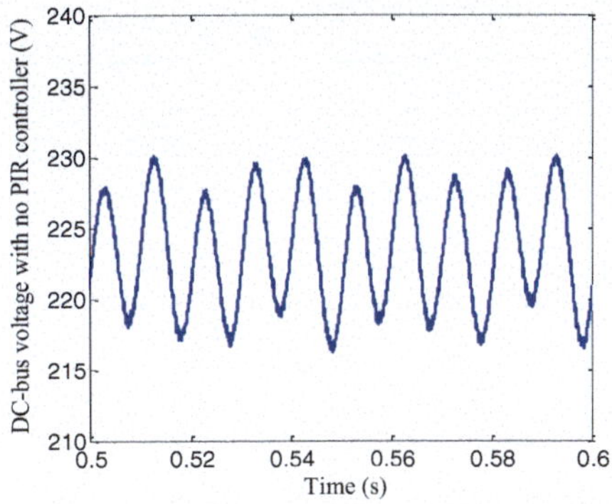

(a) DC-bus voltage with no PIR

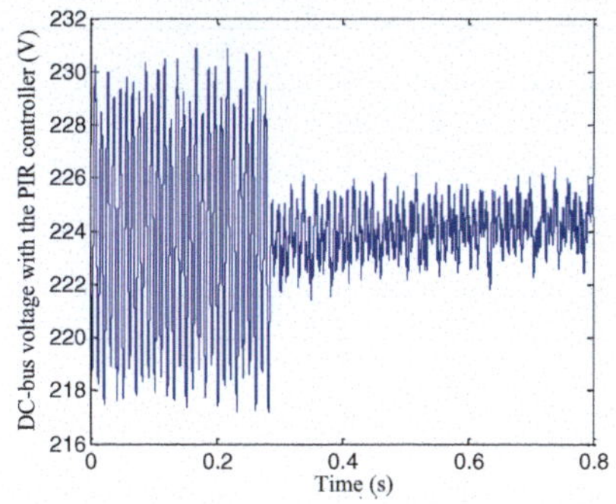

(b) DC-bus voltage with PIR

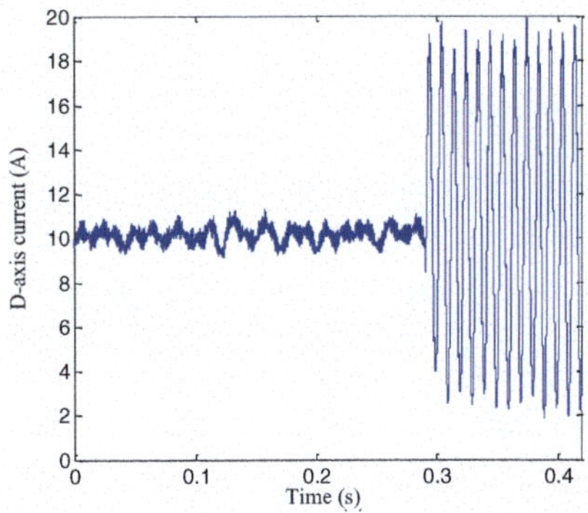

(c) D-axis current

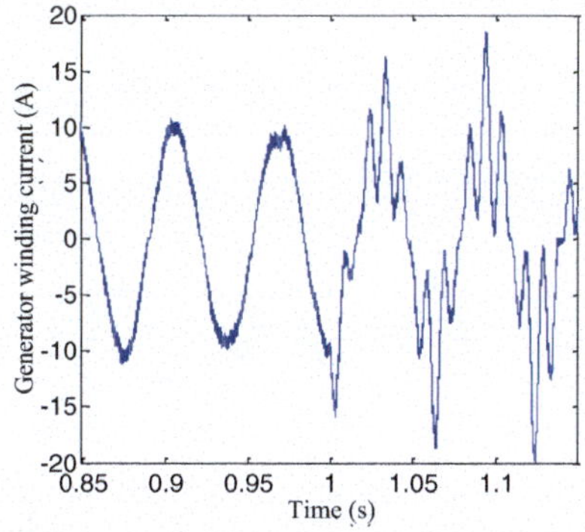

(d) Generator current

Fig. 8.46 Simulation results for suppression of DC-bus voltage

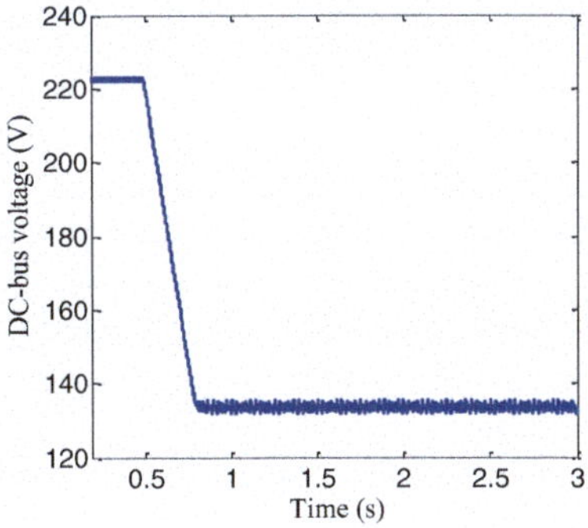

(a) DC-bus voltage of a low-voltage module

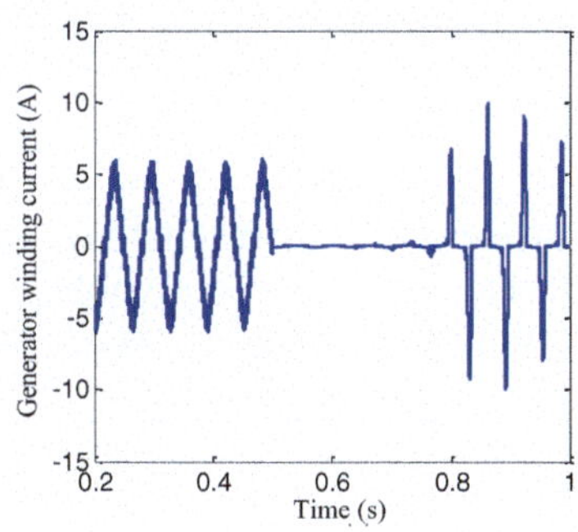

(b) Generator winding current

Fig. 8.47 Simulation results for a rectifier-side switch open-circuit fault

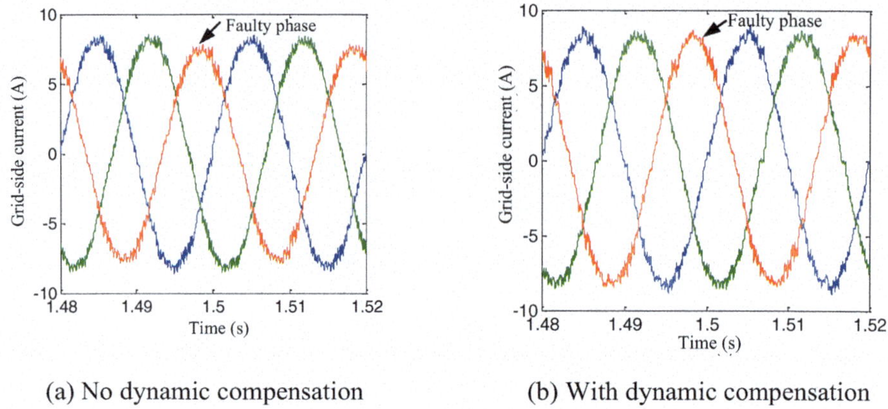

Fig. 8.48 Grid current waveform after an open-circuit fault in the PFC main switch

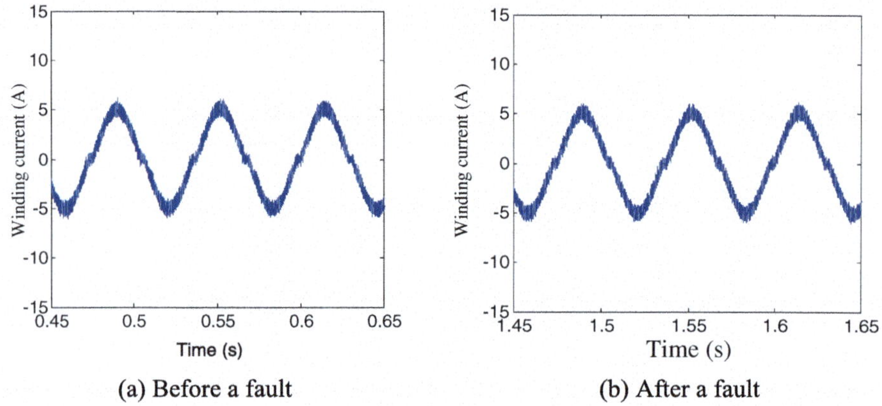

Fig. 8.49 Current waveform of the non-faulty phase winding with 30° apart from the faulted one before and after a fault occurs

cascaded H-bridge inverter is applied to the cascaded NPC inverter. The modulation scheme is as follows: according to the control system shown in Fig. 8.35, when the reference voltage u^*_{abc} is obtained from the vector control algorithm, the required output voltage for each stage module is calculated (as $1/N$ of the total output voltage), N being the number of stages of the inverter module. Half of the DC-bus voltage of each module is then selected as the base value, and the desired output voltage

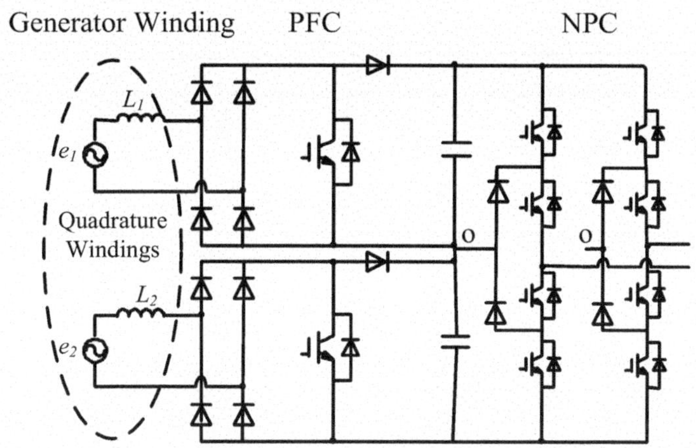

Fig. 8.50 Low-voltage module with two rectifier modules connected in series and the NPC structure inverter

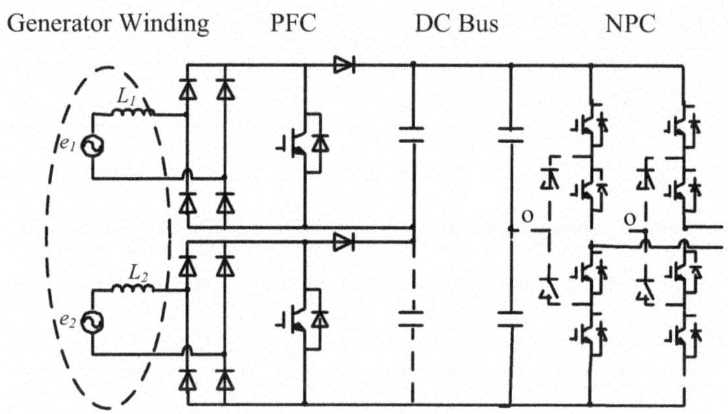

Fig. 8.51 NPC inverter with the separate DC-bus capacitor

is normalised to a value between 0 and 2, as shown in Fig. 8.52. The advantage of this normalisation is that the integer part of the output voltage represents the output voltage layer, while the fractional part represents the duty cycle of the output, the concept of the output voltage layer being the same as in Sect. 8.1. The modulation of the first stage NPC module can be completed according to Fig. 8.52. The PWM pulses for the rest NPC modules are obtained by shifting the phase of the first stage NPC module pulses, with each stage shifted by an angle of π/N sequently, which increases the harmonics of the switching frequency to $2N\omega_c$, ω_c being the switching frequency.

The experimental results of the above structure and algorithm are shown in Fig. 8.53.

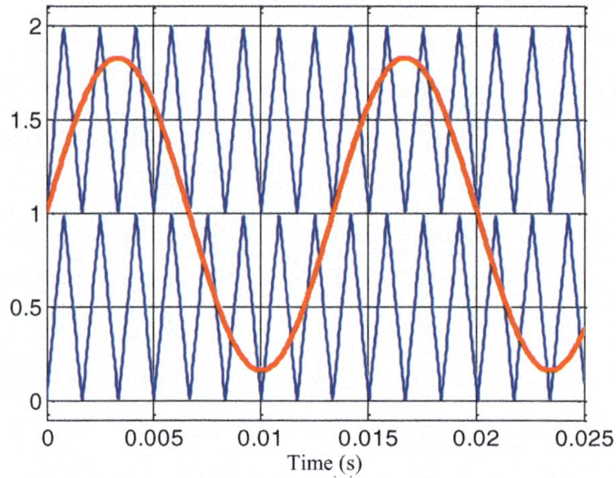

Fig. 8.52 Modulation method for the NPC inverter module

Fig. 8.53 Experimental results of the low-voltage module with two rectifiers connected in series and the cascaded NPC inverter

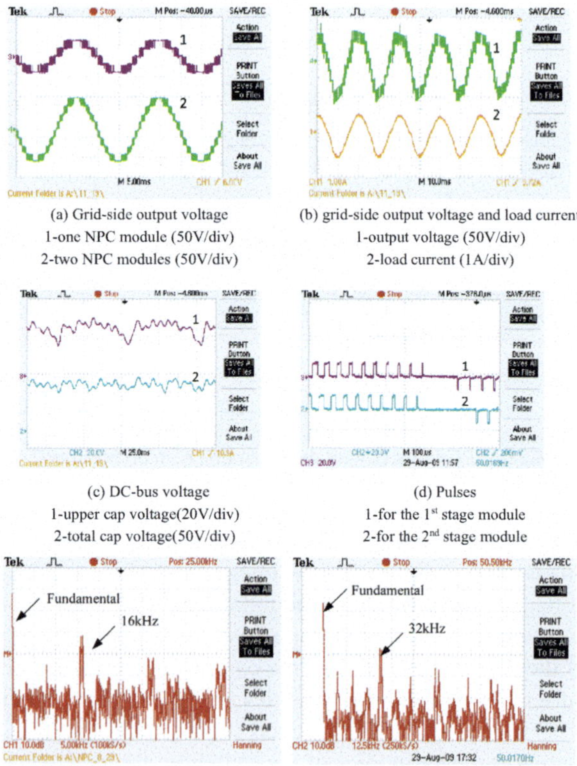

(a) Grid-side output voltage
1-one NPC module (50V/div)
2-two NPC modules (50V/div)

(b) grid-side output voltage and load current
1-output voltage (50V/div)
2-load current (1A/div)

(c) DC-bus voltage
1-upper cap voltage(20V/div)
2-total cap voltage(50V/div)

(d) Pulses
1-for the 1st stage module
2-for the 2nd stage module

(e) FFT of one-stage NPC (5-level)

(f) FFT of two-stage NPC (9-level)

References

1. H. Xiangning, C. Alian, *Theory and Applications of Multi-level Converters* (China Machine Press, Beijing, China, 2006)
2. W. Chenchen, Research on Multilevel Converter Topologies, PWM Algorithms and Midpoint Voltage Control. Ph.D. thesis (Tsinghua University, Beijing, China, 2008)
3. S. Ogasawara, H. Akagi, Analysis of variation of neutral point potential in neutral-point-clamped voltage source PWM inverters, in *IEEE IAS* (1993), pp. 965–970
4. N. Celanovic, D. Boroyevich, A fast space-vector modulation algorithm for multilevel three-phase converters. IEEE Trans. Ind. Appl. **37**(2), 637–641 (2001)
5. J.M. Erdman, R.J. Kerkman, D.W. Schlegel et al., Effect of PWM inverters on AC motor bearing currents and shaft voltages. IEEE Trans. Ind. Appl. **32**(2), 250–259 (1996)
6. H. Zhang, A. Von, S. Dai et al., Multilevel inverter modulation schemes to eliminate common-mode voltages. IEEE Trans. Ind. Appl. **36**(6), 1645–1653 (2000)
7. L. Lifei, *Research on the Rectifier-boost Circuit of a Novel High-voltage Permanent Magnet Synchronous Direct-drive Wind Power Generation System [Master Thesis]* (Tsinghua University, Beijing, China, 2008)
8. Y. Lo, S. Ou, H. Chiu, On evaluating the current distortion of the single-phase switch-mode rectifiers with current slope maps. IEEE Trans. Industr. Electron. **49**(5), 1128–1137 (2002)
9. G. Koen, D.M. Van, P.M. Van et al., Digitally controlled boost power-factor-correction converters operating in both continuous and discontinuous conduction mode. IEEE Trans. Ind. Electron. **52**(1), 88–97 (2005)
10. X. Shan, Research on multi-phase permanent magnet synchronous motor drive technology. Ph.D. thesis (Institute of Electrical Engineering of the Chinese Aca, Beijing, China, 2005)
11. H. Zhiling, R. Zhaohua, Q. Ming, Analysis of open circuit malfunction for H-bridge cascaded multilevel high voltage converter. Electric Drive **36**(7), 19–21 (2006)
12. J. Peng, Protection configuration of cell-cascaded multilevel high voltage converter. Electric Drive **39**(5), 10–14 (2009)
13. B. Ge, F. Peng, An effective SPWM control technique for 1 MVA 6000 V cascaded neutral point clamped inverter, in *IEEE IAS* (2008), pp. 1–6

Chapter 9
Offshore Wind Power Systems Based on High-Voltage DC Transmission

In large-capacity offshore wind power systems, flexible high-voltage direct current (VSC-HVDC) systems, with their advantages of no reactive power compensation, long-distance transmission, low cable costs, flexible control, bi-directional energy flow and so on, have become the most important form of transmission for offshore wind power. Wherein, the series VSC-HVDC system does not require transformers and large-capacity centralised converters in the offshore platform, with low cost, low complexity, high efficiency, high reliability, good maintainability and so on.

The first VSC-HVDC-based offshore wind farm was BorWin1 in Germany with a system capacity of 400 MW, built by ABB. BorWin1 aggregates the output of 80 sets of 5 MW turbines on an offshore HVDC platform via a 36 kV AC cable and a 36/154 kV transformer, and transmits it via HVDC cables at 300 kVdc to an onshore platform 200 km away. BorWin2 is another large-capacity offshore wind farm with a system capacity of 800 MW and an HVDC rating of 300 kV. With the connection to BorWin1, BorWin (BorWin1 + BorWin2) will be the largest HVDC offshore wind farm with the longest subsea transmission distance.

This chapter will discuss two novel series-connected VSC-HVDC systems [1], namely the Modular Multilevel Converter (MMC)-based HVDC system and the Medium Frequency Transformer (MFT)-based HVDC system, including their topologies and their control methods in detail.

H. Ma et al., *The Control Principle of Wind Power Generation System*,
https://doi.org/10.1007/978-981-99-6083-5_9

9.1 MMC-Based HVDC Transmission System for Offshore Wind Farms

9.1.1 Modular Multi-level Converter

9.1.1.1 Basic Structure of the Modular Multi-level Converter

A typical three-phase five-level MMC converter structure is shown in Fig. 9.1a. The converter contains three bridge arms, each forming a phase and three forming a three-phase output; each bridge arm contains two half-bridge arms, called the upper half-bridge arm (or positive half-bridge arm) and the lower half-bridge arm (or negative half-bridge arm), respectively; each half-bridge arm consists of four sub-modules (SM) connected in series to obtain a five-level phase voltage output; each half-bridge arm contains one inductor L_s for current control and fault current limiting [2]; the DC side consists of two equivalent capacitors connected in series.

A SM consists of a half bridge (consisting of one pair of complementary IGBTs) and an equivalent energy storage capacitor, as shown in Fig. 9.1b. The switching states of the SM are as follows:

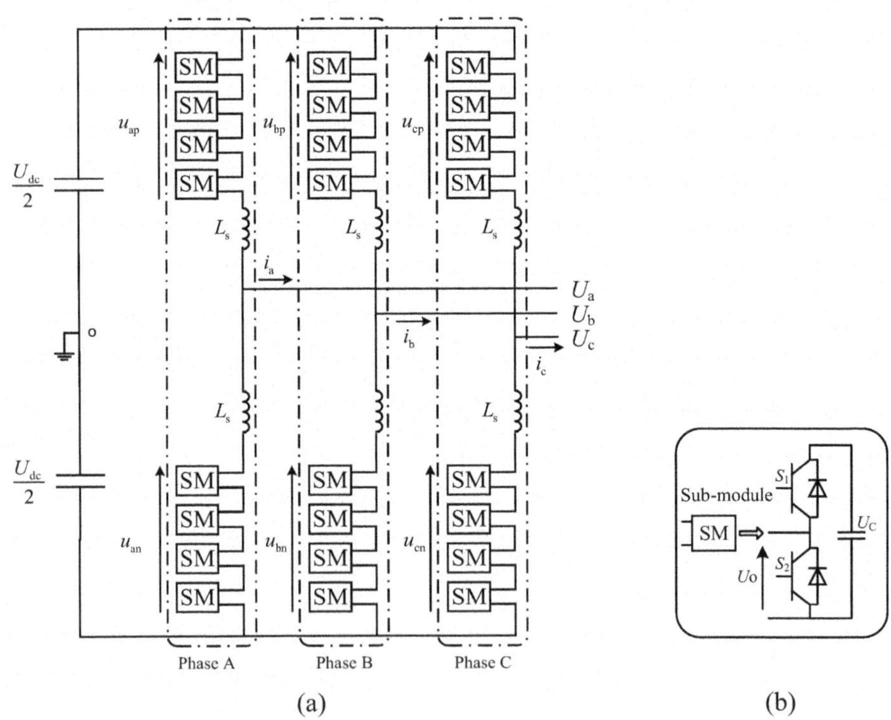

Fig. 9.1 Topology of a typical three-phase five-level MMC converter

Table 9.1 SM output voltage and capacitor state

	S_1	S_2	U_o	dU_o/dt
$i > 0$	1	0	V_c	>0
	0	1	0	0
$i < 0$	1	0	V_c	<0
	0	1	0	0

- On state: S_1 is on, S_2 is off. In the on state, the voltage at the SM terminal (U_o) is equal to the capacitor voltage and the capacitor charging and discharging state depends on the current direction.
- Off state: S_1 is off, S_2 is on. In the off state, the capacitor is bypassed and the voltage at the SM terminal is 0.
- Blocked state: Both S_1 and S_2 are switched off. This occurs in the start-up or fault state.

During the operation of the MMC converter, S_1 and S_2 in each SM are generally in complementary switching states and the number of SMs in the open state is generally equal to half the number of SMs in each phase. The SM output voltage and capacitor charging/discharging state are related to the switching state and current direction, as shown in Table 9.1.

9.1.1.2 Mathematical Model of the Modular Multi-level Converter

The functional diagram of the three-phase five-level MMC converter is shown in Fig. 9.2. Here, ideal switches are used instead of complementary IGBTs in each SM to achieve switching control of the SM output voltage. Where S_{upji} and S_{lowji} indicate the switches of the upper and lower half-bridge arms respectively, C_{upji} and C_{lowji} indicate the energy storage capacitors of the upper and lower half-bridge arms respectively, j indicates the phase sequence where the current switch is located ($j =$ a, b, c), and i indicates the sequence number where the current switch is located ($i =$ 1, 2, …, 6). Therefore, when $S_{upji} = 1$ (or $S_{lowji} = 1$), it means that the S_{upji} (or S_{lowji}) switch is connected to the positive terminal of its capacitor, i.e. the output voltage of the ith SM of the upper half-bridge arm (or lower half-bridge arm) in Phase j is equal to U_{cupji} (or U_{clowji}); when $S_{upji} = 0$ (or $S_{lowji} = 0$), it means that the S_{upji} (or S_{lowji}) switch is connected to the negative terminal of its capacitor, i.e. the output voltage of the ith SM of the upper half-bridge arm (or lower half-bridge arm) in Phase j is equal to 0.

Considering that the output of each half-bridge arm can be understood as a controllable voltage source, the output voltage satisfies Eq. (9.1).

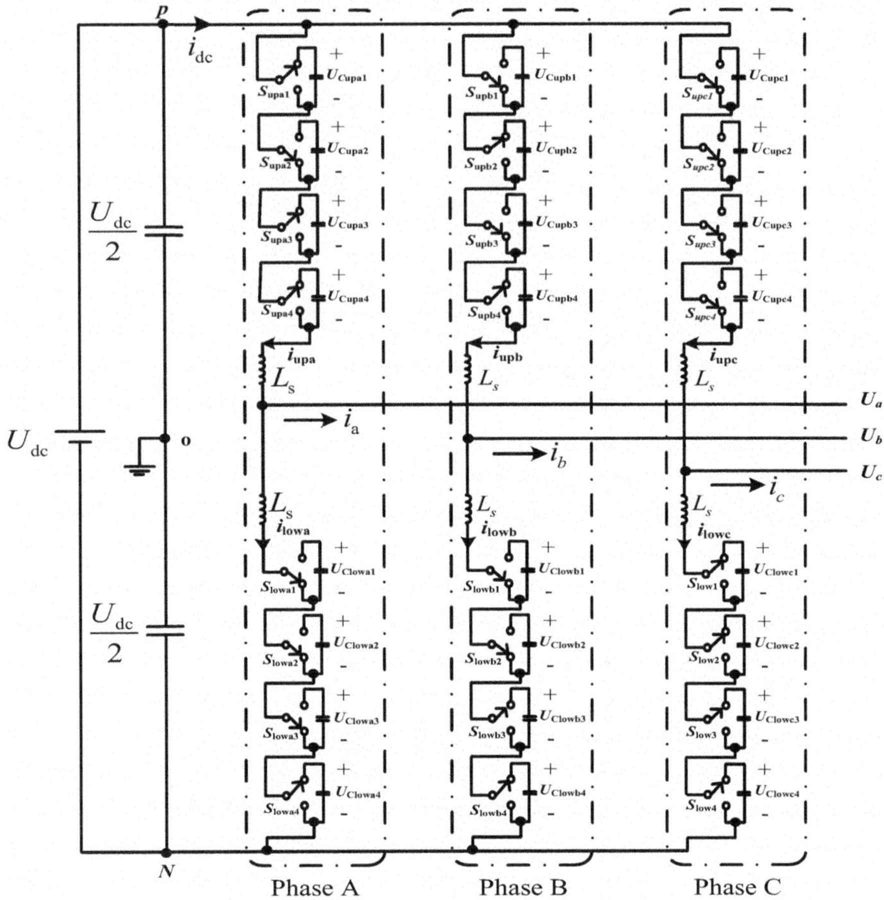

Fig. 9.2 Equivalent topology of a typical three-phase five-level MMC converter

$$
\begin{aligned}
u_{\mathrm{upj}} &= \sum_{i-1}^{4} (S_{\mathrm{upji}} U_{\mathrm{Cupji}}) + L_{\mathrm{S}} \frac{di_{\mathrm{upj}}}{dt} \\
u_{\mathrm{lowj}} &= \sum_{i-1}^{4} (S_{\mathrm{lowji}} U_{\mathrm{Clowji}}) + L_{\mathrm{S}} \frac{di_{\mathrm{lowj}}}{dt}
\end{aligned}
\qquad \mathrm{j} = \mathrm{a, b, c} \tag{9.1}
$$

where $U_{\mathrm{up}j}$ and $U_{\mathrm{low}j}$ indicate the output voltages of the upper and lower half-bridge arms respectively.

The current equation for each half-bridge arm can be expressed as

$$
\begin{aligned}
i_{\mathrm{upj}} &= \frac{i_{\mathrm{j}}}{2} + i_{\mathrm{diff}} \\
i_{\mathrm{lowj}} &= -\frac{i_{\mathrm{j}}}{2} + i_{\mathrm{diff}}
\end{aligned} \tag{9.2}
$$

where $i_{\text{diff}i}$ indicates the current deviation, which consists of two components, the AC component (i_{zj}) and the DC component (i_{yj}), as shown in Eq. (9.3) [3].

$$i_{\text{diffj}} = \frac{i_{\text{upj}} + i_{\text{lowj}}}{2} = i_{\text{zj}} + i_{\text{yj}} \tag{9.3}$$

where i_{zj} is the inter-phase loop current and $i_{\text{za}} + i_{\text{zb}} + i_{\text{zc}} = 0$. This loop current has no effect on both the AC and DC sides of the converter but can have an impact on the voltage ripple of the SM storage capacitors, thus affecting the selection of the rated parameters of the converter devices [2].

Taking idc to as the DC current on the DC side of the converter, and in balanced state, there is

$$i_{\text{yj}} = \frac{i_{\text{dc}}}{3} \tag{9.4}$$

It can be known from Eqs. (9.2) and (9.4) that

$$\begin{aligned} i_{\text{upj}} &= \tfrac{i_{\text{j}}}{2} + \tfrac{i_{\text{dc}}}{3} + i_{\text{zj}} \\ i_{\text{lowj}} &= -\tfrac{i_{\text{j}}}{2} + \tfrac{i_{\text{dc}}}{3} + i_{\text{zj}} \end{aligned} \tag{9.5}$$

The phase voltage output of AC side of the converter can be expressed as

$$U_{\text{jo}} = \frac{U_{\text{dc}}}{2} - U_{\text{jp}} = -\frac{U_{\text{dc}}}{2} + U_{\text{jn}} \tag{9.6}$$

The output voltages of the upper and lower half-bridge arms of the converter can be controlled by controlling the switching states of the SMs, so that the entire converter output voltage on the AC side can be controlled.

The DC-side voltage of the converter can be expressed as

$$\begin{aligned} U_{\text{dc}} = u_{\text{upj}} + u_{\text{lowj}} &= \sum_{i-1}^{4} (S_{\text{upji}} U_{\text{Cupji}}) + \sum_{i-1}^{4} (S_{\text{lowji}} U_{\text{Clowji}}) \\ &+ L_{\text{S}} \frac{di_{\text{lowj}}}{dt} + L_{\text{S}} \frac{di_{\text{upj}}}{dt} \end{aligned} \tag{9.7}$$

9.1.1.3 PWM Method Considering Capacitor Voltage Balance

Here a phase-disposed carrier modulation method is used to modulate the above three-phase five-level MMC converter.

In the MMC function diagram shown in Fig. 9.2, the IGBTs in each SM can be controlled by the switching functions $S_{\text{up}ji}$ and $S_{\text{low}ji}$, and for one bridge arm the number of switching functions being 1 must be 4 at any given moment, which means that 4 SMs in any bridge arm must be in the on state and the other 4 SMs in the off state, i.e.

$$n_{\mathrm{upj}} + n_{\mathrm{lowj}} = 4 \tag{9.8}$$

where $n_{\mathrm{up}j}$ is the number of SMs in the upper half-bridge arm of Phase j that are in the on state and $n_{\mathrm{low}j}$ is the number of SMs in the lower half-bridge arm of Phase j that are in the on state.

Here, the number of on-state SMs with the upper and lower half-bridge arms in Phase j depends on the magnitude of the reference voltage in Phase j. Assuming that the capacitor voltage in each SM is $U_{\mathrm{c}} = U_{\mathrm{dc}}/4$, the five-level voltage output between Phase j and the neutral point O is

- Voltage Level 1: $U_{\mathrm{o}j} = -U_{\mathrm{dc}}/2$, all 4 SMs of the upper half-bridge arm are on and all 4 SMs of the lower half-bridge arm are off, e.g. $n_{\mathrm{up}j} = 4$ and $n_{\mathrm{low}j} = 0$;
- Voltage Level 2: $U_{\mathrm{o}j} = -U_{\mathrm{dc}}/4$, 3 SMs of the upper half-bridge arm are on (1 SM is off) and 1 SM of the lower half-bridge arm is on (3 SMs are off), e.g. $n_{\mathrm{up}j} = 3$ and $n_{\mathrm{low}j} = 1$;
- Voltage Level 3: $U_{\mathrm{o}j} = 0$, 2 SMs of the upper half-bridge arm are on (2 SMs are off) and 2 SMs of the lower half-bridge arm are on (2 SMs are off), e.g. $n_{\mathrm{up}j} = 2$ and $n_{\mathrm{low}j} = 2$;
- Voltage Level 4: $U_{\mathrm{o}j} = +\mathrm{U}_{\mathrm{dc}}/4$, 1 SM of the upper half-bridge arm is on (3 SMs are off) and 3 SMs of the lower half-bridge arm are on (1 SM is off), e.g. $n_{\mathrm{up}j} = 1$ and $n_{\mathrm{low}j} = 3$;
- Voltage Level 5: $U_{\mathrm{o}j} = +U_{\mathrm{dc}}/2$, all 4 SMs of the upper half-bridge arm are off and all 4 SMs of the lower half-bridge arm are on, e.g. $n_{\mathrm{up}j} = 0$ and $n_{\mathrm{low}j} = 4$;

To further confirm which SMs are on and which are off in each of the upper and lower half-bridge arms of a phase during MMC modulation, measurements of the SM capacitor voltage and bridge arm current are required.

When the current direction in the upper and lower half-bridge arms is positive, the SM in the on state will be charged and its capacitor voltage will rise; when the current direction in the upper and lower half-bridge arms is negative, the SM in the on state will be discharged and its capacitor voltage will fall. The SM in the off state will be bypassed and its capacitor voltage will remain unchanged, independent to the current direction of the half-bridge arm.

To maintain a balanced capacitor voltage for each sub-module in the MMC converter, before each control of the switching function of each SM, the capacitor voltages of all SMs are measured for the current phase and arranged in descending order by grouping the upper and lower half-bridge arms, and then: (1) If the current direction of the upper half-bridge arm is positive, the switching function of the $n_{\mathrm{up}j}$ SMs with the lowest capacitor voltage in the upper half-bridge arm group is set to 1; (2) If the current direction of the upper half-bridge arm is negative, the switching function of the $n_{\mathrm{up}}j$ SMs with the highest capacitor voltage in the upper half-bridge arm group is set to 1; (3) If the current direction of the lower half-bridge arm is positive, the switching function of the $n_{\mathrm{low}}j$ SMs with the lowest capacitor voltage in the lower half-bridge arm group is set to 1; (4) If the current direction of the lower

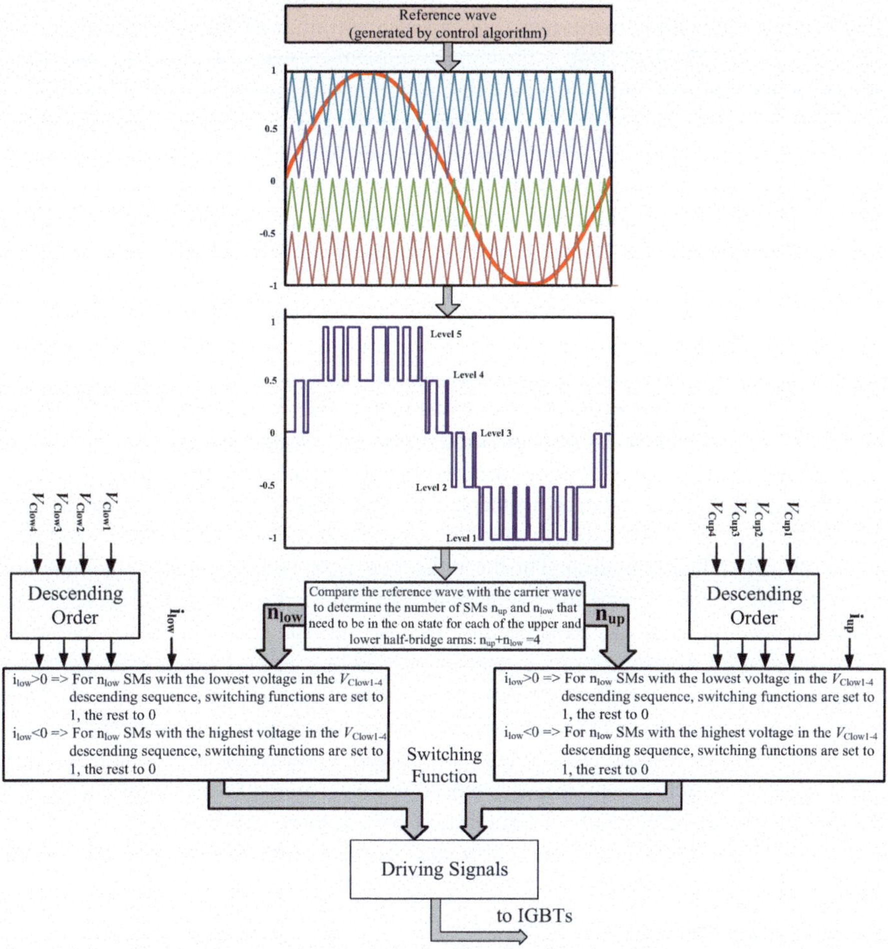

Fig. 9.3 Modulation method for 5-level MMC with voltage balance capability

half-bridge arm is negative, the switching function of the n_{lowj} SMs with the highest capacitor voltage in the lower half-bridge arm group is set to 1.

In summary, the three-phase five-level MMC converter PWM method described in this section is obtained as shown in Fig. 9.3.

9.1.2 Topology of the MMC-Based HVDC Transmission System for Offshore Wind Farms

In the typical topology of a series VSC-HVDC as shown in Fig. 1.29, offshore wind farms obtain the high DC voltage through series connection of the DC side of the wind turbine and achieve transformerless transmission, eliminating the need for large and bulky step-up transformers, thus improving system efficiency and reliability, reducing system weight and cost, but there is also the problem that individual machine and line faults are difficult to handle [4–6]. Multi-phase high power wind turbines have also started to be used in offshore wind power systems, where each symmetrical three-phase winding in a multi-phase wind turbine is connected to the AC side of a two-level AC/DC converter and the DC side of these AC/DC converters are connected in series to obtain the required high DC voltage. This topology reduces the total DC-side capacitance, uses a two-level structure to reduce the voltage and current requirements of the power switching devices, and each AC/DC converter can be independently protected and isolated, improving the system's ability to operate uninterruptedly in the event of a fault, but there are also problems, namely: the DC-side capacitor array is not effectively protected in the event of a short circuit, and the stability of the DC-side voltage depends on the regulation capability of all converters.

The topology of the MMC-based offshore wind farm HVDC transmission system discussed in this section is shown in Fig. 9.4. The offshore section consists of five generation units, each consisting of a wind turbine and its associated PMSG and a 3-Level NPC (3L-NPC) converter. The DC side of each 3L-NPC converter is connected to a Half-Bridge (H-B), and the output side of each H-B is connected in series to obtain the required high DC voltage and transmit the energy from each generation unit to the onshore section via a DC transmission cable. The onshore section uses an MMC converter to convert the high voltage DC power into AC power and feed it into the grid, while controlling the total voltage stabilisation on the DC side (Fig. 9.5).

The advantages of this topology are: (1) each wind turbine gets independent MPPT control, which means that the whole wind farm can work in global MPPT when the wind speed at each turbine varies over a large offshore area; (2) direct access to high DC voltage without additional boosting parts, thus improving system

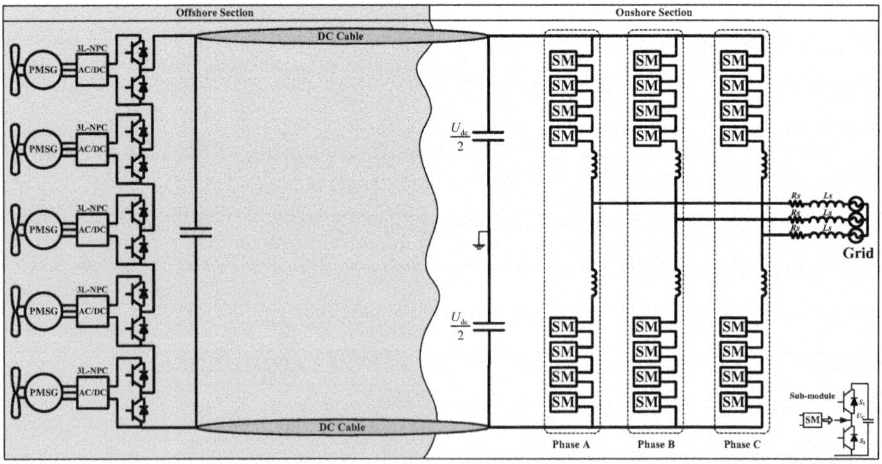

Fig. 9.4 Topology of the MMC-based HVDC transmission system for offshore wind farms

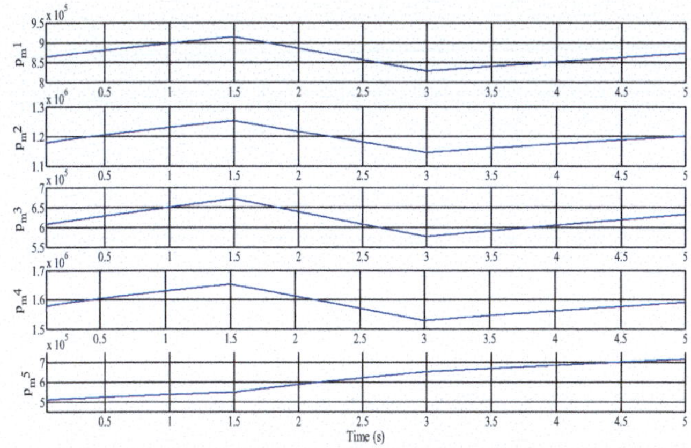

(a) Output power of each generation unit

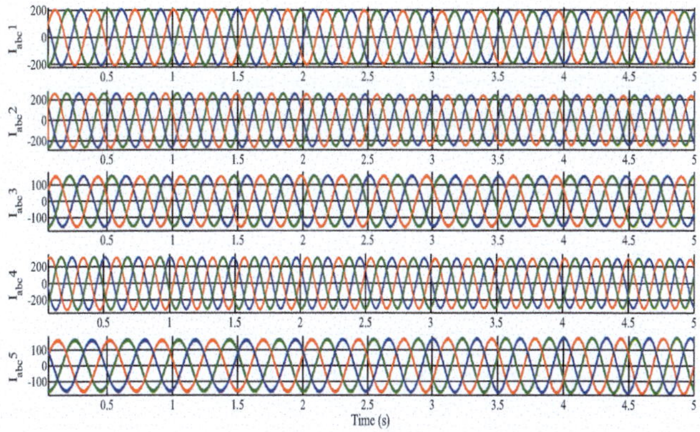

(b) Phase currents of each PMSG

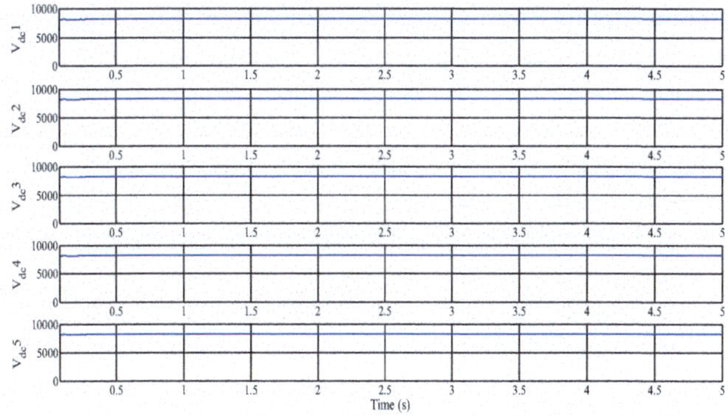

(c) DC-bus voltages of each 3L-NPC Converter

Fig. 9.5 Simulation results of MMC-based HVDC transmission system for offshore wind farms

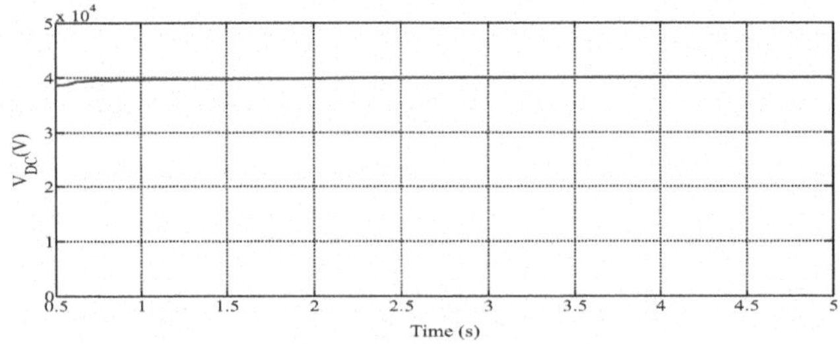

(d) High DC voltage between DC cables

Fig. 9.5 (continued)

efficiency, reducing system cost and increasing system integration. However, the 3L-NPC converter increases the weight of the generation unit to a certain extent, and its capacitor array can also affect the system lifetime to a certain extent. In order to improve the reliability of this topology, and in particular to enhance the uninterrupted operation of the series branches of the generation units, the array of generation units can be connected in series and parallel, i.e. a part of the generation units can be connected in series and then all the series branches can be connected in parallel to form a hybrid connected array of generation units; in addition, crowbar units can also be added to the DC side of each 3L-NPC converter for the bypass of the generation units under fault and online maintenance.

9.1.3 Control of the MMC-Based HVDC Transmission System for Offshore Wind Farms

In the topology shown in Fig. 9.4, the offshore section consists of the generation units and the H-B units. In particular, each generation unit contains a 3L-NPC converter for controlling the corresponding PMSG in each generation unit. The control method can be found in Chap. 7, Sect. 7.1.2, where the MPPT control of each PMSG is achieved by controlling the rotational speed; the output side of each H-B unit is serially connected for obtaining the required high DC voltage for high voltage DC transmission and, at the same time, for balancing the DC bus voltage of each 3L-NPC converter.

In the topology shown in Fig. 9.4, a 5-level MMC converter is used in the onshore section to control the high DC voltage between the HVDC transmission cables as its reference value and to convert and feed the power from the offshore section into the grid. The control method can be referred to the two-level grid-side converter control method in Chap. 7, Sect. 7.1.2, and the reference voltage generated by this control method will be used as reference for the phase-disposed carrier PWM method (shown in Fig. 9.3), which drives the MMC converter and implements the relevant control process.

9.2 MFT-Based HVDC Transmission System for Offshore Wind Farms

9.2.1 Matrix Converter

The Matrix Converter (MC) is a direct AC/AC converter that uses bidirectional controllable switches to control the amplitude and frequency of the output AC voltage without the need for storage capacitors, also known as an all-silicon solution. It features adjustable power factor, bi-directional power delivery, compact construction, high power density and has a voltage modulation ratio of 86% [7].

9.2.1.1 Basic Structure of the Matrix Converter

An $m \times n$ matrix converter with an m-phase input voltage and an n-phase output voltage requires $m \times n$ bi-directional power switches. The most widely used 3×3 matrix converter is shown in Fig. 9.6. It contains nine bi-directional power switches arranged in an array to ensure that any phase input can be connected to any phase output at any time, where an L_C filter is used to avoid overvoltage on the input side and to filter out high frequency currents on the input side. In general, the input of a matrix converter is mostly a voltage source, so its input is not allowed to be short-circuited, while the matrix converter is mostly with a resistive-inductive load, so its output side is not allowed to be open-circuited.

Bi-directional power switches in a matrix converter require bi-directional blocking capability of current and voltage, and are mostly composed of a group of power semiconductor switches (e.g. IGBTs, MOSFETs, etc.) connected in some series-parallel form [8]. Four common types of bidirectional switches are shown in Fig. 9.7, including: diode bridge parallel IGBT, common emitter reverse series IGBT, common

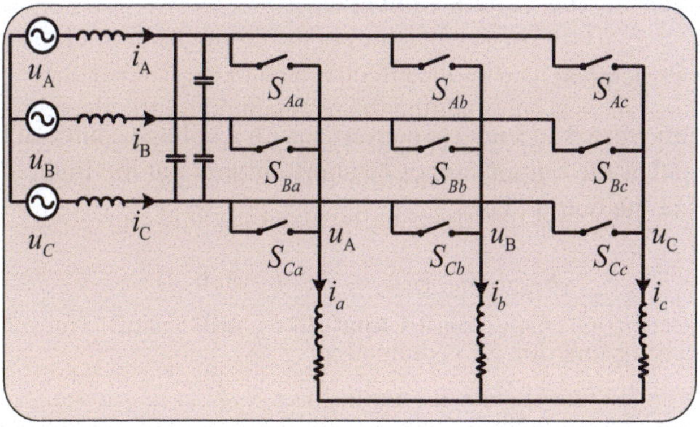

Fig. 9.6 Topology of a 3×3 matrix converter

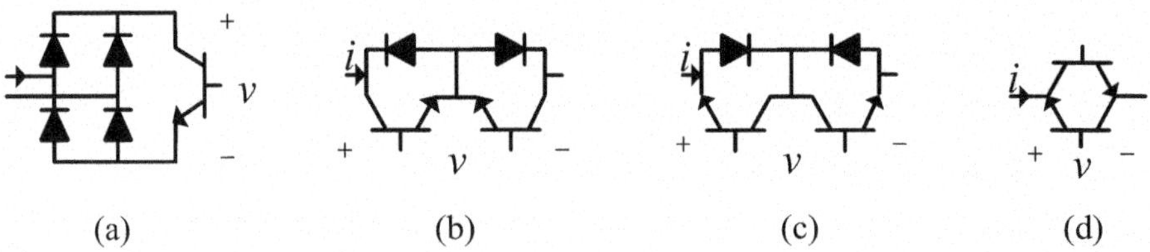

Fig. 9.7 Common Bi-directional power switches, **a** diode bridge parallel IGBT, **b** common emitter reverse series IGBT, **c** common collector reverse series IGBT, **d** reverse parallel IGBT

Table 9.2 Comparison among Bi-directional power switches

Type	Number of tubes	Advantages	Disadvantages
Diode bridge parallel IGBT	4 diodes 1 IGBT	Only 1 drive signal required	3 switches on simultaneously, increasing conduction losses
Common emitter reverse series IGBT	2 diodes 2 IGBTs	Separate control of currents in different directions	Too many isolated power supplies required for driving
Common collector reverse series IGBT	2 diodes 2 IGBTs	Separate control of currents in different directions	Too many isolated power supplies required for driving and commutation problem in large-capacity system
Reverse parallel IGBT	2 IGBTs	Compact and inexpensive	Poor reverse voltage tolerance

collector reverse series IGBT and reverse parallel IGBT. Table 9.2 provides a comparison of these common bidirectional power switches.

9.2.1.2 SVPWM of the Matrix Converter

1. Switching Principles

As mentioned above, a 3 × 3 matrix converter with a voltage source at the input and an inductive load at the output cannot be short-circuited at the input and cannot be open-circuited at the output, i.e.

$$S_{Aj} + S_{Bj} + S_{Cj} = 1,\quad j=(\mathrm{a, b, c}) \tag{9.9}$$

where the switching function S_{ij} is defined as

$$S_{ij} = \begin{cases} 1, \text{ switch } S_{ij} \text{ is on} \\ 0, \text{ switch } S_{ij} \text{ is off} \end{cases} \quad i = (\mathrm{A, B, C}) \;\; j = (\mathrm{a, b, c}) \tag{9.10}$$

Equation (9.9) shows that there is one and only one switch on in any output phase of the matrix converter at any given moment, so there are 27 switching states in a 3 × 3 matrix converter, corresponding to 27 output voltage vectors.

In Fig. 9.6, u_A, u_B and u_C are defined as the input phase voltage, i_A, i_B and i_C as the input phase current, u_a, u_b and u_c as the output phase voltage, i_a, i_b and i_c as the output phase current, U_{in} as the input voltage amplitude and I_o as the output current amplitude. For a three-phase sinusoidal system, the output voltage and input current can be expressed as

$$\begin{bmatrix} u_A(t) \\ u_B(t) \\ u_C(t) \end{bmatrix} = U_{in} \begin{bmatrix} \cos(\omega_{in}t) \\ \cos(\omega_{in}t - \frac{2\pi}{3}) \\ \cos(\omega_{in}t + \frac{2\pi}{3}) \end{bmatrix} \quad \text{and} \quad \begin{bmatrix} i_a(t) \\ i_b(t) \\ i_c(t) \end{bmatrix} = I_o \begin{bmatrix} \cos(\omega_o t) \\ \cos(\omega_o t - \frac{2\pi}{3}) \\ \cos(\omega_o t + \frac{2\pi}{3}) \end{bmatrix} \tag{9.11}$$

$$u_A + u_B + u_C = 0 \quad \text{and} \quad i_a + i_b + i_c = 0 \tag{9.12}$$

In a matrix converter, the output voltage reference vector $\boldsymbol{U}_o$ and the input current reference vector $\boldsymbol{I}_{in}$ can be expressed as

$$\begin{aligned} \boldsymbol{U}_o(t) &= \frac{2}{3}\left(u_a(t) + u_b(t)e^{j2\pi/3} + u_c(t)e^{j4\pi/3}\right) \\ &= \mathrm{Re}(\boldsymbol{U}_o(t)) + j\mathrm{Im}(\boldsymbol{U}_o(t)) + \\ &= \frac{2}{3}\left(u_a + \cos\left(\frac{2\pi}{3}\right) \cdot u_b + \cos\left(\frac{4\pi}{3}\right) \cdot u_c\right) \\ &+ j\frac{2}{3}\left(\sin\left(\frac{2\pi}{3}\right) \cdot u_b + \sin\left(\frac{4\pi}{3}\right) \cdot u_c\right) \\ &= \frac{2}{3}\left(u_a - \frac{1}{2} \cdot u_b - \frac{1}{2} \cdot u_c\right) + j\frac{1}{\sqrt{3}}(u_b - u_c) \end{aligned} \tag{9.13}$$

$$\begin{aligned} \boldsymbol{I}_{in}(t) &= \frac{2}{3}\left(i_A(t) + i_B(t)e^{j2\pi/3} + i_C(t)e^{j4\pi/3}\right) \\ &= \mathrm{Re}(\boldsymbol{I}_{in}(t)) + j\mathrm{Im}(\boldsymbol{I}_{in}(t)) \\ &= \frac{2}{3}\left(i_A + \cos\left(\frac{2\pi}{3}\right) \cdot i_B + \cos\left(\frac{4\pi}{3}\right) \cdot i_C\right) \\ &+ j\frac{2}{3}\left(\sin\left(\frac{2\pi}{3}\right) \cdot i_B + \sin\left(\frac{4\pi}{3}\right) \cdot i_C\right) \\ &= \frac{2}{3}\left(i_A - \frac{1}{2} \cdot i_B - \frac{1}{2} \cdot i_C\right) + j\frac{1}{\sqrt{3}}(i_B - i_C) \end{aligned} \tag{9.14}$$

where

$$|\boldsymbol{U}_o(t)| = \sqrt{(\mathrm{Re}(\boldsymbol{U}_o(t)))^2 + (\mathrm{Im}(\boldsymbol{U}_o(t)))^2}$$

$$= \sqrt{\left(\frac{2}{3}\left(u_a - \frac{1}{2}\cdot u_b - \frac{1}{2}\cdot u_c\right)\right)^2 + \left(\frac{1}{\sqrt{3}}(u_b - u_c)\right)^2}$$

$$\angle \boldsymbol{U}_o(t) = \sin^{-1}\frac{\mathrm{Im}(\boldsymbol{U}_o(t))}{|\boldsymbol{U}_o(t)|} = \cos^{-1}\frac{\mathrm{Re}(\boldsymbol{U}_o(t))}{|\boldsymbol{U}_o(t)|} \tag{9.15}$$

$$|\boldsymbol{I}_{in}(t)| = \sqrt{(\mathrm{Re}(\boldsymbol{I}_{in}(t)))^2 + (\mathrm{Im}(\boldsymbol{I}_{in}(t)))^2}$$

$$\angle \boldsymbol{I}_{in}(t) = \sin^{-1}\frac{\mathrm{Im}(\boldsymbol{I}_{in}(t))}{|\boldsymbol{I}_{in}(t)|} = \cos^{-1}\frac{\mathrm{Re}(\boldsymbol{I}_{in}(t))}{|\boldsymbol{I}_{in}(t)|} \tag{9.16}$$

2. Switching States

According to the amplitude and phase of the output voltage, the switching states of a 3 × 3 matrix converter can be divided into three categories:

(1) The first category, containing a total of 18 switching states, will result in the converter having two and only two output phases connected to the same input phase, where the converter output voltage and input current have a fixed direction and their amplitude varies with the phase of the input voltage and output current respectively, so the output voltage vectors corresponding to the first category of switching states are called Stationary/Active Vectors. An example is given in Fig. 9.8, where the switches S_{Aa}, S_{Ab} and S_{Cc} are on, the output voltage $u_a = u_b = u_A$ and $u_c = u_C$, and the input current $i_A = i_a + i_b$ and $i_B = 0, i_C = i_c$.

In the switching state shown in Fig. 9.8, it can be known from Eqs. (9.13) and (9.15) that

$$\begin{aligned} \mathrm{Re}(\boldsymbol{U}_o(t)) &= \tfrac{2}{3}\left[u_a + \cos\left(\tfrac{2\pi}{3}\right)u_b + \cos\left(\tfrac{4\pi}{3}\right)u_c\right] = \tfrac{2}{3}\left[u_a - \tfrac{1}{2}\cdot u_b - \tfrac{1}{2}\cdot u_c\right] = \tfrac{1}{3}u_{AC} \\ \mathrm{Im}(\boldsymbol{U}_o(t)) &= \tfrac{2}{3}\left[\sin\left(\tfrac{2\pi}{3}\right)u_b + \sin\left(\tfrac{4\pi}{3}\right)u_c\right] = \tfrac{1}{\sqrt{3}}[u_b - u_c] = \tfrac{1}{\sqrt{3}}u_{AC} = \sqrt{3}\mathrm{Re}(\boldsymbol{U}_o(t)) \end{aligned} \tag{9.17}$$

$$\begin{aligned} |\boldsymbol{U}_o(t)| &= \sqrt{(\mathrm{Re}(\boldsymbol{U}_o(t)))^2 + (\mathrm{Im}(\boldsymbol{U}_o(t)))^2} = \frac{2}{3}|U_{CA}| = \left|\frac{2}{\sqrt{3}}U_{in}\cos\left(\omega_{in}t + \frac{\pi}{6}\right)\right| \\ \angle \boldsymbol{U}_o(t) &= \sin^{-1}\frac{\mathrm{Im}(\boldsymbol{U}_o(t))}{|\boldsymbol{U}_o(t)|} = 60°\text{或}240° \end{aligned} \tag{9.18}$$

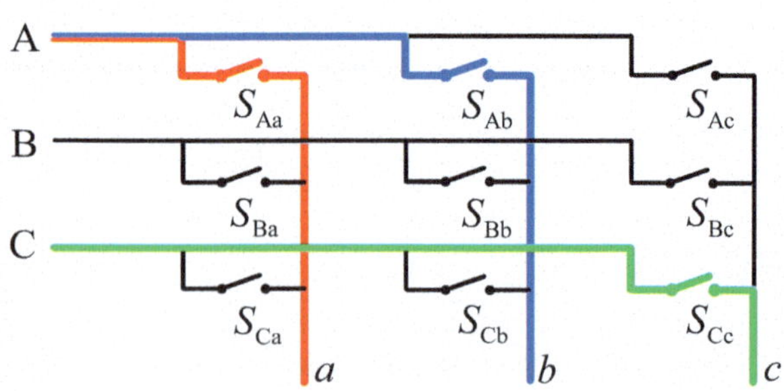

Fig. 9.8 An example of a switching state in category I

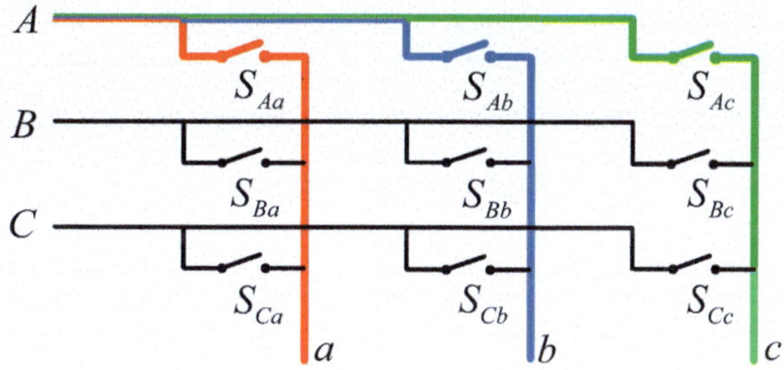

Fig. 9.9 An example of a switching state in category II

In the switching state shown in Fig. 9.8, it can be known from Eqs. (9.14) and (9.16) that

$$
\begin{aligned}
\mathrm{Re}(\boldsymbol{I}_{\mathrm{in}}(t)) &= \frac{2}{3}\left[i_{\mathrm{A}} - \frac{1}{2}i_{\mathrm{B}} - \frac{1}{2}i_{\mathrm{C}}\right] = -i_{\mathrm{c}} \\
\mathrm{Im}(\boldsymbol{I}_{\mathrm{in}}(t)) &= \frac{1}{\sqrt{3}}[i_{\mathrm{B}} - i_{\mathrm{C}}] = -\frac{1}{\sqrt{3}}i_{\mathrm{c}}
\end{aligned}
\tag{9.19}
$$

$$
\begin{aligned}
|\boldsymbol{I}_{\mathrm{o}}(t)| &= \sqrt{\left(\mathrm{Re}(\boldsymbol{I}_{\mathrm{o}}(t))\right)^2 + \left(\mathrm{Im}(\boldsymbol{I}_{\mathrm{o}}(t))\right)^2} = \frac{2}{\sqrt{3}}i_{\mathrm{c}} \\
\angle \boldsymbol{I}_{\mathrm{o}}(t) &= \sin^{-1}\frac{\mathrm{Im}(\boldsymbol{I}_{\mathrm{o}}(t))}{|\boldsymbol{I}_{\mathrm{o}}(t)|} = 30^{o}\text{或}210^{o}
\end{aligned}
\tag{9.20}
$$

(2) The second category, containing a total of 3 switching states, will result in the converter having all output phases connected to the same input phase. An example is shown in Fig. 9.9, where switches S_{Aa}, S_{Ab} and S_{Ac} are on, the output voltage $u_{\mathrm{a}} = u_{\mathrm{b}} = u_{\mathrm{c}} = u_{\mathrm{A}}$ and the input current $i_{\mathrm{A}} = i_{\mathrm{a}} + i_{\mathrm{b}} + i_{\mathrm{c}} = 0$.

The output voltage vectors corresponding to this type of switching states are located at the origin of the vector space and are known as Zero Vectors, which produce zero output voltage and zero input current, i.e. $\mathrm{Re}(\boldsymbol{U}_{\mathrm{o}}(t)) = \mathrm{Im}(\boldsymbol{U}_{\mathrm{o}}(t)) = 0$ and $\mathrm{Re}(\boldsymbol{I}_{\mathrm{in}}(t)) = \mathrm{Im}(\boldsymbol{I}_{\mathrm{in}}(t)) = 0$.

(3) The third category, containing a total of six switching states, will result in the converter output phases being connected to different input phases, where the direction and amplitude of the converter output voltage and input current are both variable, so the output voltage vectors corresponding to the third category of switching states are called Rotating Vectors. An example is shown in Fig. 9.10, where the switches S_{Aa}, S_{Bb} and S_{Cc} are on, the output voltage $u_{\mathrm{a}} = u_{\mathrm{A}}$, $u_{\mathrm{b}} = u_{\mathrm{B}}$ and $u_{\mathrm{c}} = u_{\mathrm{C}}$, and the input current $i_{\mathrm{A}} = i_{\mathrm{a}}$, $i_{\mathrm{B}} = i_{\mathrm{b}}$ and $i_{\mathrm{C}} = i_{\mathrm{c}}$.

In the switching state shown in Fig. 9.10, it can be known from Eqs. (9.13) and (9.15) that

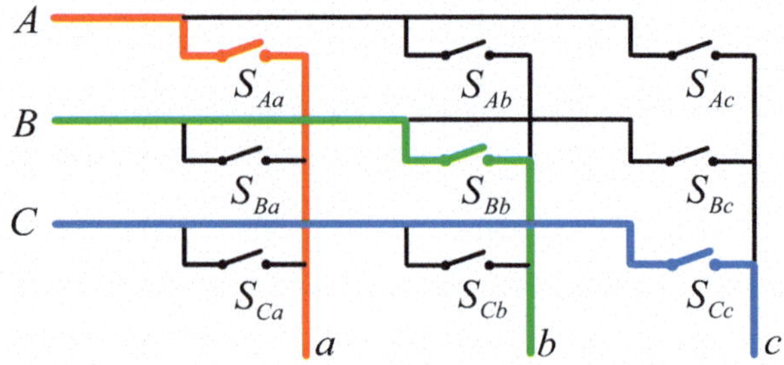

Fig. 9.10 An example of a switching state in category III

$$
\begin{aligned}
\mathrm{Re}[\boldsymbol{U}_{\mathrm{o}}(t)] &= \frac{2}{3}\left[u_{\mathrm{a}} - \frac{1}{2}u_{\mathrm{b}} - \frac{1}{2}u_{\mathrm{c}}\right] = \frac{2}{3}\left[u_{\mathrm{A}} - \frac{1}{2}u_{\mathrm{B}} - \frac{1}{2}u_{\mathrm{C}}\right] \\
&= \frac{2}{3}U_{\mathrm{in}}\left[\cos(\omega_{\mathrm{in}}t) - \frac{1}{2}\cos(\omega_{\mathrm{in}}t - \frac{2\pi}{3}) - \frac{1}{2}\cos(\omega_{\mathrm{in}}t - \frac{4\pi}{3})\right] \\
&= U_{\mathrm{in}}\cos(\omega_{\mathrm{in}}t) \\
\mathrm{Im}[\boldsymbol{U}_{\mathrm{o}}(t)] &= \frac{1}{\sqrt{3}}[u_{\mathrm{b}} - u_{\mathrm{c}}] = \frac{1}{\sqrt{3}}[u_{\mathrm{B}} - u_{\mathrm{C}}] \\
&= \frac{1}{\sqrt{3}}U_{\mathrm{in}}\left[\cos(\omega_{\mathrm{in}}t - \frac{2\pi}{3}) - \cos(\omega_{\mathrm{in}}t - \frac{4\pi}{3})\right] \\
&= U_{\mathrm{in}}\sin(\omega_{\mathrm{in}}t)
\end{aligned}
\tag{9.21}
$$

$$
\begin{aligned}
|\boldsymbol{U}_{\mathrm{o}}(t)| &= \sqrt{(\mathrm{Re}(\boldsymbol{U}_{\mathrm{o}}(t)))^2 + (\mathrm{Im}(\boldsymbol{U}_{\mathrm{o}}(t)))^2} = U_{\mathrm{in}} \\
\angle\boldsymbol{U}_{\mathrm{o}}(t) &= \sin^{-1}\frac{\mathrm{Im}(\boldsymbol{U}_{\mathrm{o}}(t))}{|\boldsymbol{U}_{\mathrm{o}}(t)|} = \omega_{\mathrm{in}}t
\end{aligned}
\tag{9.22}
$$

In the switching state shown in Fig. 9.10, it can be known from Eqs. (9.14) and (9.16) that

$$
\begin{aligned}
\mathrm{Re}[\boldsymbol{I}_{\mathrm{in}}(t)] &= \tfrac{2}{3}\left[i_{\mathrm{A}} - \tfrac{1}{2}\cdot i_{\mathrm{B}} - \tfrac{1}{2}\cdot i_{\mathrm{C}}\right] = \tfrac{2}{3}\left[i_{\mathrm{a}} - \tfrac{1}{2}i_{\mathrm{b}} - \tfrac{1}{2}i_{\mathrm{c}}\right] \\
&= \tfrac{2}{3}I_{\mathrm{o}}\left[\cos(\omega_{\mathrm{o}}t) - \tfrac{1}{2}\cos\left(\omega_{\mathrm{o}}t - \tfrac{2\pi}{3}\right) - \tfrac{1}{2}\cos\left(\omega_{\mathrm{o}}t + \tfrac{2\pi}{3}\right)\right] \\
&= I_{\mathrm{o}}\cos(\omega_{\mathrm{o}}t) \\
\mathrm{Im}[\boldsymbol{I}_{\mathrm{in}}(t)] &= \tfrac{1}{\sqrt{3}}[i_{\mathrm{B}} - i_{\mathrm{C}}] = \tfrac{1}{\sqrt{3}}[i_{\mathrm{b}} - i_{\mathrm{c}}] \\
&= \tfrac{1}{\sqrt{3}}I_{\mathrm{o}}\left[\cos(\omega_{\mathrm{o}}t - \tfrac{2\pi}{3}) - \cos(\omega_{\mathrm{o}}t - \tfrac{4\pi}{3})\right] \\
&= I_{\mathrm{o}}\sin(\omega_{\mathrm{o}}t)
\end{aligned}
\tag{9.23}
$$

$$
\begin{aligned}
|\boldsymbol{I}_{\mathrm{in}}(t)| &= \sqrt{(\mathrm{Re}(\boldsymbol{I}_{\mathrm{in}}(t)))^2 + (\mathrm{Im}(\boldsymbol{I}_{\mathrm{in}}(t)))^2} = I_{\mathrm{o}} \\
\angle\boldsymbol{I}_{\mathrm{in}}(t) &= \sin^{-1}\frac{\mathrm{Im}(\boldsymbol{I}_{\mathrm{in}}(t))}{|\boldsymbol{I}_{\mathrm{in}}(t)|} = \omega_{\mathrm{o}}t
\end{aligned}
\tag{9.24}
$$

Table 9.3 gives all 27 of the switching states of a 3 × 3 matrix converter, where the active vectors are numbered +1, −1, ..., +9, −9.

As can be seen from the table above, the first category of switching states can be subdivided into three subcategories, with 6 space vectors in the same direction in each subcategory, where the space vectors in different subcategories differ in phase by 120° from each other. The space vectors in the first and second categories can be used as the basic space vectors in the modulation method to synthesise the reference vector, while the space vectors in the third category change direction in real time and therefore cannot be used as the basic vectors in the modulation method [9].

3. SVPWM Technology of a 3-phase Matrix Converter

A three-phase matrix converter modulation requires to use active and zero vectors to synthesise the output voltage reference vector and the input current reference vector, as shown in Fig. 9.11, where k_v is the sector number of voltage vector space and k_i is the sector number of current vector space.

It can be seen that the output voltage reference vector and the input current reference vector in any sector can be synthesised from the two basic vectors located at the boundary of the sector in which the synthesised vector is located. The output voltage reference vector $\boldsymbol{U}_o$ in Fig. 9.11a is located in Sector 1 and can be synthesised from the basic output voltage vectors $\boldsymbol{U}_{01}$ (corresponding to the vector number of ±7, ±8, ±9) and $\boldsymbol{U}_{02}$ (corresponding to the vector number of ±1, ±2, ±3); the input current reference vector $\boldsymbol{I}_i$ in Fig. 9.11b is located in Sector 1 and can be synthesised from the basic input current vectors $\boldsymbol{I}_{01}$ (corresponding to the vector number of ±3, ±6, ±9) and $\boldsymbol{I}_{02}$ (corresponding to the vector number of ±1, ±4, ±7).

The space vector modulation process of a three-phase matrix converter needs to consider both the input voltage reference vector and the input current reference vector synthesis process, i.e. the switching states used in the modulation should be those that are common to both the input voltage reference vector synthesis and the input current reference vector synthesis processes. For the modulation processes of the voltage and current reference vectors as shown in Fig. 9.11, the switching states corresponding to the vector number of ±1, ±3, ±7, ±9 should be used, the determination of the sign for which will be discussed subsequently. Similarly, when the output voltage reference vector and the input current reference vector are located in different sectors of their respective spaces, the switching states selected in the modulation method are shown in Table 9.4, where "sequence" is used to mark the switching states in the table [9].

According to Tables 9.4 and 9.5, the output voltage amplitude in the space vector modulation of a three-phase matrix converter is shown in Table 9.5.

In the coordinate frame shown in Fig. 9.11, in order to make sense of the calculated duty cycle: when $k_v = k_i = 1$, the switch states are selected as +9, −7, −3 and +1; when $k_v = 1$ and $k_i = 4$, the switch states are selected as −9, +7, +3 and −1; when $k_v = 4$ and $k_i = 1$, the switch state numbers are selected as −9, +7, +3 and −1; and when $k_v = 4$ and $k_i = 4$, the switch state numbers are selected as +9, −7, −3 and +1. Similarly, the switch state numbers selected for any output voltage vector and input current vector can be obtained as shown in Table 9.6.

Table 9.3 All switching states of a 3 × 3 matrix converter

Category	No	On-state switch	U_o			I_i			Output volt phase angle	Input curr phase angle
			U_a	U_b	U_c	I_A	I_B	I_C		
(I) Active vectors	+1	$S_{Aa}\ S_{Bb}S_{Bc}$	U_A	U_B	U_B	I_a	I_b+I_c	*0*	0°/180°	*330°/150°*
	−1	$S_{Ba}S_{Ab}S_{Ac}$	U_B	U_A	U_A	I_b+I_c	I_a	*0*	0°/180°	*330°/150°*
	+2	$S_{Ba}\ S_{Cb}S_{Cc}$	U_B	U_C	U_C	*0*	I_a	I_b+I_c	0°/180°	*90°/270°*
	−2	$S_{Ca}S_{Bb}\ S_{Bc}$	U_C	U_B	U_B	*0*	I_b+I_c	I_a	0°/180°	*90°/270°*
	+3	$S_{Ca}S_{Ab}S_{Ac}$	U_C	U_A	U_A	I_b+I_c	*0*	I_a	0°/180°	*210°/30°*
	−3	$S_{Aa}\ S_{Cb}S_{Cc}$	U_A	U_C	U_C	I_a	*0*	I_b+I_c	0°/180°	*210°/30°*
	+4	$S_{Ba}S_{Ab}S_{Bc}$	U_B	U_A	U_B	I_b	I_a+I_c	*0*	120°/300°	*330°/150°*
	−4	$S_{Aa}\ S_{Bb}S_{Ac}$	U_A	U_B	U_A	I_a+I_c	I_b	*0*	120°/300°	*330°/150°*
	+5	$S_{Ca}\ S_{Bb}S_{Cc}$	U_C	U_B	U_C	*0*	I_b	I_a+I_c	120°/300°	*90°/270°*
	−5	$S_{Ba}S_{Cb}S_{Bc}$	U_B	U_C	U_B	*0*	I_a+I_c	I_b	120°/300°	*90°/270°*
	+6	$S_{Aa}S_{Cb}S_{Ac}$	U_A	U_C	U_A	I_a+I_c	*0*	I_b	120°/300°	*210°/30°*
	−6	$S_{Ca}\ S_{Ab}S_{Cc}$	U_C	U_A	U_C	I_b	*0*	I_a+I_c	120°/300°	*210°/30°*
	+7	$S_{Ba}\ S_{Bb}S_{Ac}$	U_B	U_B	U_A	I_c	I_a+I_b	*0*	240°/60°	*330°/150°*
	−7	$S_{Aa}\ S_{Ab}S_{Bc}$	U_A	U_A	U_B	I_a+I_b	I_c	*0*	240°/60°	*330°/150°*
	+8	$S_{Ca}S_{Cb}S_{Bc}$	U_C	U_C	U_B	*0*	I_c	I_a+I_b	240°/60°	*90°/270°*
	−8	$S_{Ba}\ S_{Bb}\ S_{Cc}$	U_B	U_B	U_C	*0*	I_a+I_b	I_c	240°/60°	*90°/270°*
	+9	$S_{Aa}S_{Ab}S_{Cc}$	U_A	U_A	U_C	I_a+I_b	*0*	I_c	240°/60°	*210°/30°*
	−9	$S_{Ca}S_{Cb}S_{Ac}$	U_C	U_C	U_A	I_c	*0*	I_a+I_b	240°/60°	*210°/30°*
(II) Zero vectors	O_1	$S_{Aa}S_{Ab}S_{ac}$	U_A	U_A	U_A	*0*	*0*	*0*	–	–
	O_2	$S_{Ba}S_{Bb}S_{Bc}$	U_B	U_B	U_B	*0*	*0*	*0*	–	–
	O_3	$S_{Ca}\ S_{Cb}\ S_{Cc}$	U_C	U_C	U_C	*0*	*0*	*0*	–	–

(continued)

Table 9.3 (continued)

Category	No	On-state switch	U_o			I_i			Output volt phase angle	Input curr phase angle
			U_a	U_b	U_c	I_A	I_B	I_C		
(III) Rotating vectors	R_1	$S_{Aa}S_{Bb}S_{Cc}$	U_A	U_B	U_C	I_a	I_b	I_c	Variable	Variable
	R_2	$S_{Aa}S_{Cb}S_{Bc}$	U_A	U_C	U_B	I_a	I_c	I_b	Variable	Variable
	R_3	$S_{Ba}S_{Cb}S_{Ac}$	U_B	U_C	U_A	I_b	I_c	I_a	Variable	Variable
	R_4	$S_{Ba}S_{Ab}S_{Cc}$	U_B	U_A	U_C	I_b	I_a	I_c	Variable	Variable
	R_5	$S_{Ca}S_{Ab}S_{Bc}$	U_C	U_A	U_B	I_c	I_a	I_b	Variable	Variable
	R_6	$S_{Ca}S_{Bb}S_{Ac}$	U_C	U_B	U_A	I_c	I_b	I_a	Variable	Variable

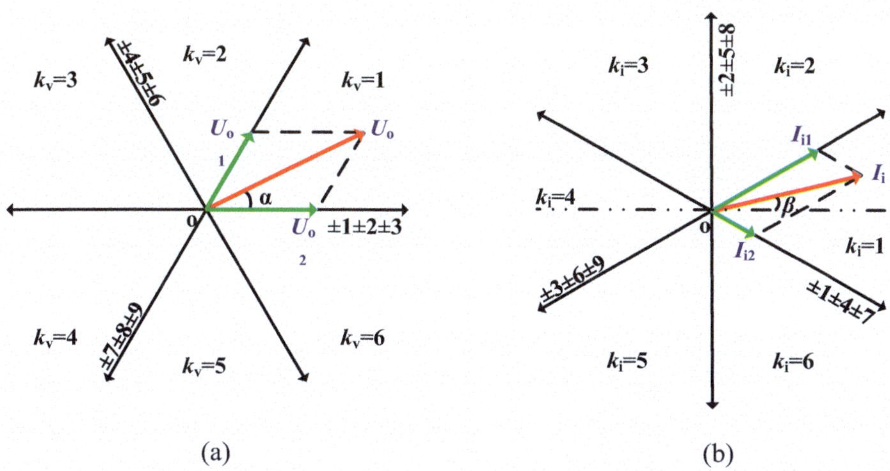

Fig. 9.11 Synthesis of the output voltage vector and input current vector in a 3-phase matrix converter

Table 9.4 Switching state selection for space vector modulation of a 3 × 3 matrix converter

Sector	$k_v = 1$ or 4				$k_v = 2$ or 5				$k_v = 3$ or 6			
$k_i = 1$ or 4	±9	±7	±3	±1	±6	±4	±9	±7	±3	±1	±6	±4
$k_i = 2$ or 5	±8	±9	±2	±3	±5	±6	±8	±9	±2	±3	±5	±6
$k_i = 3$ or 6	±7	±8	±1	±2	±4	±5	±7	±8	±1	±2	±4	±5
Sequence	I	II	III	IV	I	II	III	IV	I	II	III	IV

4. Calculation of Duty Cycle

From the previous section, it can be seen that the arbitrary reference vector is obtained from the synthesis of four active vectors and one zero vector. Taking the voltage and current reference vectors in Fig. 9.11 as an example, at this time the voltage and

Table 9.5 Active Vector Amplitude in SVPWM of a 3 × 3 matrix converter (Amp_{I}, Amp_{II}, Amp_{III} and Amp_{IV} are the active vector amplitudes in sequences)

k_{i}	Amplitude	
	sequence I or III	Sequence II or IV
$k_{\mathrm{i}} = 1$ or 4	$Amp_{\mathrm{I,III}} = U_{\mathrm{in}} \cdot \left\| \frac{2}{\sqrt{3}} \sin\left(\frac{\pi}{3} + \omega_{\mathrm{in}} t\right) \right\|$	$Amp_{\mathrm{II,IV}} = U_{\mathrm{in}} \cdot \left\| \frac{2}{\sqrt{3}} \sin\left(\frac{\pi}{3} - \omega_{\mathrm{in}} t\right) \right\|$
$k_{\mathrm{i}} = 2$ or 5	$Amp_{\mathrm{I,III}} = U_{\mathrm{in}} \cdot \left\| \frac{2}{\sqrt{3}} \sin(\omega_{\mathrm{in}} t) \right\|$	$Amp_{\mathrm{II,IV}} = U_{\mathrm{in}} \cdot \left\| \frac{2}{\sqrt{3}} \sin\left(\frac{\pi}{3} + \omega_{\mathrm{in}} t\right) \right\|$
$k_{\mathrm{i}} = 3$ or 6	$Amp_{\mathrm{I,III}} = U_{\mathrm{in}} \cdot \left\| \frac{2}{\sqrt{3}} \sin\left(\frac{\pi}{3} - \omega_{\mathrm{in}} t\right) \right\|$	$Amp_{\mathrm{II,IV}} = U_{\mathrm{in}} \cdot \left\| \frac{2}{\sqrt{3}} \sin(\omega_{\mathrm{in}} t) \right\|$

Table 9.6 Switching state selection for space vector modulation of a 3 × 3 matrix converter (with consideration of the sign)

	$k_{\mathrm{v}} = 1$				$k_{\mathrm{v}} = 2$				$k_{\mathrm{v}} = 3$			
$k_{\mathrm{i}} = 1$	+9	−7	−3	+1	−6	+4	+9	−7	+3	−1	−6	+4
$k_{\mathrm{i}} = 2$	−8	+9	+2	−3	+5	−6	−8	+9	−2	+3	+5	−6
$k_{\mathrm{i}} = 3$	+7	−8	−1	+2	−4	+5	+7	−8	+1	−2	−4	+5
$k_{\mathrm{i}} = 4$	−9	+7	+3	−1	+6	−4	−9	+7	−3	+1	+6	−4
$k_{\mathrm{i}} = 5$	+8	−9	−2	+3	−5	+6	+8	−9	+2	−3	−5	+6
$k_{\mathrm{i}} = 6$	−7	+8	+1	−2	+4	−5	−7	+8	−1	+2	+4	−5
	$k_{\mathrm{v}} = 4$				$k_{\mathrm{v}} = 5$				$k_{\mathrm{v}} = 6$			
$k_{\mathrm{i}} = 1$	−9	+7	+3	−1	+6	−4	−9	+7	−3	+1	+6	−4
$k_{\mathrm{i}} = 2$	+8	−9	−2	+3	−5	+6	+8	−9	+2	−3	−5	+6
$k_{\mathrm{i}} = 3$	−7	+8	+1	−2	+4	−5	−7	+8	−1	+2	+4	−5
$k_{\mathrm{i}} = 4$	+9	−7	−3	+1	−6	+4	+9	−7	+3	−1	−6	+4
$k_{\mathrm{i}} = 5$	−8	+9	+2	−3	+5	−6	−8	+9	−2	+3	+5	−6
$k_{\mathrm{i}} = 6$	+7	−8	−1	+2	−4	+5	+7	−8	+1	−2	−4	+5
序列	I	II	III	IV	I	II	III	IV	I	II	III	IV

current reference vectors are all located in Sector I, i.e. $k_{\mathrm{v}} = k_{\mathrm{i}} = 1$, at which point there are:

$$\boldsymbol{U}_{\mathrm{o}} = \boldsymbol{U}_{\mathrm{o}1} + \boldsymbol{U}_{\mathrm{o}2} \tag{9.25}$$

$$\begin{array}{l} |\boldsymbol{U}_{\mathrm{o}}| \cos(\alpha) = |\boldsymbol{U}_{\mathrm{o}1}| \cos(\pi/3) + |\boldsymbol{U}_{\mathrm{o}2}| \\ |\boldsymbol{U}_{\mathrm{o}}| \sin(\alpha) = |\boldsymbol{U}_{\mathrm{o}1}| \sin(\pi/3) \end{array} \Rightarrow \begin{array}{l} |\boldsymbol{U}_{\mathrm{o}1}| = \frac{2}{\sqrt{3}} |\boldsymbol{U}_{\mathrm{o}}| \sin(\alpha) \\ |\boldsymbol{U}_{\mathrm{o}1}| = \frac{2}{\sqrt{3}} |\boldsymbol{U}_{\mathrm{o}}| \sin\left(\frac{\pi}{3} - \alpha\right) \end{array} \tag{9.26}$$

$$\boldsymbol{I}_{\mathrm{i}} = \boldsymbol{I}_{\mathrm{i}1} + \boldsymbol{I}_{\mathrm{i}2} \tag{9.27}$$

$$\begin{aligned} |\boldsymbol{I}_{\mathrm{i}}|\cos(\beta+\pi/6) &= |\boldsymbol{I}_{\mathrm{i1}}|\cos(\pi/3)+|\boldsymbol{I}_{\mathrm{i2}}| \\ |\boldsymbol{I}_{\mathrm{i}}|\sin(\beta+\pi/6) &= |\boldsymbol{I}_{\mathrm{i1}}|\sin(\pi/3) \end{aligned} \Rightarrow \begin{aligned} |\boldsymbol{I}_{\mathrm{i1}}| &= \tfrac{2}{\sqrt{3}}|\boldsymbol{I}_{\mathrm{i}}|\cos\left(\tfrac{\pi}{3}-\beta\right) \\ |\boldsymbol{I}_{\mathrm{i2}}| &= \tfrac{2}{\sqrt{3}}|\boldsymbol{I}_{\mathrm{i}}|\cos\left(\tfrac{\pi}{3}+\beta\right) \end{aligned} \tag{9.28}$$

where α ($\alpha \in [0, \pi/3]$) and β ($\beta \in [-\pi/6, +\pi/6]$) are the angle of the output voltage reference vector in its sector with the corresponding sector boundary and the angle of the input current reference vector in its sector with the sector bisector, respectively.

The voltage and current vector components in Eqs. (9.25) and (9.27) can in turn be expressed as

$$\begin{aligned} &\boldsymbol{U}_{\mathrm{o1}} = \boldsymbol{U}_{\mathrm{I}}d_{\mathrm{I}} + \boldsymbol{U}_{\mathrm{II}}d_{\mathrm{II}} \\ &\boldsymbol{U}_{\mathrm{o2}} = \boldsymbol{U}_{\mathrm{III}}d_{\mathrm{III}} + \boldsymbol{U}_{\mathrm{IV}}d_{\mathrm{IV}} \\ \Rightarrow &\begin{aligned} |\boldsymbol{U}_{\mathrm{o1}}| &= |\boldsymbol{U}_{\mathrm{I}}|d_{\mathrm{I}} + |\boldsymbol{U}_{\mathrm{II}}|d_{\mathrm{II}} = Amp_{\mathrm{I}}d_{\mathrm{I}} + Amp_{\mathrm{II}}d_{\mathrm{II}} \\ |\boldsymbol{U}_{\mathrm{o2}}| &= |\boldsymbol{U}_{\mathrm{III}}|d_{\mathrm{III}} + |\boldsymbol{U}_{\mathrm{IV}}|d_{\mathrm{IV}} = Amp_{\mathrm{III}}d_{\mathrm{III}} + Amp_{\mathrm{IV}}d_{\mathrm{IV}} \end{aligned} \end{aligned} \tag{9.29}$$

$$\begin{aligned} &\boldsymbol{I}_{\mathrm{i1}} = \boldsymbol{I}_{\mathrm{I}}d_{\mathrm{I}} + \boldsymbol{I}_{\mathrm{III}}d_{\mathrm{III}} \\ &\boldsymbol{I}_{\mathrm{i2}} = \boldsymbol{I}_{\mathrm{II}}d_{\mathrm{II}} + \boldsymbol{I}_{\mathrm{IV}}d_{\mathrm{IV}} \\ \Rightarrow &\begin{aligned} |\boldsymbol{I}_{\mathrm{i1}}| &= |\boldsymbol{I}_{\mathrm{I}}|d_{\mathrm{I}} + |\boldsymbol{I}_{\mathrm{III}}|d_{\mathrm{III}} = Amp_{\mathrm{I}}d_{\mathrm{I}} + Amp_{\mathrm{III}}d_{\mathrm{III}} \\ |\boldsymbol{I}_{\mathrm{i2}}| &= |\boldsymbol{I}_{\mathrm{II}}|d_{\mathrm{II}} + |\boldsymbol{I}_{\mathrm{IV}}|d_{\mathrm{IV}} = Amp_{\mathrm{II}}d_{\mathrm{II}} + Amp_{\mathrm{IV}}d_{\mathrm{IV}} \end{aligned} \end{aligned} \tag{9.30}$$

So there is

$$\frac{Amp_{\mathrm{I}}d_{\mathrm{I}}}{Amp_{\mathrm{II}}d_{\mathrm{II}}} = \frac{Amp_{\mathrm{III}}d_{\mathrm{III}}}{Amp_{\mathrm{IV}}d_{\mathrm{IV}}} = \frac{\cos\left(\frac{\pi}{3}-\beta\right)}{\cos\left(\frac{\pi}{3}+\beta\right)} \tag{9.31}$$

It can be obtained from Eqs. (9.26), (9.29) and (9.31) that

$$\begin{aligned} d_{\mathrm{I}} &= (-1)^{k_v+k_i}\frac{2}{\sqrt{3}}|\boldsymbol{U}_o|\frac{\sin(\alpha)\cos\left(\frac{\pi}{3}-\beta\right)}{\cos(\beta)\cdot Amp_{\mathrm{I}}} \\ d_{\mathrm{II}} &= (-1)^{k_v+k_i+1}\frac{2}{\sqrt{3}}|\boldsymbol{U}_o|\frac{\sin(\alpha)\cos\left(\frac{\pi}{3}+\beta\right)}{\cos(\beta)\cdot Amp_{\mathrm{II}}} \\ d_{\mathrm{III}} &= (-1)^{k_v+k_i+1}\frac{2}{\sqrt{3}}|\boldsymbol{U}_o|\frac{\sin\left(\frac{\pi}{3}-\alpha\right)\cos\left(\frac{\pi}{3}-\beta\right)}{\cos(\beta)\cdot Amp_{\mathrm{III}}} \\ d_{\mathrm{IV}} &= (-1)^{k_v+k_i}\frac{2}{\sqrt{3}}|\boldsymbol{U}_o|\frac{\sin\left(\frac{\pi}{3}-\alpha\right)\cos\left(\frac{\pi}{3}+\beta\right)}{\cos(\beta)\cdot Amp_{\mathrm{IV}}} \end{aligned} \tag{9.32}$$

It further gets

$$d_{\mathrm{zero}} = 1 - d_{\mathrm{I}} - d_{\mathrm{II}} - d_{\mathrm{III}} - d_{\mathrm{IV}}\big|_{d_{\mathrm{I}}+d_{\mathrm{II}}+d_{\mathrm{III}}+d_{\mathrm{IV}}\in[0,1]} \tag{9.33}$$

When $k_{\mathrm{v}} = k_{\mathrm{i}} = 1$, from Table 9.5, Eqs. (9.32) and (9.33) it is obtained that

$$d_{\text{zero}} = 1 - \frac{2\sqrt{3}U_o}{U_{\text{in}}} \cdot \frac{\cos\left(\frac{\pi}{6} - \alpha\right)\left(2\cos^2\beta - 1\right)}{\cos(\beta)\left(4\cos^2\beta - 1\right)} \geq 0 \tag{9.34}$$

Solving Eq. (9.34) gives

$$\frac{U_o}{U_{in}} \leq \frac{k}{l} \quad \text{其中} \begin{cases} k = \dfrac{\cos(\beta)\cdot\left(4\cos^2\beta - 1\right)}{\left(2\cos^2\beta - 1\right)} \\ l = 2\sqrt{3}\cdot\cos\left(\dfrac{\pi}{6} - \alpha\right) \end{cases} \tag{9.35}$$

If the output voltage vector and the input current vector can be controlled independently, then the following results can be obtained.

$$\begin{matrix} 3 \leq k \leq 2\sqrt{3} \\ 3 \leq l \leq 2\sqrt{3} \end{matrix} \Rightarrow \begin{cases} \frac{3}{2\sqrt{3}} \leq \frac{k}{l} \leq \frac{2\sqrt{3}}{3} \\ \frac{U_o}{U_{in}} \leq \left(\frac{k}{l}\right)_{\min} = \frac{\sqrt{3}}{2} \approx 86.6\% \end{cases} \tag{9.36}$$

The above equation shows that the maximum modulation ratio of the matrix converter is approximately 86.6%.

5. Sequences of the Switching States

According to the above discussion, when $k_v = k_i = 1$, the switching sequence is +9(I), −7(III), −3(III), +1(IV), O1, O2 and O3, where the zero vectors are used to ensure that each switching state sequence has the same time period. In order to minimise switching losses, the zero vector is selected in such a way that only one bidirectional switch is switched during each switching state change. Thus, the sequence of switching states obtained here is shown in Fig. 9.12 [9].

In Fig. 9.12 zero vectors are used, which will introduce 12 switching commutation processes per sequence cycle, where the zero vector action time is calculated as follows:

$$d_{O1} + d_{O2} + d_{O3} = 1 - d_1 - d_3 - d_7 - d_9 \Rightarrow d_{O1} = d_{O2} = d_{O3} = d_{\text{zero}}/3 \tag{9.37}$$

The method of determining the sequence of switching states in the above way is known as Symmetrical Space Vector Modulation (SSVM), in contrast there is also Asymmetrical Space Vector Modulation (ASVM). In the ASVM method, there is only one zero vector in the middle of the cycle (e.g. the O_1 vector in Fig. 9.12) per sequence period, where S_{Aa}, S_{Ba} and S_{Ca} in Fig. 9.12 will no longer change their states (S_{Aa} always on, S_{Ba} always off, S_{Ca} always off) and the switching commutation process is reduced to eight.

6. Current Commutation

Two conditions need to be met during the commutation control of a matrix converter: (1) there is no simultaneous conduction of two bidirectional switches as shown in

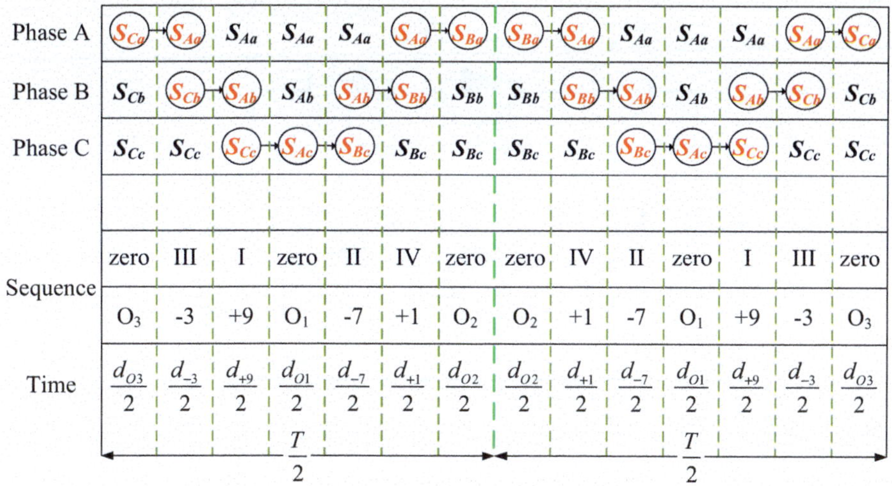

Fig. 9.12 Double-side 3-zero-vector switching sequence ($k_v = k_i = 1$)

Fig. 9.13a to avoid a phase-to-phase short circuit on the output side, and (2) there is no simultaneous switching off of two bidirectional switches as shown in Fig. 9.12b to avoid an open circuit on the output side [8].

Four-step commutation is a typical commutation method for a matrix converter, which is characterised by safety, reliability and controlled current direction, etc. [8]. The commutation process is shown in Figs. 9.14, 9.15 and 9.16.

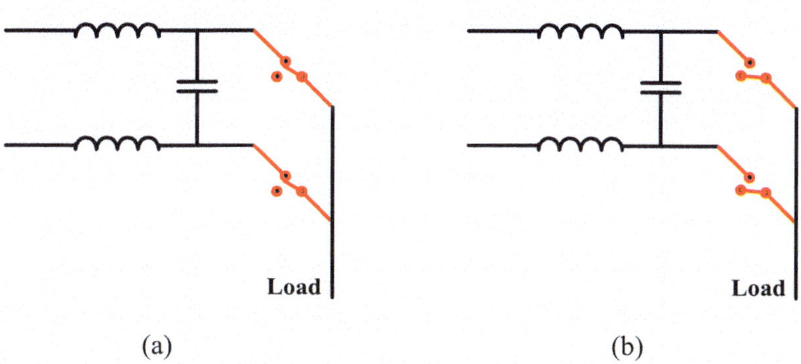

Fig. 9.13 Limits on matrix converter commutation

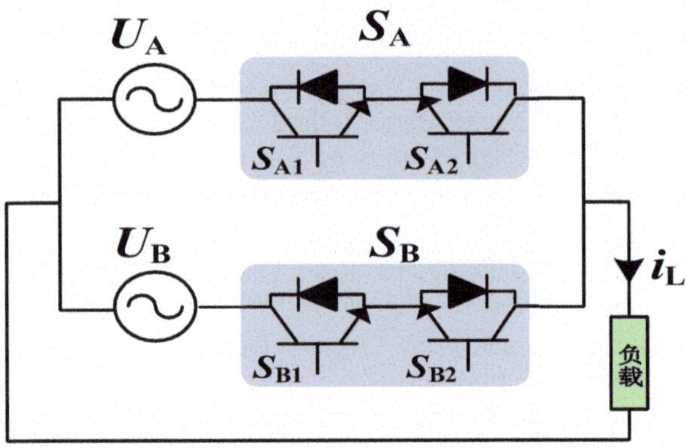

Fig. 9.14 A two-phase to single-phase matrix converter for explaining the four-step current commutation

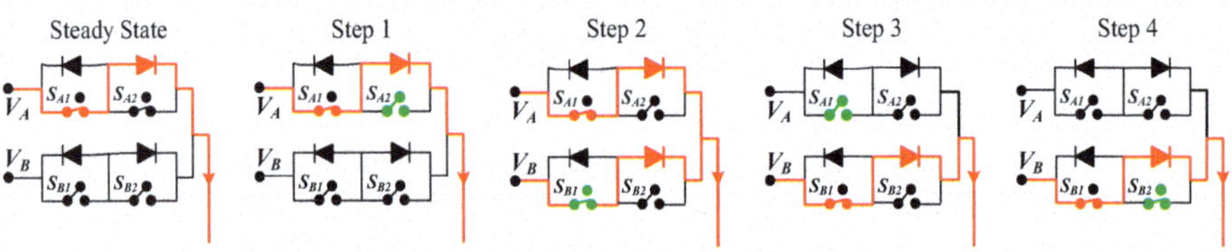

Fig. 9.15 Four-step current commutation process

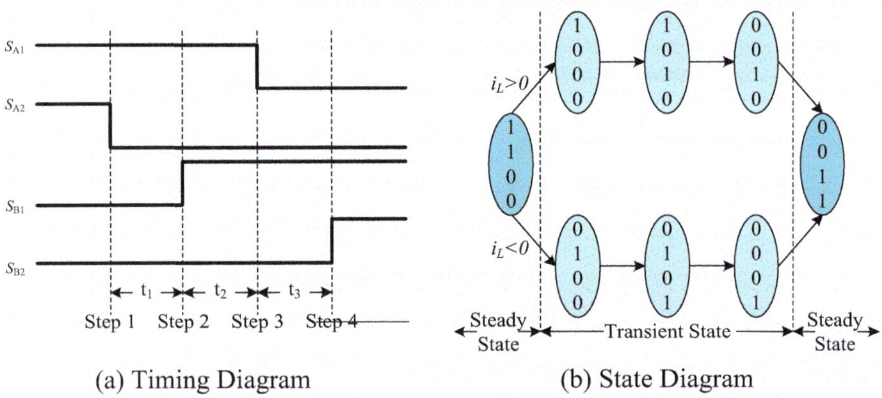

Fig. 9.16 Timing and state diagrams of four-step current commutation

9.2.2 *Topology of the MFT-Based HVDC Transmission System for Offshore Wind Farms*

The topology of the MFT-based offshore wind farm HVDC transmission system discussed in this section is shown in Fig. 9.17. The offshore section contains five power generation units, each consisting of a wind turbine, a PMSG and a 3 × 3 matrix converter; the primary side of the MFT adopts a multi-winding structure, with each winding connected to a matrix converter in the corresponding power generation unit for collecting power from the power generation unit; the secondary side of the MFT adopts a single-winding structure and is connected to a three-phase uncontrolled rectifier bridge for converting high-frequency AC power into high-voltage DC power, which is connected to the onshore section via a high-voltage DC transmission cable; its onshore section directly uses an MMC converter to convert high-voltage DC power into grid-frequency AC power and feed it into the grid.

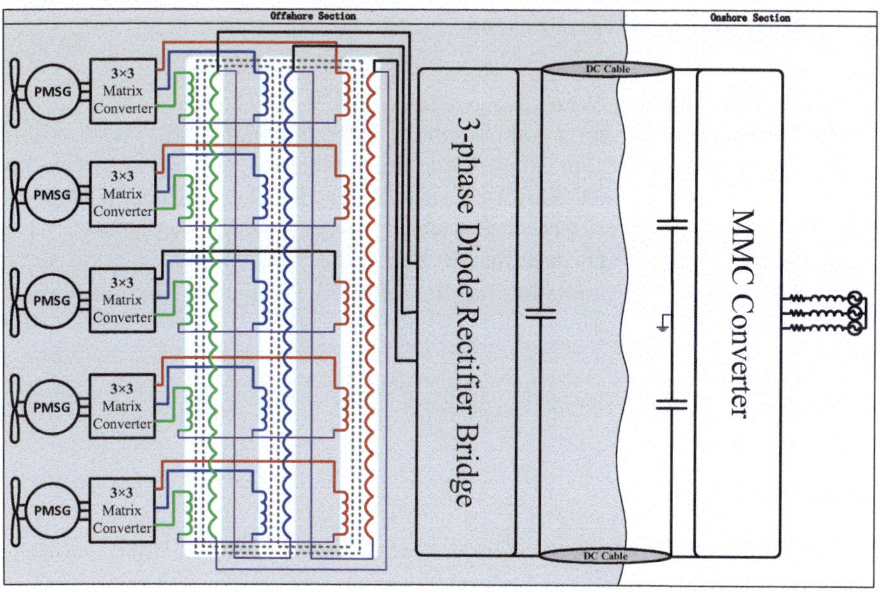

Fig. 9.17 Topology of the MFT-based HVDC transmission system for offshore wind farms

The main features of the topology shown in Fig. 9.17 are: a more compact modular structure for each generation unit, fewer steps required to convert from high- frequency AC power to high-voltage DC power, increased power density due to the MFT structure, and no balancing control between the generation units in the offshore section. The MFT has a Y/Δ connection, which effectively prevents internal transformer commutation and protects the offshore section in the event of a branch cable or onshore converter failure.

Matrix converters require overvoltage and overcurrent protection due to the lack of a self-commutation path. Among the many protection schemes, the clamp circuit is considered to be one of the most effective and reliable protection schemes for matrix converters. It requires the introduction of additional circuit structures in the original system, but the rated parameters of these protection circuit structures are usually small.

9.2.3 Control of the MFT-Based HVDC Transmission System for Offshore Wind Farms

In the system shown in Fig. 9.17, each generation unit is used to convert wind energy into electrical energy, the vector control and implementation of the MPPT algorithm can be referred to Sect. 7.1.2 in Chap. 7, and the modulation method is described in Sect. 9.2.1. The onshore MMC converter is used to feed the electrical energy from the DC bus into the grid, the vector control of which can be found in Chap. 7, Sect. 7.1.2, and the modulation method is described in Sect. 9.1.1.

Figure 9.18 shows the simulation results of the MFT-based HVDC transmission system for offshore wind farms.

Fig. 9.18 Simulation results of MFT-based HVDC transmission system for offshore wind farms

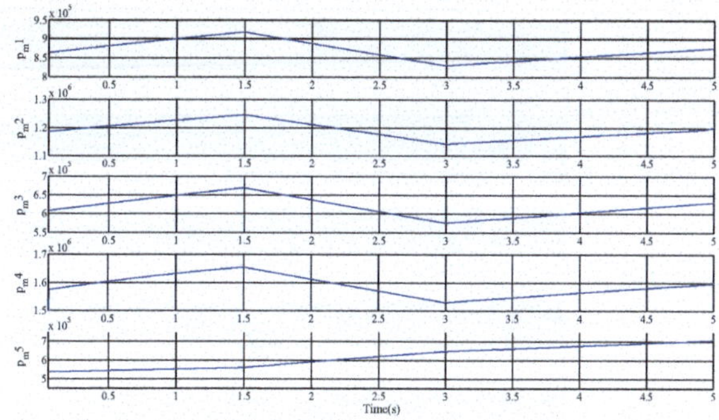

(a) Output power of each generation unit

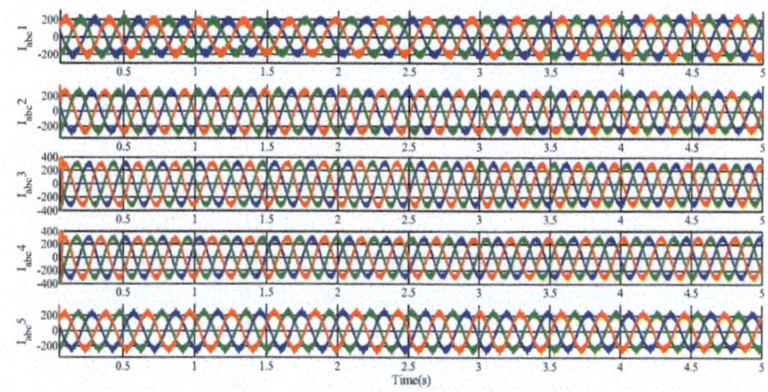

(b) Phase currents of each PMSG

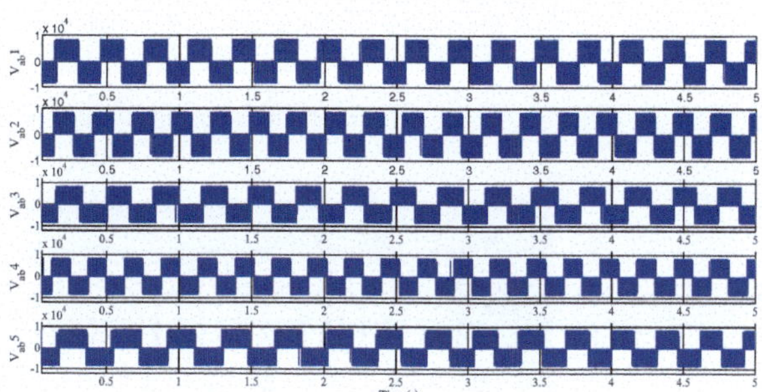

(c) Line voltagesbetween PMSGs and MCs

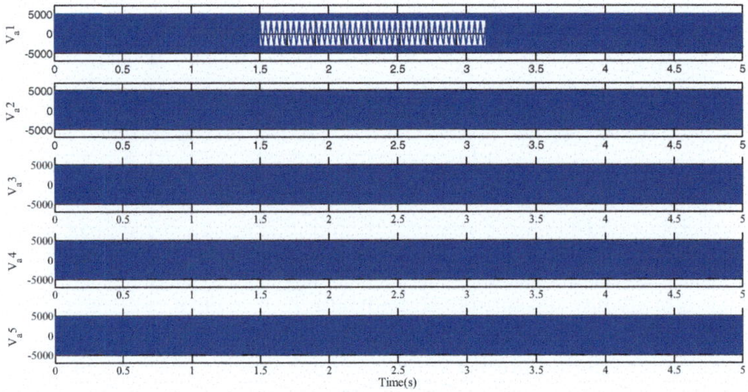

(d) Phase voltage between MFT and MCs

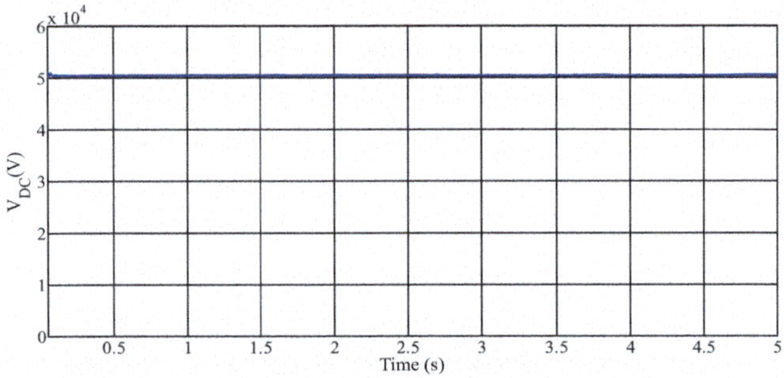

(e) High DC voltage between DC cables

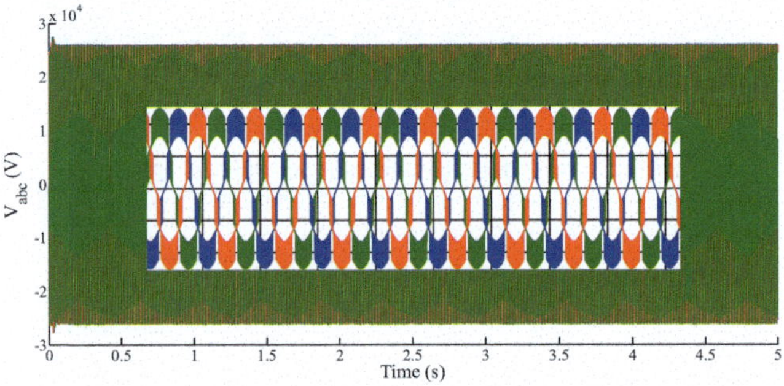

(f) AC-side voltage of MMC

Fig. 9.18 (continued)

References

1. Firas, Control of medium voltage wind turbine with 3L-NPC and integration of series DC offshore wind farm into the grid. Ph.D. thesis (Tsinghua University, Beijing, China, 2013)
2. M. Saeedifard, R. Iravani, Dynamic performance of a modular multilevel back-to-back HVDC system. IEEE Trans. Power Deliv. (2010)
3. A. Antonopoulos, L. Angquist, H.-P. Nee, On dynamics and voltage control of the modular multilevel converter, in *13th European Conference in Power Electronics and Applications, EPE'09* (IEEE, 2009)
4. M.S. Carmeli, F. Castelli-Dezza, D. Rosati, G. Marchegiani, M. Mauri, MVDC connection of offshore wind farms to the transmission system, in *International Symposium on Power Electronics, Electrical Drives, Automation and Motion* (2010)
5. M.S. Carmeli, F. Castelli-Dezza, G. Marchegiani, M. Mauri, D. Rosati, Design and analysis of a medium voltage DC wind farm with a transformer-less wind turbine generator, in *XIX International Conference on Electrical Machines (ICEM)* (2010)
6. S.S. Gjerde, T. Undeland, Power conversion system for transformer-less offshore wind turbine, in *Proceedings of the 14th European Conference on Power Electronics and Applications (EPE 2011)* (IEEE, 2011)
7. P.W. Wheeler, J. Rodríguez, J.C. Clare, L. Empringham, A. Weinstein, Matrix converters: a technology review. IEEE Trans. Ind. Electron. (2002)
8. P. Wheeler, J. Clare, L. Empringham, M. Apap, M. Bland, Matrix converters. Power Eng. J. (2002)
9. D. Casadei, G. Serra, A. Tani, L. Zarri, Matrix converter modulation strategies: a new general approach based on space-vector representation of the switch state. IEEE Trans. Ind. Electron. (2002)

SPRINGER NATURE

GPSR Compliance

The European Union's (EU) General Product Safety Regulation (GPSR) is a set of rules that requires consumer products to be safe and our obligations to ensure this.

If you have any concerns about our products, you can contact us on ProductSafety@springernature.com

In case Publisher is established outside the EU, the EU authorized representative is:

Springer Nature Customer Service Center GmbH
Europaplatz 3
69115 Heidelberg, Germany

The manufacturer's authorised representative in the EU is Springer Nature Customer Service Centre GmbH, Europaplatz 3, 69115 Heidelberg, Germany. If you have any concerns regarding our products, please contact ProductSafety@springernature.com

Printed and bound by CPI Group (UK) Ltd, Croydon, CR0 4YY
30/09/2026
02225298-0012